Deep Learning Strategies for Security Enhancement in Wireless Sensor Networks

K. Martin Sagayam
Karunya Institute of Technology and Sciences, India

Bharat Bhushan
HMR Institute of Technology and Management, New Delhi, India

A. Diana Andrushia
Karunya Institute of Technology and Sciences, India

Victor Hugo C. de Albuquerque
Universidade de Fortaleza, Brazil

A volume in the Advances in Information Security, Privacy, and Ethics (AISPE) Book Series

Published in the United States of America by
IGI Global
Information Science Reference (an imprint of IGI Global)
701 E. Chocolate Avenue
Hershey PA, USA 17033
Tel: 717-533-8845
Fax: 717-533-8661
E-mail: cust@igi-global.com
Web site: http://www.igi-global.com

Library of Congress Cataloging-in-Publication Data

Names: Sagayam, K. Martin, 1987- editor.
Title: Deep learning strategies for security enhancement in wireless sensor networks / K. Martin Sagayam, Bharat Bhushan, Diana Andrushia, Victor Hugo C. de Albuquerque, editors.
Description: Hershey, PA : Information Science Reference, an imprint of IGI Global, [2020] | Includes bibliographical references and index. | Summary: "This book explores the theoretical and practical advancements of security protocols in wireless sensor networks using artificial intelligence-based techniques"-- Provided by publisher.
Identifiers: LCCN 2020002718 (print) | LCCN 2020002719 (ebook) | ISBN 9781799850687 (hardcover) | ISBN 9781799852759 (paperback) | ISBN 9781799850694 (ebook)
Subjects: LCSH: Wireless sensor networks--Security measures. | Machine learning.
Classification: LCC TK7872.D48 D39 2020 (print) | LCC TK7872.D48 (ebook) | DDC 006.2/558--dc23
LC record available at https://lccn.loc.gov/2020002718
LC ebook record available at https://lccn.loc.gov/2020002719

This book is published in the IGI Global book series Advances in Information Security, Privacy, and Ethics (AISPE) (ISSN: 1948-9730; eISSN: 1948-9749)

British Cataloguing in Publication Data
A Cataloguing in Publication record for this book is available from the British Library.

For electronic access to this publication, please contact: eresources@igi-global.com.

Advances in Information Security, Privacy, and Ethics (AISPE) Book Series

Manish Gupta
State University of New York, USA

ISSN:1948-9730
EISSN:1948-9749

Mission

As digital technologies become more pervasive in everyday life and the Internet is utilized in ever increasing ways by both private and public entities, concern over digital threats becomes more prevalent.

The **Advances in Information Security, Privacy, & Ethics (AISPE) Book Series** provides cutting-edge research on the protection and misuse of information and technology across various industries and settings. Comprised of scholarly research on topics such as identity management, cryptography, system security, authentication, and data protection, this book series is ideal for reference by IT professionals, academicians, and upper-level students.

Coverage

- Access Control
- Global Privacy Concerns
- Data Storage of Minors
- Privacy Issues of Social Networking
- Information Security Standards
- Security Information Management
- Tracking Cookies
- Risk Management
- Privacy-Enhancing Technologies
- Cyberethics

IGI Global is currently accepting manuscripts for publication within this series. To submit a proposal for a volume in this series, please contact our Acquisition Editors at Acquisitions@igi-global.com or visit: http://www.igi-global.com/publish/.

The Advances in Information Security, Privacy, and Ethics (AISPE) Book Series (ISSN 1948-9730) is published by IGI Global, 701 E. Chocolate Avenue, Hershey, PA 17033-1240, USA, www.igi-global.com. This series is composed of titles available for purchase individually; each title is edited to be contextually exclusive from any other title within the series. For pricing and ordering information please visit http://www.igi-global.com/book-series/advances-information-security-privacy-ethics/37157. Postmaster: Send all address changes to above address.

Titles in this Series

For a list of additional titles in this series, please visit: www.igi-global.com/book-series

Personal Data Protection and Legal Developments in the European Union
Maria Tzanou (Keele University, UK)
Information Science Reference • ©2020 • 375pp • H/C (ISBN: 9781522594895) • US $225.00

Legal Regulations, Implications, and Issues Surrounding Digital Data
Margaret Jackson (RMIT University, Australia) and Marita Shelly (RMIT University, Australia)
Information Science Reference • ©2020 • 240pp • H/C (ISBN: 9781799831303) • US $195.00

Large-Scale Data Streaming, Processing, and Blockchain Security
Hemraj Saini (Jaypee University of Information Technology, India) Geetanjali Rathee (Jaypee University of Information Technology, India) and Dinesh Kumar Saini (Sohar University, Oman)
Information Science Reference • ©2020 • 300pp • H/C (ISBN: 9781799834441) • US $225.00

Advanced Security Strategies in Next Generation Computing Models
Shafi'i Muhammad Abdulhamid (Federal University of Technology Minna, Nigeria) and Muhammad Shafie Abd Latiff (Universiti Teknologi, Malaysia)
Information Science Reference • ©2020 • 300pp • H/C (ISBN: 9781799850809) • US $215.00

Advancements in Security and Privacy Initiatives for Multimedia Images
Ashwani Kumar (Vardhaman College of Engineering, India) and Seelam Sai Satyanarayana Reddy (Vardhaman College of Engineering, India)
Information Science Reference • ©2020 • 300pp • H/C (ISBN: 9781799827955) • US $215.00

Cyber Security and Safety of Nuclear Power Plant Instrumentation and Control Systems
Michael A. Yastrebenetsky (State Scientific and Technical Centre for Nuclear and Radiation Safety, Ukraine) and Vyacheslav S. Kharchenko (National Aerospace University KhAI, Ukraine)
Information Science Reference • ©2020 • 345pp • H/C (ISBN: 9781799832775) • US $195.00

Internet Censorship and Regulation Systems in Democracies Emerging Research and Opportunities
Nikolaos Koumartzis (Aristotle University of Thessaloniki, Greece) and Andreas Veglis (Aristotle University of Thessaloniki, Greece)
Information Science Reference • ©2020 • 272pp • H/C (ISBN: 9781522599739) • US $185.00

701 East Chocolate Avenue, Hershey, PA 17033, USA
Tel: 717-533-8845 x100 • Fax: 717-533-8661
E-Mail: cust@igi-global.com • www.igi-global.com

Table of Contents

Detailed Table of Contents

From its advent in mid-20th century, machine learning constantly improves the user experience of existing systems. It can be used in almost every field such as weather, sports, business, IoT, medical care, etc. Wireless sensor networks are often placed in hostile environments to observe change in surroundings. Since these communicate wirelessly, many problems such as localisation of nodes, security of data being routed create barriers for proper functioning of system. Extending the horizon of machine learning to WSN creates wonders and adds credibility to the system. This chapter aids to present various machine learning aspects applied on wireless sensor networks and the benefits and drawbacks of applying machine learning to WSN. It also describes various data aggregation and clustering techniques that aim to reduce power consumption and ensure confidentiality, authentication, integrity, and availability amongst sensor nodes. This could contribute to design and alter pre-existing ML algorithms to improve overall performance of wireless sensor networks.

Wireless sensor networks consist of unattended small sensor nodes having low energy and low range of communication. It has been observed that if there is any system to periodically start and stop the sensors sensing activities, then it saves some energy, and thus, the network lifetime gets extended. According to the current literature, security and energy efficiency are the two main concerns to improve the quality of service during transmission of data in wireless sensor networks. Machine learning has proved its efficiency in developing efficient processes to handle complex problems in various network aspects. Routing in wireless sensor network is the process of finding the route for transmitting data among different sensor

WSNs collects a huge amount of heterogeneous information and have been widely applied in various applications. Most of the information collecting techniques for WSNs can't keep away from the hotspot trouble. This problem affects the connectivity of the community and reduces the life of the entire community. Hence, an efficient speed control based data collection algorithm (SCBDCA) using mobile sink in WSNs is proposed. A tree construction technique is introduced based on weight such that the origin nodes of the built timber are described as rendezvous points (RPs). Also, other unique type of nodes named sub or additional rendezvous points (ARPs or SRPs) is chosen in accordance to their traffic consignment and hops to origin nodes in the network. Mobile sink gathers the information from the cluster head (CH) and is controlled by speed control mechanism. Simulation results among different existing procedures exhibit that the SCBDCA can extensively stabilize the consignment of the communication network, decrease the energy consumption, and extend the lifetime in the communication network.

In wired and wireless communication, providing security is extremely important. It is a very challenging issue. But the flying evolution in communication technology has triggered to sturdy research interest in wireless networks. The characteristics of wireless networks make this issue even more challenging. In Ad-hoc networks, there is a huddle of autonomous nodes, which dynamically form a temporary multi-hopped, peer-to-peer radio network, without any use of predefined infrastructure. These nodes are generally mobile in nature, and to connect these nodes, the connectionless links are used. These nodes have the potential to self-organize, self-configure, and self-arrangement. Ad-hoc networks do not have fixed structure due to their dynamic nature. Ad-hoc networks are inherently prone to a number of security threats. Lack of fixed infrastructure, use of wireless link for communication, and mobility of nodes make Ad-hoc networks extremely receptive to hostile attacks, blackhole attack being one among them, which can be implemented effortlessly.

Preface

This comprehensively edited advanced book will cover machine and deep learning strategies for security enhancement in Wireless Sensor Networks (WSNs) and Internet of Things (IoT). WSNs have gained much significance from both industry and academia owing to its wide range of applications in military and civilian scenarios. In such hostile environment, protection of WSNs from malicious attacks is of utmost concern. Design of security for such networks is a challenging task due to resource constrained nature of WSNs. Several computational intelligence paradigms have been successfully used to address problems such as data aggregation, energy aware routing, localizations, task scheduling and security. Virtualization of WSNs can enable efficient utilization of WSN deployments, as numerous applications can co-exist on the same virtualized WSN. Clustering techniques enhances energy efficiency and extends network longevity. Quality of Service (QoS) in IoT could be hugely benefitted by enhancement of clustering schemes with smart network selection solutions. This book aims to provide the state of the art of sensor networks and IoT in terms of security, vulnerabilities, virtualization, clustering, and intrusion detection as well as the involved deep learning techniques and give an insight of the major comprehensive study in the use of deep learning strategies such as machine learning, artificial intelligence and computational intelligence in sensor networks.

Owing to the scope and diversity of topics covered, the book will be of interest not only to researchers and theorists but also to professionals, material developers, technology specialists and methodologists dealing with the multifarious aspects of data privacy and security enhancement in Wireless Sensor Networks. The book also aims to provide a roadmap to recent research areas, vulnerabilities and their countermeasures in in the field of sensor networks. Also, the book is designed to be the first reference choice at research and development centers, academic institutions, university libraries and any institutions interested in analyzing wireless sensor networks and the use of deep learning techniques therein. Academicians and research scholars are other projected audience who identify applications, tools and methodologies through discussions on qualitative/quantitative results, literature reviews and reference citations.

The following describes the chapters included in this book.

Chapter 1: Introducing Machine Learning to Wireless Sensor Networks – Requirements and Applications

From its advent in mid-20th century, machine learning constantly improves the user experience of existing systems. It can be used in almost every field such as weather, sports, business, IOT, medical care etc. Wireless sensor networks are often placed in hostile environments, to observe change in surroundings. Since these communicate wirelessly, many problems such as localization of nodes, security of data being routed create barriers for proper functioning of system. Extending the horizon of machine learning to WSN creates wonders and adds credibility to the system. This chapter aids to present various Machine Learning aspects applied on Wireless Sensor Networks. The benefits and drawbacks of applying machine learning to WSN. It also describes various data aggregation and clustering techniques that aim to reduce power consumption and ensure confidentiality, authentication, integrity and availability amongst sensor nodes. This could contribute to design and alter pre-existing ML algorithms to improve overall performance of Wireless Sensor Networks.

Chapter 2: Secured Energy Efficient Routing in Wireless Sensor Networks Using Machine Learning Algorithm – Fundamentals and Applications

Wireless sensor networks consist of unattended small sensor nodes having low energy and low range of communication. It has been observed that if there is any system to periodically start and stop the sensors sensing activities, then it saves some energy and thus, the network lifetime gets extended. According to the current literatures, the security and energy efficiency are the two main concerns to improve the quality of service during transmission of data in Wireless Sensor Network. Machine learning has proved its efficiency in developing efficient processes to handle complex problems in various network aspects. Routing in wireless sensor network is the process of finding the route for transmitting data among different sensor nodes according to the requirement. Machine learning has been used in a broad way for designing energy efficient routing protocols and this chapter reviews the existing works in the said domain which can be the guide to someone who wants to explore the area further.

Chapter 3: Security and Privacy Challenges of Deep Learning – A Comprehensive Survey

Deep learning is the buzz word in recent times in the research field due to its various advantages in the field of healthcare, medicine, automobiles, etc. Huge amount of data is required for deep learning to achieve better accuracy thus it is important to protect the data from security and privacy breaches. In this article a comprehensive survey of security and privacy challenges in deep learning is presented. The security attacks such as poisoning attacks, evasion attacks, and black-box attacks are explored with its prevention and defence techniques. A comparative analysis is done on various techniques to prevent the data from such security attacks. Privacy is another major challenge in deep learning. In this chapter we presented a in-depth survey on various privacy-preserving techniques for deep learning such as differential privacy, homomorphic encryption, secret sharing and secure multi-party computation. A detailed comparison table to compare the various privacy-preserving techniques and approaches are also presented.

Chapter 4: Wireless Environment Security – Challenges and Analysis Using Deep Learning

The communication through the wireless environment is open which is completely differs from the wired. This open environment communication can be accessed by users including illegitimate and thus increases the vulnerability for malicious attacks. For that reason motivation comes to study about the different possible security challenges, threats and to devise powerful, efficient, and improved required solution to improve the various security vulnerabilities. This letter presents the challenges regarding security and the security requirement in the wireless type communication. Along with, the research performs the analysis of deep learning for detecting malicious website. These websites are responsible to disrupt normal system working and can control the complete system and its resources by installing malware on to the respective machine. To elucidate the constructive and effective way towards the detection of malicious URL, the study uses convolutional neural network.

Chapter 5: Modelling a Deep Learning-Based Wireless Sensor Network Task Assignment Algorithm – An Investigative Approach

The twenty-first century is witnessing the emergence of a wide variety of Wireless Sensor Network (WSN) applications ranging from simple environmental monitoring to complex satellite monitoring applications. The advent of complex WSN applications has led to a massive transition in the development, functioning and capabilities of wireless sensor nodes. The contemporary nodes have multi-functional capabilities enabling the heterogeneous WSN applications. The future of WSN task assignment envisions WSN to be heterogeneous network with minimal human interaction. This led to the investigative model of a deep learning based task assignment algorithm. The algorithm employs a Multilayer Feed Forward Neural Network (MLFFNN) trained by (Particle Swarm Optimization) PSO for solving task assignment problem in a dynamic centralized heterogeneous WSN. The analyses include the study of hidden layers, effectiveness of the task assignment algorithms. The chapter would be highly beneficial to a wide range of audiences employing the machine and deep learning in WSN

Chapter 6: The Threat of Intelligent Attackers Using Deep Learning – The Backoff Attack Case

The significant increase in the number of devices interconnected has brought new services and applications, as well as new network vulnerabilities. The increasing hardware capacities of these devices and the developments in the Artificial Intelligent field mean that new and complex attack methods are being developed. This Chapter focuses on the backoff attack in a wireless network using CSMA/CA multiple access, and it shows that an intelligent attacker, making use of control theory, can successfully exploit a Sequential Probability Ratio Test based defense mechanism. Also, recent developments in the Deep Reinforcement Learning field allows that attackers that do not have full knowledge of the defense mechanism are able to successfully learn to attack it. Thus, this Chapter illustrates by means of the backoff attack the possibilities that the recent advances in the Artificial Intelligence field bring to intelligent attackers and highlights the importance of researching in intelligent defense methods able to cope with such attackers.

Chapter 7: Attacks, Vulnerabilities, And Their Countermeasures in Wireless Sensor Networks

Wireless sensor network (WSN) comprises sensor nodes which have the capability to sense and compute. Due to their availability and minimal cost than the other traditional networks, WSN is used broadly. The need for sensor networks increases quickly, they are more likely to security attacks. There are many attacks and vulnerabilities in WSN. The sensor nodes have issues like limited resources of memory and power, undependable communication medium which is further complicated in unattended environments, secure communication and data transmission issues. Due to the complexity in establishing and maintaining the wireless sensor networks, the traditional security solutions if implemented will prove to be inefficient for the dynamic nature of the wireless sensor networks. Since recent times, the advance of smart cities and everything smart, Wireless sensor nodes have become an integral part of the internet of things and their related paradigms. This chapter discusses the known attacks, vulnerabilities and countermeasures existing in wireless sensor networks.

Chapter 8: Recent Trends in Channel Assignment Techniques in Wireless Mesh Networks

Wireless mesh networks have numerous advantages in terms of connectivity as well as reliability. Traditionally the nodes in wireless mesh networks are equipped with a single radio, but the limitations are lower throughput and limited use of the available wireless channel. To overcome this, the recent advances in wireless mesh networks are based on a multi-channel multi-radio approach. Channel assignment is a technique that selects the best channel for a node or to the entire network just to increase the network capacity. To maximize the throughput and the capacity of the network, multiple channels with multiple radios were introduced in these networks. In this work, algorithms are developed to improve throughput, minimize delay, reduce average energy consumption and increase the residual energy for multi-radio multi-channel wireless mesh networks.

Chapter 9: Integrated Intrusion Detection System (IDS) For Security Enhancement in Wireless Sensor Networks

Rapidly expanding application areas of wireless sensor networks (WSNs) that includes critical civilian and military applications intensifies the security concerns especially in hostile unattended environments. In order to ensure the dependability and security of continued WSN services, there is a need of alternate defense line called Intrusion Detection system (IDS). WSNs are comprised of huge number of energy constrained nodes whose battery replacement or recharging is a challenging task after their deployment. In this present work, we develop a Simple Integrated IDS scheme (SIIS) that integrates the concept of clustering along with RC4 and Digital signature. For heterogeneous systems, cluster-based protocols perform best as they work on divide and conquer strategy. In order to safeguard the transmitted data, a secured, lightweight digital signature scheme that employs symmetric cryptography is used along with RC4, a synchronous stream cipher that satisfies both efficiency and security for lightweight algorithms.

Chapter 10: A Speed Control Based Big Data Collection Algorithm (SCBDCA) Using Clusters and Portable Sink WSNs

WSNs collects a huge amount of heterogeneous information and have been widely applied in various applications. Most of the information collecting techniques for WSNs can't keep away from the hotspot trouble. This problem affects the connectivity of the community and reduces the life of the entire community. Hence, an efficient Speed Control based Data Collection Algorithm (SCBDCA) using mobile sink in WSNs is proposed. A tree construction technique is introduced based on weight such that the origin nodes of the built timber are described as Rendezvous Points (RPs). Also other unique type of nodes named sub or additional Rendezvous Points (ARPs or SRPs) is chosen in accordance to their traffic consignment and hops to origin nodes in the network. Mobile sink gathers the information from the Cluster Head (CH) and is controlled by speed control mechanism. Simulation results among different existing procedures exhibit that the SCBDCA can extensively stability the consignment of the communication network, decrease the energy consumption and extend the lifetime in the communication network.

Chapter 11: Analysis of Black-Hole Attack With Its Mitigation Techniques in Ad-Hoc Network

In wired and wireless communication, providing security is extremely important. It is a very challenging issue. But the flying evolution in communication technology has triggered to sturdy research interest on Wireless Networks. The characteristics of wireless network make this issue even more challenging. In Ad-hoc network, there is a huddle of autonomous nodes, which dynamically form a temporary multi-hopped, peer-to-peer radio network, without any use of predefined infrastructure. These nodes are generally mobile in nature and to connect these nodes the connectionless links are used. These nodes have the potential to self-organize, self-configure and self-arrangement. Ad-hoc network do not have fixed structure due to their dynamic nature Ad-hoc networks are inherently very prone to number of security threats. Lack of fixed infrastructure, use of wireless link for communication and mobility of nodes make Ad-hoc networks extremely receptive to rival's hostile attacks black-hole attack being one among them, which can be implementing effortlessly.

Chapter 12: Fundamentals of Wireless Sensor Networks Using Machine Learning Approaches – Advancement in Big Data Analysis Using Hadoop for Oil Pipeline System With Scheduling Algorithm

Big Data and analytics may be new to some industries, but the oil and gas industry has long dealt with large quantities of data to make technical decisions. Oil producers can capture more detailed data in real-time at lower costs and from previously inaccessible areas, to improve oil-field and plant performance. Stream computing is a new way of analyzing high-frequency data for real-time complex-event-processing and scoring data against a physics-based or empirical model for predictive analytics, without having to store the data. Hadoop Map/Reduce and other NoSQL approaches are a new way of analyzing massive volumes of data used to support the reservoir, production, & facilities engineering. Hence this chapter enumerates the routing organization of IoT with smart applications aggregating real-time oil pipeline sensor data as big data subjected to machine learning algorithms using the Hadoop platform.

Chapter 13: Security in Rail IoT Systems – An IoT Solution for New Rail Services

Indra Sistemas S.A. have designed and developed a safety and secure solution system for the Rail Transportation environment based on a distributed architecture under the domain of the Industrial IoT that enables V2V, V2I and I2I communications, allowing peer-to-peer data sharing. UPM has designed and implemented a HW based security infrastructure for extreme edge devices in IoT. The implementation takes advantage of HW accelerator to enhance security in low resources devices with a very low overhead in cost and memory footprint. Current security solutions are problematic due to centralized control entity. The complexity of this kind of system resides in the management, in a decentralized way, of the security at each point of the distributed architecture. This document describes how the system secures all the infrastructure based on a distributed architecture without affecting the throughput and the high availability of the data in order to get a top-performance, in compliance with the strengthen safety and security constrains of the rail environment's regulations.

Chapter 14: The Internet of Things in the Russian Federation – Integrated Security

The chapter presents a study on ensuring the information security of the Russian Federation in the field of the Internet of things (IoT); an analysis of Russian state policy in this field and methods for its implementation in terms of technical and legal regulation; the directions of state development in field creating a system of regulation of relations in the field of IoT. To present the general picture of the state's opposition to the risks and threats arising from the use of the IoT, a comparison is made of the risks and threats used by the FSTEC of Russia and ENISA. The authors disclose Russian approaches to ensuring information security, reflected in state strategic documents, including the strategy adopted in 2015 in the field of ensuring information security by switching to their own information technologies. In conclusion, recommendations are made for government bodies and users to ensure integrated information security.

Chapter 15: Artificial Intelligence and SHGs – Enabling Financial Inclusion in India

For inclusive growth and sustainable development of SHG and women, empowerment there is a need to provide an environment to access quality services from financial and non-financial agencies. While banks cannot reach all people through a 'brick and mortar' model, new and advanced banking technology has enabled financial inclusion through branchless banking. By using artificial intelligence in banking, banks have a cost-effective and efficient solution to provide access to services to the financially excluded. Digital technology improves the accessibility and affordability of financial services for the previously unbanked or underbanked individuals and MSMEs. A big data-driven model can also, be helpful for psychometric evaluations. Several psychometric tools help evaluate the applicant's answers which aid to capture information that can help to predict loan repayment behavior, comprising applicants' beliefs, performance, attitudes, and integrity.

Chapter 16: Applicability of WSN and Biometric Models in the Field of Healthcare

Health is considered as the most important ingredient in human life. Health is wealth, most frequent used proverb used in day to day lives. A healthy person can perform its entire task with full enthusiasm and great energy and can solve all problems as mind is a powerful weapon which controls all our functioning. But now due to change in our life styles, we are becoming prone to all kinds of health hazards. Due to unhealthy mind we are not able to perform any task. Humans are becoming victims of many diseases and one of the most common reason for our degradation in health is stress. In this paper, we present role of WSN & biometric models such as two factor remote authentication, verifying fingerprint operations for enhancing security, privacy preserving in healthcare, Healthcare data by cloud technology with biometric application, & Validation built hybrid trust computing perspective for confirmation of contributor's profile in online healthcare data. A comparison table is formulated listing all the advantages and disadvantages of various biometric based models used in healthcare.

Chapter 17: Augmented Data Prediction Efficiency for Wireless Sensor Network Application by AI-ML Technology

Practical Wireless Sensor Network (WSN) demands cutting edge Artificial Intelligence (AI) technology like Deep Learning (DL) which is the subset of AI paradigm to impart intelligence to end devices or nodes. Innovation of AI in WSN aid the enhanced connected world of Internet of Things (IoT). AI is an evolving area of intelligent learning methodologies by computers via Machine Learning Algorithms (MLA). This research paper entirely deals on the implementation of AI technologies in the areas of advanced Machine Learning, Language Recognition using Natural Language Processing (NLP) and Image Recognition through live example of Machine Learning. MLA are constructed to predict optimized output by giving training dataset inputs. In image recognition, an outcome model utilizing the existing reference model, to predict DL based AI prediction. Complex DL AI services is achieved by Bluemix sole power-driven Watson studio and Watson Assistant Service. Application Programming Interface keys are designated to connect Watson and Node Red Starter (NRS) to provide the web interface.

The readers of this book will be benefited about the evolution, usage, challenges and the proposed countermeasures associated with the adoption of WSNs and IoTs for wider applications. The book aims to enable the readers to realize the existing security facilities, their limitations and future possibilities. The proposed book focusses to publish original research outcomes towards industries and healthcare using various technological developments. Therefore, the readers will gain insights to taxonomy of challenges, issues and research directions in this regard. The book aims to showcase the basics of both WSNs and IoT for beginners as well as their integration and challenge discussions for existing practitioner. The readers will gain exposure to a novel paradigm where IoT, Big-data and Cloud are merged together to solve several real-time problems owing to the existence of inter-dependency between these mutually exclusive technologies. Further, the book aims to bring together state-of-the-art innovations, research activities (both in academia and industry), and the corresponding standardization impacts of machine learning and deep learning so as to make the readers aware of the requirements and promising technical options to enrich and boost research activities in this area.

Chapter 1
Introducing Machine Learning to Wireless Sensor Networks:
Requirements and Applications

Vidit Gulyani
https://orcid.org/0000-0002-8962-6098
HMR Institute of Technology and Management, Delhi, India

Tushar Dhiman
HMR Institute of Technology and Management, Delhi, India

Bharat Bhushan
https://orcid.org/0000-0002-9345-4786
HMR Institute of Technology and Management, Delhi, India

ABSTRACT

From its advent in mid-20th century, machine learning constantly improves the user experience of existing systems. It can be used in almost every field such as weather, sports, business, IoT, medical care, etc. Wireless sensor networks are often placed in hostile environments to observe change in surroundings. Since these communicate wirelessly, many problems such as localisation of nodes, security of data being routed create barriers for proper functioning of system. Extending the horizon of machine learning to WSN creates wonders and adds credibility to the system. This chapter aids to present various machine learning aspects applied on wireless sensor networks and the benefits and drawbacks of applying machine learning to WSN. It also describes various data aggregation and clustering techniques that aim to reduce power consumption and ensure confidentiality, authentication, integrity, and availability amongst sensor nodes. This could contribute to design and alter pre-existing ML algorithms to improve overall performance of wireless sensor networks.

DOI: 10.4018/978-1-7998-5068-7.ch001

INTRODUCTION

Cluster of sensor nodes deployed to observe different environment scenarios by sharing sensed information to other nodes and central body called base station (BS) collectively attribute to Wireless sensor networks. These nodes operate on low power and are stationed, specifically to observe change in surroundings. Sensors have limited resource and are stationary. These are posted in various scenarios to sense and report any change in environment. WSN's are used in field of healthcare, hostile and military grounds, weather forecasting etc.

Sensor nodes observe numerous contextual changes such as pressure, weather, thermal and optical. These are arranged in a specific fashion. Sensor nodes report change in event of their interests. If more than one node in the network sense the same event, they all work in collaboration by gathering data from the event (Lan et. al, 2017). Only one node is allowed to generate the final report. Report is delivered to Base station either directly or by hopping the information on other sensor nodes through wireless medium. Base station unit is a gateway between nodes and external world. Each node in the network has following three components:

a) Sensor subsystem: senses any change in environment.
b) Processing subsystem: it generates report of the change recorded from environment
c) Communication system: this unit sends the report to base station and receive any instruction from base station.

Figure 1 depicts a block diagram for Wireless sensor Networks.

Figure 1. A typical representation of Wireless sensor networks

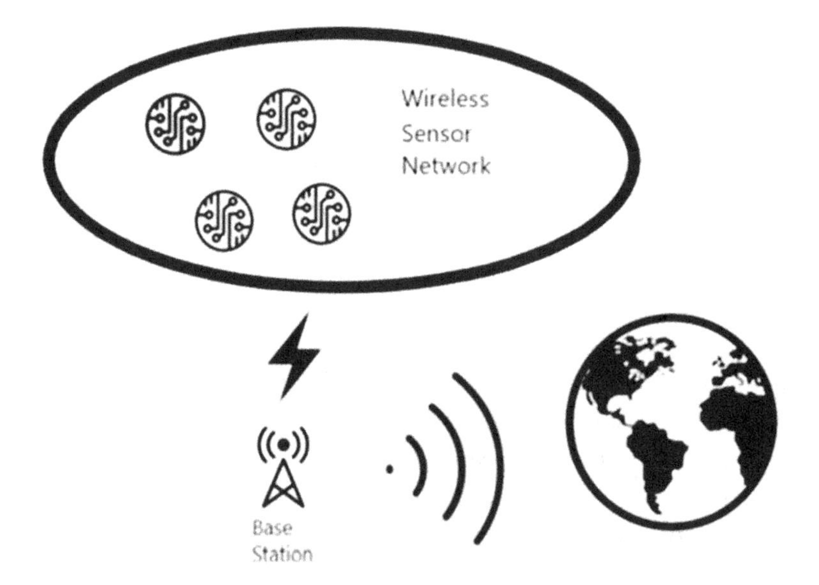

Requirements for WSN mainly comprise non-functional requirements of the system. Major non-functional requirements are security, less power consumption, reliability of the network, fault and intrusion detection. Inducing Machine Learning algorithms to WSN increases reliability as any malicious activity such as tampering of sensor data, and intrusions are easily detected and removed from network by clustering the previous data and removing outliers from each cluster (Zhao et al, 2015; Wang et al, 2019).

Machine learning emerged in mid-20th century to serve as ground for Artificial Intelligence (AI). Increased use of software's worldwide generated enormous data in clouds, hence machine learning expanded its horizons to almost every field. Different researchers define ML in varied tones as per the application ground. Combining all these definitions, machine learning can be defined as: Training a machine model to predict future outcomes by analysing old patterns and learn new ways of behaviour using statistical data as experience. Machine learning improves experience of prebuild software systems. E.g. spam detection in Gmail, AI based opponent players in games. Voice assistants and many more. Machine learning also serves as a ground for deep learning systems such as autonomous vehicles, self-trained robots etc. With robust algorithms ML analyses previously occurred patterns in various fields e.g. business, sports, weather predictions to predict future outcomes with high confidence level .

Coupling WSN and ML adds credibility to network. Ad-hoc networks have increased the use of Machine Learning algorithms from past few decades. Non-functional requirements in WSN are attained by applying machine learning algorithms on data generated by sensor networks some of which are listed below:

1. WSN are mostly positioned by armed forces in hostile environments to keep an eye on enemy's activities. It becomes important to ensure security as the system can be attacked and data can be modified. ML algorithms play their parts by recognizing any change in behaviour of nodes by using k-NN and other clustering algorithms. Details about these algorithms are discussed in Section III.
2. Furthermore, Intrusion and anomalies in the network are eliminated. It improves QoS (Quality of Service) thereby increasing reliability on the network.
3. Sensor networks are resource limited. Various clustering algorithms use very less computational power and proves to be an asset for wireless sensor networks.

The major contribution of our work is as follows. This work defines wireless sensor networks and their requirements, achieved by applying machine learning. It also summarizes three branches of machine learning namely supervised, unsupervised and reinforcement learning applied in WSNs. This article also aids to present two important concepts of ML *data aggregation* and *data clustering* which improves overall system performance. Finally, it summarizes ML aspect introduced in WSN, that benefits the application of WSN in hostile scenarios also opening horizons for further research avenues in respective fields of technology.

The remainder of the paper is organised as follows: Section II discusses Security and Non-functional requirements for WSN. Section III gives a brief introduction to various branches of machine learning and algorithms used in WSNs. Section IV describes data clustering and data aggregation with their effects on WSNs. Section V present benefits of machine learning algorithms applied on WSNs and their advantages. Section VI concludes overall report and describes future of WSNs coupled with machine learning.

REQUIREMENTS IN WSN

Security Requirements in Wireless Sensor Networks

The presence of risks and the rigid environments requests makes security considerations in the architecture of WSN an important concern. Generally, following security services are required in WSNs:

Confidentiality

Maintaining the confidentialness of crucial data passed amongst sensor nodes is important. Crucial sections of a packet are ciphered before transferring data from sending node, the sections are deciphered at the receiver node. Without the analogous decryption clefs, invaders are inhibited from approaching the crucial information. Fraction of data that has to be encrypted relies upon the application. In few cases, only the data section of a packet which is crucial is encrypted, whereas some cases involve encrypting packet header to hide the node identity.

Authenticity

Authenticity is the belief of the identity recognition of communicating nodes. Each node should investigate that a received data or information comes from an actual sender. If authentication is not ensured, attackers can bluff node identities to send false message or data to ad-hoc networks. Generally, a fitted MAC (message authentication code) may be utilized to authenticate the genesis of a signal or message.

Integrity

To avoid the modification of information by attackers, Integrity should be given to the transmitted information which confirms that no harm is caused to sensor data by attackers. Interference can be introduced by attackers in few bits of transferred packets to variate the polarities. A malevolent routing node may also variate essential information in packets, prior to sending. Like a CRC (cyclic redundancy checksum) consumed to find out random defects during packets transfer, can save packets against modification, a keyed checksum, such as a MAC.

Availability

It implies another essential ability of a wireless sensor network to endue services whenever they wanted. In spite of, attackers can launch invasion to disparage the network abilities or even dissipate the whole network. A DoS (Denial of Service) invasion is an extremely harmful threat to network (Wood and Stankovic, 2002). This happens when adversary attacks the network to mislay the overall efficiency by enduing services by decreasing the strength of nodes through different tricky procedures, outraging network protocols, transmitting radio interference.

Non-Functional Requirements

Non-functional requirements also termed as Non-behavioural requirements signify parameter which do imply to operational behaviour of the model. Some of the Non-functional requirements are data integrity, time taken to process the queries, intrusion detection, quality of service and security. This section highlights non-functional requirements achieved due to recent advancements in machine learning algorithms.

Intrusion Detection and Security

Domain of security is major challenge for WSN due to constraints on resources. Attack strategies aims at generating false, misbehaving values for sensor nodes. Anomaly detection problem is solved by classification and clustering algorithms. Sensor nodes readings are classified into cluster of correct and incorrect readings. The data points lying in cluster of expected behaviour are taken for calculations. The outlying region contains misleading values and are eliminated. WSN security advancements helps at generating the following rewards:

- Removing misbehaving values increases the network reliability.
- Saves nodes energy by eliminating outlier values.
- Increased life of WSN network due to saving of nodes energy.
- Network learns to eliminate these attacks over time without human involvement.

Quality of Service, Fault Detection and Data Integrity

QoS (Quality of service) ensure high-priority transportation of data and real-time events. Wireless Sensor Networks suffer from bandwidth and energy constraints which limits the amount of information that shall be delivered to the destination nodes from a source. Furthermore, dissemination and data aggregation in WSN can be unreliable and faulty (Paradis and Han, 2007). When random network topologies are coupled with these issues an important challenge is introduced for designing a reliable algorithm for these types of networks. The general QoS requirements and state of the art in Wireless Sensor Networks have been reviewed in (Chen and Varshney, 2004).

The adoption of machine learning methods to reach data aggregation and Quality of Service (QoS) results in advantages as follow:

- Different types of streams are recognized by using different machine learning classifiers, thus removing the need of flow-aware management techniques.
- The requirement for data integrity, fault detection, QoS rely on the application and network service. While ensuring capability of resource utilization machine learning methods are capable to manage power utilization and bandwidth.

ML ALGORITHMS HIERARCHY FOR WSN

Various machine learning algorithms (including supervised, unsupervised and reinforcement learning) employed in the realm of WSNs are detailed in the subsections below.

Supervised Learning

Supervised learning involves training of machine model using named labelled data. Teaching becomes effective when data is well defined hence new data can be easily classified using the previous data (labelled) used in training the model. This section discusses most frequently used algorithms, used to train WSN model. Supervised algorithms are used for various classification and regression applications .Supervised learning solves various WSN issues, some of which include query processing and event detection (Zhu et al., 2018), object targeting and localization (Ullah et al., 2019), media access control, data integrity (Zhang et al., 2015), Quality of Service(QoS), intrusion detection and security (Butun et al, 2014).

k-Nearest Neighbour (k-NN)

K-NN is a supervised classification algorithm which divides data into groups called clusters. Data points in a cluster have similar nature. Whenever, new data is added to the set, it is compared with clustered data points to check to which cluster it belongs. No of comparisons depend on value of **k**. Points are compared on the base of nearest distance. Consider 3 clusters A, B, C and a new data point is compared k times where k is equal to 15. The number of data points equal to cluster A, B, C are 6, 8, 1 respectively. The result would be assigned as cluster B. **Figure 2** depicts working of k-NN algorithm

Figure 2. Working of k-NN algorithm

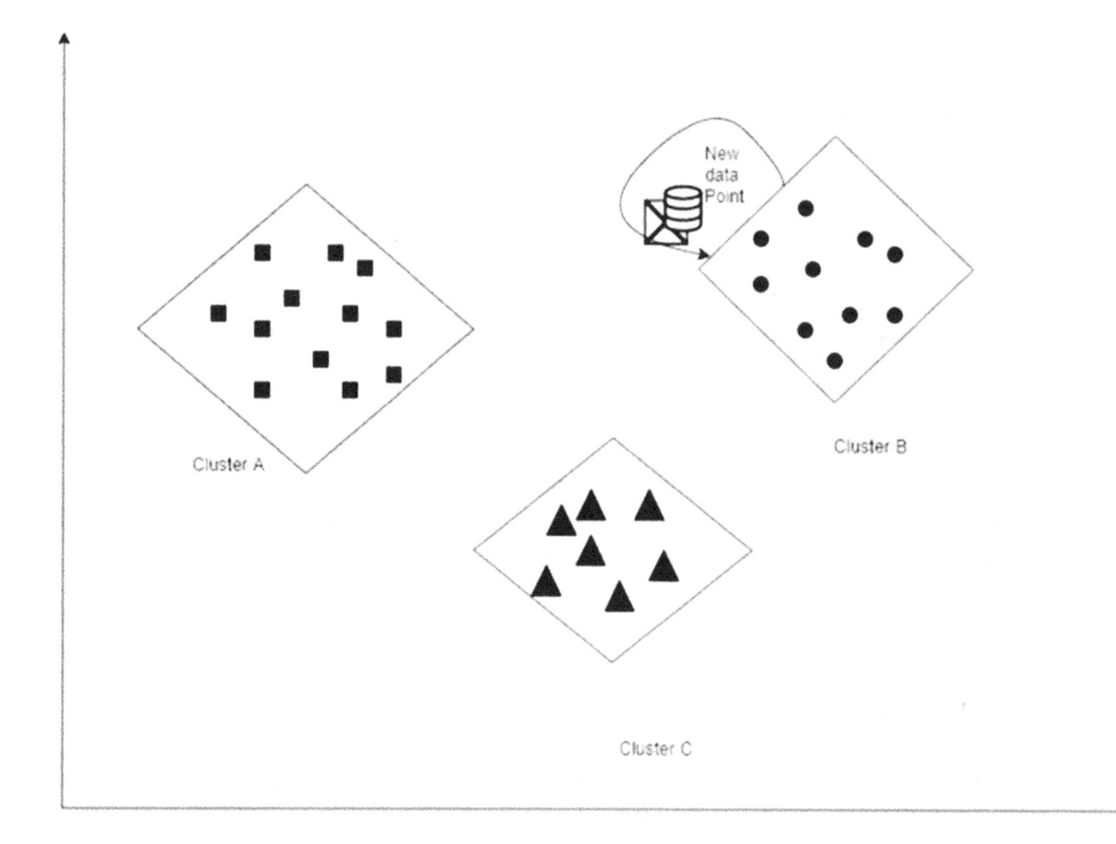

Liqiang and Jianzhong (2010) shows application of k-NN to predict missing data points in WSN's. K-NN require moderate CPU usage but results are very accurate. Large no. of clusters leads to inaccurate results as distance from data points of various clusters become similar, hence accuracy is decreased (Beyer et al., 1999). Query processing subsystem is also one of the major applications of k-NN algorithm in the domain of WSN (Fu et al., 2010). Ahmed et al. (2018) have presented increase in lifetime of wireless sensor networks using k-NN algorithm. k-NN is used for intrusion detection is WSN (Wenchao et al., 2014).

Decision Tree (DT)

Another classification algorithm that uses *linearly separable* labelled data. One goal of this algorithm is to classify a new entry of data from pre-existing data by taking sequence of Boolean decisions. The Randomness in the data is called the *entropy*. The condition for decision is taken such that the entropy for next level is decreased maximum e.g.- For classifying a banyan tree, apple tree and bamboo tree, the first condition would be taken by comparing heights of trees. If the height of tree is greater than 40 feet, then it maybe a banyan or bamboo tree else it is apple tree. **Figure 3** shows classification using decision tree algorithm. The information obtained by classifying at each step is called *information gain*. Entropy is decreased after classifying at each step and the procedure is repeated till all the different items in the data become *leaf nodes* of the decision tree.

Figure 3. Decision Tree Classifier

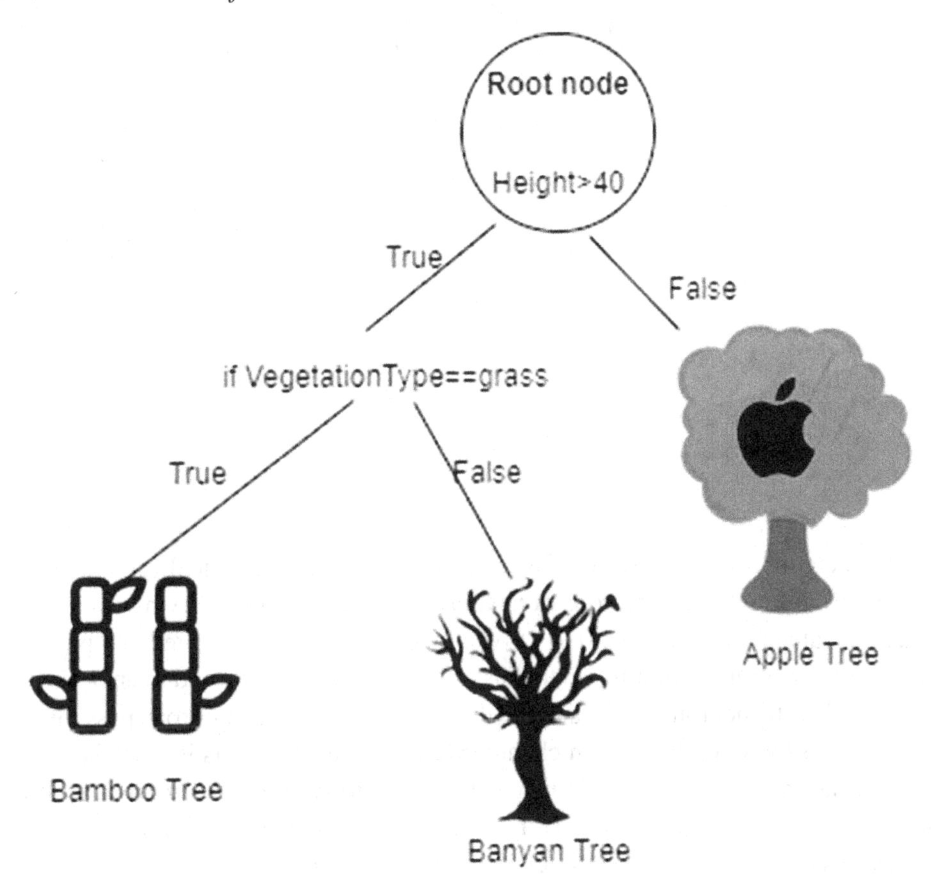

WSN's apply DT algorithm to check the link reliability of sensor nodes in network by comparing several parameters e.g. loss rate, data corruption rate etc. DT is also used to increase lifetime of WSN by identifying cluster head (Ahmed et al., 2009).

Neural Networks (NN)

Neural networks is a classification and prediction (regression) algorithm which uses sequence of decision units called neurons serving as core processing units. It has 3 layered architecture the input layer, hidden layer and output layer. **Figure 4** shows a typical neural network with all its processing units.

Figure 4. Neural Network

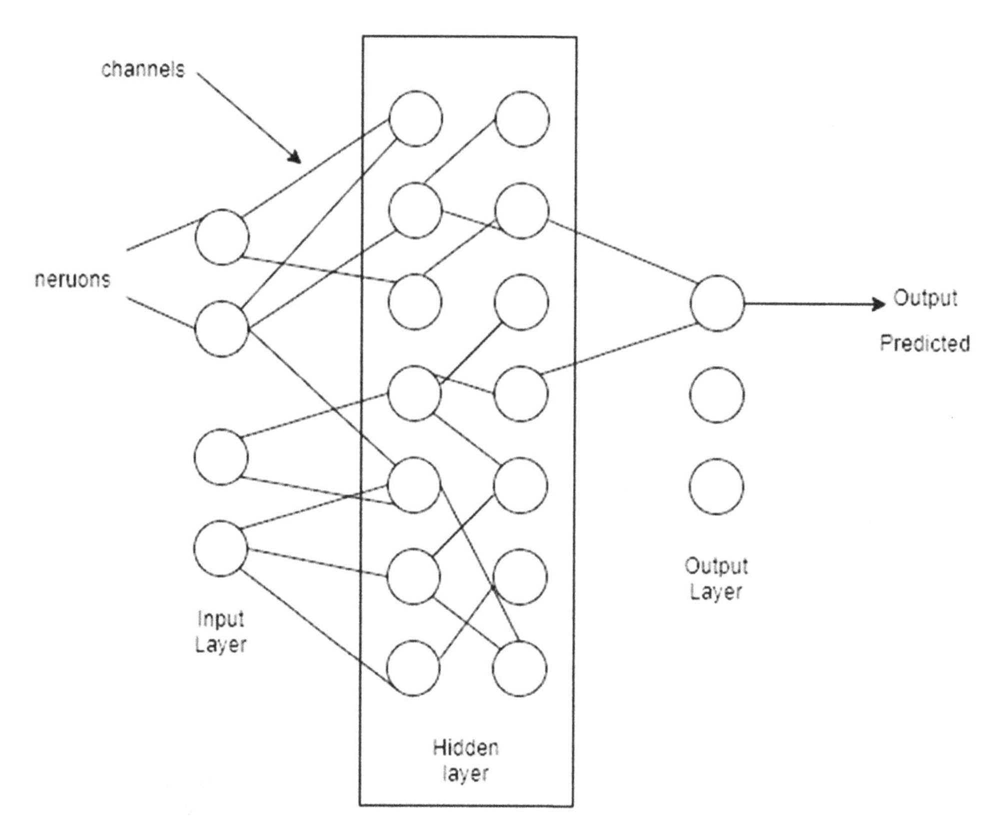

Each layer has neurons arranged in linear fashion. Neurons have a numerical value with it called *bias*. These neurons are connected to the neurons of next layer through channels having some weights which is also a numerical value ranging from 0.0 to 1. Well labelled input is bifurcated and passed to neurons of input layer. Input in neurons is multiplied to bias, this output value is passed to an *activation function* which decides whether the neuron will be activated or not. The activated neurons pass the value to next layer through channels by multiplying with channel weights, this process is iteratively performed till it reaches the output layer. This process is known as *forward propagation.* The output is compared with actual output and error rate is calculated. Error rate is used to change weights of channels. Change in weight of network is done until model starts giving correct output.

For WSN's there are only limited applications of this algorithm because of its high computation power and it takes a lot of time, probably months to train the model. Kumar et al. (2016) showed localization of wireless sensor networks using NN.

Support Vector Machines (SVM)

Support vector machines classifies new data entry only between two classes of data. Labelled data is divided into two groups. These clusters are separated by a line known as *hyper-plane.* Hyper plane is taken at a point where maximum distance from *support vector* occurs. Support vectors are corner-most data points from each cluster. The gap between support-vectors of two clusters is called the *distance margin (D).* hyper plane lies exactly at middle of distance margin. The behaviour of new data point determines to which side of hyper-plane it lies. Gap between clusters and hyper-plane is taken maximum to classify the set accurately.

SVM is very useful and efficient algorithm to classify two set of data. It doesn't require high computational power, hence makes space for its application in WSN. SVM detects any malicious activity or misbehaviour of node by comparing it with cluster of actual behaviour, making it simple and efficient to eliminate or replace any misbehaving node from the network. **Figure 5** shows SVM in action to eliminate misbehaving values. Energy efficient routing in WSN is achieved using SVM (Khan et al., 2016). Furthermore, geographical location of nodes is a concern which is also solved using SVM (Tran and Nguyen, 2008).

Figure 5. SVM to detect malicious node readings

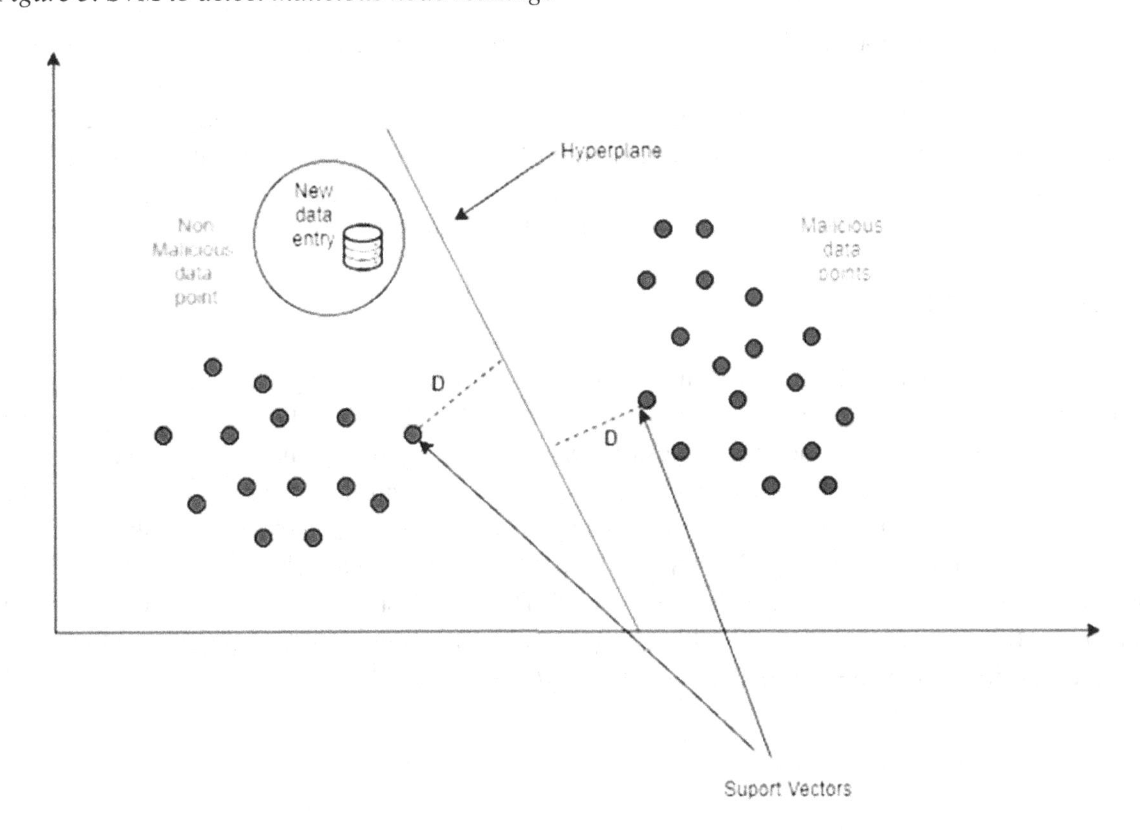

Bayesian Learning

Bayesian learning uses probability at its base to predict the unknown value from the set of clearly defined known values. Small data set is required to train machine model using Bayesian learning. Probability distribution is used to learn certain concepts on its own. Prior statistics are used to find and validate posterior statistics. At the heart of Bayesian learning lies bayes theorem, used to calculate probability of a hypothesis(h) from of given set of data(D) i.e. P(h|D)Where P(h|D) is calculated as

$$P(h|D) = \frac{p(D|h) * p(h)}{p(D)} \quad (1)$$

For hypothesis to be estimated, *maximum likelihood value* i.e. the maximum value of probability from different scenarios for the hypothesis is considered as the posterior belief (prediction). The main advantage of Bayesian statistics lies in the fact, that it gives an accurate estimate using posterior belief before reaching to the final conclusion.

WSN's use Bayesian inference to evaluate event consistence by using missing information in data sets and comparing them with prior environment statistics. It is also used to maintain trust between the sensor nodes (Momani et al., 2010). This statistical knowledge about prior environment restricts the extensive adoption of Bayesian learning for WSN's

Unsupervised Learning

Opposite to supervised, the data in this case is not well labelled. There is no supervision for the algorithm to be controlled if the algorithm goes wrong. Unsupervised algorithm learns on its own over time to resolve unknown patterns. Overtime they become known patterns, which then becomes base of that model to solve problems. Unsupervised learning works by categorizing unknown data into the sets of similar behaviour but biggest disadvantage is you cannot control the algorithm if it starts giving wrong result. You have to start from scratch due to unlabelled data being used for training the model.

K-Means Clustering

Falling in unsupervised category of machine learning it works by representing similar data values into a group called *cluster.* The number of clusters depend on the value of k. This algorithm starts by plotting k initial random centroids for k number of clusters. Distance of every data point is calculated from these k centroids. Closest distance from data point to the centroid is assigned that cluster by evaluating the Euclidean distance. After assigning each data point to a cluster, reallocation of centroid takes place by using the data points of that cluster. The repositioning is performed iteratively until the centroid stops changing its position in accordance with data points. WSN's use clustering to solve the node clustering problem by the steps stated above. It's efficient to use algorithm due to its less requirement of CPU resource and simple implementation even underwater (Harb et al., 2015).

Principle Component Analysis (PCA)

Large data sets contain numerous dimensions, some of which are redundant. Principle component analysis is dimension reduction method to modify data set with large dimensions into set of smaller dimensions. The main focus is to reduce attributes without losing any informative part of data set by eliminating the correlated data.

The data is standardized or scaled to a comparable range to reduce chances of partial output. Covariance of dimensions is calculated followed by evaluating the Eigen values and Eigen vectors. Covariance gives the degree of association among different dimensions. Eigen values and Eigen vectors tell the variance. The dimension having the highest value of variance (Eigen value) has the information content, thus considered principle component1, which is followed by other principle components. The last step is to reduce the data by taking only the principle components as the dimension. A detailed review of PCA procedure can be viewed in (Mishra et al., 2017). PCA reduces redundant data for WSN nodes, thereby reducing the calculation time and resource requirement for WSN algorithms. Detailed use of PCA in WSN can be reviewed in (Borgne et al., 2008).

Reinforcement Learning

Reinforcement learning follows a distinctive approach as compared to other counterparts of machine learning algorithms. Reinforcement learning doesn't consume wide set of data to train the model, rather it learns by its own mistakes. Most combat games use reinforcement learning to train the computer bot to play against human opponent. Reinforcement learning is almost identical to deep learning where the machine models learn on its own. The major difference between these two learning techniques is the deep learning uses well labelled data to train the model by seeing its mistakes. Expected behaviour is defined in deep learning, thus the model can be transformed according to output but the reinforcement model learns by using each of its output as training input. Algorithm's final behaviour is undefined. Sometimes deep learning and reinforcement learning go hand in hand to train various models e.g. autonomous cars. Reinforcement learning is the first and a major step towards Artificial Intelligence.

This algorithm involves an agent (i.e. the machine model on which algorithm is applied) which learns by performing some actions to get the reward. The action taken by the agent is to maximize the reward The agent changes its state after receiving the award from the environment. Agent uses a policy to decide the next action to move to next state. The approach to solve reinforcement learning algorithm is called *Markov's decision process (MDP).* MDP is order pair represented by (S, A, $\{P_{sa}\}$, γ, R).Where,

- Si is finite set of states i.e. no of positions an agent can hold.
- {A} is set of actions that agent can perform to change the position from one state to the next state
- P_{sa} is the transition probability or chance of policy that agent will incorporate to move to next state
- γ is the discount factor that lies between 0 and 1. If the value of discount factor lies near to 0, the agent will not take a risk by compromising on smaller value of reward. Discount factor near to 1 implies agent taking risk to catch a higher value of reward.
- R is the reward function R:SxA-> R signifies at state S taking action A will generate a reward R

There is no fix behaviour of this model. For changing the behaviour, the algorithm must be redesigned from scratch. Adaptive routing for WSN underwater is presented in [24]. Reinforcement learning is also applied to save energy in communication between the nodes (Savaglio et. al, 2019).

FACETS OF WIRELESS SENSOR NETWORKS

Data Clustering

In extensively energy-constrain WSNs, transmission of entire data straight to the sink is inefficient. An economical way is to send data to a cluster head well-known as local aggregator which adds each of the sensors data inside its cluster and forwards it to the sink. This will generally lead in energy saving. The most favourable picking of the cluster head has been discussed in (Crosby et al., 2006). Comparison and taxonomy of elegant clustering algorithms is given in (Abbasi and Younis, 2007).

Researchers have classified clustering in two types: *Soft* and *Hard*. In hard clustering, every data point is allied to a cluster fully or not. Data points in soft cluster can belong to more than one cluster. Liu et al. (2012) used soft clustering to identify dissimilarity in shapes.

The cluster-based collection of data towards a base station from sources within Wireless Sensor Networks is represented in Figure 6. In figure, there can be a few defective nodes that should be erased from the WSNs. These defective nodes may produce inaccurate results which adversely influence the correctness of the complete network. The operations of data collection and node clustering is improved as data is compressed generally at cluster heads by taking out dissimilarity. Lee et. al. (2016) presented three layered adaptive clustering hierarchy for WSN.

Data Aggregation

Aggregation of data in WSNs affects different parameters like computational overhead, communication overhead, memory and power. Minimizing communication overhead and number of transfers, data aggregation plays a valuable role in wireless sensor networks. An accomplished data aggregation procedure equilibrates the energy usage of the sensor nodes and increases the network lifetime. There are various types of data aggregation procedures: centralized data aggregation, in-network aggregation, tree-based data aggregation and cluster-based data aggregation. Selection depends upon the network fabrication. Various ways to approach data aggregation in wireless sensor networks have proposed in (Ambigavathi and Sridharan, 2018). Benefits of data aggregation are as follows:

- ML accessibility select accomplished cluster heads(CH) in the circuits for data aggregation which equilibrates the power of the sensor nodes.
- ML methods are very useful for dimensionality minimization of information at the sensor node stages thereby minimizing the communication overhead in the system. The minimization can be executed at cluster heads or sensor nodes to reduce the latency in transfer of information.
- ML siphon the atmosphere and work according to it without reprogramming or reconfiguring in the reference of data aggregation.

Song et al (2013) proposed a DLRDG (distributed linear regression-based data gathering) method has been introduced to increase the CHs operativity in wireless sensor networks. Based upon the previous data at every CH in wireless sensor networks, the regression algorithm predicts the genuine monitoring measurements and transmit to the base station. Linear regression procedures also select the incorrect nodes in the system network during the information consolidation process. An energy accomplished multivariate information minimization model has been evolved by Atoui et al. (2016) based upon periodical data aggregation by polynomial regression functions. The decrease of information in sensor level will decrease the communication overhead all along the transfer of the information in between the CHs or nodes to the base station. Euclidean distance is not solely adequate in every cases of the wireless sensor networks, it's also needs to assess a spatial and temporal correlation within the sensor node in the complicated wireless sensor networks formation. A linear regression method is developed by Gispan et al. (2017) to get a rough calculation of the non-random parameters while data consolidation. The procedure of collection and aggregation of data from sensor nodes is shown in **Figure 6**.

Figure 6. Cluster based data aggregation

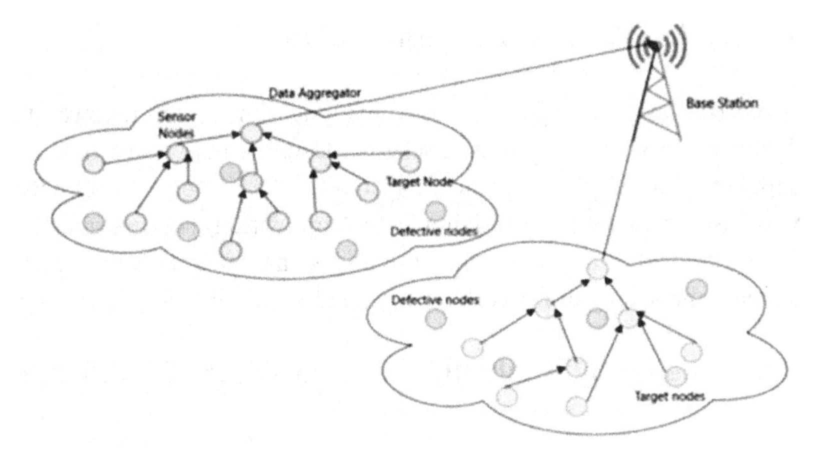

BENEFITS TO WIRELESS SENSOR NETWORKS ON APPLYING MACHINE LEARNING ALGORITHMS

Clustering Large Scale Network Using Neural Network

A neural network establishes the relation between distinct levels of received signal power and the position of wireless sensors in an indoor habitat. Hongmei et al. (2009) discussed how neural networks are used for the development of self-managing clusters. The clustering problems on huge scale network with small transmission radii are targeted by this scheme where consolidation algorithms may or may not work precisely. However, for large transmission radii, the performance of the algorithm is nearly same as centralized algorithms.

Decision Tree to Elect Cluster Overhead

Ahmed et al. [35] used the decision tree algorithm to elect cluster head. Several critical features such as battery level, cluster centroids, the vulnerability indications, and degree of mobility are used for this approach by repeating the input vector via decision tree.

Gaussian Process Models for Sensor Readings

The collaboration of random variables that are parameterized by covariance and mean methods are called Gaussian process (GP). A scheme is presented by Ertin (2007) to initialize probabilistic models of the readings, formed on Gaussian process regression. Kho et al. (2009) expanded Gaussian process regression to adjust sensor information depending on its significance. Study was to define trade-off between optimality and computational cost by focusing on energy consumption. For predicting smooth functions with limited training data sets Gaussian process models are preferred. However, when dealing with huge scale networks WSN designers must recognize the immense computational complexity for the methods.

Self Organizing Map (SOM) for Data Aggregation

The competitive memorization, for mapping spaces from higher dimensions to lower dimensions is self-organizing map algorithm. It is an unsupervised scheme. Usually, map units form a two-dimensional lattice, thus the mapping is from high dimensional spaces onto a plane. A novel network architecture known as "CODA" (Cluster-based self-Organizing Data Aggregation) is proposed by Lee et al. (2005). Using a self-organizing algorithm in this scheme, the nodes can categorize the aggregated data. Data aggregation will increase the saving of network energy, quality of data, decreasing the network traffic.

Online Data Compression by Applying Learning Vector Quantization

While an entire knowledge is required regarding the network topology in the above methods, some of the algorithms does not have such restriction. For example, a scheme called "ALVQ" (Adaptive Learning Vector Quantization) introduced by Lin et al. (2006) to retrieve constricted versions of data accurately by the sensor nodes. Using historical patterns and data correlation, the LVQ learning algorithm is used by ALVQ to forecast the codebook by using the previous training samples. The ALVQ scheme increases the efficiency of recovering authentic reading from the constricted data, minimizes the bandwidth required during the transmission. The dead neurons are vital disadvantage of practicing LVQ for online data aggregation as they are very far from training samples. Algorithms which are robust against outliers are important to be developed. And it is relevant for representing large datasets by very less vectors.

Principal Component Analysis for Data Aggregation

To enhance data aggregation there are two important algorithms (Compressive sensing, Expectation-maximization) which is accurately used in combination with PCA (principal component analysis) in Wireless Sensor Networks.

- **Expectation-maximization algorithm (EM)** is a ceaseless method in which we calculate maximum a posteriori (MAP) or maximum likelihood estimates of criterion in statistical models, where the model is depended on unobserved latent variables. It is made up of two steps, i.e., an E (expectation step) and M (maximization step). Along with its E-step, while fixing the current expectation EM formulates the cost function of the system's criterion. After that, the M-step recomputes the criterion using which the cost function's approximation error is minimized.
- **Compressive sensing algorithm (CS)**: it is an efficient method of acquiring images or signals with minimum numbers of samples. It is searched to change the traditional method "sample then compress" by "sample while compressing". From few random measurements to recover the original signal Compressive sensing explores sparsity property of signals.

Using few collected samples from Wireless sensor Networks a method is developed by Masiero et al. (2009). for measuring distributed observations. This method is based on PCA scheme to develop orthogonal factors used by CS to regenerate the authentic readings. In addition, due to its capability to measure temporal correlations and spatial data, this method is independent from the routing agreement. Similarly, to enhance the transportation of reading directly to the base station PCA is applied by Rooshenas et al. (2010). By combining nodes PCA concludes in a considerable traffic deduction from aggregated data into small packets. In intermediate nodes this distributed scheme is performed to combine incoming packets rather than sending them to destinations. The main concern of PCA-based data collection is high computational requirement for the solution which magnificently cope up with the higher data dimensionality reduction (dimensionality of aggregated data by preserving only crucial information).

Collaborative Data Processing Using K-Means Algorithm

Li et al. (2002) addressed about tracking a single target and distributed detection by the fundamental concept using sensor networks. Collection of information from the monitored environment is a framework of "Collaborative Signal Processing" (CSP). In addition, using classification techniques like K-nearest neighbours and SVM, this algorithm can track numerous targets.

Role-Free Clustering

Murphy and Foster (2009) introduced the Wireless Sensor Network cluster formulation scheme "CLIQUE" (Role-Free Clustering with Q-Learning for Wireless Sensor Networks). CLIQUE empowers every node to examine the abilities of itself to function as head of the cluster node.

Decentralized Learning for Data Latency

Mihaylov et al. (2010) talked about the concern of higher data latency using reinforcement learning in arbitrary topology sensor networks. To optimize the data aggregation, each node locally executes the learning algorithm without the requirement for a central control station. Therefore, the accuracy of the network is increased with small learning transportation overhead. This approach preserves the budget of node energy during data collection process, and hence increase the network lifeline.

Outlier Detection

Using k-NN (k-Nearest Neighbour)

The system of outlier detection using k-nearest neighbours was developed by Branch et al. (2013). Furthermore, missing readings of nodes can be replaced by mean value of neighbour nodes. Drawback to this is, it needs extra memory space to store readings of neighbour nodes gathered from monitoring environment.

Using Bayesian Belief Network (BBN's)

The scheme was developed by Janaki-ram et al. (2006). Maximum number of neighbour nodes have similar readings due to spatial and temporal correlations. Conditional relationship among node readings is evaluated and conditional dependency is used to deduce outliers among data. Furthermore, the missing readings of sensor nodes are deduced using the same technique.

Using Quarter Sphere SVM

Heavy computational requirement of one-class SVM can be reduced with the application of quarter sphere SVM (Laskov et al., 2004). Quartered sphere SVM encloses correct data points in a sphere, all the misbehaving nodes lie out of spherical region. Eliminating these tampered nodes reduces communication overhead. Zhang et al. (2013) deduced an outlier detection scheme by investigating spatial and temporal correlation of gathered readings using traditional SVM. Yang (2008) used unsupervised quarter sphere SVM to reduce computational complexity of one-class SVM based detection of outliers.

These SVM algorithms offer efficient learning but lag due to scalability of data values in large data sets.

Detecting Grey Hole and Black Hole Attacks Using SVM

In black hole attacks malicious nodes refuse to transfer the packets, whenever routing request (RREQ) is received, instead they start giving routing reply to source nodes that initiated RREQ (Tseng et al., 2011). Source nodes are misguided and assume that packet was delivered to the destination but there is no routing reply. This is also termed as packet drop attack. In grey-hole attacks malicious nodes transfer the packets on selective basis by deliberately dropping some packets leading into transferring of incomplete information. This is known as selective forwarding attacks. Kaplantzis et al. (2007) proposed a technique to detect grey and black hole attacks by using traditional SVM. It takes into account routing information, hop count and bandwidth to detect malicious nodes of network.

Self Organizing Map for Analysing Attacks

Avram et al. (2007) researched to detect attacks using Self Organizing Maps. SOM analyses the properties of data fed as input using neural networks at its core. Complexity of algorithm is to determine input weights. SOM technique is not very useful to detect attack in a huge network of nodes.

QoS Estimation Using Neural Networks

Recently, there is dissemination of interest to improve performance of WSN. For example, Snow et al. (2005) proposed a scheme for estimation of network dependability benchmark of a sensor using neural network. Dependability is a metric that shows survivability, maintainability, reliability and availability of a sensor network.

Metric Map

Link quality measurement tools can provide unstable and inaccurate readings across varying environments because of different conditions such as interference and signal variations. As a result, using supervised learning method to make a link quality estimation framework, MetricMap is presented by Wang et al. (2008). MetricMap increases the MintRoute agreement by acquiring offline and online learning methods, like decision tree which are used to gain link quality indicators. This type of frameworks uses local characteristic to construct classification tree such as transmission buffer size, RSSI (received signal strength indicator), the backward and forward probabilities and channel load. the ratio of the collected packets to the transferred packets over the link l is known as forward probability whereas backward probability is deduced over converse path. Three times development in data transmission rate can be achieved over basic MintRoute method.

Assessing Correctness and Credibility of Nodes Using Multi-Output Gaussian Processes

A real-time algorithm is presented by Osborne et al. (2008) to ascertain a set of nodes which are capable of managing data processing tasks such as predicting the missing reading and judge the accuracy of gathered sensor readings. A probabilistic Gaussian process based on repitive implementation is provided by this algorithm which is trained to maintain a rational training data size and re-use the prior experience.

Delivering QoS Using Reinforcement Learning

A QoS task scheduler is introduced by Ouferhat and Mellouk (2007) based on Q-learning technique for adaptive multimedia sensor networks. By reducing the transmission delay this scheduler significantly increases the network productivity. Comparatively, a QoS metric in WSN's are considered by Seah et al. (2007) which represents how the area of intrigue will be perceived efficiently. To find weakly monitored sites a distributed learner is developed by using Q-learning method. In the above QoS mechanisms energy harvesting is not considered. Conversely, a QoS-aware power management method is introduced by Hsu et al. (2009) for Wireless Sensor Networks with energy harvesting capacities, namely "RLPM" (Reinforcement Learning based QoS-aware Power Management). With energy harvesting capabilities this method is able to adapt to the changing levels of nodes energy in system. Liang et al. (2009) designed "MRL-CC" (Multi-agent Reinforcement Learning based Multi-hop mesh Cooperative Communication) to be structured as modelling tool for QoS provisioning in Wireless Sensor Networks.

CONCLUSION

This work summarizes every working aspect of machine learning inculcated by Wireless Sensor Networks by clearly defining requirements and applications. Non behavioural requirements that are important for ad-hoc networks is *security* and *less power consumption.* Grey and black hole attacks are eliminated using SVM thereby ensuring integrity. Various clustering algorithms such as k-NN, k- means use very less energy which increases scope of their application. Electing the cluster head saves network power in various scenarios as communication overhead is removed to a great extent. Neural network is implanted in WSN to ensure QoS. This gives rise to new techniques of quality assessment such as metric map and self-organizing map.

Machine learning is an enormous asset for the society. With its possibility to fit almost everywhere in IT applications, it improves the user experience of systems that too without continuous human involvement. Wireless Sensor Networks although placed in adverse environment feel safe under the guidance of Machine Learning Algorithms applied on the them. This extends applicability of WSN in IOT, Threat detection and medical applications such as implanting sensors in human body.

REFERENCES

Abbasi, A. A., & Younis, M. (2007). A survey on clustering algorithms for wireless sensor networks. *Computer Communications*, *30*(14-15), 2826–2841. doi:10.1016/j.comcom.2007.05.024

Ahmed, G., Khan, N., Khalid, Z., & Ramer, R. (2009). *Cluster head selection using decision trees for Wireless Sensor Networks*. doi:10.1109/ISSNIP.2008.4761982

Ahmed, G., Khan, N. M., Khalid, Z., & Ramer, R. (2008). Cluster head selection using decision trees for Wireless Sensor Networks. *2008 International Conference on Intelligent Sensors, Sensor Networks and Information Processing*. 10.1109/ISSNIP.2008.4761982

Ahmed, M., Taha, A., Hassanien, A. E., & Hassanien, E. (2018). *An Optimized K-Nearest Neighbor Algorithm for Extending Wireless Sensor Network Lifetime*. . doi:10.1007/978-3-319-74690-6_50

Ambigavathi, M., & Sridharan, D. (2018). Energy-Aware Data Aggregation Techniques in Wireless Sensor Network. *Advances in Power Systems and Energy Management Lecture Notes in Electrical Engineering*, 165–173. doi:10.1007/978-981-10-4394-9_17

Atoui, I., Makhoul, A., Tawbe, S., Couturier, R., & Hijazi, A. (2016). Tree-Based Data Aggregation Approach in Periodic Sensor Networks Using Correlation Matrix and Polynomial Regression. *2016 IEEE Intl Conference on Computational Science and Engineering (CSE) and IEEE Intl Conference on Embedded and Ubiquitous Computing (EUC) and 15th Intl Symposium on Distributed Computing and Applications for Business Engineering (DCABES)*. 10.1109/CSE-EUC-DCABES.2016.267

Avram, T., Oh, S., & Hariri, S. (2007). Analyzing Attacks in Wireless Ad Hoc Network with Self-Organizing Maps. *Fifth Annual Conference on Communication Networks and Services Research (CNSR 07)*. 10.1109/CNSR.2007.15

Beyer, K., Goldstein, J., Ramakrishnan, R., & Shaft, U. (1999). When Is "Nearest Neighbor" Meaningful? *Lecture Notes in Computer Science Database Theory — ICDT'99*, 217–235. doi: . doi:10.1007/3-540-49257-7_15

Borgne, Y.-A. L., Raybaud, S., & Bontempi, G. (2008). Distributed Principal Component Analysis for Wireless Sensor Networks. *Sensors (Basel)*, *8*(8), 4821–4850. doi:10.33908084821 PMID:27873788

Branch, J. W., Giannella, C., Szymanski, B., Wolff, R., & Kargupta, H. (2013). In-network outlier detection in wireless sensor networks. *Knowledge and Information Systems*, *34*(1), 23–54. doi:10.100710115-011-0474-5

Butun, I., Morgera, S. D., & Sankar, R. (2014). A Survey of Intrusion Detection Systems in Wireless Sensor Networks. *IEEE Communications Surveys and Tutorials*, *16*(1), 266–282. doi:10.1109/SURV.2013.050113.00191

Chen, D., & Varshney, P.K. (2004). QoS Support in Wireless Sensor Networks: A Survey. *Proceedings of the International Conference on Wireless Networks, ICWN'04*, *1*, 227-233.

Crosby, G., Pissinou, N., & Gadze, J. (n.d.). A Framework for Trust-based Cluster Head Election in Wireless Sensor Networks. *Second IEEE Workshop on Dependability and Security in Sensor Networks and Systems*. 10.1109/DSSNS.2006.1

Ertin, E. (2007). Gaussian Process Models for Censored Sensor Readings. *2007 IEEE/SP 14th Workshop on Statistical Signal Processing*, 665–669. doi: 10.1109sp.2007.4301342

Forster, A., & Murphy, A. L. (2009). CLIQUE: Role-Free Clustering with Q-Learning for Wireless Sensor Networks. *2009 29th IEEE International Conference on Distributed Computing Systems*. doi: 10.1109/icdcs.2009.43

Fu, T.-Y., Peng, W.-C., & Lee, W.-C. (2010). Parallelizing Itinerary-Based KNN Query Processing in Wireless Sensor Networks. *IEEE Transactions on Knowledge and Data Engineering*, *22*(5), 711–729. doi:10.1109/TKDE.2009.146

Gispan, L., Leshem, A., & Beery, Y. (2017). Decentralized estimation of regression coefficients in sensor networks. *Digital Signal Processing*, *68*, 16–23. doi:10.1016/j.dsp.2017.05.005

Harb, H., Makhoul, A., & Couturier, R. (2015). An Enhanced K-Means and ANOVA-Based Clustering Approach for Similarity Aggregation in Underwater Wireless Sensor Networks. *IEEE Sensors Journal*, *15*(10), 5483–5493. doi:10.1109/JSEN.2015.2443380

He, H., Zhu, Z., & Makinen, E. (2009). A Neural Network Model to Minimize the Connected Dominating Set for Self-Configuration of Wireless Sensor Networks. *IEEE Transactions on Neural Networks*, *20*(6), 973–982. doi:10.1109/TNN.2009.2015088 PMID:19398401

Hsu, R. C., Liu, C.-T., Wang, K.-C., & Lee, W.-M. (2009). QoS-Aware Power Management for Energy Harvesting Wireless Sensor Network Utilizing Reinforcement Learning. *2009 International Conference on Computational Science and Engineering*. 10.1109/CSE.2009.83

Janakiram, D., Kumar, A., & V., A. M. R. (2006). Outlier Detection in Wireless Sensor Networks using Bayesian Belief Networks. *2006 1st International Conference on Communication Systems Software & Middleware*. doi:10.1109/comswa.2006.1665221

Kaplantzis, S., Shilton, A., Mani, N., & Sekercioglu, Y. A. (2007). Detecting Selective Forwarding Attacks in Wireless Sensor Networks using Support Vector Machines. *2007 3rd International Conference on Intelligent Sensors, Sensor Networks and Information*. doi: 10.1109/issnip.2007.4496866

Khan, F., Memon, S., & Jokhio, S. H. (2016). Support vector machine based energy aware routing in wireless sensor networks. *2016 2nd International Conference on Robotics and Artificial Intelligence (ICRAI)*. doi: 10.1109/icrai.2016.7791218

Kho, J., Rogers, A., & Jennings, N. R. (2009). Decentralized control of adaptive sampling in wireless sensor networks. *ACM Transactions on Sensor Networks*, *5*(3), 1–35. doi:10.1145/1525856.1525857

Kumar, S., Sharma, R., & Vans, E. (2016). Localization for Wireless Sensor Networks: A Neural Network Approach. *International Journal of Computer Networks & Communications, 8*, 61-71. doi:10.5121/ijcnc.2016.8105

Lan, K.-C., & Wei, M.-Z. (2017). A Compressibility-Based Clustering Algorithm for Hierarchical Compressive Data Gathering. *IEEE Sensors Journal*, *17*(8), 2550–2562. doi:10.1109/JSEN.2017.2669081

Laskov, P., Schäfer, C., Kotenko, I., & Müller, K.-R. (2004). Intrusion Detection in Unlabeled Data with Quarter-sphere Support Vector Machines. *PIK - Praxis Der Informationsverarbeitung Und Kommunikation, 27*(4), 228–236. doi:10.1515/piko.2004.228

Lee, J.-S., & Kao, T.-Y. (2016). An Improved Three-Layer Low-Energy Adaptive Clustering Hierarchy for Wireless Sensor Networks. *IEEE Internet of Things Journal*, *3*(6), 951–958. doi:10.1109/JIOT.2016.2530682

Lee, S., & Chung, T. (2005). Data Aggregation for Wireless Sensor Networks Using Self-organizing Map. *Lecture Notes in Computer Science Artificial Intelligence and Simulation*, 508–517. doi:10.1007/978-3-540-30583-5_54

Li, D., Wong, K., Hu, Y. H., & Sayeed, A. (2002). Detection, classification, and tracking of targets. *IEEE Signal Processing Magazine*, *19*(2), 17–29. doi:10.1109/79.985674

Li, W., Yi, P., Wu, Y., Pan, L., & Li, J. (2014). A New Intrusion Detection System Based on KNN Classification Algorithm in Wireless Sensor Network. *Journal of Electrical and Computer Engineering*, 1–8. doi:10.1155/2014/240217

Liang, X., Chen, M., Xiao, Y., Balasingham, I., & Leung, V. C. (2009). A novel cooperative communication protocol for QoS provisioning in wireless sensor networks. *2009 5th International Conference on Testbeds and Research Infrastructures for the Development of Networks & Communities and Workshops*. doi: 10.1109/tridentcom.2009.4976244

Lin, S., Kalogeraki, V., Gunopulos, D., & Lonardi, S. (2006). Online Information Compression in Sensor Networks. *2006 IEEE International Conference on Communications*. 10.1109/ICC.2006.255237

Liu, M., Vemuri, B. C., Amari, S.-I., & Nielsen, F. (2012). Shape Retrieval Using Hierarchical Total Bregman Soft Clustering. *IEEE Transactions on Pattern Analysis and Machine Intelligence*, *34*(12), 2407–2419. doi:10.1109/TPAMI.2012.44 PMID:22331859

Masiero, R., Quer, G., Munaretto, D., Rossi, M., Widmer, J., & Zorzi, M. (2009). Data Acquisition through Joint Compressive Sensing and Principal Component Analysis. *GLOBECOM 2009 - 2009 IEEE Global Telecommunications Conference*. doi: 10.1109/glocom.2009.5425458

Mihaylov, M., Tuyls, K., & Nowé, A. (2010). Decentralized Learning in Wireless Sensor Networks. Adaptive and Learning Agents Lecture Notes in Computer Science, 60–73. doi:10.1007/978-3-642-11814-2_4

Mishra, S., Sarkar, U., Taraphder, S., Datta, S., Swain, D., Saikhom, R., Panda, S., & Laishram, M. (2017). Principal Component Analysis. *International Journal of Livestock Research*, *1*. Advance online publication. doi:10.5455/ijlr.20170415115235

Momani, M., Challa, S., & Al-Hmouz, R. (2010). Bayesian Fusion Algorithm for Inferring Trust in Wireless Sensor Networks. *JNW*, *5*(7), 815–822. doi:10.4304/jnw.5.7.815-822

Osborne, M. A., Roberts, S. J., Rogers, A., Ramchurn, S. D., & Jennings, N. R. (2008). Towards Real-Time Information Processing of Sensor Network Data Using Computationally Efficient Multi-output Gaussian Processes. *2008 International Conference on Information Processing in Sensor Networks (Ipsn 2008)*. 10.1109/IPSN.2008.25

Ouferhat, N., & Mellouk, A. (2007). A QoS Scheduler Packets for Wireless Sensor Networks. *2007 IEEE/ACS International Conference on Computer Systems and Applications*. 10.1109/AICCSA.2007.370885

Pan, L., & Li, J. (2010). K-Nearest Neighbor Based Missing Data Estimation Algorithm in Wireless Sensor Networks. *Wireless Sensor Network.*, *2*(02), 115–122. doi:10.4236/wsn.2010.22016

Paradis, L., & Han, Q. (2007). A Survey of Fault Management in Wireless Sensor Networks. *Journal of Network and Systems Management*, *15*(2), 171–190. doi:10.100710922-007-9062-0

Rooshenas, A., Rabiee, H. R., Movaghar, A., & Naderi, M. Y. (2010). Reducing the data transmission in Wireless Sensor Networks using the Principal Component Analysis. *2010 Sixth International Conference on Intelligent Sensors, Sensor Networks and Information Processing*. 10.1109/ISSNIP.2010.5706781

Savaglio, C., Pace, P., Aloi, G., Liotta, A., & Fortino, G. (2019). Lightweight Reinforcement Learning for Energy Efficient Communications in Wireless Sensor Networks. *IEEE Access: Practical Innovations, Open Solutions*, *7*, 29355–29364. doi:10.1109/ACCESS.2019.2902371

Seah, M. W. M., Tham, C.-K., Srinivasan, V., & Xin, A. (2007). Achieving Coverage through Distributed Reinforcement Learning in Wireless Sensor Networks. *2007 3rd International Conference on Intelligent Sensors, Sensor Networks and Information*. doi: 10.1109/issnip.2007.4496881

Snow, A., Rastogi, P., & Weckman, G. (2005). Assessing dependability of wireless networks using neural networks. *MILCOM 2005 - 2005 IEEE Military Communications Conference*. doi: . doi:10.1109/milcom.2005.1606090

Song, X., Wang, C., Gao, J., & Hu, X. (2013). DLRDG: Distributed linear regression-based hierarchical data gathering framework in wireless sensor network. *Neural Computing & Applications*, *23*(7-8), 1999–2013. doi:10.100700521-012-1248-z

Tran, D., & Nguyen, T. (2008). Localization In Wireless Sensor Networks Based on Support Vector Machines. Parallel and Distributed Systems. *IEEE Transactions on.*, *19*, 981–994. doi:10.1109/TPDS.2007.70800

Tseng, F.-H., Chou, L.-D., & Chao, H.-C. (2011). A survey of black hole attacks in wireless mobile ad hoc networks. *Human-Centric Computing and Information Sciences*, *1*(1), 4. doi:10.1186/2192-1962-1-4

Ullah, I., Chen, J., Su, X., Esposito, C., & Choi, C. (2019). Localization and Detection of Targets in Underwater Wireless Sensor Using Distance and Angle Based Algorithms. *IEEE Access: Practical Innovations, Open Solutions*, *7*, 45693–45704. doi:10.1109/ACCESS.2019.2909133

Valerio, V. D., Presti, F. L., Petrioli, C., Picari, L., Spaccini, D., & Basagni, S. (2019). CARMA: Channel-Aware Reinforcement Learning-Based Multi-Path Adaptive Routing for Underwater Wireless Sensor Networks. *IEEE Journal on Selected Areas in Communications*, *37*(11), 2634–2647. doi:10.1109/JSAC.2019.2933968

Wang, Y., Martonosi, M., & Peh, L.-S. (2007). Predicting link quality using supervised learning in wireless sensor networks. *Mobile Computing and Communications Review*, *11*(3), 71–83. doi:10.1145/1317425.1317434

Wang, Z., Song, G., & Gao, C. (2019). An Isolation-Based Distributed Outlier Detection Framework Using Nearest Neighbor Ensembles for Wireless Sensor Networks. *IEEE Access: Practical Innovations, Open Solutions*, *7*, 96319–96333. doi:10.1109/ACCESS.2019.2929581

Wood, A., & Stankovic, J. (2002). Denial of service in sensor networks. *Computer*, *35*(10), 54–62. doi:10.1109/MC.2002.1039518

Yang, Z., Meratnia, N., & Havinga, P. (2008). An online outlier detection technique for wireless sensor networks using unsupervised quarter-sphere support vector machine. *2008 International Conference on Intelligent Sensors, Sensor Networks and Information Processing*. 10.1109/ISSNIP.2008.4761978

Zhang, J., Ren, F., Gao, S., Yang, H., & Lin, C. (2015). Dynamic Routing for Data Integrity and Delay Differentiated Services in Wireless Sensor Networks. *IEEE Transactions on Mobile Computing*, *14*(2), 328–343. doi:10.1109/TMC.2014.2313576

Zhang, Y., Meratnia, N., & Havinga, P. J. (2013). Distributed online outlier detection in wireless sensor networks using ellipsoidal support vector machine. *Ad Hoc Networks*, *11*(3), 1062–1074. doi:10.1016/j.adhoc.2012.11.001

Zhao, C., Zhang, W., Yang, Y., & Yao, S. (2015). Treelet-Based Clustered Compressive Data Aggregation for Wireless Sensor Networks. *IEEE Transactions on Vehicular Technology*, *64*(9), 4257–4267. doi:10.1109/TVT.2014.2361250

Zhu, W., Cao, J., & Raynal, M. (2018). Energy-Efficient Composite Event Detection in Wireless Sensor Networks. *IEEE Communications Letters*, *22*(1), 177–180. doi:10.1109/LCOMM.2017.2764458

Chapter 2
Secured Energy-Efficient Routing in Wireless Sensor Networks Using Machine Learning Algorithm:
Fundamentals and Applications

Ahona Ghosh
Brainware University, Kolkata, India

Chiung Ching Ho
Department of Computing and Information Systems, School of Science and Technology, Sunway University, Malaysia

Robert Bestak
Department of Telecommunication Engineering, Czech Technical University in Prague, Czech Republic

ABSTRACT

Wireless sensor networks consist of unattended small sensor nodes having low energy and low range of communication. It has been observed that if there is any system to periodically start and stop the sensors sensing activities, then it saves some energy, and thus, the network lifetime gets extended. According to the current literature, security and energy efficiency are the two main concerns to improve the quality of service during transmission of data in wireless sensor networks. Machine learning has proved its efficiency in developing efficient processes to handle complex problems in various network aspects. Routing in wireless sensor network is the process of finding the route for transmitting data among different sensor nodes according to the requirement. Machine learning has been used in a broad way for designing energy efficient routing protocols, and this chapter reviews the existing works in the said domain, which can be the guide to someone who wants to explore the area further.

DOI: 10.4018/978-1-7998-5068-7.ch002

INTRODUCTION

The rise of artificial intelligence (AI) has influenced every field largely; from entertainment to education or from agriculture to manufacturing. Healthcare and computer network are not outside the list which have witnessed a great impact with the magical touch of AI and Machine Learning (ML). In computer network, Wireless Sensor Network (WSN) denotes a collection of spatially isolated and dedicated sensors for monitoring and recording the physical situations of the circumstances and shaping the collected data at a central position. Apart from the different factors like time consumption, energy efficiency, security and cost which contribute to the process of routing in WSN, the active research areas include different Quality of service parameters like Packet Delivery Ratio (PDR), efficiency, robustness, reliability etc. The process of data transmission between different sensor nodes and communication between them is called routing and the goal always remains to reduce the energy consumption during this routing (Pathan et al., 2007) and increase the lifetime of the sensor nodes as much as possible. In this chapter, the design issues of WSN have been addressed first and then the applications of machine learning are described in the concerned domain.

In the next section, related background study and currently available applications of machine learning algorithms in the concerned domain are highlighted. In section 3, the advantages and drawbacks for the existing ML approaches in Energy efficient routing scenario are discussed. Section 4 describes the recent algorithms or techniques used to develop energy efficient routing in WSN and compares them after performance evaluation. Segment 5 provides an overall discussion about the scope and limitation of the chapter and future direction.

Motivation of the Chapter

This chapter collects views of different researchers worldwide from different perspectives. The survey outline presented here is definitely going to help and guide the present researchers in the concerned domain. WSNs can be used to track and monitor the dangerous and unreachable areas where exploration of locations having irregular behaviours like volcanic eruption, forest fire etc. The initial configurations should have the capability of changing its nature to adopt with circumstances, because anytime anything can happen. Machine learning algorithms are capable of calibrating itself to newly acquired knowledge, so application of machine learning in these types of systems will be really useful. The sensor devices are often capable of collecting large data, but sometimes they cannot find the correlation between them. Machine learning can be applied to them for exploring the correlation for better deployment and wide area of coverage which is always desired for the systems. This chapter summarizes the existing systems where limited resource and diversity in the learning patterns have been considered. However, areas like development of distributed and lightweight message transmission system and using machine learning in resource scheduling and management have still remained unexplored. Further experiments and researches can be undertaken in the domain mentioned.

Background of WSN

Wireless Sensor Networks control and monitor rapidly changing environments efficiently. This dynamic nature is sometimes due to some external factors affecting the system and sometimes it is due to some requirement from designer's perspective. To cope up with such ever changing scenarios, machine learning is applied to WSN implementation so that the network learns the nature and trend by its own.

Contribution of the Chapter

In this chapter, we have presented a review of the machine learning methods applied in various literatures over the period 2004-2020 related to WSN, especially in security and energy efficient routing methods of WSN. Various problems of the existing works have been addressed along with experimental finding comparison and advantages of the methods. We have also provided a guidance to current and future WSN designers for their concerned application challenges aiming to maximize resource utilization and lifespan of sensor nodes.

Literature Survey

Exploration of machine learning in different WSN protocols and applications have become the reason behind increasing attention of researchers in recent years. The set of rules applied to design a routing technique is called routing protocol and, in this context, different protocols (Osisanwo et al.,2017) have been taken into account in different scenarios. Low Energy Adaptive Clustering Hierarchy (LEACH) (Dhawan et al.,2014) is a traditional protocol which has taken energy efficiency as one of its considerable factors, most of the machine learning based routing techniques have been compared with LEACH to evaluate their performance in the concerned area of research. Pathan et al. have proposed an approach in (Pathan et al.,2007) to design an energy efficient routing protocol where two security aspects, i.e. authenticity and confidentiality have been considered by adopting one-way hash chain and preloaded shared secret keys. Experimental results show that the method is promising one in its concerned domain, but here the concept of optimization of interval value for network restructuring purpose is missing. Artificial neural network, swarm intelligence and artificial immune system have been used to improve the existing routing processes of WSN in (Kumar et al.,2014) by making the network 'cognitive' in nature. The attacks in wsn to capture nodes have been tried to reduce by introducing game theory to model the movements of victim node and attacker node and the nature of the attacker has been learnt through neural network approach. After reviewing the existing works, Bhanderi et al. in (Bhanderi et al., 2014) have concluded that machine learning is an appropriate solution for the optimization problems in WSN and truthfully utilize the complex characteristics of distributed systems. Two stage machine learning approach has been designed by Vimalapriya et al. (Vimalapriya et al., 2019) to verify node adaptability depending on the energy, delay and the capacity of accepting packets. The nodes with deficiency are identified and the routes are refurbished after pruning with lesser delay and latency. The autonomy of sensor nodes results to threats and malicious node injection in the network leads to performance degradation (Ishmanoy et al., 2017), path loss etc. Contrary, Security, end to end latency, alive node, average energy utilization and performance analysis show that the model can retain energy in an appreciable level and the active nodes are capable of better survival than the traditional methods.

Protocol function-based routing is of five types according to Sharawi et al. (Sharawi et al.,2013) which include negotiation-based routing, query-based routing, QoS based routing, coherent based routing. Among different paradigms of soft computing, reinforcement learning has been the most appropriate as its memory consumption is the lowest, flexibility is the highest and processing power consumption is the lowest compared to the other methods, i.e. Swarm Intelligence, Evolutionary Algorithm, Fuzzy logic, Neural network and artificial immune system. The hybrid computing paradigms which are inspired by biology are also promising to solve optimization problems in WSN.

Alsheikh et al. have presented a literature review where the new application areas of WSN have been explored, like cyber physical system, Internet of Things, machine to machine communication (Alsheikh et al., 2017). With restricted human intervention, the machine learning algorithms extract the different abstraction levels. Specialized machine learning based WSN systems like development of outlier detection system, task scheduling and optimal deployment systems have been also reviewed here. Apart from different security aspects like anomaly and intrusion detection, Quality of Service enhancement and data integrity have been considered as non-operational characteristics in (Khan et al., 2017) as because of random network topologies, faulty and unreliable data in WSNs. The challenges in WSNs are most of the time application specific like selection of sensors, management of resource where future trends and estimations are required. Deep learning strategies have been very effective in existing unsophisticated networking systems where intelligent and autonomous characteristics of network are required. Scattered applications of deep learning in WSN have been summarized in (Fadlullah et al., 2017) where area of prediction and classification of network flow and traffic, mobility prediction have been explored for the newly emerging technology called deep learning. Open research challenges like smart network have been discussed and one researcher can be guided well through this.

Machine Learning Methods Applied in WSN Routing

With the rapid increase in applications of machine learning, the area of computer network has also been flourished with different formation techniques having flexible and localized communication mediums. Machine learning methods can be broadly categorized into three types, i.e. supervised, unsupervised and reinforcement learning. All three of these have been applied to design secured energy efficient routing for WSN in existing literatures and comparison among the methods has been shown in Table 1. The most popular methods include Support vector machine (SVM), K-Means Clustering, Decision tree (Zhong et al.,2016), Ant colony optimization (Dorigo et al.,2006), Fuzzy logic (Zadeh et al.,1999), Evolutionary algorithm (Vikhar et al.,2016), swarm intelligence (Kennedy et al.,2006), Artificial Neural Network (ANN), Convolutional Neural Network (CNN) of Deep Learning as shown in Figure 1. In (Egrova et al., 2006), the main contribution of Forster et al is in considering multiple destinations for the individual sensor nodes instead of targeting only one destination for multiple sources in WSN. In each iteration the route fitness feedback is provided and learning better routes is accomplished accordingly. Estimation of link costs for routing optimization has been performed in (Singh et al.,2017). Particle swarm optimization-based routing protocol by the authors has shown better performance in terms of consumed energy, end to end delay and data transmitted than the traditional LEACH protocols designed for the purpose of routing in WSN. The main advantage of LEACH as no need of maintaining routing information and forwarding the collected data from every sensor node to its neighbour nodes and further forwarding of data to base station node to its neighbour nodes using multi hop fashion has been discussed in (Ali et al.,2019). Data redundancy, Route updating cost and end to end delay can be caused due to blackholes in network for quick draining of energy of sensor nodes. To overcome these issues a cyclic neural network-based approach of data fusion has been proposed by the authors here. Fission fusion social structure-based spider monkey optimization and Termite colony optimization-based clustering in WSN have been proposed in (Gui et al.,2016) and several meta heuristics-based research efforts have been reviewed here. Improving SMO algorithms by reducing space and time complexity and development of SMO based routing optimization technique can be done in future for better understanding of the concept.

Figure 1. Categories of Machine learning algorithms used in WSN

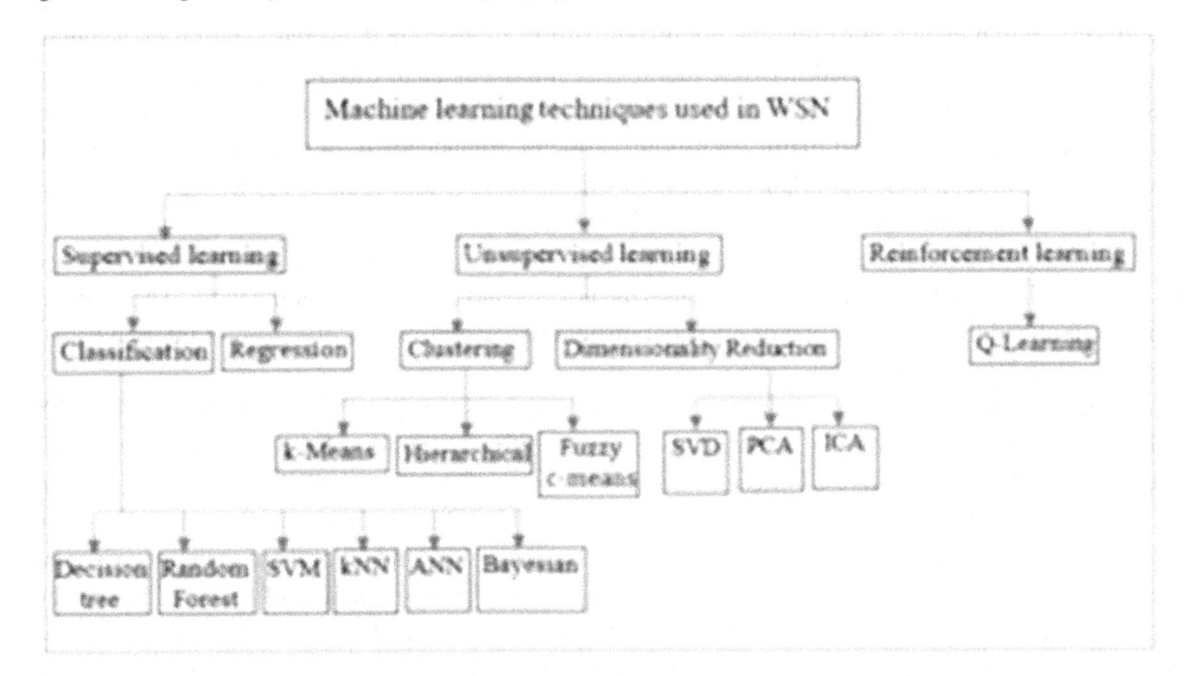

Fuzzy logic is used to deal with uncertainty of some relative terms Arabi et al. have used fuzzy logic in (Arabi et al.,2010) to select the cluster head based on some predefined condition and to shift between two techniques, namely Earliest Fast tree and Source initiated dissemination for designing a hybrid routing method having flexibility and network lifetime increase as its features. Transmission capability of nodes, link quality within a range and battery capacity have been the three main considered factors in the work of (Jaradat et al.,2013) to find out the next hop of relay node towards the target node where the routing decision of next hop has been made by fuzzy logic controller.

Initial concept of artificial neural network comes from the working pattern of biological neurons. Some nodes in the form of neurons are connected to each other and some function according to the application area is invoked for assigning weights to the connections. The objective remains to adjust the weights to get the output exactly as the mapping function. This technique has been used in different functional aspects of Wireless sensor network in the recent literatures. Authors (Nehra et al.,2009) have presented the problem of energy efficient routing as linear programming model with some specified restrictions. Experimental results show that the number of active nodes and the residual energy are better than the LEACH protocol in WSN. Authors (Zhao et al.,2009) have used self-organizing neural network on a sensor node called MODABER to solve the problem of frequency interference. Authors (Bin et al.,2006) have discussed a system where dynamic reduction of information to be forwarded has taken place to reduce the energy consumption. Trunk of a genetic tree supported by the network structure gets updated time to time for analysing the trade-off for the full system's energy consumption and simulation results show that the proposed method is promising one.

Natural evolution of adaptation to ever changing environment and existence capability is modelled using one of the widely used machine learning approaches called evolutionary method. For solution of routing in two tier sensor networks by the authors (Chakraborty et al.,2012) have proposed a differential evolution based memetic algorithm. Local search algorithm has been modified by combining it with differential evolution to find the optimized path from every node to base station by achieving energy efficiency and increased network lifetime. A multi objective differential evolution approach has been introduced by the authors (Sharma et al.,2012) where latency for a linkage in communication has been a factor in single and multipath routing problem. Enhancement of data quality has been the major concern of data fusion problem addressed by the authors (Pinto et al.,2014) where different user defined parameters have designed a trade-off in between using the genetic machine learning algorithm of proposed approach. The ability to automatically adjust the communication rate of specific sensor nodes according to the dynamic topology proves the novelty and significance of the work. The proposed model's application to optimization of other parameters like evolution and check point interval has not been tested and thus missing here.

The uncontrolled systems are distributed by the concept of swarm intelligence (Okdem et al.,2009, Saleem et al.,2019, Zungeru et al.,2012) based on some effective routing formula of shortest path in ant's colony (Sarangi et al.,2012). Basic Ant Based Routing (BABR), Flooded Forward Ant Routing (FFAR), Sensor Driven and Cost Aware Ant Routing, Flooded Piggybacked Ant Routing are some of the well-known techniques in this domain. The common concept behind all of these techniques is transmission from one node to its neighbour nodes depends on the energy function of the neighbour nodes and the present pheromone in the connection between them. When the data gets delivered to the proper target node, a backward path tracking takes place for updating the node number and energy consumed. A new idea of introducing dominant node has been expressed in (Venkataramana et al.,2019) where the proposed approach has been compared with LEACH, LEACH-Two Level Group Head, Energy Efficient Dynamic Clustering Algorithm (Sarella et al.,2017) and Energy-Efficient Level based and Time-based Clustering algorithm (EELTC) (Tashtarian et al.,2007) and PEGASIS (Kim et al.,2006) for performance evaluation and shown a better performance than the conventional ones.

Forster has reviewed presently used ML techniques in WSN and Mobile Ad Hoc Network (MANET) (Forster et al.,2007). ML approaches have definitely performed better than the non-learning techniques in WSN. Reinforcement learning has been the most effective one in mostly static WSN whereas swarm intelligence has worked well in non-energy restricted, highly mobile MANET environment. Ant colony optimization as a heuristic way for energy consumption reduction purpose has been proposed by (Yan et al.,2011). The process of optimization here is divided into several stages, among which, the first one is Ant System (AS), the second one is Elitist technique for the Ant System (EAS) and the later ones are Ant System Rank, Max-Min Ant System, Ant Colony System (ACS). The simulation results shown in six different scales, compare the performances of three algorithms, namely ACS, AS and an enhanced AS called ASW and show that ACS consumes less energy than the other two.

Authors (Prajapati et al.,2018) have explored a new aspect of WSN where according to them, the quantity of the sensor nodes should not affect Quality of Service of WSN as the traffic gets unbalanced and packet transmission rate falls due to continuous change in the number of nodes for addition and deletion of nodes. Artificial neural network methods like multilayer perceptron (MLP) have been very useful in distributed and parallel network environment, whereas Naïve Bayes classifier has shown little bit poor performance compared to SVM and MLP. Explicit programming of conventional designs of WSN are not capable of handling its dynamic nature. Synchronization, congestion control and avoidance have been discussed in (Kumar et al.,2019) apart from the common discussed factors of dealing with WSN challenges using machine learning. Malicious nodes named as Hello flooding, selective forwarding have been classified and modelled using Bayesian classifier and a malicious node detection system has

Table 1. Comparison of the existing literatures based on ML methods

	Ref.	Objective	Methodology	ML method	Performance	Drawback(s)
Supervised	(Khan et al., 2016)	-To assign sensor nodes to the nearest cluster and balance the energy dissipation in the cluster heads	-Designing a WSN routing protocol for clustering sensor nodes	SVM	-Performance is better than the LEACH by comparing the PDR, processing overhead, energy dissipation & memory overhead	--
	(Wang et al., 2006)	-To predict link quality of wsn for optimization of routing	-Feature extraction & o/p labelling -Sample collection -Offline training	Decision Tree & Rule learner	-300% improved data delivery rate evaluated by MetricMap & MintRoute -Less complexity	-Reduction of labelling cost not considered -No online incremental training approach considered
Unsupervised	(Preeth et al., 2018)	-To design energy efficient cluster & immune inspired routing protocol	-Cluster head selection based on QoS impact & energy status	Adaptive Fuzzy multi criteria decision making	-OoS parameters, i.e. channel load, PDR 99%, Jitter BER, throughput 0.95Mbps, network lifetime 5500 rounds	-No optimal route with high residual energy for increasing the sensor lifetime is searched
	(Thangaramya et al., 2019)	-Cluster formation in WSN for energy efficient routing of packets	-Weight adjustment using CNN to prolong network lifetime	-Neuro fuzzy rule-based clustering	-Better network lifetime than the LEACH, FLCFP &HEED protocol	-Assumption of every node to be trustful
	(Townsend, 2018)	-Node clustering in near optimal configuration for energy efficiency	Two chromosome repair methods Arepair and BFSrepair to design gateway nodes	-Genetic algorithm -4 selection methods, i.e. elite, roulette wheel, linear rank, tournament	-Less time & energy consumption than traditional genetic algorithm	-Consideration of only single sink -No variation considered in transmission range -Consideration of only static nodes in WSN
Reinforcement	(Förster et al., 2008)	-Routing on Real WSN peripherals using QLearning method of Reinforcement learning	FROMS (multi-source multicast routing protocol) & DD (Directed Diffusion in a test case of ScatterWeb nodes	QLearning	-Improvement of delivery rate in different network scenarios	-Small size of testbed -No consideration of changing topologies, new cost functions and mobile sinks
	(Kadam et al., 2012)	-Energy efficient routing in heterogenous network	-Propagation of energy beyond the direct neighbours	Enhanced version of Qlearning method in [57]	Improvements in n/w lifetime after load balancing also	--
	(Liang et al., 2008)	-Computation of QoS route using distributed value function-distributed reinforcement learning method	-Identification of optimal routing policy through previous rewards & experience	Qlearning algorithm	-Packet routing more efficient than AODV routing	-No energy consumption parameter is considered for QoS route determination
	(Boyan et al., 1994)	-To track which routing decision leads to minimum delivery time	-Shortest path measurement by Bellman Ford algorithm	Qrouting algorithm	Qrouting based algorithms perform better in the packet routing domain	-Only table-based representation -no concept of function approximator
	(Dong et al., 2007)	-To design a geographic routing protocol for providing suitable localization and high data rate at low cost in WSN	-capable of adapting energy variation	Reinforcement learning	-Robustness & network lifetime is 75% to 213% better than GPSR [67]	--

been proposed and simulated for performance evaluation in (Zou et al.,2012). The error rate of 0.6% in classification proves the effectiveness and appropriateness of the method. Network health abnormality prediction based on regression algorithm and network object detection and monitoring as well as state transition analysis using pattern matching machine learning approach have been described in (Vijayakumar et al,2019) where the need of revolutionary technology like Internet of Things in sensor and cloud environment have been presented also apart from the WSN. Three stages, namely pre-processing, data aggregation and inference of information processing in WSN have been highlighted in (Di et al.,2007) where inference refers to the process of extracting hidden information from the aggregated data. Most of the machine learning applications nowadays focus on classification of moving object in WSN and can be integrated throughout these three steps for optimal information processing.

Supervised Learning Methods

Supervised learning methods (Asusanwo et al., 2017) are the tasks of learning where the inputs are mapped to some outputs based on some predefined sample input output pairs. Analysis of training sample is required to get the learning function and to trace the output of the testing sample. Supervised methods, like classification and regression have been applied in several aspects of Wireless Sensor Network (Khan et al.,2016, Wang et al.,2006) where classification include SVM, artificial neural network, deep learning, Random forest, Decision tree, Bayesian network and K-nearest neighbour as well. Decision tree has been applied in (Rajagopalan et al.,2006) to choose the cluster head and important features like the distance between decision tree node and the cluster centroid, the degree of movement, battery level and identification of vulnerability have been considered during the iteration of input vector through decision tree which results to better performance of the model than traditional LEACH protocol.

Janakiram et al. have proposed an outlier detecting scheme (Janakiram et al.,2006) using Bayesian network where the first assumption is most of the neighbours of a particular node always have the similar readings with the said node. Based on this phenomenon, the correlations or the dependencies among several nodes are calculated and the outliers are identified accordingly. K-Nearest Neighbour has been used in (Branch et al.,2013) where if there is any lost reading of some node, then it gets replaced by the average value of readings provided by k number of nearest neighbours to that specific node, but the limitation here is the large memory requirement for storing and monitoring each and every reading for all the nodes of a network.

Support Vector Machine

Support vector machine is a widely popular binary classifier where data points belonging to two different classes get separated by a line called, hyperplane. The working mechanism of SVM has been shown in Figure 2.

Decision Tree

The classification and regression method works here in the form of a tree structure where the dataset is broken down into sub datasets at first and at the same time an associated decision tree gets built. The decision nodes contain two or more branches and finally the leaf nodes contain the decision or classification. The mechanism is shown in Figure 3.

Figure 2. Working mechanism of SVM

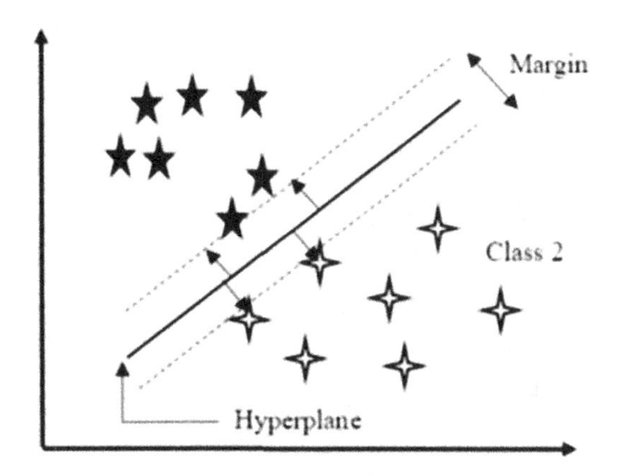

Figure 3. Working mechanism of Decision tree

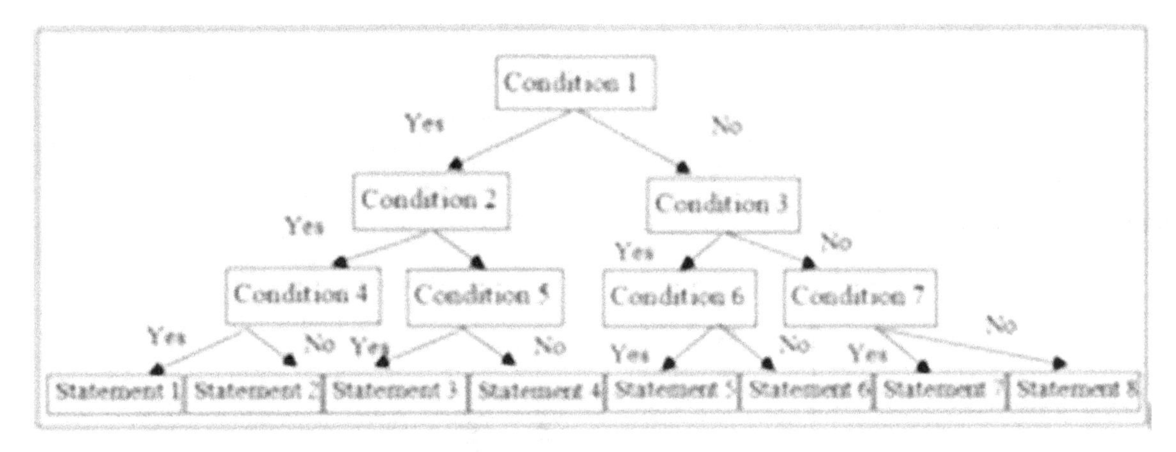

Random Forest

The random forest is basically a collection of decision trees where ensemble method of classification and regression is integrated. During the training time, it inputs some classes in the form of decision trees and the output class becomes the mode of the classes in case of classification and mean prediction in case of regression. The functionality of Random Forest algorithm is shown in Figure 4.

Naïve Bayes Classifier

Naïve bayes classifiers are simple families of probabilistic classifiers in machine learning which applies Bayes theorem having strong (naïve) assumptions among the considered features. Its accuracy and speed are good when it is applied on large datasets and the mechanism is shown in Figure 5.

Figure 4. Working mechanism of Random forest

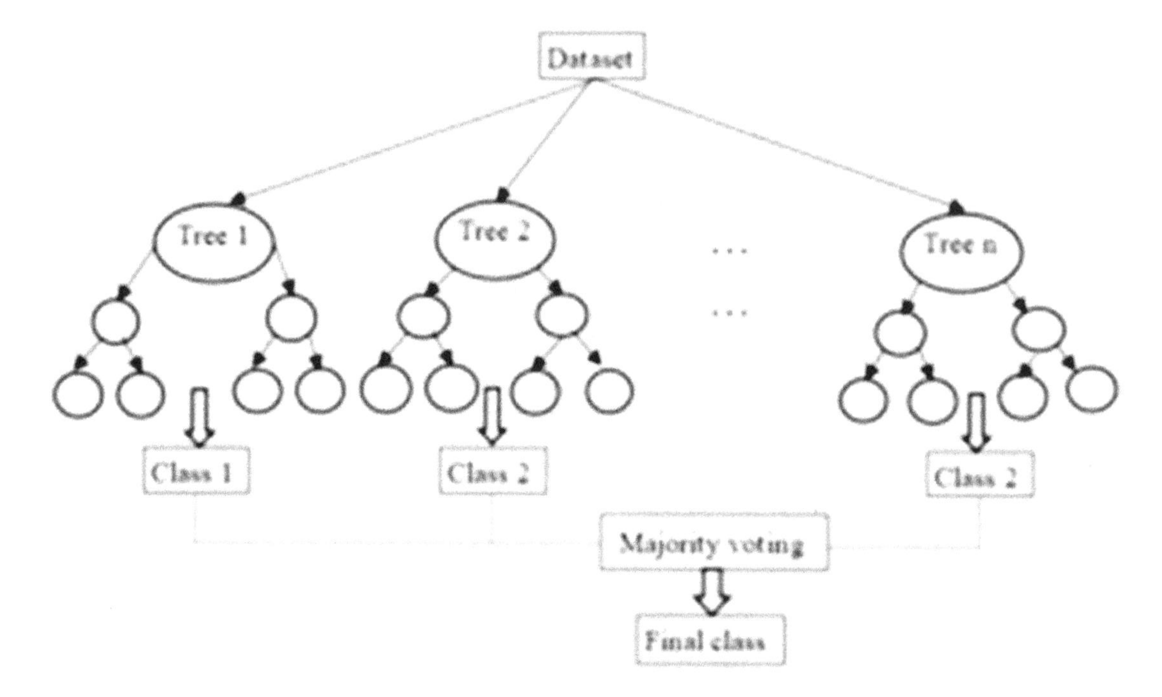

Figure 5. Working mechanism of Naïve Bayes Classifier

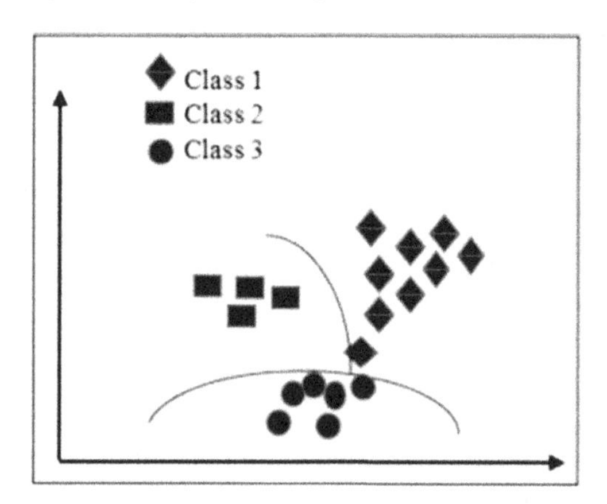

Unsupervised Learning Methods

Unsupervised learning approaches namely clustering, dimensionality reduction have been widely used (Preeth et al.,2018, Thangaramya et al.,2019, Townsend et al.,2018, Fakhet et al.,2017) in Wireless Sensor Network. Clustering is a technique of data analysis where an intuition about the structure of a collection of data points is analysed and groups/clusters of similar data points based on some application specific conditions like Euclidean distance or correlation-based distance etc are formed. It is inefficient to transmit each and every data directly to sink in case of a large energy restricted network, if data from every node gets transmitted to its local aggregator (also known as cluster head) and the aggregator after aggregating all the data received from its cluster nodes, passes it to the sink finally, here lies the concept of clustering the similar nodes in WSN. The popular clustering algorithms which have been applied in WSN are k-means clustering, hierarchical clustering and Fuzzy C-means clustering, whereas dimensionality reduction algorithms like principal component analysis, independent component analysis and singular value decomposition have been widely used in various aspects of WSN. Principal Component Analysis (PCA) is a technique to extract the most important features from some large feature set and present them as a set of orthogonal variables called principal components. In WSN, the PCA reduces the data amount among sensor nodes by eliminating the un correlated data from the nodes and keeps only the important portion of data to reduce the overall complexity of the system. Two data aggregation techniques (Li et al.,2004, Masiero et al.,2009, Mekua et al.,2010, Rajagopalan et al.,2006), namely Compressive Sensing (CS) (Duarte et al.,2011) and Expectation Maximization (EM) (Dempster et al.,1977) are combinedly used with Principal component analysis to achieve a better performance with enhanced data aggregation in WSN.

K-Means Clustering

K means clustering is the most well-known unsupervised learning method which has been applied in several literatures for determining the wireless sensor nodes cluster and in an efficient way to make the system effective and reliable to the network user. The working mechanism of K-Means clustering has been shown in Figure 6. The steps followed in K-means clustering are:

Step 1: Choose k number of nodes randomly to be the initial centroids of a cluster
Step 2: Mark each node using the nearest centroid measured using distance function
Step 3: Re-evaluate the centroid using current node's membership
Step 4: Stop when the predefined convergence condition gets satisfied

The addition of the distances between every node and its corresponding centroid is compared with a pre-defined threshold value every time and if the condition is satisfied then the problem converges, otherwise the flow goes back to step 2. A cluster is a collection of similar data points aggregated together. Typically unsupervised methods create inferences from the concerned datasets using only input vectors without having any known or labelled output.

Figure 6. Working mechanism of K means clustering

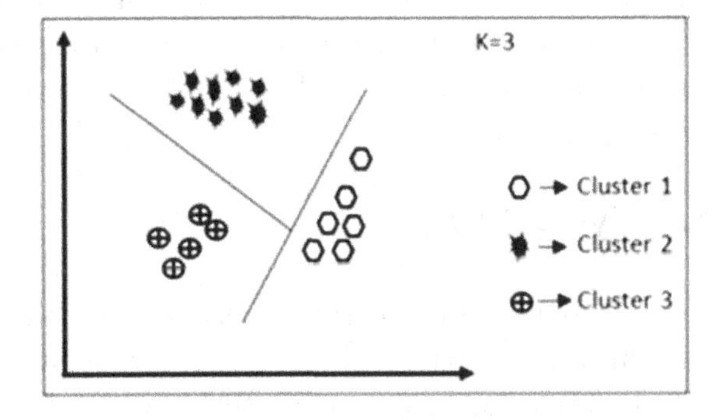

Hierarchical Clustering

In this method (also called hierarchical cluster analysis) similar features are grouped with each other based on some pre-defined condition and the output given is a set of distinct clusters where the objects within each cluster are similar to each other. The working mechanism (Rani et al.,2013) is shown in Figure 7.

Figure 7. Working mechanism of Hierarchical clustering algorithm

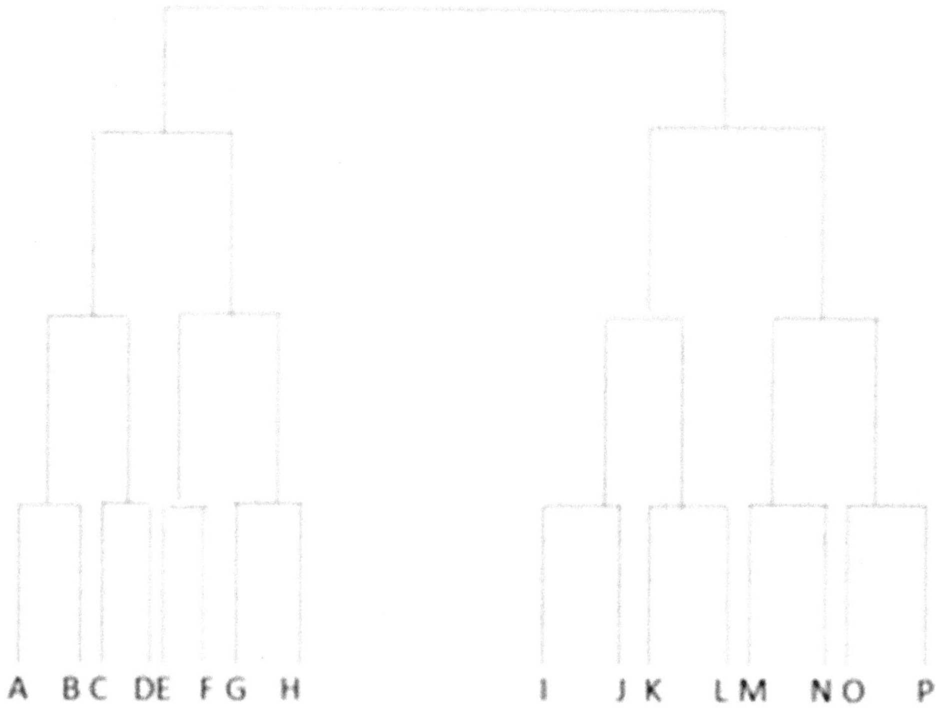

Fuzzy Clustering

Fuzzy clustering is also known as soft clustering and soft k means (Yang et al.,1993). The speciality of this method is here one data point can belong to more than one clusters. The points in a class are similar to each other and in different classes, the points are dissimilar to among them. The data points are assigned some values between 0 and 1 according to their degree of membership which avoids the distortions in the final solution of regular clustering algorithms. The working mechanism of Fuzzy clustering is shown in Figure 8.

Principal Component Analysis (PCA)

It is a linear transformation process for reduction of dimensionality for a large dataset having datapoints at two, three or more than that dimensional space. A line which is having the least average squared distance from a point is said to be best fitted and the next best fitted line can be defined as the line which

Figure 8. Working mechanism of Fuzzy clustering

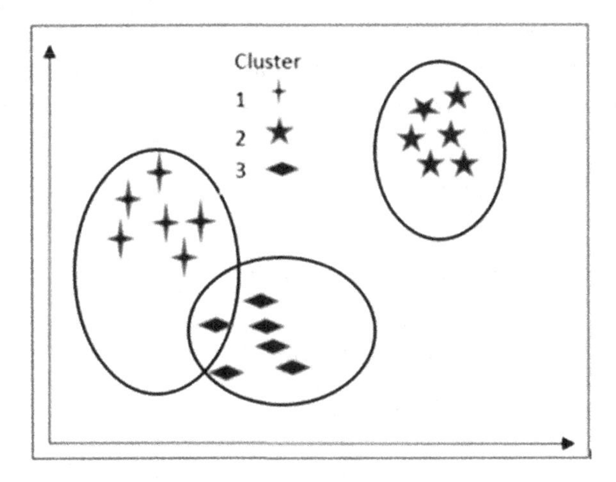

Figure 9. Working mechanism of PCA

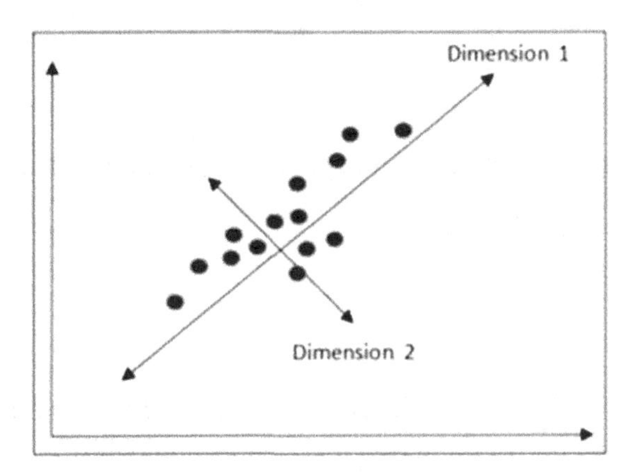

is having perpendicular directions from the first. By repetition of this process, some orthogonal basis gets formed where the data are not correlated, these vectors are called as principal components and the process is called principal component analysis. The working mechanism of PCA is shown in Figure 9.

Reinforcement Learning Methods

Reinforcement learning methods have strong contributions (Forster et al.,2010, Kadam et al.,2012, Boyan et al.,1994, Dong et al.,2007) in the domain of computer network and especially in wireless sensor network. The working mechanism of Reinforcement learning as shown in Figure 10 describes the process of learning what to do next and mapping process of different scenarios to actions for maximizing the reward signal for acting perfectly upon corresponding environment. The decision of which action to take is not available to the learner at the very beginning, by time, he/she discovers the actions which will result to the most rewarding statements and these actions affect not only the next step, but also all the subsequent stages of rewards. Existing approaches to Q-learning method of WSN sometimes face difficulties in convergence, to overcome this problem, Sharma et al. have (Sharma et al.,2012) proposed an enhanced model of Q-learning method, but Network Simulator version 2 provided the simulation result of enhanced model as same as the traditional model. Authors (Dorigo et al.,2006) have presented an approach to manage power dynamically by using a learning algorithm which does not have any prior model, application of neural network having multi-layer provides the workload information and helps to take improved decision depending on the varying workload. Based on the estimated workload, the learning method is formed to use optimized time out value and idle time is minimized.

Figure 10. Working mechanism of Reinforcement Learning

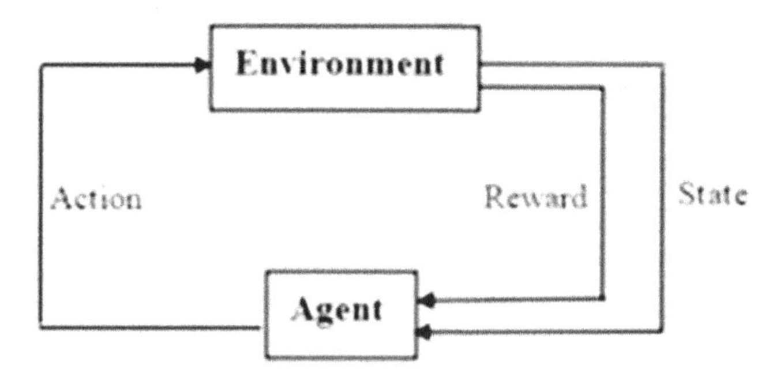

Reinforcement learning is also used in scheduling, service provisioning, medium access control apart from the routing in WSN.

CONCLUSION

This chapter explains various techniques applied to design secured energy efficient routing schemes for wireless sensor networks using one of the most emerging technologies in current scenario, called machine learning. To meet the demands of the latest technologies and the research trend, this chapter will be able to provide guidance to those who are interested to conduct research in the concerned domain. The machine learning approach helps to enhance the overall performance by making the network intelligent and effective. Collaboration with Internet of Things (IoT) based technology and environment will bring more advanced and real-world features in WSN to proceed further. Future research may lead to a solution of more matured protocol and applications of it in practical scenario.

REFERENCES

Ahmed, G., Khan, N. M., Khalid, Z., & Ramer, R. (2008, December). Cluster head selection using decision trees for wireless sensor networks. In *2008 International Conference on Intelligent Sensors, Sensor Networks and Information Processing* (pp. 173-178). IEEE. 10.1109/ISSNIP.2008.4761982

Ali, B., Mahmood, T., Abbas, M., Hussain, M., Ullah, H., Sarker, A., & Khan, A. (2019). LEACH Robust Routing Approach Applying Machine Learning. *IJCSNS*, *19*(6), 18–26.

Alsheikh, M. A., Lin, S., Niyato, D., & Tan, H. P. (2014). Machine learning in wireless sensor networks: Algorithms, strategies, and applications. *IEEE Communications Surveys and Tutorials*, *16*(4), 1996–2018. doi:10.1109/COMST.2014.2320099

Arabi, Z. (2010, June). HERF: A hybrid energy efficient routing using a fuzzy method in wireless sensor networks. In *2010 International Conference on Intelligent and Advanced Systems* (pp. 1-6). IEEE. 10.1109/ICIAS.2010.5716145

Beyens, P., Peeters, M., Steenhaut, K., & Nowe, A. (2005). Routing with compression in wireless sensor networks: a q-learning approach. In *Fifth European workshop on adaptive agents and multi-agent systems (AAMAS 05)* (*Vol. 8*). Academic Press.

Bhanderi, M. V., & Shah, H. B. (2014). Machine Learning for Wireless Sensor Network: A Review, Challenges and Applications. *Adv. Electron. Electr. Eng*, *4*, 475–486.

Bin, G., Zhe, L., & Ze-Jun, W. (2005, September). A dynamic-cluster energy-aware routing algorithm based on neural structure in the wireless sensor networks. In *The Fifth International Conference on Computer and Information Technology (CIT'05)* (pp. 401-405). IEEE. 10.1109/CIT.2005.8

Boyan, J. A., & Littman, M. L. (1994). Packet routing in dynamically changing networks: A reinforcement learning approach. In Advances in neural information processing systems (pp. 671-678). Academic Press.

Branch, J. W., Giannella, C., Szymanski, B., Wolff, R., & Kargupta, H. (2013). In-network outlier detection in wireless sensor networks. *Knowledge and Information Systems*, *34*(1), 23–54. doi:10.100710115-011-0474-5

Chakraborty, U. K., Das, S. K., & Abbott, T. E. (2012, June). Energy-efficient routing in hierarchical wireless sensor networks using differential-evolution-based memetic algorithm. In *2012 IEEE Congress on Evolutionary Computation* (pp. 1-8). IEEE. 10.1109/CEC.2012.6252985

Dempster, A. P., Laird, N. M., & Rubin, D. B. (1977). Maximum likelihood from incomplete data via the EM algorithm. *Journal of the Royal Statistical Society. Series B. Methodological*, *39*(1), 1–22. doi:10.1111/j.2517-6161.1977.tb01600.x

Dhawan, H., & Waraich, S. (2014). A comparative study on LEACH routing protocol and its variants in wireless sensor networks: A survey. *International Journal of Computers and Applications*, *95*(8).

Di, M., & Joo, E. M. (2007, December). A survey of machine learning in wireless sensor networks from networking and application perspectives. In *2007 6th international conference on information, communications & signal processing* (pp. 1-5). IEEE.

Dong, S., Agrawal, P., & Sivalingam, K. (2007, November). Reinforcement learning based geographic routing protocol for UWB wireless sensor network. In *IEEE GLOBECOM 2007-IEEE Global Telecommunications Conference* (pp. 652-656). IEEE. 10.1109/GLOCOM.2007.127

Dorigo, M., Birattari, M., & Stutzle, T. (2006). Ant colony optimization. *IEEE Computational Intelligence Magazine*, *1*(4), 28–39. doi:10.1109/MCI.2006.329691

Duarte, M. F., & Eldar, Y. C. (2011). Structured compressed sensing: From theory to applications. *IEEE Transactions on Signal Processing*, *59*(9), 4053–4085. doi:10.1109/TSP.2011.2161982

Egorova-Förster, A., & Murphy, A. L. (2006, May). A feedback-enhanced learning approach for routing in WSN. In *Communication in Distributed Systems-15. ITG/GI Symposium* (pp. 1-12). VDE.

Fadlullah, Z. M., Tang, F., Mao, B., Kato, N., Akashi, O., Inoue, T., & Mizutani, K. (2017). State-of-the-art deep learning: Evolving machine intelligence toward tomorrow's intelligent network traffic control systems. *IEEE Communications Surveys and Tutorials, 19*(4), 2432–2455. doi:10.1109/COMST.2017.2707140

Fakhet, W., El Khediri, S., Dallali, A., & Kachouri, A. (2017, October). New K-means algorithm for clustering in wireless sensor networks. In *2017 International Conference on Internet of Things, Embedded Systems and Communications (IINTEC)* (pp. 67-71). IEEE. 10.1109/IINTEC.2017.8325915

Forster, A. (2007, December). Machine learning techniques applied to wireless ad-hoc networks: Guide and survey. In *2007 3rd international conference on intelligent sensors, sensor networks and information* (pp. 365-370). IEEE.

Förster, A., Murphy, A. L., Schiller, J., & Terfloth, K. (2008, October). An efficient implementation of reinforcement learning based routing on real WSN hardware. In *2008 IEEE International Conference on Wireless and Mobile Computing, Networking and Communications* (pp. 247-252). IEEE. 10.1109/WiMob.2008.99

Gui, T., Ma, C., Wang, F., & Wilkins, D. E. (2016, March). Survey on swarm intelligence based routing protocols for wireless sensor networks: An extensive study. In 2016 IEEE international conference on industrial technology (ICIT) (pp. 1944-1949). IEEE.

Ishmanov, F., & Bin Zikria, Y. (2017). Trust mechanisms to secure routing in wireless sensor networks: Current state of the research and open research issues. *Journal of Sensors, 2017*, 2017. doi:10.1155/2017/4724852

Janakiram, D., Reddy, V. A., & Kumar, A. P. (2006, January). Outlier detection in wireless sensor networks using Bayesian belief networks. In *2006 1st International Conference on Communication Systems Software & Middleware* (pp. 1-6). IEEE. 10.1109/COMSWA.2006.1665221

Jaradat, T., Benhaddou, D., Balakrishnan, M., & Al-Fuqaha, A. (2013, July). Energy efficient cross-layer routing protocol in wireless sensor networks based on fuzzy logic. In *2013 9th International Wireless Communications and Mobile Computing Conference (IWCMC)* (pp. 177-182). IEEE. 10.1109/IWCMC.2013.6583555

Kadam, K., & Srivastava, N. (2012, March). Application of machine learning (reinforcement learning) for routing in Wireless Sensor Networks (WSNs). In *2012 1st International Symposium on Physics and Technology of Sensors (ISPTS-1)* (pp. 349-352). IEEE.

Karp, B., & Kung, H. T. (2000, August). GPSR: Greedy perimeter stateless routing for wireless networks. In *Proceedings of the 6th annual international conference on Mobile computing and networking* (pp. 243-254). 10.1145/345910.345953

Kennedy, J. (2006). Swarm intelligence. In *Handbook of nature-inspired and innovative computing* (pp. 187–219). Springer. doi:10.1007/0-387-27705-6_6

Khan, F., Memon, S., & Jokhio, S. H. (2016, November). Support vector machine based energy aware routing in wireless sensor networks. In *2016 2nd International Conference on Robotics and Artificial Intelligence (ICRAI)* (pp. 1-4). IEEE. 10.1109/ICRAI.2016.7791218

Khan, Z. A., & Samad, A. (2017). A study of machine learning in wireless sensor network. *Int. J. Comput. Netw. Appl*, *4*(4), 105–112. doi:10.22247/ijcna/2017/49122

Kim, D. S., & Chung, Y. J. (2006, June). Self-organization routing protocol supporting mobile nodes for wireless sensor network. In First International Multi-Symposiums on Computer and Computational Sciences (IMSCCS'06) (Vol. 2, pp. 622-626). IEEE. doi:10.1109/IMSCCS.2006.265

Kumar, D. P., Amgoth, T., & Annavarapu, C. S. R. (2019). Machine learning algorithms for wireless sensor networks: A survey. *Information Fusion*, *49*, 1–25. doi:10.1016/j.inffus.2018.09.013

Kumar, S. (2014). Improving WSN Routing and Security with an Artificial Intelligence Approach. In DWAI@ AI* IA (pp. 62-71). Academic Press.

Lee, S., & Chung, T. (2004, October). Data aggregation for wireless sensor networks using self-organizing map. In *International Conference on AI, Simulation, and Planning in High Autonomy Systems* (pp. 508-517). Springer.

Liang, X., Balasingham, I., & Byun, S. S. (2008, October). A multi-agent reinforcement learning based routing protocol for wireless sensor networks. In *2008 IEEE International Symposium on Wireless Communication Systems* (pp. 552-557). IEEE. 10.1109/ISWCS.2008.4726117

Macua, S. V., Belanovic, P., & Zazo, S. (2010, June). Consensus-based distributed principal component analysis in wireless sensor networks. In *2010 IEEE 11th International Workshop on Signal Processing Advances in Wireless Communications (SPAWC)* (pp. 1-5). IEEE.

Masiero, R., Quer, G., Munaretto, D., Rossi, M., Widmer, J., & Zorzi, M. (2009, November). Data acquisition through joint compressive sensing and principal component analysis. In *GLOBECOM 2009-2009 IEEE Global Telecommunications Conference* (pp. 1-6). IEEE. 10.1109/GLOCOM.2009.5425458

Nehra, N. K., Kumar, M., & Patel, R. B. (2009, December). Neural network based energy efficient clustering and routing in wireless sensor networks. In *2009 First International Conference on Networks & Communications* (pp. 34-39). IEEE. 10.1109/NetCoM.2009.56

Okdem, S., & Karaboga, D. (2009). Routing in wireless sensor networks using an ant colony optimization (ACO) router chip. *Sensors (Basel)*, *9*(2), 909–921. doi:10.339090200909

Osisanwo, F. Y., Akinsola, J. E. T., Awodele, O., Hinmikaiye, J. O., Olakanmi, O., & Akinjobi, J. (2017). Supervised machine learning algorithms: Classification and comparison. *International Journal of Computer Trends and Technology*, *48*(3), 128–138. doi:10.14445/22312803/IJCTT-V48P126

Pathan, A. S. K., & Hong, C. S. (2007, August). A secure energy-efficient routing protocol for WSN. In *International symposium on parallel and distributed processing and applications* (pp. 407-418). Springer. 10.1007/978-3-540-74742-0_38

Patil, M., & Biradar, R. C. (2012, December). A survey on routing protocols in wireless sensor networks. In *2012 18th IEEE International Conference on Networks (ICON)* (pp. 86-91). IEEE. 10.1109/ICON.2012.6506539

Pinto, A. R., Montez, C., Araújo, G., Vasques, F., & Portugal, P. (2014). An approach to implement data-fusion techniques in wireless sensor networks using genetic machine learning algorithms. *Information Fusion*, *15*, 90–101. doi:10.1016/j.inffus.2013.05.003

Prajapati, J., & Jain, S. C. (2018, April). Machine learning techniques and challenges in wireless sensor networks. In *2018 Second International Conference on Inventive Communication and Computational Technologies (ICICCT)* (pp. 233-238). IEEE. 10.1109/ICICCT.2018.8473187

Preeth, S. S. L., Dhanalakshmi, R., Kumar, R., & Shakeel, P. M. (2018). An adaptive fuzzy rule based energy efficient clustering and immune-inspired routing protocol for WSN-assisted IoT system. *Journal of Ambient Intelligence and Humanized Computing*, 1–13. doi:10.100712652-018-1154-z

Rajagopalan, R., & Varshney, P. K. (2006). *Data aggregation techniques in sensor networks: A survey*. Academic Press.

Rani, Y., & Rohil, H. (2013). A study of hierarchical clustering algorithm. *ter S & on Te SIT-2*, 113.

Saleem, M., Di Caro, G. A., & Farooq, M. (2011). Swarm intelligence based routing protocol for wireless sensor networks: Survey and future directions. *Information Sciences*, *181*(20), 4597–4624. doi:10.1016/j.ins.2010.07.005

Sarangi, S., & Thankchan, B. (2012). A novel routing algorithm for wireless sensor network using particle swarm optimization. *IOSR Journal of Computer Engineering*, *4*(1), 26–30. doi:10.9790/0661-0412630

Sarella, V. R., Reddy, P. P., Rao, S. K., & Padala, P. (2017). EEECARP: Efficient Energy Clustering Adaptive Routing Procedure for Wireless Sensor Networks. *Journal of Global Information Management*, *25*(4), 125–138. doi:10.4018/JGIM.2017100108

Sharawi, M., Saroit, I. A., El-Mahdy, H., & Emary, E. (2013). Routing wireless sensor networks based on soft computing paradigms: Survey. *International Journal on Soft Computing Artificial Intelligence and Applications (Commerce, Calif.)*, *2*(4), 21–36.

Sharma, V. K., Shukla, S. S. P., & Singh, V. (2012, December). A tailored Q-Learning for routing in wireless sensor networks. In *2012 2nd IEEE International Conference on Parallel, Distributed and Grid Computing* (pp. 663-668). IEEE. 10.1109/PDGC.2012.6449899

Singh, A., Rathkanthiwar, S., & Kakde, S. (2016, April). Energy efficient routing of WSN using particle swarm optimization and V-LEACH protocol. In *2016 International Conference on Communication and Signal Processing (ICCSP)* (pp. 2078-2082). IEEE. 10.1109/ICCSP.2016.7754544

Singh, K., & Kaur, J. (2017). Machine learning based link cost estimation for routing optimization in wireless sensor networks. *Adv. Wirel. Mob. Commun*, *10*, 39–49.

Tashtarian, F., Haghighat, A. T., Honary, M. T., & Shokrzadeh, H. (2007, September). A new energy-efficient clustering algorithm for wireless sensor networks. In *2007 15th International Conference on Software, Telecommunications and Computer Networks* (pp. 1-6). IEEE.

Thangaramya, K., Kulothungan, K., Logambigai, R., Selvi, M., Ganapathy, S., & Kannan, A. (2019). Energy aware cluster and neuro-fuzzy based routing algorithm for wireless sensor networks in IoT. *Computer Networks*, *151*, 211–223. doi:10.1016/j.comnet.2019.01.024

Townsend, L. (2018). *Wireless Sensor Network Clustering with Machine Learning*. Academic Press.

Venkataramana, S., Sekhar, B. V. D. S., Deshai, N., Chakravarthy, V. V. S. S. S., & Rao, S. K. (2019, May). Efficient time reducing and energy saving routing algorithm for wireless sensor network. *Journal of Physics: Conference Series*, *1228*(1), 012002. doi:10.1088/1742-6596/1228/1/012002

Vikhar, P. A. (2016, December). Evolutionary algorithms: A critical review and its future prospects. In *2016 International conference on global trends in signal processing, information computing and communication (ICGTSPICC)* (pp. 261-265). IEEE. 10.1109/ICGTSPICC.2016.7955308

Vimalapriya, M. D., Vignesh, B. S., & Sandhya, S. (2019). *Energy-Centric Route Planning using Machine Learning Algorithm for Data Intensive Secure Multi-Sink Sensor Networks. International Journal of Innovative Technology and Exploring Engineering*. IJITEE.

Wang, Y., Martonosi, M., & Peh, L. S. (2006, May). A supervised learning approach for routing optimizations in wireless sensor networks. In *Proceedings of the 2nd international workshop on Multi-hop ad hoc networks: from theory to reality* (pp. 79-86). 10.1145/1132983.1132997

Yan, J. F., Gao, Y., & Yang, L. (2011, July). Ant colony optimization for wireless sensor networks routing. In *2011 International Conference on Machine Learning and Cybernetics* (Vol. 1, pp. 400-403). IEEE. 10.1109/ICMLC.2011.6016670

Yang, M. S. (1993). A survey of fuzzy clustering. *Mathematical and Computer Modelling*, *18*(11), 1–16. doi:10.1016/0895-7177(93)90202-A

Zadeh, L. A. (1999). Fuzzy logic= computing with words. In *Computing with Words in Information/ Intelligent Systems 1* (pp. 3–23). Physica. doi:10.1007/978-3-7908-1873-4_1

Zhao, W., Liu, D., & Jiang, Y. (2009, January). Distributed neural network routing algorithm based on global information of wireless sensor network. In *2009 WRI International Conference on Communications and Mobile Computing* (Vol. 1, pp. 552-555). IEEE. 10.1109/CMC.2009.103

Zhong, Y. (2016, August). The analysis of cases based on decision tree. In *2016 7th IEEE International Conference on Software Engineering and Service Science (ICSESS)* (pp. 142-147). IEEE.

Zou, K., Ouyang, Y., Niu, C., & Zou, Y. (2012, September). Simulation of Malicious Nodes Detection Based on Machine Learing for WSN. In *International Conference on Information Computing and Applications* (pp. 492-499). Springer. 10.1007/978-3-642-34038-3_68

Zungeru, A. M., Ang, L. M., & Seng, K. P. (2012). *Performance evaluation of ant-based routing protocols for wireless sensor networks.* arXiv preprint arXiv:1206.5938

Chapter 3
Security and Privacy Challenges of Deep Learning:
A Comprehensive Survey

J. Andrew Onesimu
https://orcid.org/0000-0003-3592-6543
Karunya Institute of Technology and Sciences, India & Vellore Institute of Technology, India

Karthikeyan J.
Vellore Institute of Technology, India

D. Samuel Joshua Viswas
https://orcid.org/0000-0002-6792-7391
Karunya Institute of Technology and Sciences, India

Robin D Sebastian
Karunya Institute of Technology and Sciences, India

ABSTRACT

Deep learning is the buzz word in recent times in the research field due to its various advantages in the fields of healthcare, medicine, automobiles, etc. A huge amount of data is required for deep learning to achieve better accuracy; thus, it is important to protect the data from security and privacy breaches. In this chapter, a comprehensive survey of security and privacy challenges in deep learning is presented. The security attacks such as poisoning attacks, evasion attacks, and black-box attacks are explored with its prevention and defence techniques. A comparative analysis is done on various techniques to prevent the data from such security attacks. Privacy is another major challenge in deep learning. In this chapter, the authors presented an in-depth survey on various privacy-preserving techniques for deep learning such as differential privacy, homomorphic encryption, secret sharing, and secure multi-party computation. A detailed comparison table to compare the various privacy-preserving techniques and approaches is also presented.

DOI: 10.4018/978-1-7998-5068-7.ch003

INTRODUCTION

In recent industry revolution, customer oriented services are getting popular every day. In order to provide customer oriented services huge amount of data is being collected from the users. The data collection is either done actively with the consent of the user or it is done passively without the knowledge of the user. However, such data contains personal information about individuals which is necessary to be protected. It is important to protect the collected data in the data storage, data processing and data transmission. Various security mechanisms are available to protect the data, nevertheless they differ based on the computational and time complexity.

Users generates the data through online and offline. Every electronic transaction is collected as data. This in turn helps the researchers to collect information about the various events and predict the future. Especially in the area of sciences and medicines a lot of research is being done to invent something new, to predict and treat diseases, to develop drugs and medicines, etc., So the researchers will always be looking for huge amount of data generated by the users. When such data is collected it is necessary to protect the personally identifiable information (PII) (Krishnamurthy & Wills, 2009) from the data. As the data are collected from the individuals it contains PII. Analyzing the data without removing PII could lead to privacy breach (J, Karthikeyan, & Jebastin, 2019). Thus it is essential to remove PII before publishing the data various researches.

There is a huge rise in artificial intelligence, (Russell & Norvig, 2016) machine learning, (Andrieu, De Freitas, Doucet, & Jordan, 2003) and deep learning (Arel, Rose, Karnowski, & others, 2010) in recent times because of the huge availability of the data and improved accuracy of the models. AI gives the decision making ability to the system based the previous records. Machine learning (ML) is a subset of AI, it can process large amount of data and provide accurate results than traditional approaches(Bhushan & Sahoo, 2018). Machine learning consists of supervised, unsupervised and reinforcement learning techniques. It can work with various types of data and can provide promising results in the field of stock market prediction, drug discovery, disease prediction, etc. (Andrew, Mathew, & Mohit, 2019) Deep learning is a subset of ML (Jindal, Gupta, & Bhushan, 2020), it is used when the dataset is huge and have complex structures. Deep learning mimics the human brain neurons to learn patterns from the data. Deep Learning gives promising results in the field of computer vision, audio processing, video processing, pattern recognition etc.

Contributions

Though the recent advancements provide greater advantages in terms of accuracy and processing time, it has considerable risks in securing the data during the model training and testing. Also, the private data has to be prevented from privacy leaks. These are the two challenges we have identified for this book chapter. In this book chapter, the various security and privacy challenges on deep learning models are presented. At first, the security challenges of training data and model data are presented. Then the various security attacks on deep learning models are analyzed. Secondly, the privacy issues of deep learning models presented along with various privacy attacks. Finally a comparative analysis is given on security and privacy attacks.

Chapter Organization

The remainder of the chapter goes as follows, section 2 discusses about the security challenges on machine learning or deep learning based model training and testing security. We also presented in depth survey on various security attacks on deep learning model. A comparative analysis table is presented for security attacks. Section 3 describes about various privacy issues in the deep learning models. This section also describes the various privacy preserving mechanisms to protect the data from privacy breaches. A comparative analysis table presented to compare the various privacy preserving techniques and approaches. Finally, the conclusion of the chapter and references are presented.

SECURITY CHALLENGES

As we look towards solving more and more problems in medicine, science and technology, agriculture, communication and cyber security (Andrew & Kathrine, 2018) through Machine learning/Deep Learning we need to take a careful look at the security challenges (Bhushan & Sahoo, 2020a) which are currently threaten these approached and also at the challenges which will threaten such approaches in the near future.

In the following sections we take look at the various security challenges faced by machine learning and deep learning, namely, training data security (Bhushan & Sahoo, 2020b), model security, adversarial attacks, poisoning attacks, evasion attacks and black box attacks.

Training Data Security

Training Data security is an important aspect which is generally overlooked by machine learning model developers. Training data security though right now is slowly gaining prominence it will play a major role in the near future as the data we collect gets more and more sophisticated.

Development of tools and testing sets for detection of SQL attacks has also been done. Bayes classifier system has been developed to detect SQL Injections attacks with accuracy varying between 97% - 99% (Lodeiro-Santiago, Caballero-Gil, & Caballero-Gil, 2017).The paper also proposes of an idea to create a distributed system which can classify multiple clients.

Regression models have also been proposed which does not assume feature independence and sub-Gaussian noise with low variance and takes into account that the feature matrix can be well approximated (C. Liu, Li, Vorobeychik, & Oprea, 2017).

Model Data Security

Model security is one of the primary aims of businesses worldwide which provide Machine Learning as a service. As increasing business tend to look towards machine learning and deep learning for solutions to modern problems, model security plays a prominent role.

Machine Learning as a service or MLaaS models have also be looked at which are mostly Support Vector Machine and Support Vector Regression(SVR) Machines which used Secured Multi-Party Communication protocols (Reith, Schneider, & Tkachenko, 2019). Model Extraction attacks were performed on Homomorphic based SVR. Assuming that the attacked has access to a substantial set of unlabelled training data. Lowd-Meek attack was implemented on Linear SVM (Hsieh, Chang, Lin, Keerthi, &

Sundararajan, 2008). It has been found that extraction takes 17 . n queries for n features. Simulation of extraction of SVM using RBF kernel was also done. To extract 20 features 420 queries were required. For extraction of linear SVR, first equation solving attack was employed for extracting the model parameters, adaptive retraining was found beneficial with lesser error rates for budgetary factor set as one. In case of extraction of Quadratic SVR's it was found that the error rates were high were using retraining attacks on extraction of natural datasets and also quadratic kernel SVR were found to be slow. Extraction of Radial Basis Function (RBF) SVR's it was found that for artificial datasets the Mean Squared Error decreases as we increase the budgetary factor.

Practical attacks were also demonstrated on Wi-Fi Localization SVR's which used Wi-Fi fingerprints using SVRs to predict client's location. An explicit model was created with RBF Kernel with 100 training samples. Models were attacked using adaptive retraining. It was found that lower number of features result in faster extraction. It was also found that not every Sigmoid Kernel is a valid kernel, it behaves similar to RBF kernel. The countermeasures suggested from this study were to round of the actual prediction in regression problems as this would decrease the accuracy of the equation solving attack. It was also noted that cutting off a client with more number of queries would be a strategic all was for MLaaS, more queries would mean more revenue. Monitoring for Suspicious Queries such as zero vectors and unit vectors would also be effective as it would help in mitigating equation solving attacks. Server side feature translation would also be an effective method.

Furthermore, research has been done to prevent acquiring of the structure of Deep Neural Network model running on accelerator was done (Wang, Hou, Zhu, Zhang, & Meng, 2019). A secure architecture 'NPUFort' was proposed which aims to prevent side-channel information leakage and protect the parameters of model. NPUFort contains unit dedicated for security and a unit dedicated for the security of instructions. The initial configuration suggested was read of configurable registers and interrupt registers has to be disabled. Verification of instructions should be done through bypass and link check which use cyclic redundancy check algorithm. AES-CTR which is designed to run as a pipeline is used to encrypt and decrypt data running on DNN accelerator. Clocks on the security unit should be synchronized with DNN clock accelerator. DNN accelerator obtains the key through RSA. Encryption of only a selected features based on the critical feature map which has been noted to reduce power consumption and obscures the boundary which is present between layers. Making it impossible for an attacker to obtain accurate execution time of each layer by timing side channel as there is read-write latency between encrypted feature map and unencrypted feature map. The proposed architecture was found to increase security and boost performance by 1.112X (approximately).

Security Attacks on Deep Learning

This section describes the various security attacks on deep learning such as adversarial attacks, poisoning attack, evasion attack, and black box attack.

Adversarial Attack

Adversarial attacks are generally performed by adding specially crafted noise to mis-classify a target class. The noise crafted is generally non recognizable to the human eye but can cause the reliable model with a high accuracy percentage to predict incorrect results.

Figure 1. Sample architecture of a medical image prediction model

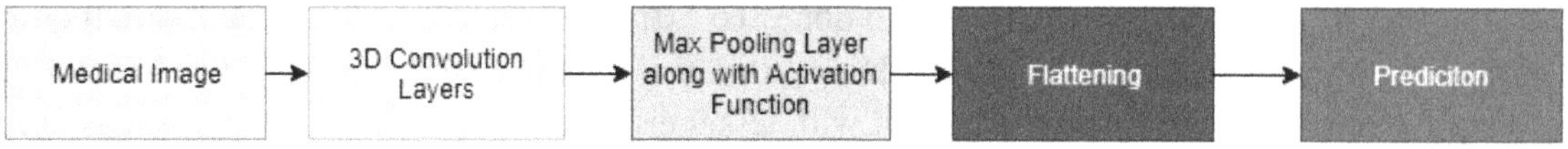

Recent research has shown that the prediction of Conventional Deep Neural Network Models are susceptible by adversarial modifications. The model fails to predict correctly when crafted minimal noise is placed in the data and has been proved by (Yi Li et al., 2019) on medical imaging data where the attack parameters where limited by modification of number of pixels, extent of modification of each independent pixel and Euclidean norm. Fig. 1. shows a sample architecture of a medical image prediction model.

Generally, there is an obfuscation masking where we try to make the information about the gradients of the model harder to obtain and Non-obfuscation masking whereby making gradients farther than the decision boundary we make the model robust against adversarial samples

It has also be found out that defensive distillation does not increase the robustness of target classifier. Focus has been to leverage two intact model in parallel to detect adversarial examples. Transferable Prediction Difference is a method proposed to detect adversarial examples (Guo et al., 2019). Network architectures which have been trained on MNIST and CIFAR were used along with TPD method. High level of accuracy was found against adversarial examples however the accuracy of the models dropped by a slight percentage. Detection rate of 98% was achieved using TPD.

A different paradigm approach was also considered while dealing with adversarial attacks. Consideration of adversarial examples not drawn from the same distribution as that of original data was also put forth (Grosse, Manoharan, Papernot, Backes, & McDaniel, 2017). A method of detection of adversarial examples using statistical tests was also proposed. Statistical metrics, 'Maximum Mean Discrepancy' and 'Energy Distance' were used to computed and compared to the threshold. This method was called as 'Two-sample hypothesis testing'. An acceptance of 95% was obtained. A sample size of 50 adversarial examples was found sufficient for statistical tests. However it should be noted that JSMA was an exception requiring 100 adversarial samples.

Focus then shifted to train Deep Neural Networks against evasion attacks from adversarial samples generated from Noise-GAN (Hashemi & Mozaffari, 2019). Production of perturbation is done using Noise-GAN, which is them sent to the Discriminator along with the original image. The Discriminator here is a multi-class network. Multi-Layer Perceptron and DeConvolution Neural Networks are used as generators. Evaluation has been performed on MNIST and CIFAR-10 with attack success rate on adversarial images being more than 91.6%.

Poisoning Attack

Poisoning Attacks (Abad & Bonilla, 2007) are generally white box attacks where the attacker tends to poison the dataset so that the model miss-classifies a specific class or an instance. Poisoning attacks are generally performed by well-intended attackers who might want to evade detection for gains. Poisoning attacks have a potential to cause massive damage if unnoticed as it aims to attack the training dataset of the model itself.

Poisoning Attacks are generally used to reduce accuracy in prediction in DNN models for the benefit of the attacker. Selective poisoning attack which reduces the accuracy of only a chosen class was proposed in (Kwon, Yoon, & Park, 2019). This method's intended target neural network applications include image recognition, face recognition and autonomous vehicles. The method also assumes that the attack is a white box attack. Malicious data is generated with the least probability to be recognized as a particular class by a model. Common convolution neural networks were pretrained on MNIST, Fashion-MNIST and CIFAR10. Reduction of accuracy by 43.2%, 41.7% and 55.3% was achieved on MNIST, Fashion-MNIST and CIFAR10 respectively.

Poisoning attacks wherein the attacker doesn't require control over the training data were also proposed (Shafahi et al., 2018). The attacks are targeted with a specific objective in mind. This attack is intended to make the target instance to be classified as the base class. Pretrained Inception V3 model is attacked first, then AlexNet architecture modified for CIFAR-10 is attacked. "One-Shot Kill" attack was found 100% effective. The reasons stated by the author for obtaining this efficiency is due to the availability of more trainable weights than training examples. Poisoning attacks on transfer learning models though difficult were found achievable through watermarking and multiple poison instances. Parameters were tweaked so that the target and the poison instances overlap. Care must be taken such that the target and the poison instance do not get separated in the feature space. To prevent this a low opacity watermark of the target instance is placed on the poisoning instance. In multiple instance poisoning attack, targeting data outliers has been found effective.

Training Data Poisoning has been specifically targeted against feature selection. Past research has shown that methods such as LASSO (least absolute shrinkage and selection operator) are vulnerable to training data poisoning by adding less than 5 percentage of poisoned samples (Xiao et al., 2015). This emphasises the importance of training data security. The successful of attacking feature selection by inserting very few poisoned samples in training data increases when the feature selection algorithm enforces sparsity. It was suggested that by looking at the bias variance decomposition of MSE (Mean Squared Error) we would be able to obtain useful insights into the impact of not only poisoning but also evasion attacks.

Algorithms have been developed which find the worst attack point and update the model with the direction of the attack point (Steinhardt, Koh, & Liang, 2017). Defenders which are dependent on data are much more vulnerable as poisoned data samples can change the region of clean dataset. By addition of poisoned data the test error for IMDB sentiment dataset error rate increase by 11% .The framework developed emphasizes in maximizing linear functions for SVM's and Logistic Regression.

Evasion Attack

Evasion attacks (Varghese, Fingerhut, & Bonomi, 2006) are a key obstacle for Deep Neural Networks. Proposal to develop new DNN's which are immune to latest evasion attacks was proposed by (Cao & Gong, 2017). It was noted that adversarial examples tend to lie in the classification boundary, for this reason, region-based classification system was proposed which does not reduce the accuracy of the model and intern makes the model more robust against evasion attacks. Region-based classification system creates a hypercube around the testing sample. Data points are then sampled from the hypercube. White-box settings was applied in this case. Evaluation of the model on MNIST and CIFAR-10 datasets to CW attacks showed a sharp decline to 7%(CW-L_0) and 16%(CW-L_0) respectively.

Black-box Attack

Black box attacks (Y. Liu, Chen, Liu, & Song, 2016) have been an interesting field for security research in deep learning. Target models in deep learning have generally been black box models (Milton, 2018) Suggested Momentum Diverse Input Iterative fast gradient sign method (M-D12FGSM) which resulted in a validation score of 1.404 to generate adversarial images with specific crafted noise. Fig. 2. shows the model of black box testing.

Exploratory attacks which are data driven on black box classifiers have also seen considered as a possibility (Sethi & Kantardzic, 2018). Seed-Explore-Exploit framework was used to simulate data driven and reverse engineering attacks on classifiers. Defender's classifier type, model parameters and training data have been assumed to be kept confidential by the defender. It was found that the effective attack rate for reverse engineering attacks on Credit, Theorem and Spam base datasets was low and they could benefit from parameter tuning. The effective attack rates (Soni & Bhushan, 2019) reduces with the increase in the exploitation radius. Choice between choosing to maintain a large list of black list data and remodeling the model must be made

Figure 2. Black box testing model

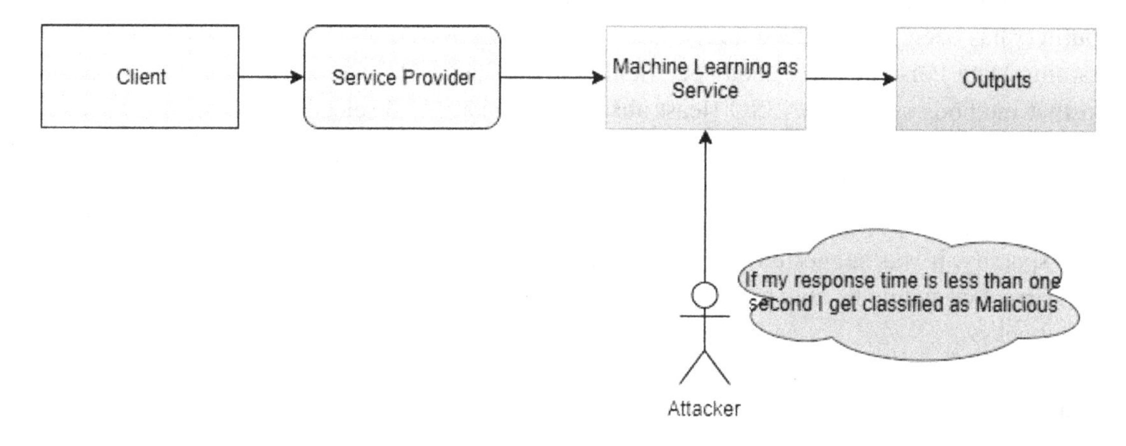

Table 1. Comparison of security attacks and it defense techniques

Reference	Year	Security Attack	Deep Learning Model	Defence Technique	Remarks
(Yi Li et al., 2019)	2019	Adversarial Attack	Hybrid Deep Learning Model	DICOM, domain knowledge, multi-alias image segmentation	The suggested techniques have to be tried on adversarial scenarios
(Guo et al., 2019)	2019	Adversarial Attack	Architectures of MNIST and CIFAR	Transferable Prediction Difference (TPD)	Proper training of the architectures is required prior to implementation of TPD
(Grosse et al., 2017)	2017	Adversarial Attack	Decision Trees, Support Vector Machine, Neural Networks		

continued on following page

Table 1. Continued

Reference	Year	Security Attack	Deep Learning Model	Defence Technique	Remarks
(Hashemi & Mozaffari, 2019)	2019	Adversarial Attack	Generative Adversarial Network	Training DNN on adversarial images	The adversarial training dataset should be large
(J. Chen, Su, Shen, Xiong, & Zheng, 2019)	2019	Black Box Adversarial Attack	Deep Neural Network(DNN)	Perturbation Optimized Black Box on Genetic Algorithm	
(Sethi & Kantardzic, 2018)	2018	Exploratory attack on black box classifier	Linear Kernel Support Vector Machine	Moving target defense strategies, Dynamic adversarial drift handling techniques	Pre-emptive strategies in training phase of classifier have to be researched and developed
(Reith et al., 2019)	2019	Attacks on Model	Linear Support Vector Machines, Support Vector Regression Machines	Server side translation, Rounding, Monitoring of Suspicious Queries and number of queries	Implementation of these methods stands with trade off in computation time and revenue
(Wang et al., 2019)	2019	Model Inversion Attack	General DNN accelerators	NPUFort	Methods such as memory isolation, information hiding through access patterns should also be considered
(Kwon et al., 2019)	2019	Selective Poisoning Attack	DNN models trained on CIFAR10, MNIST and Fashion-MNIST	Integrity of training data must be checked with number and hash value.	Defense methods suggested show promising protection
(Shafahi et al., 2018)	2018	Poisoning Attack	Inception V3, AlexNet	Exploring training of models with poisoned datasets must be done.	Side effect of adversarial training must be explored
(Cao & Gong, 2017)	2017	Evasion Attack	Region based classification DNN	Region-based classification	New evasion attacks should be evaluated against this model
(Lodeiro-Santiago et al., 2017)	2017	SQL Injection Attack	Naïve Bayes Classifier	Frequency Analysis approach	Analysis should be extended to other database engines
(C. Liu et al., 2017)	2017	Training Data Poisoning	Linear Regression		
(Mo, Shamsabadi, Katevas, Cavallaro, & Haddadi, 2019)	2019	Model Security	Deep Neural Networks	Framework to determine the memorized sensitive information	Protection of individual layers has to be considered
(T. Liu, Xu, Liu, Wang, & Wen, 2019)	2019	Model Security at a System Level	Deep Neural Networks	Static Analysis and Dynamic Analysis	Constant need to update mitigation methods
(Kwon et al., 2019)	2019	Poison Attacks	Models pretrained on MNIST, CIFAR10, Fashion-MNIST	Addition of selective malicious data	Extension of poison attacks to voice and video attacks has to be studied
(Milton, 2018)	2018	Evasion Attacks	Convolution Neural Networks	Momentum Diverse Input Iterative Fast Gradient Sign Method (M-DI2-FGSM)	Face Recognition Systems should be tested against this attack
(Xiao et al., 2015)	2017	Poisoning Attack	Feature Selection Algorithms	Secure Feature Engineering	More research has to be done in developing algorithms with security in mind
(Steinhardt et al., 2017)	2017	Poisoning Attack	Various algorithms	Defence Framework	More ways that attacks can occur should be thought

A comparative analysis of various security attacks and defense mechanisms are presented in Table 1. The table compares the security attacks on deep learning model with defense mechanisms and remarks about

PRIVACY CHALLENGES IN DEEP LEARNING

Training a deep learning model involves huge amount of personal data. Such data contains sensitive information that needs to be protected. Leakage of personal information through deep learning process lead to privacy breach. This section presents various challenges of protecting privacy in deep learning model.

Types of Privacy Breaches in Deep Learning

Privacy breaches in deep learning are possibly classified as deep learning data privacy and deep learning model privacy. Data privacy describes privacy of the data involved in training a deep learning model and the model privacy describes the privacy of deep learning model itself. The following section describes it in more detail with state-of-the-art research.

Deep Learning Data Privacy

Large data is transmitted from user to cloud or it is used for training the model. There are several privacy issues and an even chance of the data to be leaked. Data privacy is the process to secure the data of the users which contains private information (Andrew J & Karthikeyan J, 2019). The used for training should be secured or there is a possibility of the data to be leaked or misused by others.

Usually, the data from the user is sent to the cloud by encrypting it and while the data is retrieved it is decrypted into its original form. To encrypt or decrypt there is a key through which the data can be accessed.

Figure 3. Data encryption model with Cloud Server

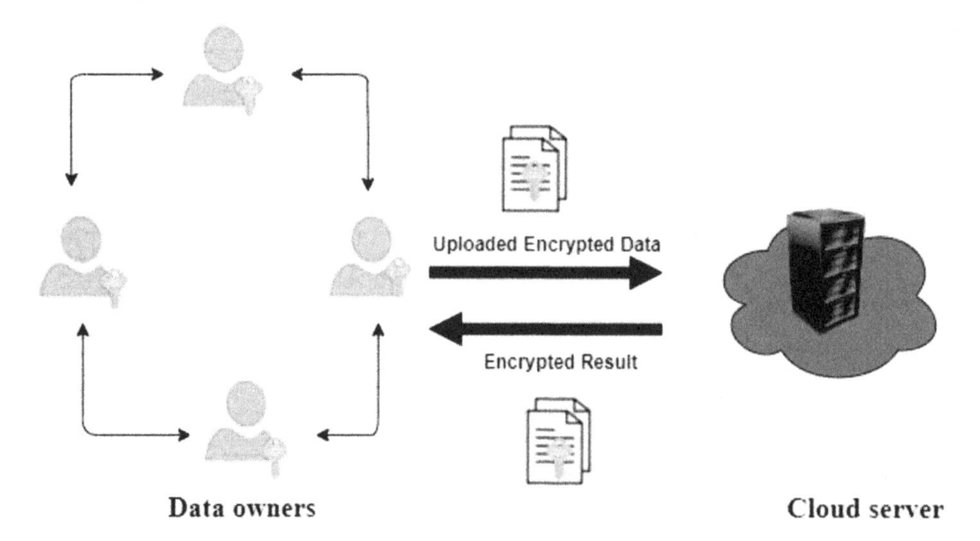

Below listed are some of the privacy-preserving methods implemented in several papers, each paper has implemented different privacy methods for data privacy.

Due to a large amount of data and insufficient storage large amount of data has been stored in the cloud. However, the cloud cannot be trusted as it is a third-party tool. Hence the paper (P. Li et al., 2018) has introduced public-key encryption with a double decryption algorithm that will encrypt the data of the users that have been uploaded in the cloud. The cloud server is to add the data into the cloud after encrypting it under a single public key. When the data analyst downloads the data from the cloud he gets the noise added dataset, he decrypts the dataset using his key and then uses it for training the machine learning model.

Data privacy is the process of securing the data by adding some noise or other in to keep the data private. To maintain the privacy of the data paper (Yuancheng Li, Wang, & Li, 2019) has proposed an algorithm called triplet loss that will generate data and some of the additional noise that will disturb the original data hence the data remains private. Triplet loss is an algorithm that can be used for individual level recognition which will improve the accuracy of the model. LFW, YTF and SFC datasets were implemented and the privacy was 98.5%, 96.2%, 97.8% respectively.

The collecting and storing of data and training are done using conventional Deep Neural Network (DNN). The training of the data is done centrally. Hence the paper (L. Chen et al., 2018) proposed a method called Crowd learning. The heavy load training procedure is decentralized and training is done in computation restricted devices that can generate training data. The computation cost and the minimization of total communication costs are checked by SliceNet. The sensitive data generated by mobile devices is calculated by crowd learning and SliceNet. MNIST and CIFAR-10 have experimented and the overall communication cost is calculated to be 0.77KB and 3KB.

Deep Learning Model Privacy

One of the major issues in deep learning is the model privacy. If the privacy of the model not maintained then private data from the model may be released and misused. There are several techniques to improve privacy in model privacy and thus leading to a secure model.

Figure 4. Privacy Preserved Deep Learning Model

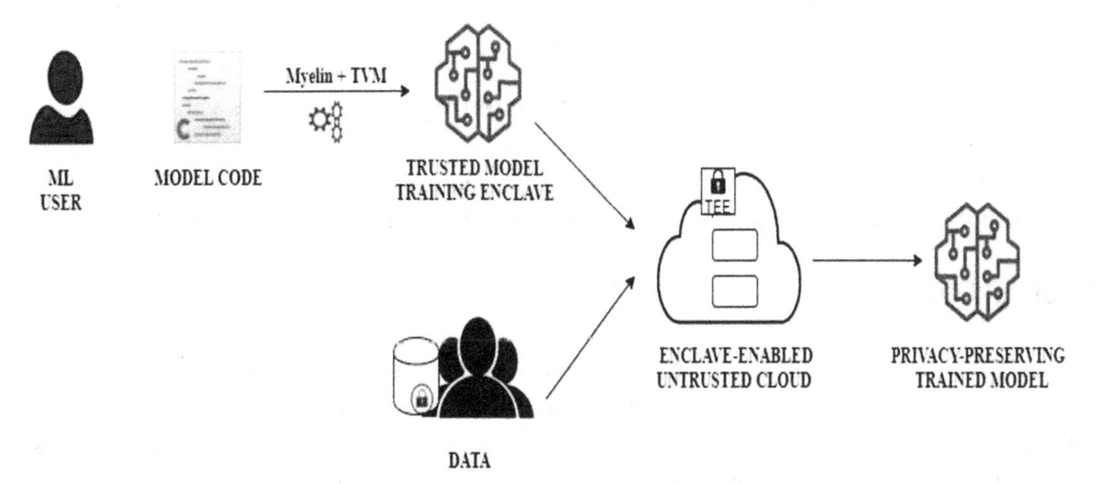

Some of the types of model privacy implemented in papers are listed below,

Model privacy (Shokri & Shmatikov, 2015) is one of the major issues in deep learning. During the training of the model the sensitive data in the dataset can be leaked which leads to the disclosure of private data. Two-stage of architecture have been implemented in (Alguliyev, Aliguliyev, & Abdullayeva, 2019). One is modified sparse denoising that performs the transformation of data and the other is CNN that classifies the transformed data. To achieve the low loss in data transformation the sparsification parameter is added to the objective function of the autoencoder by the Kullback–Leibler divergence function. Mean square error is used to calculate the loss function. The smaller the error and better the performance. The implementation of the proposed sparse denoising autoencoder is conducted on the Keras library of the Python program package. The experiments were carried out over various epochs and the best result was obtained when the epoch value was 250.

The training of the data becomes high as it requires more computation as a reason everyone uses the cloud. But the data in the cloud may not be safe and can be used by others or the private data can revealed. To overcome this problem this paper (Sharma & Chen, 2018) has come up with a solution of disguising the image and generating a secret image transformation key for each of the outsourced data. MNIST and CIFAR-10 have experimented. The disguise image has performed 95.6% in MNIST and 89.3% in CIFAR-10 which is very close to the accuracy of the data that has not been disguised. The image disguising mechanism observes high visual privacy as 94.4% in MNIST and 89.7% in CIFAR-10.

There are many security attacks due to which the data may be leaked from the model. The paper (Ma, Chen, & Zhang, 2019) comes up with a solution of an outsourcing model that uses non-interactive Privacy-Preserving Network Prediction (pp – NNP) that will increase the security of the model. The PP- NNP consists of four algorithms with it they are Setup, Query, Predict and Recover. The experiments have been done on pairing-based cryptography and found to be secured. The computation cost of input size and output size has been set to 1000 and 50 respectively.

When multiple users use the Same model for training the data can be leaked. To make the model secure the paper (Phong & Phuong, 2019) has implemented Stochastic Gradient Descent (SGD). There is no restriction for the activation function. The main advantage is that the weight parameters are shared instead of sharing the gradients computed by SGD. UCI, MNIST, CIFAR-10, CIFAR-100 dataset have experimented. One of the datasets at UCI had an accuracy of 99.95%(Skin or non-skin). Using ResNet-32v1 in CIFAR-10 the testing accuracy was 92.28% when the epoch was 97. Using ResNet-110v2 in CIFAR-100 the testing accuracy was 74.75% when the central epoch was 74.75%. Using CNN in CIFAR-10 the testing accuracy was 79% after 50 epochs.

Privacy-Preserving Techniques in Deep Learning

In this section various privacy preserving techniques for deep learning such as differential privacy, homomorphic encryption, secure multi-party computation, and secret sharing are presented.

Differential Privacy

Differential Privacy (Biryukov et al., 2011) is an algorithm that will analyze and check whether the individual's data is not determined or identified once the output is released. The main aim of this differential privacy is that it reduces the chance of an individual's private data being leaked misused by any others while the model is being trained or once it completes its training.

To achieve differential privacy in these dataset noises is added in a certain amount and even certain algorithms can add noise differently and enhance more privacy. When the dataset is less the noise added is more and when the dataset is huge the noise added is less. These noise variations are done to achieve constant privacy and the constant result of all the training models. There are two types of differential privacy they are listed below.

Laplace Mechanism

The Laplace mechanism adds Laplacian-distributed noise to a function which is also known as a double exponential distribution

$$P_r\left[A_{q,d} \in r\right] \leq e^{\varepsilon} P_r\left[A_{q,d'} \in r\right] \quad (1)$$

Gaussian Mechanism

The Gaussian mechanism is one of the methods to preserve privacy by adding some noise. This is used to minimize the risk

$$\sigma \geq \sqrt{2\log\left(\frac{1.25}{\delta}\right)} r^2 f / \varepsilon \quad (2)$$

Below listed are some of the papers where differential privacy is implemented to secure the privacy in deep learning models,

The private data of the individuals may be leaked during the model is trained. To avoid this issue, a method known as Learning Vector Quantization in (Brinkrolf, Göpfert, & Hammer, 2019) is introduced, this will limit the sensitivity of Learning Vector Quantization to a single data point and thus preserve privacy. Three datasets have been used such as MNIST, Motion, and Segment. In MNIST dataset non-private Stochastic Gradient Descent is used and the privacy loss is calculated to be 0.1857. When the privacy value goes higher the loss gets reduced.

The private data of the users may be leaked while training. Hence a method in differential privacy is used. To control this, a new method called a Heterogeneous Gaussian Mechanism is proposed in (N. H. Phan et al., 2019) and implemented. To prove the model is more secure and efficient Gaussian noise is being injected by Heterogeneous Gaussian Mechanism in the first hidden layer. This enables the model to be more secure without any loss of private data. MNIST and CIFAR-10 have experimented and the privacy budget has been increased by 29.67% and 18.17% respectively.

To train a deep learning model we need lots of data and it may be crowdsourced. There is a chance that the data may be leaked hence the paper (Abadi et al., 2016) proposed a method called Differentially Private SGD algorithm. As there is an issue in calculating the privacy sanitizer is used that protects the privacy and privacy accountant that keeps track of the privacy spending over the model. MNIST and CIFAR-10 have experimented and the accuracy is 97% and 73%. Hence it shows that using the non-convex objects at a low cost and privacy budget the model is trained.

Figure 5. Differential Privacy Model

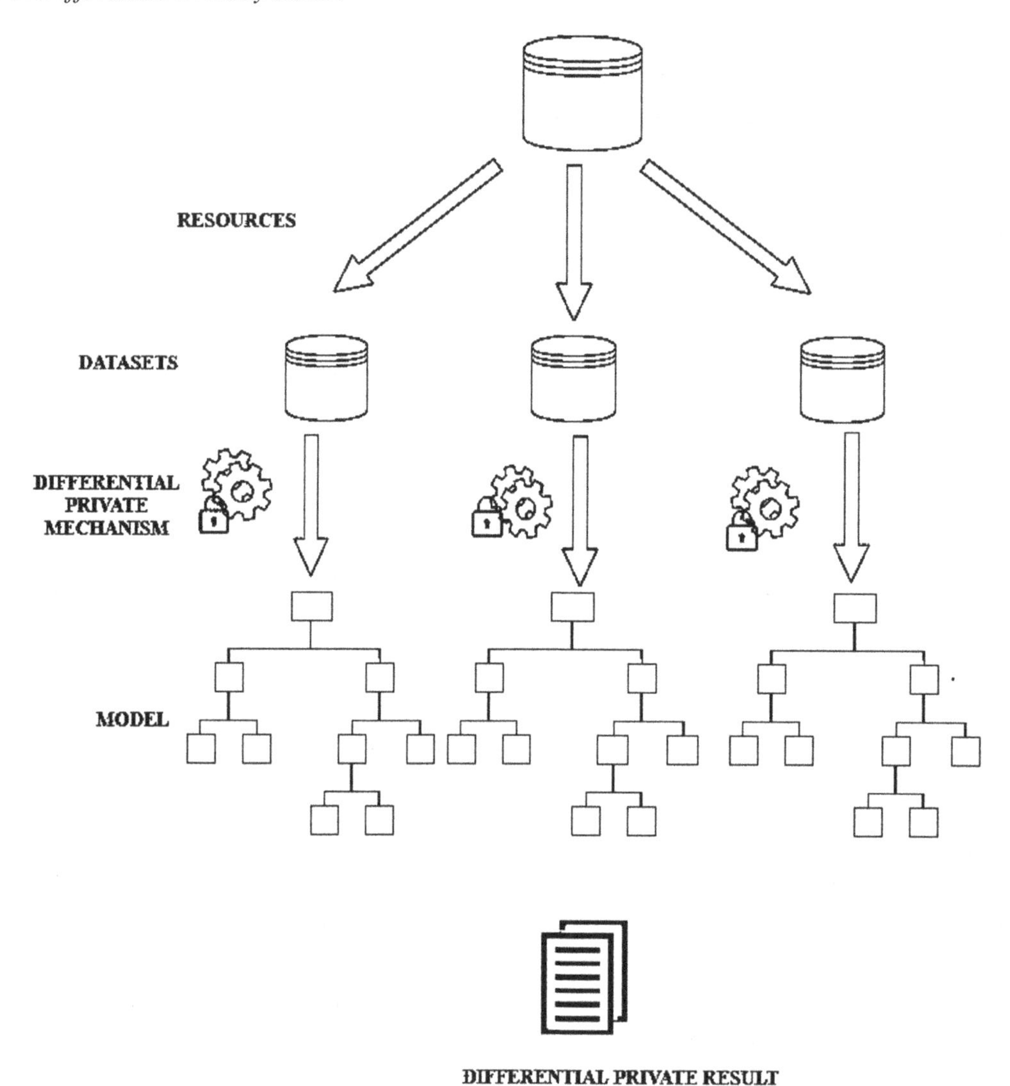

The details of the image can be exposed and can be used by others, to prevent this, the paper (N. Phan, Wu, Hu, & Dou, 2017) proposed a method Adaptive Laplace Mechanism (ADLM). In the Adaptive Laplace, the Mechanism is the network is trained by optimizing the loss function. In each computation of Layer-wise Relevance Propagation, the noise is added. MNIST and CIFAR-10 have experimented. The ADLM model gives an accuracy of 88.59% with higher noise in MNIST. In CIFAR-10 the model gives an accuracy of 77% with large noise.

Homomorphic Encryption

Homomorphic encryption (Gentry, 2009) is the process of encrypting the data before adding to the cloud or using the data for training the module. The user encrypts the data and then add it to the cloud and while accessing the data it will be decrypted and the user uses it. While encrypting the data a key is generated for the data and only with the generated key the data is decrypted. A model of homomorphic encryption technique is shown in Fig. 6.

Below are some of the papers that use homomorphic encryption to make the model more secure and private

Due to large amounts of data, users store their data in a third-party cloud. Homomorphic Encryption is a method that allows the user to keep the data private while it is processed. Multiple owners will store their data in the cloud. In this paper (P. Li et al., 2017) two multi-key privacy-preserving technique has been implemented to keep these data private. The first scheme is multi-key fully homomorphic encryption that will allow owners to encrypt their data with a different public key. The second scheme called Fully Homomorphic Encryption is proposed that will not allow interaction between several data owners. The disadvantage of this paper is to implement the FHE scheme and to reduce the cost of computation and communication.

Figure 6. Homomorphic Encryption Model

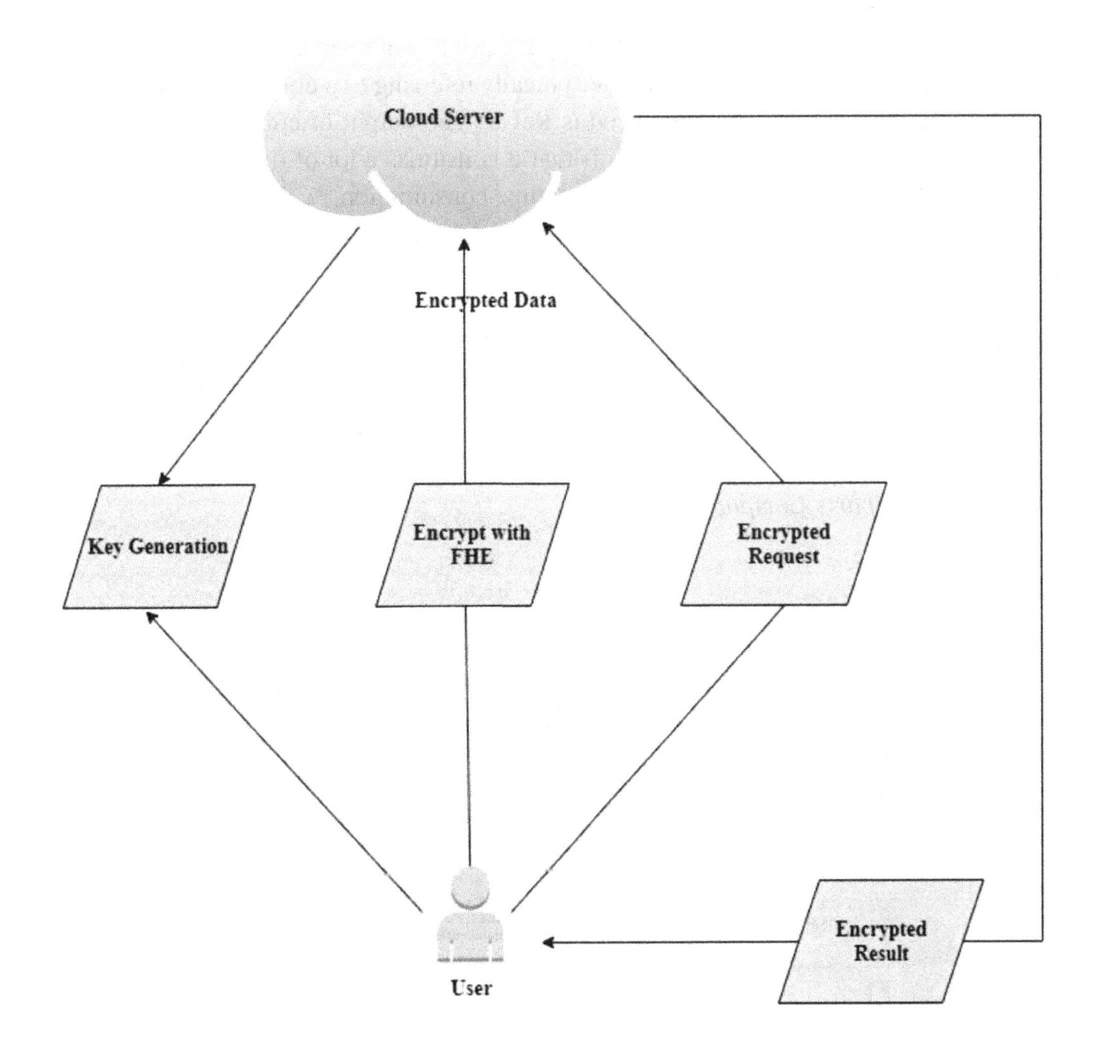

Once the machine learning model is trained and converted into a web application the administrator can view the private data, to avoid the leakage of data Homomorphic encryption is used. Homomorphic Encryption allows a user to encrypt the data before providing it as an input to the machine learning model. To avoid the leakage of data the paper (Muhammad, Sugeng, & Murfi, 2018) has implemented a model that will encrypt the input using the Paillier Encryption Scheme and produces an encrypted output that shares the same key. The MNIST dataset has experimented with 150 epochs. The model gives an accuracy of 92.2% when it reaches 122nd epoch

After the model is trained and still the private data may be leaked or stolen by others to avoid this homomorphic encryption is used that will encrypt the data while we pass it as an input for a model. Hence the paper (Arita & Nakasato, 2017) proposed a method called FHE4GT, a fully Homomorphic scheme that will encrypt the data and assign a private key to decrypt the data. The encrypted data by the client is sent to the cloud server. Another classifier called homClassify that will classify the encrypted data without decrypting it. These will classify the data in the server and the encrypted and the classified result is sent to the client to decrypt it.

Several different images may be trained such as medical images, photos of a person, etc. Hence we need to encrypt and classify the images while uploading the images or passing it for training, it can be done using homomorphic encryption. The paper (Jacobsen, 2019) implements bootstrapped Fully Homomorphic Encryption with CNN for preserving the privacy of deep learning. Bootstrap FHE is the process that runs a decrypting function homomorphically referring to a ciphertext. The MNIST dataset is used for testing, the activation function used is ReLU. The output of correctly classified images is 91.2%. The problem with Bootstrapping FHE is that it consumes a lot of time and noise. In this case, LFHE can be used which will reduce the noise and time consumption.

Secure Multiparty Computation

Secure multi-party computation (MPC) (Atallah & Du, 2001) aims to allow multiple parties to perform a computation over their input data while ensuring the data is kept private and the final result is accurate. A model of secure multiparty computation technique is shown in Figure 7.

Figure 7. Secure Multiparty Computation Model

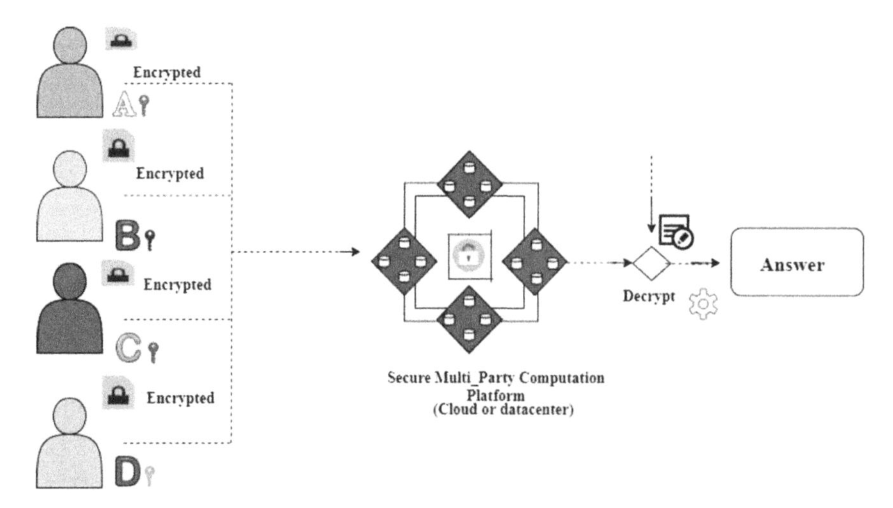

Below are some of the papers in which different techniques have been implemented to preserve privacy in deep learning,

Multi-party computation in the neural networks is a jointly trained neural network model by multiple parties' private data. Individual data can be revealed through the trained model. To avoid this, a practical multi-party computing system is designed in (Shokri & Shmatikov, 2015). In this system, the stochastic gradient descent is parallelized among the participants asynchronously. This enables the participants to train the model without sharing their private data. However, this system proposed a selective parameter sharing technique to update the model parameter for training. MNIST and SVHN datasets have experimented and it produced an accuracy of 99.14% and 98.71% accuracy with 10% of parameter sharing.

Multiparty computation in neural network analyses the private datasets without sharing the data. A method that will preserve the privacy of multiple parties by adding cryptographic tools such as Diffie–Hellman key exchange protocol, ElGamal encryption, and an aggregate signature has been introduced in (Ma, Zhang, Chen, & Shen, 2018) that will enable several parties to get trained in a same neural network model that is based on combined dataset.

Multi-party computation is the process of jointly training a deep learning model through a common medium. Due to this reason, the data may be leaked. The existing framework consumes a high total privacy budget for privacy and hence leads to data leakage. The paper (Gong, Feng, & Xie, 2020) has proposed a method called privacy-enhanced deep learning framework that can reduce the privacy issues in it. All the information of the participants that is sent to the centralized server is encrypted by the Paillier algorithm. Even if there is any problem in the central server the data is not leaked which has been proved. MNIST and SVHN dataset have experimented and it produced an accuracy of 96.18% in MNIST and 91.38% using SVHN with the values $\epsilon = 10$ and $\theta = 1$.

Multiparty computation is the process where a single model is being trained by multiple entities. To make a model effective this method is used. But there are chances that the data may be leaked, to avoid this paper (Gupta & Raskar, 2018) have implemented a method called Distributed Neural Network. To steal the data of others, the parameter is needed so that the data can be encoded and used. The parameter is unique and this parameter will not affect the model accuracy. Four datasets have been implemented and the accuracy of MNIST is 99.20%, CIFAR-10 is 92.45%, CIFAR-100 is 66.59% and ILSVRC is 57.1% which is better than the previously proposed method.

Secret Sharing

Secret sharing is the process of sharing the data or training of data without the private details getting released, the data gets split between parties and after the training, the data of both the user will join to get a result. The paper (Jarecki & Krawczyk, 1995) proposed a method called a proactive secret sharing scheme. Proactive secret sharing often changes the shares (data) and renews it, in this way the attack gets reduced. A proactive secret sharing is proved to be reconstructable every time. The update phase which consists of three stages such as private key renewal, share recovery protocol and share renewal protocol at the beginning of every period.

The paper (Calkavur, 2018) proposed a method of sharing the image privately based on two methods one is matrix projection and the other is the Shamir Secret sharing scheme. The image can be separated into k parts and the image is shared, to retrieve the image it is again converted using k parts. If the image is reconstructed using any k-n then the original image will not be restored. Hence it avoids privacy leakage while sharing the image and it is called a lossless scheme.

Table 2. Comparison of Privacy Preserving Techniques

Reference	Year	Dataset	Privacy Preserving Technique	Privacy Preserving Approach	Remarks
(P. Li et al., 2018)	2018	Clinical dataset	Data Privacy	public-key encryption with double decryption	While working with a multiparty dataset with PKE encryption there was no privacy leakage.
(Yuancheng Li et al., 2019)	2019	LFW, YTF, and SFC	Data Privacy	triplet loss	The proposed algorithm protects the privacy of the data whereas achieves high recognition accuracy
(L. Chen et al., 2018)	2018	MNIST CIFAR-10	Data Privacy	Crowd Learning	Compared to the training the overall communication is relatively large, splitting the images into more pieces will increase privacy
(Alguliyev et al., 2019)	2019	Cleveland Medical Dataset	Model Privacy	modified sparse denoising	To make the features learn directly from the raw data-sparse denoising encoder has been developed.
(Sharma & Chen, 2018)	2018	MNIST and CIFAR-10	Model Privacy	disguising the image and generating a secret image transformation key	In an outsourced setting, several image disguising has been proposed to attain privacy.
(Ma et al., 2019)	2019	Pairing-Based Cryptography (PBC) library	Model Privacy	Privacy-Preserving Network Prediction (pp – NNP)	The proposed algorithm satisfies the condition of the model privacy
(Phong & Phuong, 2019)	2019	UCI MNIST CIFAR-10 CIFAR-100	Model Privacy	Stochastic Gradient Descent	The proposed system uses weight parameters.
(Brinkrolf et al., 2019)	2019	MNIST Motion Segment	Differential Privacy	Learning Vector Quantization (LVQ)	LVQ is the proposed method which gave good and high privacy
(N. H. Phan et al., 2019)	2019	MNIST and CIFAR -10	Differential Privacy	Heterogeneous Gaussian Mechanism is proposed	Due to the proposed model promising result, there is a long term avenue to the tradeoff between privacy preservation
(Abadi et al., 2016)	2016	MNIST CIFAR-10	Differential Privacy	Differentially Private SGD algorithm	The given method will be applied directly to gradient computation and can be applied to other classical optimization methods.
(N. Phan et al., 2017)	2018	MNIST CIFAR-10	Differential Privacy	Adaptive Laplace Mechanism	The proposed mechanism can add more noise to the input feature intentionally.
(P. Li et al., 2017)	2017	Clinical dataset	Homomorphic Encryption	Multi-key Fully Homomorphic and FHE	Both the developed scheme can solve the problem of privacy-preserving and hence provide privacy
(Muhammad et al., 2018)	2018	MNIST	Homomorphic Encryption	Paillier Encryption	The weights of the simplified model the encrypted data are generated.

continued on following page

Table 2. Continued

Reference	Year	Dataset	Privacy Preserving Technique	Privacy Preserving Approach	Remarks
(Arita & Nakasato, 2017)	2017	Clinical dataset	Homomorphic Encryption	Fully Homomorphic Encryption	FHE scheme can compute encryption and homclassify is used to classify the images without decrypting it.
(Jacobsen, 2019)	2019	MNIST	Homomorphic Encryption	Bootstrap FHE with CNN	The secure prediction has been done using CNN with FHE.
(Shokri & Shmatikov, 2015)	2015	MNIST SVHN	Secure Multiparty Computation	stochastic gradient descent is parallelized	The proposed method allows the user to share their data without any concern
(Ma et al., 2018)	2018	Clinical dataset	Secure Multiparty Computation	DNN based SMC	The proposed method works for any type of deep neural network and preserves privacy and parameter privacy by retaining the accuracy of the resulting model.
(Gong et al., 2020)	2019	Clinical dataset	Secure Multiparty Computation	privacy-enhanced deep learning framework-Paillier Algorithm	A framework is proposed that will encrypt the data so the private data is not revealed.
(Gupta & Raskar, 2018)	2018	MNIST CIFAR-10 CIFAR-100 ILSVRC	Secure Multiparty Computation	Distributed Neural Network	A method to train deep neural networks has been implemented
(Jarecki & Krawczyk, 1995)	1995	Clinical dataset	Secret Sharing	proactive secret sharing scheme	The proactive sharing scheme is proposed to be efficient and secured.
(Calkavur, 2018)	2018	Image Dataset	Secret Sharing	Matrix projection and Secret Sharing	Secret sharing scheme and Shamir scheme is proved to be a lossless scheme

A comparative analysis of various privacy preserving techniques on deep learning models are presented in Table 2. The table compares the privacy preserving techniques and approaches on different datasets and remarks about the techniques.

CONCLUSION

Deep learning is the buzz word in recent times in the research field due to its various advantages in the field of healthcare, medicine, automobiles, etc. However, it has many security and privacy challenges. Huge amount of data is required for deep learning to achieve better accuracy thus it is important to protect the data from security and privacy breaches. This article addressed the security challenges during the deep learning model training and testing. The security attacks such as poisoning attacks, evasion attacks and black box attacks are explored with its prevention and defense techniques. A comparative analysis is done on various techniques to prevent the data from such security attacks. Privacy is another major challenge in deep learning. In this article, we presented an in-depth survey on various privacy preserving techniques for deep learning such as differential privacy, homomorphic encryption, secret sharing and secure multi-party computation. We also presented a detailed comparative analysis to compare the various privacy preserving techniques and approaches.

The future research direction is to develop a novel hybrid algorithm that addresses both security and privacy challenges of deep learning models that resists different types of security and privacy attacks.

REFERENCES

Abad, C. L., & Bonilla, R. I. (2007). An analysis on the schemes for detecting and preventing ARP cache poisoning attacks. *Proceedings - International Conference on Distributed Computing Systems*. 10.1109/ICDCSW.2007.19

Abadi, M., Chu, A., Goodfellow, I., McMahan, H. B., Mironov, I., Talwar, K., & Zhang, L. (2016). Deep Learning with Differential Privacy. In *Proceedings of the 2016 ACM SIGSAC Conference on Computer and Communications Security - CCS'16* (pp. 308–318). New York: ACM Press. 10.1145/2976749.2978318

Alguliyev, R. M., Aliguliyev, R. M., & Abdullayeva, F. J. (2019). Privacy-preserving deep learning algorithm for big personal data analysis. *Journal of Industrial Information Integration*. doi:10.1016/j.jii.2019.07.002

Andrew, J., & Karthikeyan, J. (2019). Privacy-Preserving Internet of Things: Techniques and Applications. *International Journal of Engineering and Advanced Technology*, *8*(6), 3229–3234. doi:10.35940/ijeat.F8830.088619

Andrew, J., & Kathrine, G. J. W. (2018). *An intrusion detection system using correlation, prioritization and clustering techniques to mitigate false alerts* (Vol. 645). Advances in Intelligent Systems and Computing. doi:10.1007/978-981-10-7200-0_23

Andrew, J., Mathew, S. S., & Mohit, B. (2019). *A Comprehensive Analysis of Privacy-preserving Techniques in Deep learning based Disease Prediction Systems*. doi:10.1088/1742-6596/1362/1/012070

Andrieu, C., De Freitas, N., Doucet, A., & Jordan, M. I. (2003). An introduction to MCMC for machine learning. *Machine Learning*, *50*(1–2), 5–43. doi:10.1023/A:1020281327116

Arel, I., Rose, D. C., & Karnowski, T. P. (2010). Deep machine learning-a new frontier in artificial intelligence research. *IEEE Computational Intelligence Magazine*, *5*(4), 13–18. doi:10.1109/MCI.2010.938364

Arita, S., & Nakasato, S. (2017). Fully Homomorphic Encryption for Classification in Machine Learning. *2017 IEEE International Conference on Smart Computing, SMARTCOMP 2017*, 2–5. 10.1109/SMARTCOMP.2017.7947011

Atallah, M. J., & Du, W. (2001). Secure multi-party computational geometry. In Lecture Notes in Computer Science (including subseries Lecture Notes in Artificial Intelligence and Lecture Notes in Bioinformatics) (Vol. 2125, pp. 165–179). Springer Verlag. doi:10.1007/3-540-44634-6_16

Bhushan, B., & Sahoo, G. (2018). Recent advances in attacks, technical challenges, vulnerabilities and their countermeasures in wireless sensor networks. *Wireless Personal Communications*, *98*(2), 2037–2077. doi:10.100711277-017-4962-0

Bhushan, B., & Sahoo, G. (2020a). *A Hybrid Secure and Energy Efficient Cluster Based Intrusion Detection system for Wireless Sensing Environment*. Institute of Electrical and Electronics Engineers (IEEE). doi:10.1109/icspc46172.2019.8976509

Bhushan, B., & Sahoo, G. (2020b). Requirements, Protocols, and Security Challenges in Wireless Sensor Networks: An Industrial Perspective. In Handbook of Computer Networks and Cyber Security (pp. 683–713). Springer International Publishing. doi:10.1007/978-3-030-22277-2_27

Biryukov, A., De Cannière, C., Winkler, W. E., Aggarwal, C. C., Kuhn, M., Bouganim, L., … Smith, S. W. (2011). Differential Privacy. In Encyclopedia of Cryptography and Security (pp. 338–340). Boston, MA: Springer US. doi:10.1007/978-1-4419-5906-5_752

Brinkrolf, J., Göpfert, C., & Hammer, B. (2019). Differential privacy for learning vector quantization. *Neurocomputing*, *342*, 125–136. doi:10.1016/j.neucom.2018.11.095

Calkavur, S. (2018). An Image Secret Sharing Method Based on Shamir Secret Sharing. *Current Trends in Computer Sciences & Applications*, *1*(2). Advance online publication. doi:10.32474/CTCSA.2018.01.000106

Cao, X., & Gong, N. Z. (2017). Mitigating evasion attacks to deep neural networks via region-based classification. *ACM International Conference Proceeding Series, Part F1325*, 278–287. 10.1145/3134600.3134606

Chen, J., Su, M., Shen, S., Xiong, H., & Zheng, H. (2019). *POBA-GA : Perturbation optimized black-box adversarial attacks via genetic algorithm*. doi:10.1016/j.cose.2019.04.014

Chen, L., Jung, T., Du, H., Qian, J., Hou, J., & Li, X.-Y. (2018). Crowdlearning: Crowded Deep Learning with Data Privacy. In *2018 15th Annual IEEE International Conference on Sensing, Communication, and Networking (SECON)* (pp. 1–9). IEEE. 10.1109/SAHCN.2018.8397100

Gentry, C. (2009). *Fully Homomorphic Encryption Using Ideal Lattices*. Academic Press.

Gong, M., Feng, J., & Xie, Y. (2020). *Privacy-enhanced multi-party deep learning*. Academic Press.

Grosse, K., Manoharan, P., Papernot, N., Backes, M., & McDaniel, P. (2017). *On the (Statistical) Detection of Adversarial Examples*. Academic Press.

Guo, F., Zhao, Q., Li, X., Kuang, X., Zhang, J., Han, Y., & an Tan, Y. (2019). Detecting adversarial examples via prediction difference for deep neural networks. *Information Sciences*, *501*, 182–192. doi:10.1016/j.ins.2019.05.084

Gupta, O., & Raskar, R. (2018). Distributed learning of deep neural network over multiple agents. *Journal of Network and Computer Applications*, *116*, 1–8. doi:10.1016/j.jnca.2018.05.003

Hashemi, A. S., & Mozaffari, S. (2019). Secure deep neural networks using adversarial image generation and training with Noise-GAN. *Computers & Security*, *86*, 372–387. doi:10.1016/j.cose.2019.06.012

Hsieh, C. J., Chang, K. W., Lin, C. J., Keerthi, S. S., & Sundararajan, S. (2008). A dual coordinate descent method for large-scale linear SVM. In *Proceedings of the 25th International Conference on Machine Learning* (pp. 408–415). 10.1145/1390156.1390208

J, A., Karthikeyan, J., & Jebastin, J. (2019). Privacy Preserving Big Data Publication On Cloud Using Mondrian Anonymization Techniques and Deep Neural Networks. In *2019 5th International Conference on Advanced Computing & Communication Systems (ICACCS)* (pp. 722–727). IEEE. doi:10.1109/ICACCS.2019.8728384

Jacobsen, R. H. (2019). *On Fully Homomorphic Encryption for Privacy-Preserving Deep Learning*. Academic Press.

Jarecki, S., & Krawczyk, H. (1995). Or : How to Cope With Perpetual Leakage. *Communication*, 1–22.

Jindal, M., Gupta, J., & Bhushan, B. (2020). *Machine learning methods for IoT and their Future Applications*. Institute of Electrical and Electronics Engineers (IEEE). doi:10.1109/icccis48478.2019.8974551

Krishnamurthy, B., & Wills, C. E. (2009). *On the Leakage of Personally Identifiable Information Via Online Social Networks*. Retrieved from https://www.facebook.com/profile.php?

Kwon, H., Yoon, H., & Park, K. W. (2019). Selective poisoning attack on deep neural networks. *Symmetry*, *11*(7), 1–13. doi:10.3390ym11070892

Li, P., Li, J., Huang, Z., Li, T., Gao, C.-Z., Yiu, S.-M., & Chen, K. (2017). Multi-key privacy-preserving deep learning in cloud computing. *Future Generation Computer Systems*, *74*, 76–85. doi:10.1016/j.future.2017.02.006

Li, P., Li, T., Ye, H., Li, J., Chen, X., & Xiang, Y. (2018). Privacy-preserving machine learning with multiple data providers. *Future Generation Computer Systems*, *87*, 341–350. doi:10.1016/j.future.2018.04.076

Li, Y., Zhang, H., Bermudez, C., Chen, Y., Landman, B. A., & Vorobeychik, Y. (2019). Anatomical context protects deep learning from adversarial perturbations in medical imaging. *Neurocomputing*. doi:10.1016/j.neucom.2019.10.085

Li, Y., Wang, Y., & Li, D. (2019). Privacy-preserving lightweight face recognition. *Neurocomputing*, *363*, 212–222. doi:10.1016/j.neucom.2019.07.039

Liu, C., Li, B., Vorobeychik, Y., & Oprea, A. (2017). Robust linear regression against training data poisoning. *AISec 2017 - Proceedings of the 10th ACM Workshop on Artificial Intelligence and Security, Co-Located with CCS 2017*, 91–102. 10.1145/3128572.3140447

Liu, T., Xu, N., Liu, Q., Wang, Y., & Wen, W. (2019). A System-level Perspective to Understand the Vulnerability of Deep Learning Systems. *Proceedings of the Asia and South Pacific Design Automation Conference, ASP-DAC*, 544–549. 10.1145/3287624.3288751

Liu, Y., Chen, X., Liu, C., & Song, D. (2016). *Delving into Transferable Adversarial Examples and Black-box Attacks*. Retrieved from https://arxiv.org/abs/1611.02770

Lodeiro-Santiago, M., Caballero-Gil, C., & Caballero-Gil, P. (2017). Collaborative SQL-injections detection system with machine learning. *ACM International Conference Proceeding Series*. 10.1145/3109761.3158395

Ma, X., Chen, X., & Zhang, X. (2019). Non-interactive privacy-preserving neural network prediction. *Information Sciences*, *481*, 507–519. doi:10.1016/j.ins.2018.12.015

Ma, X., Zhang, F., Chen, X., & Shen, J. (2018). Privacy preserving multi-party computation delegation for deep learning in cloud computing. *Information Sciences*, *459*, 103–116. doi:10.1016/j.ins.2018.05.005

Milton, M. A. A. (2018). *Evaluation of Momentum Diverse Input Iterative Fast Gradient Sign Method (M-DI2-FGSM) Based Attack Method on MCS 2018 Adversarial Attacks on Black Box Face Recognition System*. Academic Press.

Mo, F., Shamsabadi, A. S., Katevas, K., Cavallaro, A., & Haddadi, H. (2019). *Towards Characterizing and Limiting Information Exposure in DNN Layers*. doi:10.1145/3319535.3363279

Muhammad, K., Sugeng, K. A., & Murfi, H. (2018). Machine Learning with Partially Homomorphic Encrypted Data. *Journal of Physics: Conference Series*, *1108*(1), 012112. Advance online publication. doi:10.1088/1742-6596/1108/1/012112

Phan, N., Wu, X., Hu, H., & Dou, D. (2017). Adaptive Laplace Mechanism: Differential Privacy Preservation in Deep Learning. In *2017 IEEE International Conference on Data Mining (ICDM)* (pp. 385–394). IEEE. 10.1109/ICDM.2017.48

Phan, N. H., Vu, M. N., Liu, Y., Jin, R., Dou, D., Wu, X., & Thai, M. T. (2019). Heterogeneous Gaussian mechanism: Preserving differential privacy in deep learning with provable robustness. *IJCAI International Joint Conference on Artificial Intelligence,* 4753–4759. 10.24963/ijcai.2019/660

Phong, L. T., & Phuong, T. T. (2019). Privacy-Preserving Deep Learning via Weight Transmission. *IEEE Transactions on Information Forensics and Security*, 1–1. doi:10.1109/TIFS.2019.2911169

Reith, R. N., Schneider, T., & Tkachenko, O. (2019). *Efficiently Stealing your Machine Learning Models*. doi:10.1145/3338498.3358646

Russell, S. J., & Norvig, P. (2016). *Artificial intelligence: a modern approach*. Pearson Education Limited.

Sethi, T. S., & Kantardzic, M. (2018). Neurocomputing Data driven exploratory attacks on black box classifiers in adversarial domains. *Neurocomputing*, *289*, 129–143. doi:10.1016/j.neucom.2018.02.007

Shafahi, A., Ronny Huang, W., Najibi, M., Suciu, O., Studer, C., Dumitras, T., & Goldstein, T. (2018). Poison frogs! Targeted clean-label poisoning attacks on neural networks. Advances in Neural Information Processing Systems, 6103–6113.

Sharma, S., & Chen, K. (2018). *Image Disguising for Privacy-preserving Deep Learning*. doi:10.1145/3243734.3278511

Shokri, R., & Shmatikov, V. (2015). Privacy-Preserving Deep Learning. In *Proceedings of the 22nd ACM SIGSAC Conference on Computer and Communications Security - CCS '15* (pp. 1310–1321). New York: ACM Press. 10.1145/2810103.2813687

Soni, S., & Bhushan, B. (2019). Use of Machine Learning algorithms for designing efficient cyber security solutions. In *2019 2nd International Conference on Intelligent Computing, Instrumentation and Control Technologies, ICICICT 2019* (pp. 1496–1501). Institute of Electrical and Electronics Engineers Inc. 10.1109/ICICICT46008.2019.8993253

Steinhardt, J., Koh, P. W., & Liang, P. (2017). Certified defenses for data poisoning attacks. Advances in Neural Information Processing Systems, 3518–3530.

Varghese, G., Fingerhut, J. A., & Bonomi, F. (2006). *Detecting evasion attacks at high speeds without reassembly*. Association for Computing Machinery (ACM). doi:10.1145/1159913.1159951

Wang, X., Hou, R., Zhu, Y., Zhang, J., & Meng, D. (2019). NPUFort: A secure architecture of DNN accelerator against model inversion attack. *ACM International Conference on Computing Frontiers 2019, CF 2019 - Proceedings*, 190–196. 10.1145/3310273.3323070

Xiao, H., Biggio, B., Brown, G., Fumera, G., Eckert, C., & Roli, F. (2015). Is feature selection secure against training data poisoning? *32nd International Conference on Machine Learning, ICML 2015*, *2*, 1689–1698.

Chapter 4
Wireless Environment Security: Challenges and Analysis Using Deep Learning

Vidushi
Banasthali Vidyapith, India

Manisha Agarwal
Banasthali Vidyapith, India

Aditya Khamparia
https://orcid.org/0000-0001-9019-8230
Lovely Professional University, India

Naghma Khatoon
Usha Martin University, India

ABSTRACT

The communication through the wireless environment is open, which completely differs from the wired. This open environment communication can be accessed by users including illegitimate and thus increases the vulnerability for malicious attacks. For that reason, motivation comes to study about the different possible security challenges, threats, and to devise powerful, efficient, and improved required solution to improve the various security vulnerabilities. This chapter presents the challenges regarding security and the security requirement in the wireless type communication. The research performs the analysis of deep learning for detecting malicious websites. These websites are responsible to disrupt normal system working and can control the complete system and its resources by installing malware on to the respective machine. To elucidate the constructive and effective way towards the detection of malicious URL, the study uses convolutional neural networks.

DOI: 10.4018/978-1-7998-5068-7.ch004

INTRODUCTION

The present era is timed by emphasizing on the security due to advancement in technology that exceedingly trends the devices towards the connectivity through the wireless network. Nowadays, an enormous volume of information is transmitted with the help of wireless networks that has generated a myriad of privacy problems. The wireless system enhances the technological flexibility and makes the environment more convenient to work (Aliu, 2012). Along with advantages, it also increases risk of security. It directly invites and open the security threats that may affect day to day life and breaches privacy. Wireless network provides convenience like searching the web, interconnection, etc. however, it is certain that some problems of security, data breach or privacy may come into the picture. Several complications are faced by society like academic loss, economic loss and many personal life problems. Problem of privacy breach is unavoidable, this can be disastrous and life changing event. Thus, security has paramount importance along with improvements in technology. Lack of security can be a barrier in the way of technology usage and advancement. Because security is a necessary evil, it opens the door for the academic researchers, scientists, as well as industrial researchers.

Deep learning seems as a possible solution to analyze security problems. Machine learning technique by using past data and experience will lead to improvement. It can learn without any requirement of human. Machine learning (Alpaydin, 2020) technology has shown great impact and achievement in the area of artificial intelligence. Deep learning (Shokri, 2015) is also a part and member of machine learning. Significant use and success of machine and deep learning have been observed in many applications like recognition of an object, in medical science, feature prediction or detection, jamming, etc.

The existing serious problem of malicious URL detection is understandable and has been deeply analyzed in this document using deep learning model. Because of the active adoption of deep learning in different above-mentioned fields, this letter uses this learning for detecting the malicious URL. Therefore, the objective of this paper is investigating the deep learning potential in terms of accuracy and to measure loss occurred while detecting malicious URL.

In deep learning model, this research presents a convolutional network model (Zhao, 2013) for malicious URL detection. However, this document shows the potential as well as loss in training together with validation phase. This letter presents analysis of a convolutional network that is a deep learning model for the detection of malicious URL in the wireless network. Deep learning is a technique which mimics brain learning process. It tries to apply the working technique of brain to the machine. Deep learning thus adds intelligence and makes the model more powerful. Deep learning tries to simulate behavior of the brain therefore possess capability of data processing. The human brain works smoothly even in a complex situation, the brain takes a decision based on inputs collected by the sense organs. Deep learning seems eye catching method as it achieves similar capability like brain. For accomplishing this aim numerous layers of the neural network are present in deep learning. These layers help to solve complicated and compound problems. Deep learning has the ability to learn from experience, so more experience means more learning. Learning from the past is used to frame the model to train the dataset. With the completion of the neural network training, suitable decision can be taken to get an intellectual reward. The implementation of this complete idea in the real world situations has shown remarkable success, such as face identification, interpretation of language, drug discovery, etc. The astonishing outcome have been shown by deep learning dealing with real problems.

Deep learning is not a new branch, rather it is a subcategory of machine learning. With the help of cascading layers, it retrieves useful features that are further used in a powerful decision-making task.

Prediction, analysis, detection, recognition, and many such complicated tasks need complex decision making. The only correct decision will lead to successful completion of the task. Otherwise unpredictable problems can arise. So It can be said that, success can only be achieved by the right decision. For instance, incorrect weather prediction or incorrect predictions in business can lead to huge damage in terms of not only monetary but life also. In brief, deep learning can represent the complex environment, can extract abstract information and last but not least a good decision-maker.

This document is motivated to implement the deep learning model to detect the malicious website. The paper contribution is to analyze the malicious website that can redirect the users of their demanded website to the other one that exploits the browser vulnerabilities. The user may get infected by malware and exploited because of browser vulnerabilities. Attackers lure the client and redirect them towards the malicious website or URL. The clients or users are forced to download malware from the malicious website that can further damage even the complete system. This paper elaborates the security challenges for the wireless network and detects the malicious website using convolutional neural network.

This article remaining part is organized into different sections. Section 2 elaborates related and useful work performed by researchers previously. Further, in the next section 3, the wireless environment concept, including types, and OSI layers are discussed. Section 4 mentions various challenges, parameters, attacks, possibly related to the wireless system. Section 5 describes the malicious_n_non-malicious URL dataset used in this paper. While section 6 focuses on the model called convolutional network. Studies' section 7, presents the experimental analysis work graphically and in tabular form on the dataset using the convolution model during both training and testing phase with 16 epochs. Section 8 discusses the future work along with a conclusion.

BACKGROUND

Toshiki et. al. (Shibahara, 2017) discussed the malicious URL and proposed a method to find out malicious sequences of URLs. Furthermore, this paper compares the individual-based, CNN, and EDCNN approaches to classify URL sequences. The authors of (Zou, 2016) presented a detailed survey on the security challenges faced by wireless network and also the present defense mechanism for malicious attacks to protect the transmission through the wireless network. Minho et. at. (Shin, 2006) aims to come up with a comprehensive survey related to security problems and 3G, WLAN latest technology. The thorough detail students with several jamming techniques along with James types and jamming countermeasures are presented by the study authors (Jaitly, 2017).

Sydney et. al. (Kasongo, 2019) proposes FFDNN-IDS (Feed Forward Deep Neural Network) and compare with present machine learning techniques. The paper comparison result proves FFDNN-IDS accuracy better than existing one. The authors of (Mao, 2018) comprehensively analyzed and reviewed different methodologies to enhance the performance of the wireless network by applying schemes of deep learning. The article (Mao, 2018) performed the survey related to the applications of various algorithms of deep learning for network layers. Safa et. al. (Otoum, 2019) provides the survey and analysis report provides the solution concerning the intrusion detection system for wireless sensor network.

To detect efficiently the intrusive behavior of the wireless network, (Yang, 2019) proposes a deep learning-based model for intrusion detection called combined wireless. The intension of the authors (Kavianpour, 2017) of this research is the identification of the principal elements required for network security. Along with it, this paper presents an overview of security related to the wireless LAN. The authors (Xie, 2018) present wireless sensor networks overview and also classify wireless attacks based on the layers of the protocol stack attacks are classified. The research of the authors (Tomić, 2017) culminates the attacks of network layer with deep analyses in opposition to the protocol for RPL routing. The meaning of RPL is a routing protocol for low power and lossy networks.

WIRELESS ENVIRONMENT

The communication between devices is enhanced because of the advancement in manufacturing technologies in hardware and wireless system. This enables small, low power, and multi-functional sensors for the network called wireless sensor networks (Bhushan, 2019). These networks have many issues (Bhushan, 2018) along with popularity of wireless system among users. The major issue is a network security (Han, 2019), (Hennebert, 2014). There are several reasons for this popularity; some important ones are as follows:

- To connect devices, cable is not required. It saves the cost as well as space. Because of network extendibility feature, user's mobility increase. This makes it better than the traditional one.
- It is simple to use and has comparatively high speed.
- Its coverage area is many times wider compared to wire based systems.
- A high level of scalability, efficiency, availability, accessibility, and its flexible nature increases its demand in users.
- Wireless environment opens billions of business opportunities.
- It has become possible for anyone to work anytime and from anywhere.

Above mentioned benefits are major causes for using a wireless system that can become useless if security threats are not dealt properly (Bhushan, 2019), (Sharma, 2017). Today's world is fast, everyone is busy, and wants to complete work in seconds or as early as possible, wireless communication is solution to accomplish the need of society. Today's society is incomplete without wireless system. This communication becomes the basic need of society. But security threat like malicious node presence in sensor networks of type wireless (She, 2019) makes it a serious concern. Broadly, the wireless network can be divided into three categories. These are mentioned below in **Table 1.**

Table 1.Wireless Network Types

Wireless Types	Definition
Wireless Wide Area Network (Sinha, 2017), (Zou, 2016)	Its creation is done by the signals of the mobile phone that is further maintained by service providers (Sinha, 2017), (Zou, 2016).
Wireless Local Area Network (Sinha, 2017), (Zou, 2016)	Radio waves are used by these networks. Its range limit is varying from room to campus (Sinha, 2017), (Zou, 2016).
Wireless Personal Are Network (Sinha, 2017), (Zou, 2016)	Bluetooth technology is used by these networks. Their range is near about 30 feet (Sinha, 2017), (Zou, 2016).

The above table 1 describes the types of wireless systems. These systems are used as per the requirement. The communication through the wireless environment generally adopts OSI architecture. This is comprised of seven layers. These OSL layers are described briefly in below **Table 2.**

Table 2. Layers of OSI Architecture

	OSI Layer	Description
First Layer	Application Layer	It is placed at the top. The user interacts to the system through the application layer.
Second Layer	Presentation Layer	Also known with the name translation layer. Encryption-decryption, compression, and translation are the main task of this layer
Third Layer	Session Layer	The session-dialogue control, maintenance, and termination are done at this layer
Fourth Layer	Transport Layer	Flow, as well as control of error, is performed at this layer
Fifth Layer	Network Layer	The function of routing and local addressing is performed by this layer.
Sixth Layer	Data Link Layer	This layer takes the responsibility of delivery of message between the nodes. LLC and MAC are its sub-layers.
Seventh Layer	Physical Layer	The information contained by this layer is in the form of bits. The connection between devices is present in this layer.

The above table 2 describes all the layers of the OSI model along with their description (Zou, 2016). The OSI model is the conceptual model. Means it is not implemented yet, but all other models like TCP/IP use the concept of the OSI model and are also practically implemented (Zou, 2016), (Bhushan, 2020). OSI is indeed the base for other network model, but till now it is not implemented practically.

WIRELESS ENVIRONMENT SECURITY CHALLENGES

From the last decades, the communication through the wireless system is proliferating to achieve the goal of drastically increasing demand. Along with the advantages, the biggest disadvantage also comes into the picture called a security that can diminish the complete technology if not handled properly and on time. As mentioned by the authors of (Zou, 2016) with the mere use of wireless devices also increases illicit activities related to cyber fraud or crimes. Some of the cyber-related crimes, in short, are mentioned below in Table 3.

Table 3. Important cyber crimes and their description

	Cyber Crimes	Description
1	Computer hacking (Zou, 2016), (Bhushan, 2020)	It is a technique to alter the original information of the system (Zou, 2016), (Bhushan, 2020).
2	Malicious attack (Zou, 2016), (Bhushan, 2020)	It is a way to abuse the other's system (Zou, 2016), (Bhushan, 2020).
3	Information theft (Zou, 2016), (Bhushan, 2020)	To take other's data by cheating (Zou, 2016), (Bhushan, 2020).
4	Financial theft (Zou, 2016), (Bhushan, 2020)	To cheat someone monetarily (Zou, 2016), (Bhushan, 2020).
5	Identity theft (Zou, 2016), (Bhushan, 2020)	To hide the original identity of the user (Zou, 2016), (Bhushan, 2020).
6	Phishing (Zou, 2016), (Bhushan, 2020)	It is a wrong practice of getting users personal data by sending an email (Zou, 2016), (Bhushan, 2020).

The above table 3 demonstrates different well-known cybercrime types in brief. Because of the presence of all these frauds possibilities in the wireless network, special attention towards the wireless security in the physical layer (Wang, 2017) and others comes into the picture. It is the need of today's society to work on security-related problems. The necessity of security breach attracts the attention of researchers in the world in the field of security attacks, technical serious challenges, vulnerabilities, along with their countermeasures (Bhushan, 2017). It can be said that there is no meaning of technological advancement of research or technology is meaningless if the user's data, private information is not secure. This critical concern opens the new research areas and researchers are giving their efforts in this crucial field. Thus, improving the matter of security in communication through a wireless medium has paramount importance. The main parameters needed for wireless security are depicted in the following **Figure 1.**

Figure 1. Wireless Network Security

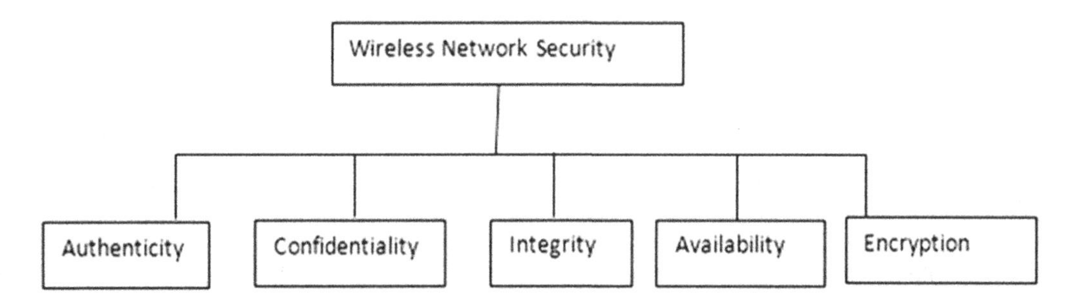

To take advantage of wireless technology without any fear the parameters mentioned in the above figure 1 have to be present in its security mechanism. In a wireless environment (Jedari, 2018), (Conti, 2016), information exchanged takes place only in between the users who are authorized for that. To protect the data from harmful hands encryption, authentication, and all mentioned in above figure 1 are needed to be present during the communication. They are described as follows:

- Authenticity can be simply defined as the process of identifying authorized from others.
- Confidentiality defined the process in such a way so that only intended clients can access data.
- Integrity means that the information transmission needs to be both reliable as well as accurate.
- Availability means that the intended user should be capable of getting a wireless network as per the need.
- Encryption is to hide the original with the other one.

Different OSI layers are mentioned in the above section and described in table 2. Each layer has problems related to security (Bhushan, 2017). Because every layer's protocol is different, which makes each layer security issue different from the others. So different types of attacks at different layers are possible in wireless system. Security (Bhushan, 2017) is the main concern problem in wireless and different kinds of attacks possibility is present in different layers. This possibility of different attacks on different layers is present because each layer has different protocols. Almost every work is done in the layers with the help of protocols. These OSI layers are also known as protocol layers. Multiple protocols are present on the layers to perform work. Each layer working is different from the others because of the presence of different protocols. Table 4 below mention some important protocols and possible attacks with the concerned layer. Layers with their important protocols and attacks possible are described below in **Table 4.**

Table 4. Layers, Protocols, and possible attacks

OSI Layer	Protocols	Possible attacks
Physical Layer	Actual transmission occurs (Haibo, 2010).	Eavesdropping (Wang, 2017), (Perrig, 2004), Jamming (Mpitziopoulos, 2009).
Data Link Layer	CDMA (Derryberry, 2002), OFDMA (Haibo, 2010), (Derryberry, 2005).	Spoofing (Mpitziopoulos, 2009), (Allman, 1999),MITM attack (Zhou, 2006), Network injection (Park, 2007), (Harris, 1999).
Transport Layer	TCP (Lee, 2001), UDP (Bruno, 2007), (Lee, 2001).	TCP flooding (Kuzmanovic, 2006), (Schuba, 1997), UDP flooding (Chang, 2002).
Network Layer	IP, ICMP (Labovitz, 2000)	IP Spoofing (Hastings, 1996), (Center, 1995),IP hijacking (Hastings, 1996).
Application Layer	HTTP, SMTP (Perrig, 2004)	Malicious URl attacks (Kieyzun, 2009), Malware attacks (Kieyzun, 2009), FTP bounce (Allman, 1999).

The above table 4 describes attack possible on layers that can breach the user's security and this breaching may lead to misuse of private data of any user involved in communication through the wireless system. Any type of communication possible through the wireless mode needs special attention because of the involvement of user private information. Online banking is one of the uprising communication that needs wireless system. This process includes monetary of the user and that makes security the highest priority. Thus, to detect the security-related problem at early stage becomes vital.

MALICIOUS_N_NON-MALICIOUS URLDATASET

The research studies the wireless sensor network (Bhushan, 2019), attacks (Bhushan, 2017) and detects the malicious URL because this is one of the uprising challenges in the wireless environment and becomes a crucial research for researchers. This document uses Malicious_n_Non-Malicious URL (Althunibat, 2016) dataset. This dataset consists of two columns and 420465 rows. It has 420465 records having URL address and associated labels. The label indicates that the URL is bad or good. The bad label indicates that there is some problem, wrong information, and the possibility of attacks may be present with the website.

User should not go to URL, labelled bad. There is a possibility of presence of malware and loss of important information about the user. Even the damage of the system in which that URL opens is also possible. This type of URL (Tandon, 2019) with a bad label is also said the malicious URL. On the other hand, the user can go ahead with the good URL indicating that no problem or malware present, that can cause any loss to the user. This type of URL with label good is also known as non-malicious URL (Tandon, 2019). For online banking, hotel booking, purchasing, etc. are some of the tasks that need wireless communication. To accomplish these tasks, the security of user information becomes highly prioritized. To redirect the user to the malicious URL or website for breaching the private information is very common. It makes it necessary to predict or detect these malicious URL. This research uses malicious_n_non-malicious URL datasets to detect the malicious. In general, datasets need preprocessing to get more potential output. But this dataset has no null value or missing value. It is the complete data, so no need to perform preprocessing task on it. This dataset is used to train the convolutional neural network model and after training the analysis work is performed. The experimental analysis work is performed for both pieces of training as well as the validation phase. Both the potential in terms of accuracy and the loss is measured using a convolutional neural network implemented on the malicious_n_non-malicious dataset.

CONVOLUTIONAL NEURAL NETWORK

A convolutional neural network is a type of deep learning model and deep learning is a subset of machine learning. All these models come under artificial intelligence. The main idea of deep learning is to learn like a human brain (Alali, 2019). This learning is giving its effort in the direction to imitate brain working. Its purpose is to process the data, getting significant patterns and finally utilize them for making the right decision. Business analysis, risk measurement, fraud-detection, malicious URL detection, and any critical task need a decision. The success or the future of work totally depends on the decision.

Deep learning helps in making the right decision based on processed data. The main and important characteristic of this learning is that it is self-adaptive (Mishra, 2019), (Mishra, 2018). As data and experience increases, its decision-making strength also improves. Deep learning to carry out the tasks adopts the basic concept of a neural network called an artificial neural network. The artificial neural network brings the achievement to accomplish the analysis work for not only the linear approach but also for non-linear. This network is structured with multiple layers and each layer with multiple neuron nodes. Layers are broadly divided into three types. One input, one or many hidden, and output layers. Its basic architecture is shown in below **Figure 2.**

Figure 2. Basic Artificial Neural Network Architecture

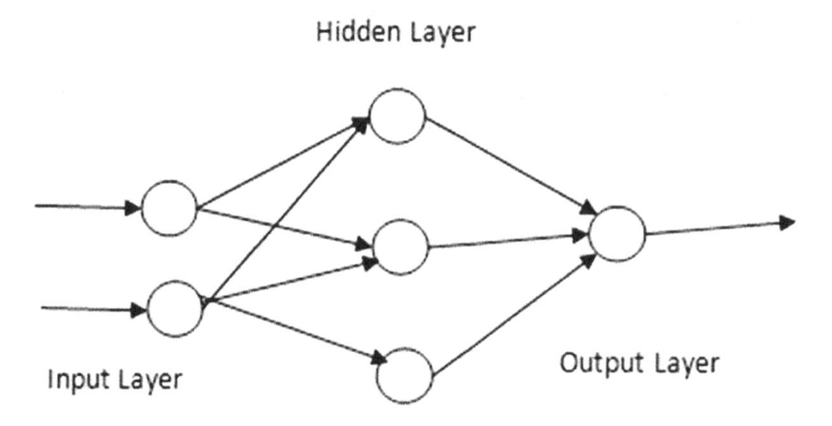

The above figure 2 shows the basic architecture of an artificial neural network. Basic three layers are shown in the above figure 2. The first layer is the input layer and its basic work is to receive the input from the user. The second layer called the hidden layer. In figure 2 only one hidden layer is shown, but there can be multiple hidden layers. Computation is performed on the hidden layers. And finally, the output is given by the output layer. This network working is represented through the following equation 1.

$$y = \sum_{i=1}^{n} (w_i x_i) + b \quad (1)$$

Above equation 1 explains the artificial neural network concept. The convolutional neural network is the type of artificial neural network. It consists of an input layer, convolution layer, pooling layer, dense or fully connected layer, and output layer. The requirement of pre-processing is less because it can filter. Its architecture is analogous to the brain's neuron connectivity. The basic architecture of the convolution network (Shomron, 2018), (Agarwal, 2019) is given in below **Figure 3.**

Figure 3. Overview of Convolutional Neural Network

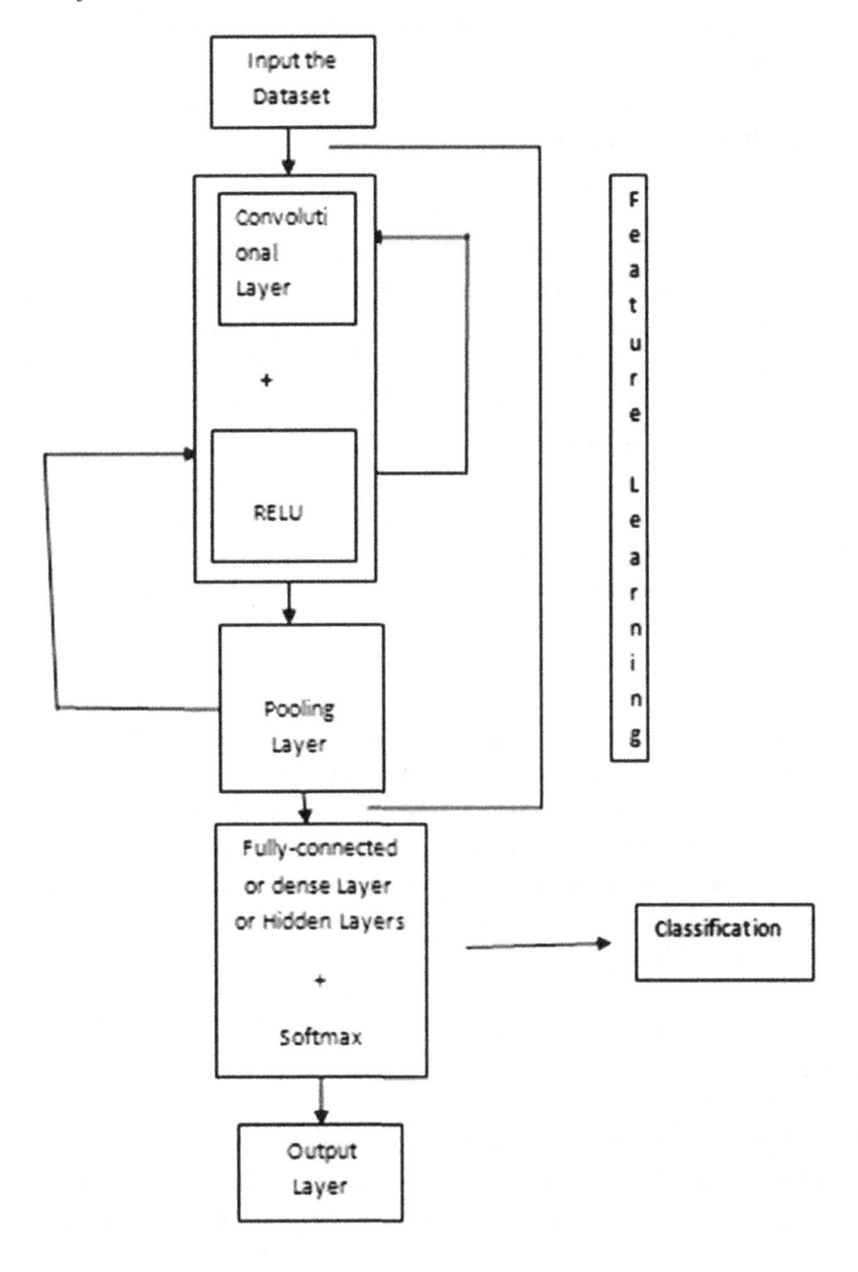

The above figure 3 depicts the basic flow of the convolutional neural network. It can solve both linear as well as a non-linear problem with less need for pre-processing. The working of this network is divided into the layers. Layers work by receiving the input data from the previous and transform it into the suitable form of the upcoming layers (Mellouli, 2019), (Yang, 2019). The basic steps taken by the convolutional neural network to accomplish the desired task are as follows:

- Assign the dataset as an input to the model.
- Input is received by the first layer called the convolutional layer that performs the dot computation of weights and input data to filter the input.
- The next layer is the RELU layer through which the activation function is applied element-wise.
- Second and third steps can be repeated as per the need.
- Next is the Pooling layer and its basic function is to perform the reduction.
- As per need repeat steps second, third, and fourth.
- The next is the fully connected layer which performs the basic neural network working and transfer the result to the last output layer.
- Last is the output layer.

In short, convolutional neural network (Sharma, 2012) can be simply defined distinct layers list. Each one gets input from previous and after performing respective task transfer the result to the next one. All these layers are independent of each other, but work as per the output received from the previous layer. The model performance depends on the working of every layer as well as on the input dataset. Briefly, the complete performance of this deep learning model depends on the following essential points and the variability of these points results also varies:

- The most important point for any model is the type of input dataset used.
- The number of layers used by the system.
- The number of neurons present in each layer.
- The optimizer used.
- The activation function used by the model.

Above mentioned points are some important parameters on which result of the model depends. This is the supervised learning model and can be used for both predictions as well as the classification task. This article uses this model to detect the malicious website after analyzing the Malicious_n_Non-Malicious URL dataset. Based on analyzing the dataset by this model and this model layer, the detection of malicious URL result is generated in terms of accuracy and loss. The following experimental analysis section shows the result of applying Malicious_n_Non-Malicious URL dataset with the convolutional neural network (Ha, 2019), (Xiao, 2019).

EXPERIMENTAL ANALYSIS

This section shows the experiment conducted with a convolutional neural network model and Malicious_n_Non-Malicious URL dataset. The potential of this deep learning model is shown in this section. The research used the experimental methodology to present the potential of the model for detecting the malicious website. The document presents the potential terms of accuracy as well as loss also. The experiment is performed two times. First, after the training of the model is completed and the second invalidation. In this paper during both training and validation phase total 16 epochs and iterations are used. The following **Table 5** shows the experimental result after training the model with 16 epochs.

Table 5. Model Performance during Training Phase

Epoch	Accuracy	Loss
1	0.8423	0.4387
2	0.8525	0.3711
3	0.8565	0.3572
4	0.8602	0.3463
5	0.8634	0.3374
6	0.8663	0.3303
7	0.8687	0.3246
8	0.8708	0.3194
9	0.8727	0.3152
10	0.8745	0.3104
11	0.8762	0.3062
12	0.878	0.302
13	0.8797	0.2975
14	0.8812	0.2949
15	0.8828	0.2902
16	0.8839	0.2874

The above table 5 shows the model potential during the 16 epochs. The following **Figure 4 and Figure 5** shows pictorially the analysis result after training.

Figure 4. Epochs v/s Accuracy (Training phase)

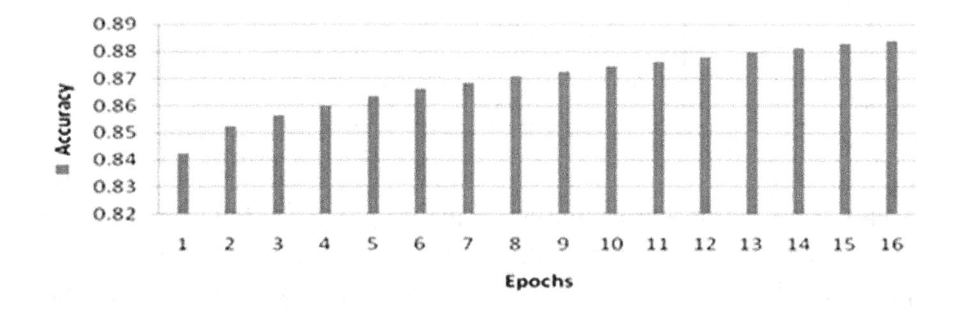

Figure 5. Epochs v/s Loss (Training Phase)

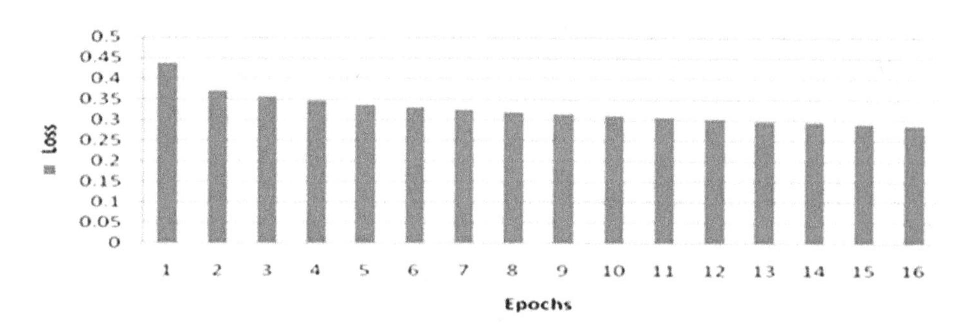

The above figure 4 and figure 5 represents the accuracy and loss, respectively during the training phase. Approximately, 88% accuracy is achieved by this model in the training phase. In the testing or validation phase accuracy and loss achieved by the model are shown below in **Table 6**:

Table 6. Model Performance during Validation Phase

Epoch	Accuracy	Loss
1	0.849	0.3799
2	0.8532	0.3662
3	0.8566	0.358
4	0.8606	0.3498
5	0.863	0.344
6	0.864	0.3413
7	0.8656	0.3381
8	0.867	0.3371
9	0.8678	0.3358
10	0.8691	0.3353
11	0.8701	0.3351
12	0.8704	0.3355
13	0.8712	0.3341
14	0.8718	0.3343
15	0.8717	0.3364
16	0.8727	0.3371

The above table 6 shows the model potential during the 16 epochs. The following **Figure 6** and **Figure 7** shows pictorially the analysis result in the validation phase. The variation of accuracy with the increase in epoch is depicted in below figure 6. It shows how the epoch increment improves accuracy. Similarly, figure 7 present the effect on loss, means how it changes, the increase of the epoch. In the validation phase how this model works with the dataset, means the potential of the model is depicted graphically in below figure 6 and figure 7 concerning the accuracy and loss respectively.

Figure 6. Accuracy v/s Epoch (during Validation Phase)

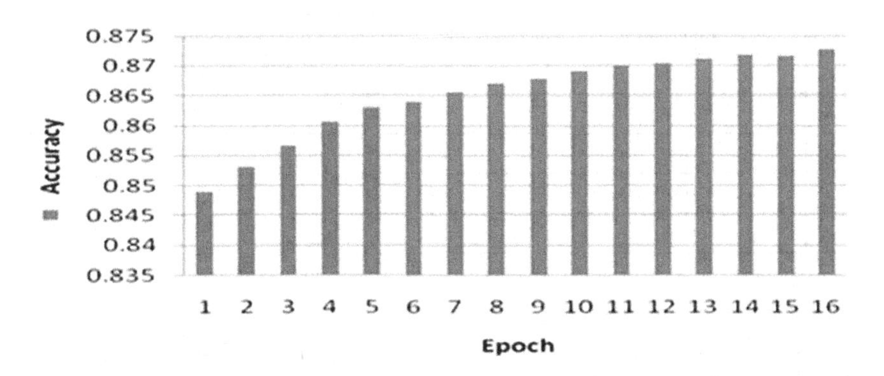

Figure 7. Loss v/s Epoch (Validation Phase)

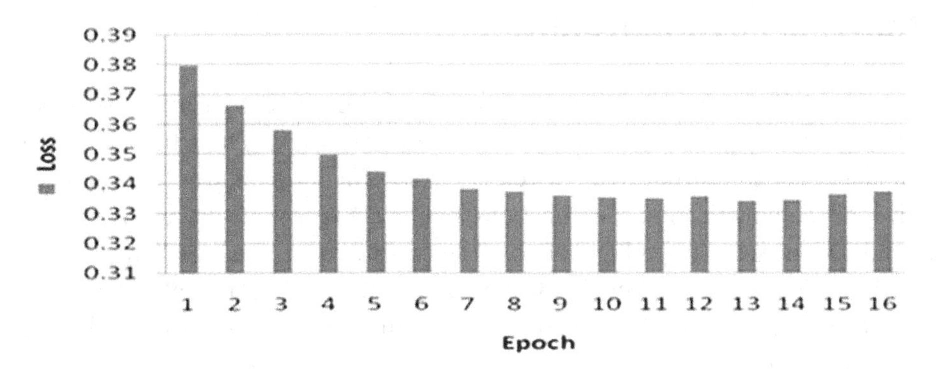

The above figure 6 and figure 7 represents the accuracy and loss, respectively during the training phase. Approximately, 87% accuracy is achieved by this model in the validation phase.

The experiment performed in this research, retrieve the classification report during the testing phase. The classification report consists of a precision, recall, f1-score, and support. These parameters show the performance measured by the model. The paper evaluates the model potential based on mentioned parameters. All mentioned parameters have different meaning and effectively show the effectiveness of the model on the dataset. The brief explanation of these parameters is as follows:

- Precision: it is defined as the ratio between the positive observations that are correctly predicted and all the positive observations predicted.
- Recall: it is defined as the ratio between the positive observations that are correctly predicted and all actual class observations.
- F1-score: the weighted average of both the precision as well as the recall is called f1-score.
- Support: it is defined as the true response sample amount present in that class.

The below **Table 7** shows the report of classification result generated during the experiment of model testing:

Table 7. Model Testing Classification Report

	Precision	Recall	F1-score	Support
0	0.88	0.97	0.92	68919
1	0.76	0.4	0.52	15174
micro avg	0.87	0.87	0.87	84093
macro avg	0.82	0.68	0.72	84093
weighted avg	0.86	0.87	0.85	84093

The classification result gets from the model testing phase is presented in above table 7. This report shows the evaluation parameters output and their micro average, macro average, and weighted average.

CONCLUSION AND FUTURE WORKS

The study discusses in detail, the wireless system, wireless environment positive points as well as also discusses regarding security problem that is being faced by the society. The article demonstrates different possible challenges or attacks due to the wireless system. On one hand where wireless facilitates the user and removes the communication complications, but on the other side due to the presence of illegal hands, major and serious security problem arises. The paper discusses the malicious website problem and also experiments to detect the malicious URL by using a conventional neural network that is the deep learning model. The research applied this model on the Malicious_n_Non-Malicious URL dataset. The experimental result evaluated and presented in the above section concerning many evaluating parameters. These parameters are accurate, loss, precision, f1-score, recall, and support. The paper also discussed them in brief. The document research can be concluded with the following points:

- The convolutional model is a deep learning model that is approximately 88% efficient in terms of accuracy to detect the malicious website in the training phase.
- During the training phase, the loss is about 28% in the detection of malicious URL by this learning model.
- In the phase of validation, this model potential in regards to accuracy is almost 87%.
- The loss occurred in the validation is nearly about 33%.

This article is useful for researchers who are working towards the security concern in wireless environments and would like to overcome this serious and crucial obstacle in the way of wireless success. The research uses the deep learning convolutional model for detecting malicious URL. In future researches this model can be used in other security challenges related to the wireless system.

REFERENCES

Alali, M., Sharef, N. M., Murad, M. A. A., Hamdan, H., & Husin, N. A. (2019). Narrow Convolutional Neural Network for Arabic Dialects Polarity Classification. *IEEE Access: Practical Innovations, Open Solutions*, *7*, 96272–96283. doi:10.1109/ACCESS.2019.2929208

Aliu, O. G., Imran, A., Imran, M. A., & Evans, B. (2012). A survey of self organisation in future cellular networks. *IEEE Communications Surveys and Tutorials*, *15*(1), 336–361. doi:10.1109/SURV.2012.021312.00116

Allman, M., & Ostermann, S. (1999). *FTP security considerations*. Academic Press.

Alpaydin, E. (2020). *Introduction to machine learning*. MIT Press.

Althunibat, S., Antonopoulos, A., Kartsakli, E., Granelli, F., & Verikoukis, C. (2016). Countering intelligent-dependent malicious nodes in target detection wireless sensor networks. *IEEE Sensors Journal*, *16*(23), 8627–8639. doi:10.1109/JSEN.2016.2606759

Bhushan, B., & Sahoo, G. (2017). Recent Advances in Attacks, Technical Challenges, Vulnerabilities and Their Countermeasures in Wireless Sensor Networks. *Wireless Personal Communications*, *98*(2), 2037–2077. doi:10.100711277-017-4962-0

Bhushan, B., & Sahoo, G. (2017). A comprehensive survey of secure and energy efficient routing protocols and data collection approaches in wireless sensor networks. *2017 International Conference on Signal Processing and Communication (ICSPC)*. 10.1109/CSPC.2017.8305856

Bhushan, B., & Sahoo, G. (2017). Detection and defense mechanisms against wormhole attacks in wireless sensor networks. *2017 3rd International Conference on Advances in Computing,Communication& Automation (ICACCA)*. DOI: 10.1109/icaccaf.2017.8344730

Bhushan, B., & Sahoo, G. (2018). Routing Protocols in Wireless Sensor Networks. *Computational Intelligence in Sensor Networks Studies in Computational Intelligence,* 215-248. DOI: . doi:10.1007/978-3-662-57277-1_10

Bhushan, B., & Sahoo, G. (2019). $$E^{2} SR^{2}$$ E 2 S R 2: An acknowledgement-based mobile sink routing protocol with rechargeable sensors for wireless sensor networks. *Wireless Networks*, *25*(5), 2697–2721. doi:10.100711276-019-01988-7

Bhushan, B., & Sahoo, G. (2019). ISFC-BLS (Intelligent and Secured Fuzzy Clustering Algorithm Using Balanced Load Sub-Cluster Formation) in WSN Environment. *Wireless Personal Communications*. Advance online publication. doi:10.100711277-019-06948-0

Bhushan, B., & Sahoo, G. (2019). Secure Location-Based Aggregator Node Selection Scheme in Wireless Sensor Networks. *Proceedings of ICETIT 2019 Lecture Notes in Electrical Engineering*, 21–35. DOI: 10.1007/978-3-030-30577-2_2

Bhushan, B., & Sahoo, G. (2020). Requirements, Protocols, and Security Challenges in Wireless Sensor Networks: An Industrial Perspective. In *Handbook of Computer Networks and Cyber Security* (pp. 683–713). Springer. doi:10.1007/978-3-030-22277-2_27

Bhushan, B., Sahoo, G., & Rai, A. K. (2017). Man-in-the-middle attack in wireless and computer networking — A review. *2017 3rd International Conference on Advances in Computing,Communication & Automation (ICACCA)*. DOI: 10.1109/icaccaf.2017.8344724

Bruno, R., Conti, M., & Gregori, E. (2007). Throughput analysis and measurements in IEEE 802.11 WLANs with TCP and UDP traffic flows. *IEEE Transactions on Mobile Computing*, *7*(2), 171–186. doi:10.1109/TMC.2007.70718

Center, C. C. (1995). IP Spoofing Attacks and Hijacked Terminal Connections. CA-95: 01.

Chang, R. K. (2002). Defending against flooding-based distributed denial-of-service attacks: A tutorial. *IEEE Communications Magazine*, *40*(10), 42–51. doi:10.1109/MCOM.2002.1039856

Conti, M., Dragoni, N., & Lesyk, V. (2016). A survey of man in the middle attacks. *IEEE Communications Surveys and Tutorials*, *18*(3), 2027–2051. doi:10.1109/COMST.2016.2548426

Derryberry, R. T., Gray, S. D., Ionescu, D. M., Mandyam, G., & Raghothaman, B. (2002). Transmit diversity in 3G CDMA systems. *IEEE Communications Magazine*, *40*(4), 68–75. doi:10.1109/35.995853

Ha, M., Byun, Y., Kim, J., Lee, J., Lee, Y., & Lee, S. (2019). Selective Deep Convolutional Neural Network for Low Cost Distorted Image Classification. *IEEE Access: Practical Innovations, Open Solutions*, *7*, 133030–133042. doi:10.1109/ACCESS.2019.2939781

Haibo, B. L., Sohraby, L. K., & Wang, C. (2010). Future internet services and applications. *IEEE Network*, *24*(4), 4–5. doi:10.1109/MNET.2010.5510911

Han, W., Tian, Z., Huang, Z., Huang, D., & Jia, Y. (2019). Quantitative Assessment of Wireless Connected Intelligent Robot Swarms Network Security Situation. *IEEE Access: Practical Innovations, Open Solutions*, *7*, 134293–134300. doi:10.1109/ACCESS.2019.2940822

Harris, B., & Hunt, R. (1999). TCP/IP security threats and attack methods. *Computer Communications*, *22*(10), 885–897. doi:10.1016/S0140-3664(99)00064-X

Hastings, N. E., & McLean, P. A. (1996, March). TCP/IP spoofing fundamentals. In *Conference Proceedings of the 1996 IEEE Fifteenth Annual International Phoenix Conference on Computers and Communications* (pp. 218-224). IEEE. 10.1109/PCCC.1996.493637

Hennebert, C., & Dos Santos, J. (2014). Security protocols and privacy issues into 6LoWPAN stack: A synthesis. *IEEE Internet of Things Journal*, *1*(5), 384–398. doi:10.1109/JIOT.2014.2359538

Jaitly, S., Malhotra, H., & Bhushan, B. (2017). Security vulnerabilities and countermeasures against jamming attacks in Wireless Sensor Networks: A survey. *2017 International Conference on Computer, Communications and Electronics (Comptelix)*. 10.1109/COMPTELIX.2017.8004033

Jedari, B., Xia, F., & Ning, Z. (2018). A survey on human-centric communications in non-cooperative wireless relay networks. *IEEE Communications Surveys and Tutorials*, *20*(2), 914–944. doi:10.1109/COMST.2018.2791428

Kasongo, S. M., & Sun, Y. (2019). A Deep Learning Method With Filter Based Feature Engineering for Wireless Intrusion Detection System. *IEEE Access: Practical Innovations, Open Solutions*, *7*, 38597–38607. doi:10.1109/ACCESS.2019.2905633

Kavianpour, A., & Anderson, M. C. (2017, June). An Overview of Wireless Network Security. In *2017 IEEE 4th International Conference on Cyber Security and Cloud Computing (CSCloud)* (pp. 306-309). IEEE. 10.1109/CSCloud.2017.45

Kieyzun, A., Guo, P. J., Jayaraman, K., & Ernst, M. D. (2009, May). Automatic creation of SQL injection and cross-site scripting attacks. In *Proceedings of the 31st International Conference on Software Engineering* (pp. 199-209). IEEE Computer Society. 10.1109/ICSE.2009.5070521

Kuzmanovic, A., & Knightly, E. W. (2006). Low-rate TCP-targeted denial of service attacks and counter strategies. *IEEE/ACM Transactions on Networking*, *14*(4), 683–696. doi:10.1109/TNET.2006.880180

Labovitz, C., Ahuja, A., Bose, A., & Jahanian, F. (2000). Delayed Internet routing convergence. *Computer Communication Review*, *30*(4), 175–187. doi:10.1145/347057.347428

Lee, S. B., Ahn, G. S., & Campbell, A. T. (2001). Improving UDP and TCP performance in mobile ad hoc networks with INSIGNIA. *IEEE Communications Magazine*, *39*(6), 156–165. doi:10.1109/35.925684

Lee, S. B., Ahn, G. S., & Campbell, A. T. (2001). Improving UDP and TCP performance in mobile ad hoc networks with INSIGNIA. *IEEE Communications Magazine*, *39*(6), 156–165. doi:10.1109/35.925684

Mao, Q., Hu, F., & Hao, Q. (2018). Deep learning for intelligent wireless networks: A comprehensive survey. *IEEE Communications Surveys and Tutorials*, *20*(4), 2595–2621. doi:10.1109/COMST.2018.2846401

Mellouli, D., Hamdani, T. M., Sanchez-Medina, J. J., Ayed, M. B., & Alimi, A. M. (2019). Morphological Convolutional Neural Network Architecture for Digit Recognition. *IEEE Transactions on Neural Networks and Learning Systems*, *30*(9), 2876–2885. doi:10.1109/TNNLS.2018.2890334 PMID:30676985

Mishra, V., & Agarwal, M. (2019). *Effect of Redundant Work on Machine.* SSRN 3444784

Mishra, V., Agarwal, M., & Puri, N. (2018). Comprehensive and Comparative Analysis of Neural Network. *International Journal of Computers and Applications*, *2*(8).

Mpitziopoulos, A., Gavalas, D., Konstantopoulos, C., & Pantziou, G. (2009). A survey on jamming attacks and countermeasures in WSNs. *IEEE Communications Surveys and Tutorials*, *11*(4), 42–56. doi:10.1109/SURV.2009.090404

Otoum, S., Kantarci, B., & Mouftah, H. T. (2019). On the feasibility of deep learning in sensor network intrusion detection. *IEEE Networking Letters*, *1*(2), 68–71. doi:10.1109/LNET.2019.2901792

Park, J. C., & Kasera, S. K. (2007, March). *Securing Ad Hoc wireless networks against data injection attacks using firewalls. In 2007 IEEE Wireless Communications and Networking Conference*. IEEE.

Perrig, A., Stankovic, J., & Wagner, D. (2004). *Security in wireless sensor networks*. Academic Press.

Schuba, C. L., Krsul, I. V., Kuhn, M. G., Spafford, E. H., Sundaram, A., & Zamboni, D. (1997, May). Analysis of a denial of service attack on TCP. In *Proceedings. 1997 IEEE Symposium on Security and Privacy (Cat. No. 97CB36097)* (pp. 208-223). IEEE. 10.1109/SECPRI.1997.601338

Sharma, M., Tandon, A., Narayan, S., & Bhushan, B. (2017). Classification and analysis of security attacks in WSNs and IEEE 802.15.4 standards: A survey. *2017 3rd International Conference on Advances in Computing, Communication & Automation (ICACCA)*. DOI: 10.1109/icaccaf.2017.8344727

Sharma, V., Rai, S., & Dev, A. (2012). A comprehensive study of artificial neural networks. *International Journal of Advanced Research in Computer Science and Software Engineering*, *2*(10).

She, W., Liu, Q., Tian, Z., Chen, J. S., Wang, B., & Liu, W. (2019). Blockchain Trust Model for Malicious Node Detection in Wireless Sensor Networks. *IEEE Access: Practical Innovations, Open Solutions*, *7*, 38947–38956. doi:10.1109/ACCESS.2019.2902811

Shibahara, T., Yamanishi, K., Takata, Y., Chiba, D., Akiyama, M., Yagi, T., ... Murata, M. (2017, May). Malicious URL sequence detection using event de-noising convolutional neural network. In *2017 IEEE International Conference on Communications (ICC)* (pp. 1-7). IEEE. 10.1109/ICC.2017.7996831

Shin, M., Ma, J., Mishra, A., & Arbaugh, W. A. (2006). Wireless network security and interworking. *Proceedings of the IEEE*, *94*(2), 455–466. doi:10.1109/JPROC.2005.862322

Shokri, R., & Shmatikov, V. (2015, October). Privacy-preserving deep learning. In *Proceedings of the 22nd ACM SIGSAC conference on computer and communications security* (pp. 1310-1321). Academic Press.

Shomron, G., & Weiser, U. (2018). Spatial Correlation and Value Prediction in Convolutional Neural Networks. *IEEE Computer Architecture Letters*, *18*(1), 10–13. doi:10.1109/LCA.2018.2890236

Sinha, P., Jha, V. K., Rai, A. K., & Bhushan, B. (2017). Security vulnerabilities, attacks and countermeasures in wireless sensor networks at various layers of OSI reference model: A survey. *2017 International Conference on Signal Processing and Communication (ICSPC)*. 10.1109/CSPC.2017.8305855

Tandon, A., Lim, T. J., & Tefek, U. (2019). Sentinel based malicious relay detection in wireless IoT networks. *Journal of Communications and Networks (Seoul)*, *21*(5), 458–468. doi:10.1109/JCN.2019.000049

Tomić, I., & McCann, J. A. (2017). A survey of potential security issues in existing wireless sensor network protocols. *IEEE Internet of Things Journal*, *4*(6), 1910–1923. doi:10.1109/JIOT.2017.2749883

V., & Agarwal, M. (2019). Intelligent Handwritten Digit Recognition Based on Multiple Parameters using CNN. *International Journal of Computer Sciences and Engineering*, *7*(5), 636–641.

Wang, K., Yuan, L., Miyazaki, T., Zeng, D., Guo, S., & Sun, Y. (2017). Strategic antieavesdropping game for physical layer security in wireless cooperative networks. *IEEE Transactions on Vehicular Technology*, *66*(10), 9448–9457. doi:10.1109/TVT.2017.2703305

Xiao, Y., Xing, C., Zhang, T., & Zhao, Z. (2019). An Intrusion Detection Model Based on Feature Reduction and Convolutional Neural Networks. *IEEE Access: Practical Innovations, Open Solutions*, *7*, 42210–42219. doi:10.1109/ACCESS.2019.2904620

Xie, H., Yan, Z., Yao, Z., & Atiquzzaman, M. (2018). Data Collection for Security Measurement in Wireless Sensor Networks: A Survey. *IEEE Internet of Things Journal*, *6*(2), 2205–2224. doi:10.1109/JIOT.2018.2883403

Yang, H., Qin, G., & Ye, L. (2019). Combined Wireless Network Intrusion Detection Model Based on Deep Learning. *IEEE Access: Practical Innovations, Open Solutions*, *7*, 82624–82632. doi:10.1109/ACCESS.2019.2923814

Yang, H., & Wang, F. (2019). Wireless Network Intrusion Detection Based on Improved Convolutional Neural Network. *IEEE Access: Practical Innovations, Open Solutions*, *7*, 64366–64374. doi:10.1109/ACCESS.2019.2917299

Zhao, P., & Hoi, S. C. (2013, August). Cost-sensitive online active learning with application to malicious URL detection. In *Proceedings of the 19th ACM SIGKDD international conference on Knowledge discovery and data mining* (pp. 919-927). 10.1145/2487575.2487647

Zhou, W., Marshall, A., & Gu, Q. (2006, April). A novel classification scheme for 802.11 WLAN active attacking traffic patterns. In *IEEE Wireless Communications and Networking Conference, 2006. WCNC 2006.* (Vol. 2, pp. 623-628). IEEE. 10.1109/WCNC.2006.1683541

Zou, Y., Zhu, J., Wang, X., & Hanzo, L. (2016). A survey on wireless security: Technical challenges, recent advances, and future trends. *Proceedings of the IEEE*, *104*(9), 1727–1765. doi:10.1109/JPROC.2016.2558521

Zou, Y., Zhu, J., Wang, X., & Hanzo, L. (2016). A survey on wireless security: Technical challenges, recent advances, and future trends. *Proceedings of the IEEE*, *104*(9), 1727–1765. doi:10.1109/JPROC.2016.2558521

Chapter 5
Modelling a Deep Learning-Based Wireless Sensor Network Task Assignment Algorithm:
An Investigative Approach

Titus Issac
https://orcid.org/0000-0002-0725-4084
Karunya Institute of Technology and Sciences, India

Salaja Silas
Karunya Institute of Technology and Sciences, India

Elijah Blessing Rajsingh
Karunya Institute of Technology and Sciences, India

ABSTRACT

The 21st century is witnessing the emergence of a wide variety of wireless sensor network (WSN) applications ranging from simple environmental monitoring to complex satellite monitoring applications. The advent of complex WSN applications has led to a massive transition in the development, functioning, and capabilities of wireless sensor nodes. The contemporary nodes have multi-functional capabilities enabling the heterogeneous WSN applications. The future of WSN task assignment envisions WSN to be heterogeneous network with minimal human interaction. This led to the investigative model of a deep learning-based task assignment algorithm. The algorithm employs a multilayer feed forward neural network (MLFFNN) trained by particle swarm optimization (PSO) for solving task assignment problem in a dynamic centralized heterogeneous WSN. The analyses include the study of hidden layers and effectiveness of the task assignment algorithms. The chapter would be highly beneficial to a wide range of audiences employing the machine and deep learning in WSN.

DOI: 10.4018/978-1-7998-5068-7.ch005

INTRODUCTION

The smart computing era is massively led by the inception of smart devices in everyday life. The smart computing era infers to a world, powered by a collective set of semi-autonomous or autonomous devices, with the ability to function with minimal or no human intervention(Friday, Wah, Al-garadi, & Rita, 2018). The smart devices compared to its legacy counterparts inherently have sensing, processing, and communication capabilities(Issac, Silas, & Rajsingh, 2019a). Nowadays, contemporary, smart devices are mounted with dedicated artificial intelligence engines to provide decision making functionality without major human interventions. The transfiguration of the everyday device into a smart device is pictorially represented in figure 1.

Figure 1. The evolution from every-day device to a smart device

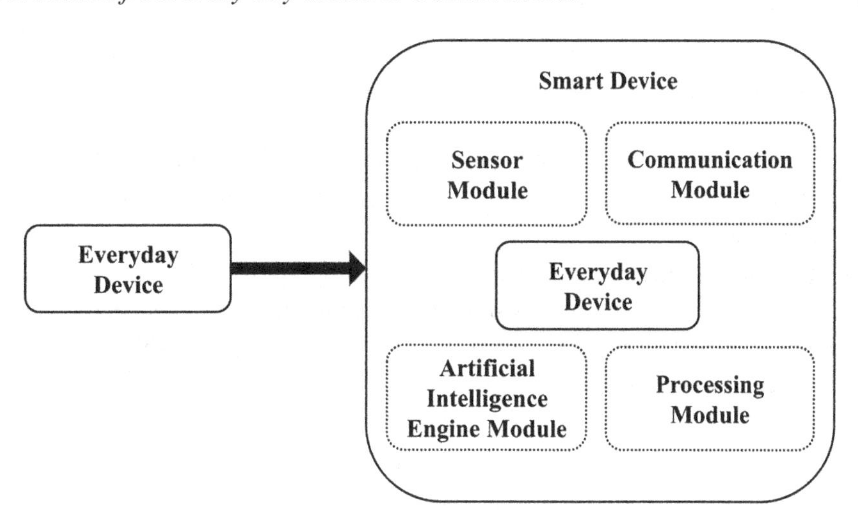

A wide range of autonomous smart devices has been visualized from time to time based on human requirements(Sohraby, Minoli, & Znati, 2007). Since the inception of smart devices, sensors have been playing a critical role in everyday devices(Sohraby et al., 2007). The inception of a wide variety of sensors along with its functionality and applications has given rise to multiple research domains such as the Wireless Sensor Network (WSN), Wireless Sensor and Actuator Network (WSAN), Body Area Network (BAN), Vehicular Ad-hoc Network (VANETS), etc(Munir, Gordon-Ross, & Ranka, 2014). However, this chapter concentrates on WSN, for the following set of reasons.

1. WSN continues as one of the prime and evolving research domain(Bhushan & Sahoo, 2019c, 2020).
2. The contemporary nodes have better capability compared to their legacy counterparts(Munir et al., 2014; Roozeboom et al., 2013).
3. The envisioned next-generation WSN applications require a set of heterogeneous wireless sensor nodes(Misra & Vaish, 2011; Tkach & Edan, 2020).

Motivation

The primary notion assuming a WSN to be a resource-constrained network still prevails, however, the contemporary nodes in the WSN have comparatively gained higher processing, storage, and energy capabilities, as well as, multi-sensing capabilities, in comparison to the legacy wireless sensor nodes(Munir et al., 2014; Roozeboom et al., 2013). To this end, the need to consider the contemporary nodes' additional capabilities during the task assignment arises, as the majority of the existing algorithms are confined to homogeneous WSN(Yu & Prasanna, 2005). The primary objective of such next-generation task assignment algorithms would be to maximize the performance and minimize the energy utilization. In a nutshell, the systematic overview of the heterogeneous WSN along with the investigation on the various existing WSN task assignment methods and the future directions of the task assignment algorithms is needed.

Major Contributions

The contributions of the chapter are manifold and are enumerated below;

1. The genesis of computing and Wireless Sensor Network has been presented.
2. Various types of task assignment in WSN along with factors influencing the task assignment were identified and presented.
3. The past and recent task assignment methods have been identified and investigated.
4. Future perspective of task assignment using deep learning methods has been investigated.
5. A PSO trained Multi-layer Feed Forward Neural Network (MLFFNN) based task assignment algorithm has been proposed and simulated.

Summary

The chapter begins with the genesis of computing and wireless sensor network (WSN). Section 3 discusses the task assignment algorithm, it types and the major factors influencing the task assignment. Investigation of the existing task assignment algorithms in WSN is discussed in section 4. Section 5 discusses the task assignment algorithms using deep learning techniques and PSO method trained Multi-layer Feed Forward Neural Network (MFFNN) based dynamic task assignment algorithm for heterogeneous WSN has been modelled. Section 6 discusses the experimental results of the proposed method. Section 7 concludes the chapter with future directions.

THE GENESIS

While considering, a wide plethora of living organisms, humans have been pioneers and experts in the pursuit of gaining intelligence(Akhand, Ayon, Shahriyar, Siddique, & Adeli, 2020; Pant, Deep, Bansal, Das, & Nagar, 2018). Intelligence, defined as the ability to gain knowledge, has been one of the key human survival instincts. The rate of gaining intelligence signifies the effectiveness of solutions for everyday problems. The humans through direct and indirect observation process, gradually commence obtaining intelligence from an early infant stage(Bansal, 2019; Păun, 2000). However, the process gets accelerated after the initial stage, through the gradual adoption of persistent learning techniques along with lateral thinking. The persistent learning infers to the cumulative knowledge obtained from the earlier human generations, as well as, the self-obtained knowledge obtained through experience. While lateral thinking is defined as the process of creative, innovative and out-of-the-box thinking. The cumulative rational intelligence obtained from time to time has led humans to provide innovative solutions to the majority of everyday problems(Akhand et al., 2020; Lalwani et al., 2019; Tkach & Edan, 2020; Yang, Zhang, Ling, Pan, & Sun, 2014). The solutions invariably include the process of dissemination of the problem into a set of indivisible tasks and obtaining a way to efficiently execute the task.

Humans have developed insightful expertise that evolved them to effortlessly perform multiple tasks simultaneously and the process is better known as multitasking. Amidst, the successful task executions, a subset of everyday tasks in their raw form was unable to be effectively carried out by humans, due to various task constraints, as well as, human limitations(Issac, Silas, & Rajsingh, 2019b). The requirements of the defaulted tasks combined with the gained intelligence motivated the inception of human-assistive technologies(Issac, Silas, & Rajsingh, 2020a; Titus, Rajsingh, & Silas, 2015).

Initially, a newly incepted human-assistive technology may try to cater to the task requirements sufficiently. However, the technologies are bound by a certain set of constraints and shortcomings and they cumulatively foster inspirations to generate a new set of requirements for next-generation technologies(Munir et al., 2014). The pursuit of excellence for an ideal technology has become a persistent process, resulting in the newer generations of never-ending new requirement cycles and it is pictorially represented in figure 2. From time to time, the higher requirements, along with the need for achieving better benchmark standards through overcoming the limitations and shortcomings of the former technologies have led to the inception of countless, innovative, next-generation, disruptive technologies.

Figure 2. The life cycle of technology - Overview

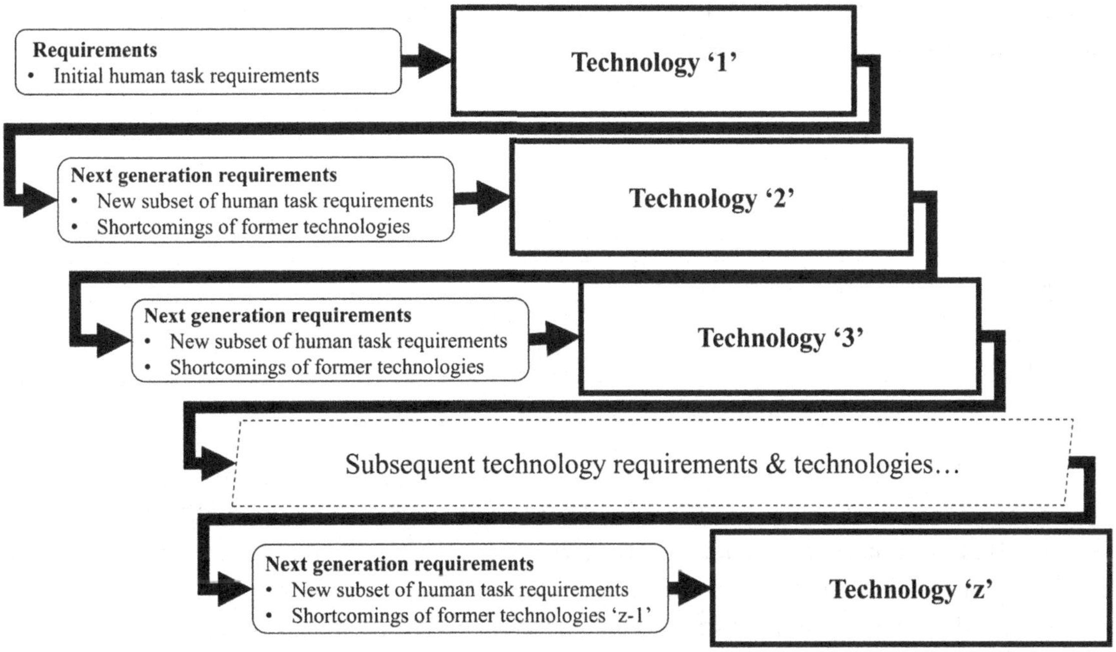

Requirements
- Initial human task requirements

Technology '1'

Next generation requirements
- New subset of human task requirements
- Shortcomings of former technologies

Technology '2'

Next generation requirements
- New subset of human task requirements
- Shortcomings of former technologies

Technology '3'

Subsequent technology requirements & technologies…

Next generation requirements
- New subset of human task requirements
- Shortcomings of former technologies 'z-1'

Technology 'z'

The nineteenth-century is a major milestone in the field of disruptive technologies while considering all other centuries in history as it witnessed the major industrial revolution inception. The first industrial revolution paved the way for the invention of mechanical devices to assist humans("Industrial Revolution," n.d.). The disruptive technology transformed various walks of life and set a milestone in the field of manufacturing. The industrialization facilitated the transition of technologies from a few, hand-manufactured products to large industrial-scale production. Electrically powered devices marked the commencement of the second industrial revolution and were found more effective than its legacy counterparts("Second Industrial Revolution," n.d.). The positive impacts of the next-generation technologies were strikingly evident in all walks of life and it started becoming an integral part of human life in a rapid phase.

Amidst, the various industrial revolutions, the field of computing got its equal share of the technological appraisals. Computing tasks have been one of the primary, every day, human requirements. The first-generation computing devices were a set of handheld primitive devices. Abacus is one such device used since its inception and has a long documented legacy starting from the ancient civilizations such as the Egyptians("Abacus," n.d.). However, the need for repetitive and persistent computing along with the higher computing requirements led to the consecutive generations of next-generation human-assisted computing technologies. The initial developments led to the era of mechanical computing.

The computing machine capabilities were further enhanced using electro-mechanical devices. Further, the transition from electro-mechanical technology to digital computing predominantly occurred after the adoption of vacuum tubes. Invented by John Ambrose Fleming in 1904, the initial years of digital computing were based on vacuum tube technology. The first fully functional electronic general-purpose computing device was funded by the US Army and it was code-named as Electronic Numerical Integrator and Computer (ENIAC)("ENIAC," n.d.). The invention of ENIAC denoted the start of the digital computing era and table 1 tabulates the properties of ENIAC that commenced in the year 1943 and concluded its operations in 1956. The ENIAC properties tabulation is made to understand the capabilities and limitations of the first electronic computing device.

Table 1. ENIAC- Properties overview

Properties	Values
Weight (kg)	30,000
Dimensions (length x breadth x height) (m)	$2.4 \times 0.9 \times 30$
Power Consumption (W)	150×10^3
No of Vacuum tubes	360
Maximum no of digits	10

Down the years, a wide range of electronic devices were invented, however, the invention of the transistor at Bell Laboratory in 1948 initiated the dawn of the modern electronic age("Information age," n.d.). The transistors paved the way to realize and achieve a wide range of new electronic devices and applications such as the wireless transmitters and receivers used for wireless communication. In 1949, a first of its kind, the integrated transistor amplifier was patented by Werner Jacob. And it triggered and proliferated the idea of integrating electronic devices better known as Integrated Circuits (IC) using Complementary Metal Oxide Semiconductor (CMOS). The IC technology was indeed a disruptive technology that facilitated the integration of multiple electronic devices on to a single chip. The concept of ICs led to the invention of a wide range of multi-functional products such as the wireless transceivers that simultaneously performed the tasks of the wireless transmitter and receiver (Munir et al., 2014).

The contemporary electronic devices were inspired, modelled and driven by the vision proposed by Gordon Moore in the year 1965. Moore's vision better known as Moore's law predicted that the next generation IC's would double its performance with no subsequent increase in power consumption every 18 months(Munir et al., 2014; Roozeboom et al., 2013). The contemporary electronic devices became quickly ubiquitous in comparison to the legacy devices due to its (i) multi-functional nature, (ii) lower power consumption rate, (iii) better performance, (iv) smaller size and (v) less bulkiness(Roozeboom et al., 2013). Despite all the advancements, the authors perform investigations on resource-constrained wireless sensor network due to its unique requirements and resource-constrained applications.

WSN Genesis

The inception of the Wireless Sensor Network (WSN) occurred at the rear end of the twentieth century. The first documented usage of the keyword "Wireless Sensor Network" was presented by the researchers Bult and his team of 22 researchers in the international symposium on Low Power Electronics and Design conducted in Monterey, California, United States of America (USA) in the year 1996 (Bult et al., 1996).

Wireless Node

A WSN is constituted by a set of wireless sensor nodes (nodes) having sensing and actuation capabilities. General characteristics that the participating nodes are resource-constrained in terms of energy, storage, memory, and communication exists. A generic component overview of a sensor node is depicted in figure 3.

Figure 3. Component wise overview of a Wireless Sensor Node

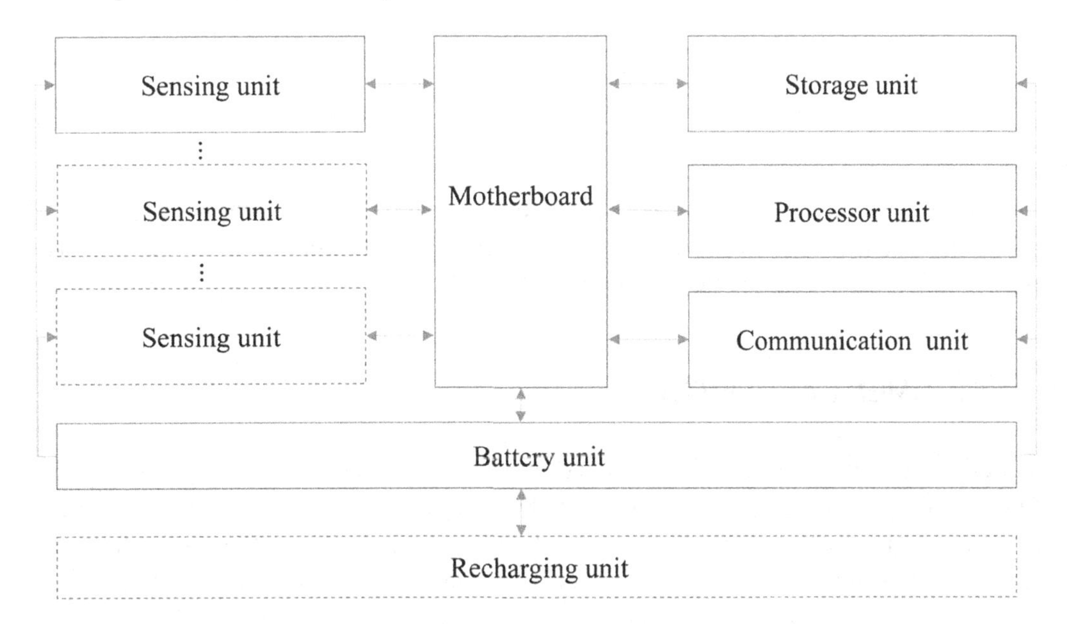

The most essential parts in the node are depicted in solid lines, while the dotted component units are additional units to a node. In the recent past as Roozeboom et al (Roozeboom et al., 2013) have successfully modelled and fabricated ten environmental sensors onto a single die, demonstrating the potential of the modern embedding and fabricating techniques. Due to the modern embedding and fabrication technique, a handheld node is much capable in terms of computing, storage, processing and communication compared to its former generations.

A WSN contains a set of wireless sensor nodes as depicted in figure 4 alias nodes collaborating with each other to attain a common goal. Over the years, WSN technologies have witnessed a significant improvement in its hardware, software, and communication modules. However, the core WSN assumption of the nodes being resource-constrained still holds true and prevails.

Figure 4. Wireless Sensor Network- Overview

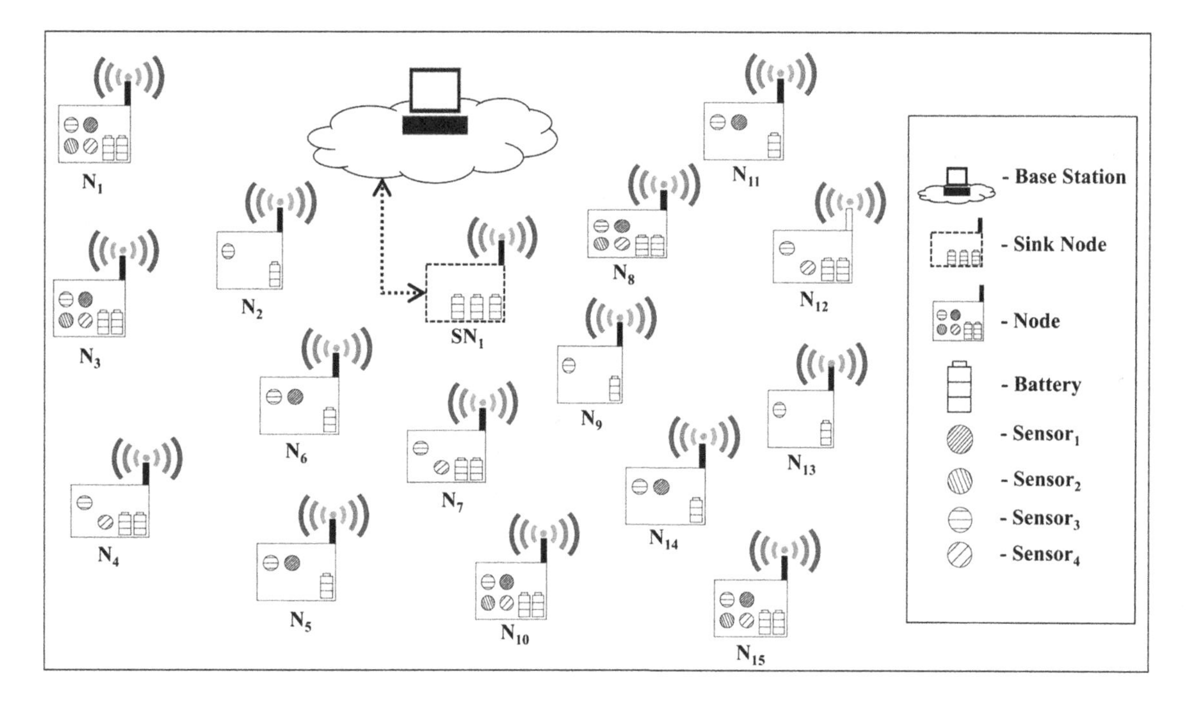

The major research challenges in the Wireless Sensor Network include security (Bhushan & Sahoo, 2018, 2020), routing(Bhushan & Sahoo, 2019a, 2019c), clustering(Bhushan & Sahoo, 2019b). However, the proposed work concentrates on task assignment in WSN.

TASK ASSIGNMENT ALGORITHMS

A task is defined as a basic operation performed on the node in the WSN (Titus, Silas, & Rajsingh, 2016). One of the major objectives is to maximize the overall performance and lifetime of the network by optimally assigning tasks to the nodes (Yang et al., 2014). The selection of the nodes for task assignment is influenced by multiple service parameters such as task priority, task deadlines, energy required and task completion time. Selecting and assigning an appropriate node for performing tasks as well as satisfying the quality of service parameters is a major research challenge. Therefore task assignment, especially with contemporary multifunctional sensors, is gaining much interest.

Task Classification

A task in a WSN could be broadly into various types and it is depicted in table 2. The major properties of task considered during the task assignment in WSN are tabulated in table 3.

Table 2. Task classification

S.no	Classification	Task types	Definition
1	Functionality	(a)Sensing	The sensing task performed by the node.
		(b)Communication	The task of transmitting and receiving messages by a node
		(c)Processing	The processing task performed by the node.
2	Repetitive Frequency	(a)Periodic	A task is said to be periodic if it repeats itself at a particular interval.
		(b)Sporadic	A task is said to be sporadic if it repeats itself and not in a particular interval.
3	Prerequisites	(a)Prerequisites dependent task	A task having a set of prerequisites
		(b)Prerequisites independent task	A task independent of any prerequisites.

Table 3. Task properties in WSN

S.no	Task properties	Task value
1	Max Task count (z)	Maximum count task count usually an integer.
2	Task identifier (t) ($1 \leq t \leq z$)	A unique identifier 't' is used to identify a task across WSN
3	Task priority (TPY_t)	Indicates the task priority of the task
4	Task preemption (TPE_t)	A boolean value property denoting whether a task could be stopped in between to carry out another task.
5	Task energy demand (TED_t)	The total energy required by a task and measured in J.
6	Task deadline (TDL_t)	Maximum deadline before the task has to be executed.
7	Task completion time (TDL_t)	The maximum time require to execute a task and it measured in s.

The Major Factors Influencing Task Assignment

The major factors influencing task assignment are broadly classified based on (i) node type, (ii) programming modes, (iii) architecture type and (iv) nature of the task assignment algorithm.

Node Type

A set of wireless sensor nodes collaborate to form a WSN. As described in the earlier section in figure 4, a node may comprise of one or more sensors as well as various components. Table 4 tabulates a set of widely used sensors in the WSN.

Table 4. Widely Used Sensors in WSN

	Sensors	Description
1	Humidity Sensor (SH)	Measure the quantity of water vapour in the observation medium.
2	Pressure Sensor (SP)	Measure the physical force exerted on to the observation medium.
3	Temperature Sensor (ST)	Measure the degree of heat in the observation medium.
4	Wind direction and speed (SW)	Measure the rate of flow of wind along with flow direction.
5	Illumination intensity sensor (SI)	Measure of perceived power per unit solid angle.
6	Acceleration sensor (SA)	Measure the vibration, shock, displacement, velocity, inclination and tilt.
7	Noise Sensor (SS)	Measure the noise level in the environment.

Until the early twenty-first century, one of the primary assumptions was that the majority of the participating nodes in WSN were considered homogeneous. The assumption of a homogeneous network resulted in requirements to consider a few node properties. Table 5 tabulates some of the major properties of homogeneous networks considered during task assignment.

Table 5. Homogeneous WSN- Properties

S.no	Properties	Representation & Values
1	Maximum number of nodes (n)	n
2	Node identifier (i)	$1 \leq i \leq n$
3	Initial Energy (IE)	IE_i
4	Remaining Energy (RE)	RE
5	No of Sensors in a node (NoS_i)	$NoS_i = j; \{1 \leq j \leq m\}$
6	Types of Sensors ($ToS_{i,y}$)	$ToS_{i,y} \in \{SH, SP, ST\ldots\}$
7	Task completion time (TCT)	TCT
8	Type of wireless communication (TOC)	Bluetooth, Zigbee, Wifi, LTE

However, for task assignment algorithms in a heterogeneous WSN, various properties have to be considered during task assignment. The major properties to be considered have been compiled into tables. The sensor properties are tabulated in table 6. Table 7 tabulates the communication properties. Storage properties of the sensor node is tabulated in Table 8. While the properties of the Operating System (OS) is tabulated in Table 9. The properties of the micro-processor of the node is tabulated in Table 10.

Table 6. Heterogeneous WSN- Sensor Properties

S.no	Properties	Representation & Values
1	Sensor Energy Requirement (SPR)	The maximum energy requirement of the sensor
2	Sensor Effective Measurement Range (SMX & SMN)	The maximum and minimum measurable unit of the sensor
3	Sensor Resolution (SRN)	The smallest measurable change detectable by the sensor.
4	Sensor Sensitivity (SSY)	The input parameter change required to produce an output change.
5	Sensor Accuracy (SAC)	The maximum difference value of the sensor measured value to the actual value.
6	Sensor Precision (SPN)	The degree of measurement reproducibility
7	Sensor Response Time (SRT)	The time required to change to the new state.

Table 7. Heterogeneous WSN- Communication Properties

S.no	Properties	Values
1	Transmission rate (TX)	Mbps
2	Reception rate (RX)	Mbps
3	Energy demand for transmission of data / bit (J)	J
4	Energy demand for receiving data / bit (ER)	J
5	Bandwidth (BW)	Mbps
6	Channels (CH)	Integer

Table 8. Heterogeneous WSN- Storage Properties

S.no	Properties	Values
1	Storage Size (SSS)	MB
2	Read Speed (SRS)	MBPS
3	Write Speed (SWS)	MBPS
4	Operating Temperature (SOT)	J

Table 9. Heterogeneous WSN- OS Properties

S.no	Properties	Values
1	OS Name (OSN)	String
2	OS version (OSV)	String
3	OS type (OST)	String
4	OS Architecture (OSA)	x86 / x64
5	Parallel operation (OSP)	Single / Multi- tasking

Table 10. Heterogeneous WSN- Processor Properties

S.no	Properties	Values
1	Clock speed (PCS)	Hertz (hz)
2	Cache memory (PCM)	Byte (MB)
3	Parallel operations (PPO)	Single/ Multi- threading
4	Processor Core (PCR)	Integer
5	Ideal Operating Temperature (POT)	Celsius

Programming Modes

The nodes are initialized and programmed based on the requirement of the WSN application. Programming includes providing the node id with a unique id, synchronization of the nodes local timer with global timing, loading with the private key and public keys, secure hashtags, and, base station and cluster head details. The programming modes for the WSN could be broadly classified into two modes: (a) Online programming, and (b) Offline programming. If the task assignment programming occurs prior deployment of the sensor nodes, it is offline programming while if it is performed after deployment and during its active stage is online programming.

Architecture

The classification is based on the execution node of the task assignment algorithm. Centralized task assignment adheres to multi-level hierarchical architecture. The top-level nodes are generally considered powerful and the decisions related to the task assignment process originate from these nodes (Base station, Cluster heads) (Yang et al., 2014). Distributed approach adheres to peer to peer architecture. Task assignment occurs via negotiation between the peer nodes (Pilloni, Navaratnam, Vural, Atzori, & Tafazolli, 2014). Adding nodes dynamically in a centralized approach involves high communication and computational overload whereas in decentralized method scalability.

Nature of the Task Assignment

Based on the modes of operation, task assignment can be classified as static and dynamic task assignment algorithm. Dynamic task assignment is an on-demand method in which the tasks are evaluated and assigned dynamically as per the demand of the WSN application in real-time, whereas in a static task assignment type, the tasks are pre-assigned for a particular period. Most of the task assignment algorithms are static. However, a wide plethora of dynamic WSN applications is in the rise (Munir et al., 2014).

INVESTIGATION ON TASK ASSIGNMENT ALGORITHMS

Investigations of the existing WSN task assignment algorithms have been carried out in this section. The genesis of WSN task assignment algorithms was initiated by Yu et al. in the year 2005(Yu & Prasanna, 2005). They proposed an energy balanced WSN task assignment algorithm by employing Dynamic Voltage Scaling (DVS) on a single hop homogeneous WSN. In 2005, Frank et al. (Römer, Frank, Marrón, & Becker, 2004) were one of the forerunners in proposing a generic role assignment scheme considering the properties of the node.

Subsequently, various task assignment algorithms were proposed from time to time with varying objectives. Tkach et al have investigated on distributed heterogeneous multi-sensor task allocation systems (Tkach & Edan, 2020). The work comprises of various task allocation and administration algorithms in a single and double layer approach. A nature-inspired distributed bee colony algorithm in a heterogeneous environment was also been proposed. However, the approaches majorly were confined to distance between the nodes, and task priority.

Haung et al. (Huang, Savkin, Ding, & Huang, 2019) surveyed on tasks performed by the mobile robots in WSN. The tasks were grouped under (i) collection (sensor data, residual energy data) (ii) delivery (energy delivery, content distribution, software update) and (iii) combination of collection and delivery. Critical evaluations have been performed and presented based on taxonomy, sensor node features, mobility, algorithm objectives and performance metrics. Hamouda (Hamouda, 2019) proposed a set of two algorithms for task mapping and scheduling. (i) Lifetime Awareness Sensor Node Selection Algorithm (LA-SNSA) was proposed to select sensor nodes for the execution of the application. (ii) A single-solution meta-heuristic modified Random Bit Climbing (λ -MRBC) algorithm was proposed to escape local optima. It uses transposition operator to improve the performance of the search. The work was limited by considering the energy consumption and application execution time (make-span) factors during task assignment. The other major factors affecting in HeWSN were not considered.

Ferjani et al. (Ferjani, Liouane, & Kacem, 2016) proposed a logic gate-based evolutionary algorithm for task allocation in WSN. It majorly considered only computational and communication factors. The task model was modelled as DAG considering task interdependencies. The algorithm did not consider the heterogeneity of the nodes. An intelligent heuristic task allocation scheme (HITAS) based on genetic algorithm was proposed by Jin et al. for a centralized multi-hop wireless sensor network(Jin, Jin, Gluhak, Moessner, & Palaniswami, 2012). The optimal task assignment solutions generated was more likely in ending up in local optima. Also, the residual energy and heterogeneity factors were not taken into consideration during the task assignment process.

Yang et al. (Yang et al., 2014) modelled task assignment based on the Modified Binary Particle Swarm Optimization (MBPSO), a centralized static task allocation scheme. Each particle was encoded as a binary matrix that consisted of the availability of the nodes for the task execution and a v-shaped family function was used to update the position of the particle. Mutation operation was performed to overcome local minima. The position updates during re-organizing often resulted in the rise of many infeasible solutions. This, in turn, led to higher computation overload. The methods (Ferjani et al., 2016; Jin et al., 2012; Yang et al., 2014) are based on nature-inspired meta-heuristic methods and it uses higher computational, storage requirements and it is not always ideal for resource-constrained WSN(Hamouda, 2019).

Guo et al (Guo, Zhang, Mi, Yang, & Obaidat, 2019) proposed a non-centralized task allocation algorithm called Min-Actor Algorithm (MIATA) for WSAN. MIATA solves network connectivity and task allocation simultaneously. Stage 1 involves MIA constructing a two-hop RCDS (Rough strong Connected Dominating Set) to solve the network connectivity issue by dividing the network into multiple clusters. After construction of the clusters, redundant nodes in RCDS are pruned by the rules. The second phase involves the actual task allocation process consisting of two phases (i) task discovery and (ii) task allocation phase. The work was limited by considering energy level, execution cost and distance factors. Honey bee colony based based distributed task assignment algorithm have been proposed by (Tkach &

Edan, 2020). It is based on swarming of bees and its modes of communication is used to employ task assignment to WSN.

An extensive literature survey of the existing task assignment algorithms proposed before July 2016 has been carried out in the previous work (Titus et al., 2016). In summary, the existing significant task assignment algorithms in WSN have been presented. The contemporary WSN applications require a composite set of heterogeneous nodes to achieve its WSN application goals. And the existing task assignment algorithms requires maximum human intervention during the during every re-organization stage. Thus, the requirement for artificial intelligence-based WSN emerges.

TASK ASSIGNMENT ALGORITHMS USING DEEP LEARNING TECHNIQUES

One of the objectives of the chapter is to model and design a futuristic, effective WSN task assignment algorithm with minimal human intervention. Deep learning is a special subset of machine learning technique based on the artificial neural network that minimizes human intervention ("Deep learning" 2020). Initially, the term 'deep learning' was coined by Rina Dechter in 1986 ("Deep learning," 2020). And down the years, it has evolved significantly to improve its performance. While the conventional approaches are predominantly feature-based learning, deep learning has the inherent ability to learn from the raw data (Friday et al., 2018). Deep learning can be classified as supervised, semi-supervised or unsupervised models based on learning models.

For the preliminary investigations, datasets for the experimentations have been generated from our previous work PSO based task assignment algorithm (Issac, Silas, & Rajsingh, 2020b). The datasets comprise of input data labels such as luminaire properties, zones, duration and the illumination intensity as target label. A supervised type of deep learning is much effective over other learning models owing to the inherent nature of the available labelled datasets to carry out the feasibility study. Deep learning methods based on architectures are classified as (i) restricted Boltzmann machine, (ii) auto-encoder, (iii) sparse coding, (iv) convolutional neural network, (v) recurrent neural network (Friday et al., 2018). The majority of the neural networks, including convolutional neural network, recurrent neural network are based on the functioning of Multi-layer Feed-forward Neural Network (MLFFNN).

The concept of artificial neural networks were first introduced by Igor Aizenberg ("Deep learning" 2020). MLFFNN is one of the first, simple yet effective, and widely employed supervised artificial neural network (HECHT-NIELSEN, 1992; Svozil, Kvasnicka, & Pospichal, 1997). The feasibility study is to be carried out using MLFFNN as it is (i) an ideal method to understand the functioning of deep learning, (ii) an ideal benchmark method for upcoming systems employing other deep learning models. To this end, MLFFNN has been chosen to carry out the feasibility study in this work.

System Modelling

System modelling of PSO trained Multi-layer Feed Forward Neural Network to solve the dynamic heterogeneous task assignment problem is presented in this section. The section is broadly categorized as follows:

1. Modelling a composite Heterogeneous Wireless Sensor Network (HeWSN)
2. Modelling the heterogeneous task assignment problem

3. Modelling a heterogeneous task assignment algorithm using Particle Swarm Optimization
4. Modelling PSO trained MLFFNN

Modelling a Composite Heterogeneous Wireless Sensor Network

A Heterogeneous Wireless Sensor Network (HeWSN) comprises of multiple sensor nodes SN_i ($1 \leq i \leq n$). The heterogeneity of the sensor nodes varies in (i) sensing, (ii) processing, (iii) storage, (iv) communication and (v) energy capabilities. For the initial investigations, all the nodes in the HeWSN are assumed to be centralized architecture in which the nodes report to the base station (BS) as depicted in figure 4.

Modelling the HeWSN Task Assignment Problem

The task assignment is modelled as a multi-criteria problem in HeWSN. The major criteria influencing the task assignment in a HeWSN are as follows: (i) initial energy (Ganeriwal, Balzano, & Srivastava, 2008), (ii) remaining energy[10], (iii) bootstrap time(Ganeriwal et al., 2008) [9], (iv) sensitivity(Brown, 2020; Carr.J & John M, 2002), (v)accuracy(Brown, 2020; Carr.J & John M, 2002) [11], (vi) response time(Ganeriwal et al., 2008), (vii)waiting time (Ganeriwal et al., 2008; Pilloni et al., 2014), (viii) processing speed (Munir et al., 2014). The criteria CR_i ($1 \leq i \leq m$) are chosen for the task assignment based on the WSN application requirements.

Modelling Particle Swarm Optimization (PSO)

The PSO adopted for this chapter has been adopted from our previous work that performs dynamic task assignment for HeWSN(Issac et al., 2020b). The fitness function of the kth particle (P(k).FF) is calculated using the weighted average (WA) of the criteria and it is calculated using equation (1) where CW_i are the corresponding weights for the criteria CR_i. And CW satisfy the equation (2). Equation (2) performs the velocity calculation by summation of the particle's best velocity and global best velocity denoted by equations (3) and (4). And the new particle identifier is generated using equation (5). The criteria weights (CW_i) for the nodes are awarded based on the priority of the factors to achieve the desired objective of the WSN application. At a given time, Task 't_i' is assigned to the node having the maximum WA value.

$$\mathbf{P(k).FF = (w1 \times cr1 + w2 \times cr2 + \ldots + wm \times crm) / (w1 + w2 + \ldots + wz)} \quad (1)$$

$$P(k).Velocity = (P(k).Velocity) + (c1 \times A) + (c2 \times B); \quad (2)$$

$$A = rand() \times (P(Best.id) P(k).id); \quad (3)$$

$$\mathrm{B = rand() \times (Globalbest.Pos - P(k).id);} \quad (4)$$

$$\mathrm{P(k).id = P(k).id + P(k).velocity;} \quad (5)$$

The pseudo code for the PSO based task assignment algorithm for HeWSN is as follows.

```
01: Get node properties p, q, c1, c2,w cr1, cr2,...crz, w1, w2,...wz
02: Randomly select candidate node(s) for P(k)
03: If task satisfying condition satisfied
04:Then Evaluate fitness function P(k).FF
05:Else Go to step 2
06: End If
07: If SN(k).FF > P(k).FF.Lbest
08:    Then  set P(k).FF.Lbest= P(k).FF
09:           If P(k).FF > P(k).FF.Gbest
10:           Then  set P(k).FF.Gbest= P(k).FF
11:    End if
12: End if
13: If the termination condition is false
14:    Then calculate new velocity & find new candidate & go to step 3
15:    Else halt the iteration and produce the FF.Gbest
16: End if
```

Modelling PSO Trained MLFFNN

MLFFNN can perform pattern recognition, clustering and fitting operations. Among the aforementioned operations, fitting the input and output solution is found ideal to solve the task assignment problem. The MLFFNN technique is found ideal to solve the dynamic task assignment problem in a heterogeneous WSN.

Multi-Layer Feed Forward Neural Network

MLFFNN was inspired by the functioning of the biological nervous system. Every neuron in the nervous system feeds forwards a signal on achieving a threshold limit. Similarly, MLFFNN is comprised of multiple parallel and feed-forward operating neurons. The MLFFNN operates by adjusting the weight and bias of the neurons in the hidden layer during feed forward. Every connection between the neurons; say x_i and h_j has a weight coefficient $w_{i,j}$. The weight coefficient signifies the degree of importance between the neurons. The weighted sum of the neuron ($\sum(x_i w_i)$) is applied to the activation function and checked with the output threshold value. The bias is reassigned based on threshold values to achieve the desired output during the training phase. Figure 5 depicts the functional overview of MLFFNN. The weights and biases are randomly initialized and are adjusted until the desired the output is achieved or stopped after a certain epoch count.

In this investigation, the output of PSO based task assignment algorithm proposed in the previous section is provided as the inputs along with every sensor node criteria and its corresponding criteria weights. The MFFNN is trained by varying (i) hidden layers, (ii) training sample size until the ideal sensor node is obtained. The training scenario is depicted in figure 6.

The overall PSO trained MLFFNN structure is depicted in figure 6. The first and last layer is the input and output layer respectively. The actual deep learning processing occurs in the hidden layer. The input and output of the neurons are fed and obtained as numbers. Every neuron in each layer is connected with all the neuron of its predecessor.

Figure 5. MLFFNN- Overview

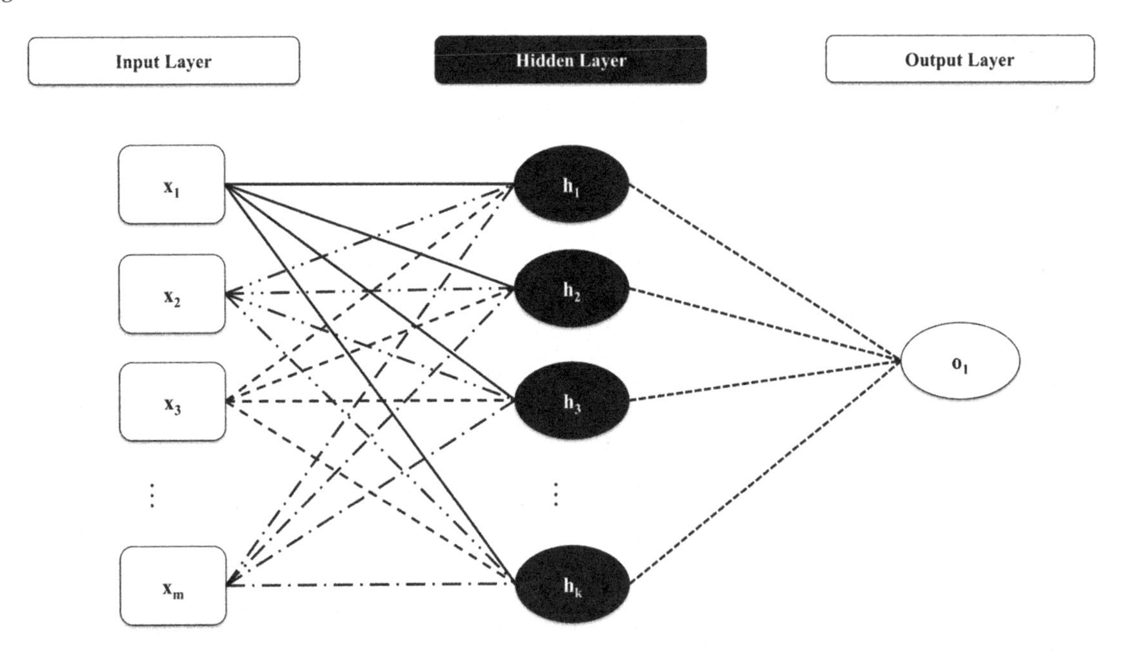

Figure 6. The training phase of PSO-MLFFNN

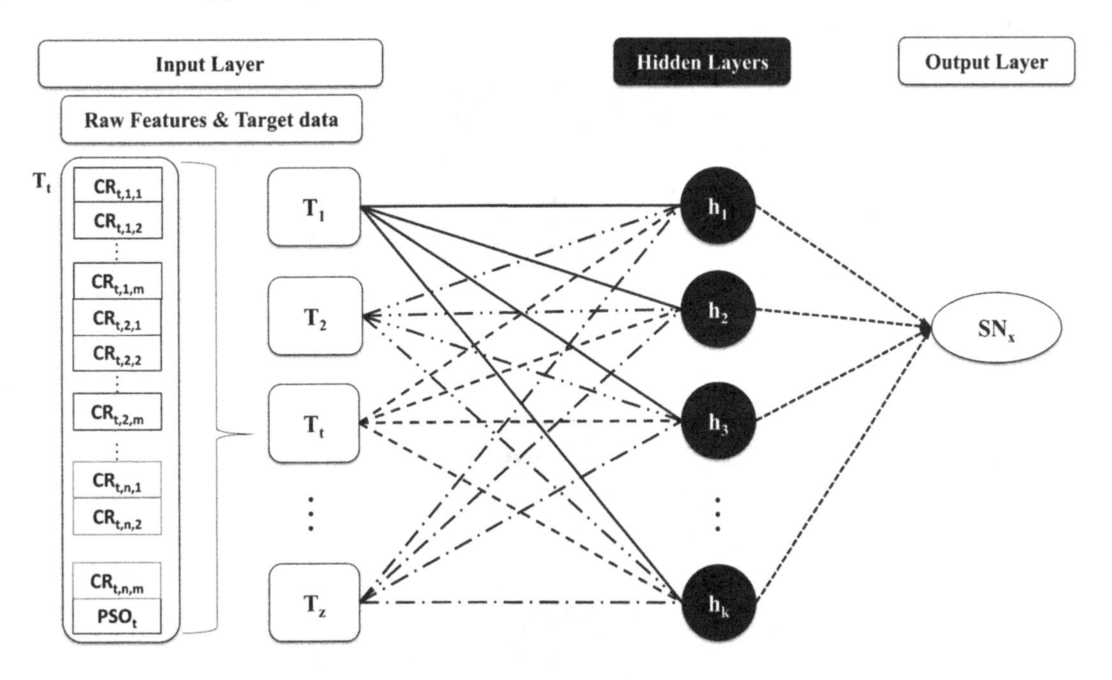

Figure 7. PSO-MLFFNN Task assignment – Workflow

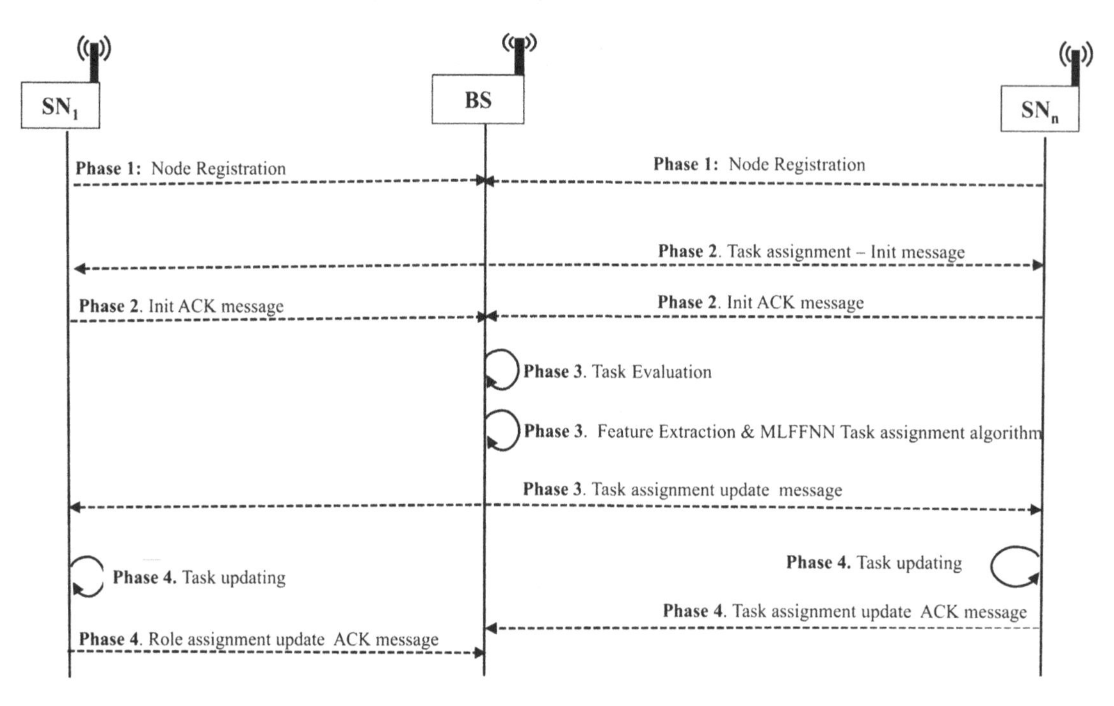

Figure 8. PSO-MLFFNN- Overview

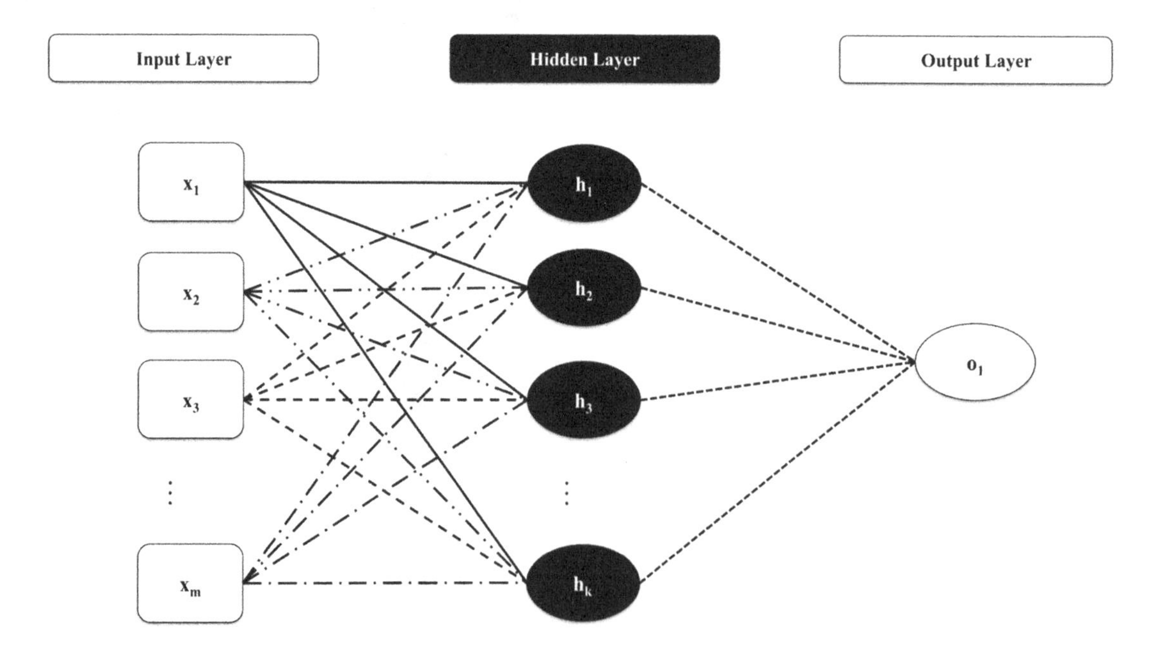

MFFNN Based Dynamic Task Assignment Algorithm

Upon successful modelling and training of the MLFFNN, task assignment algorithm is deployed in the base station (BS). The overall workflow of the task assignment algorithm is represented in figure 7. The HeWSN application is divided into multiple indivisible tasks by the base station. The generated tasks have to be mapped to the candidate node by employing MLFFNN based task assignment algorithm. On completion of the task assignment algorithm, task assignment update message is broadcasted to the corresponding nodes. The recipient nodes provide a corresponding acknowledgement message on the acceptance for the proposed task assignment message. The overall overview of the developed PSO- MLFFNN is depicted in figure 8. After the modelling of the heterogeneous WSN task assignment algorithm using PSO trained MLFFNN, the performance of the system is to be subjected to inspection.

Experimental Simulation

A Heterogeneous Wireless Network (HeWSN) is simulated by initializing a set of nodes with a random initial energy and sensing capabilities. Waspmote, a macro sensor node from libelium has been selected for simulation of heterogeneous WSN as a wide range of sensors could be plugged in the node as per the application demand The properties of waspmote node is obtained from its libelium data sheets("Waspmote Technical guide," 2018). The properties of the waspmote and sensor used for the simulation is tabulated in table 11 and table 12.

Table 11. Libelium Waspmote Properties

Parameters	Value
Max Energy	23760 J
Voltage	3.7 v
$Energy_{ON}$	0.017 J
$Energy_{SLEEP}$	0.00003 J
$Energy_{Send}$	0.1 J
$Energy_{Receive}$	0.1 J

Table 12. Libelium Waspmote – Sensor details

Sensor Type	Model	Sensing Energy(J)
Temperature	SHT75	10.8
	MCO9700A	0.0216
	L_MCP9700A	0.0216
Humidity	SHT75	10.8
	808H5V5	9.0
	L_808H5V5	9.0
Pressure	MPX4115A	192.6
	L_MPX4115A	192.6

A total of 200 nodes are randomly placed such that it is within a single-hop communication range of the Base Station (BS). ZigBee wireless communication is adapted as in the work of [6]. Heterogeneity in the network is achieved by (i) initiating the nodes with random initial energy,(ii) mounting a set of random sensors tabulated in table 12 on to the node.

The task assignment evaluation and deep learning process occur in the Base Station (BS) as depicted in figure 7. The tasks are randomly generated by the BS and assumed to be independent of other tasks. MLFFNN is trained using a data set. Data sets are generated where an individual data set contains the data of 200 nodes. The data includes the node's (i) initial energy, (ii) remaining energy, (iii) sensor energy, (iv) bootstrap time, (v) location.

A training set of 200 individual datasets has been randomly generated to train the MLFFNN. Levenberg-Marquardt training function predefined in Matlab was during training. The training, =validation and testing ratio was taken as 80, 10 and 10 respectively. Matlab inbuilt MLFFNN function called fitnet was selected to perform the investigation. Experimentation has been carried out by varying the hidden layer ranging from 1 to 200 and training set 5 to 200 with constant initial weights and biases. The experimentation led to a total of 200 x 195 individual training and evaluations have been carried out.

DISCUSSIONS

The investigations are grouped under (i) hidden layers, (ii) training sets and are discussed in the following section.

Hidden Layers

Hidden layer plays a vital in obtaining the ideal output from the neural network. The effect of minimal and maximum hidden ranges on the neural network has been carried out. Figure 9 depicts the hidden layer-wise range of maximum successful task assignment ratio. A maximum successful task assignment ratio by varying the hidden layers is obtained as 93%. The average successful task assignment ratio between various hidden layers ranges is depicted in figure 10. The ideal hidden layer range is found between 31 to 35 hidden layers as it has the maximum average successful task assignment ratio. The experimentation results of successful task assignment ratio based on hidden layers is tabulated in table 13.

Training Set

A training set contains individual data of 200 participating sensor nodes. The maximum successful task assignment ratio between various training set ranges is depicted in figure 10. The highest successful task assignment ratio achieved by varying the training sets is 95.3%. The average successful task assignment ratio between the training set ranges is depicted in figure 11. The maximum average successful task assignment ratio occurs with data set size ranging between 0 and 05. The figures 10 and 11 depict the effect of over-training that lead to the fall of the successful task assignment ratio. The experimentation results of successful task assignment ratio based on training set size is tabulated in table 14.

Figure 9. Maximum successful task assignment ratio with varying hidden layer size

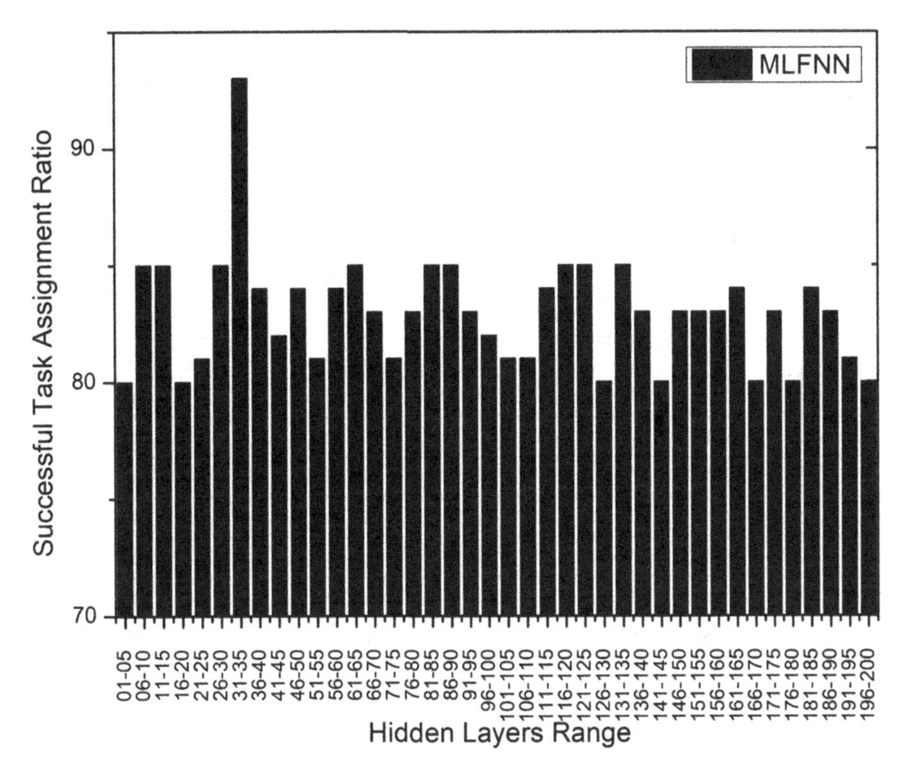

Figure 10. Average successful task assignment ratio with varying hidden layer size

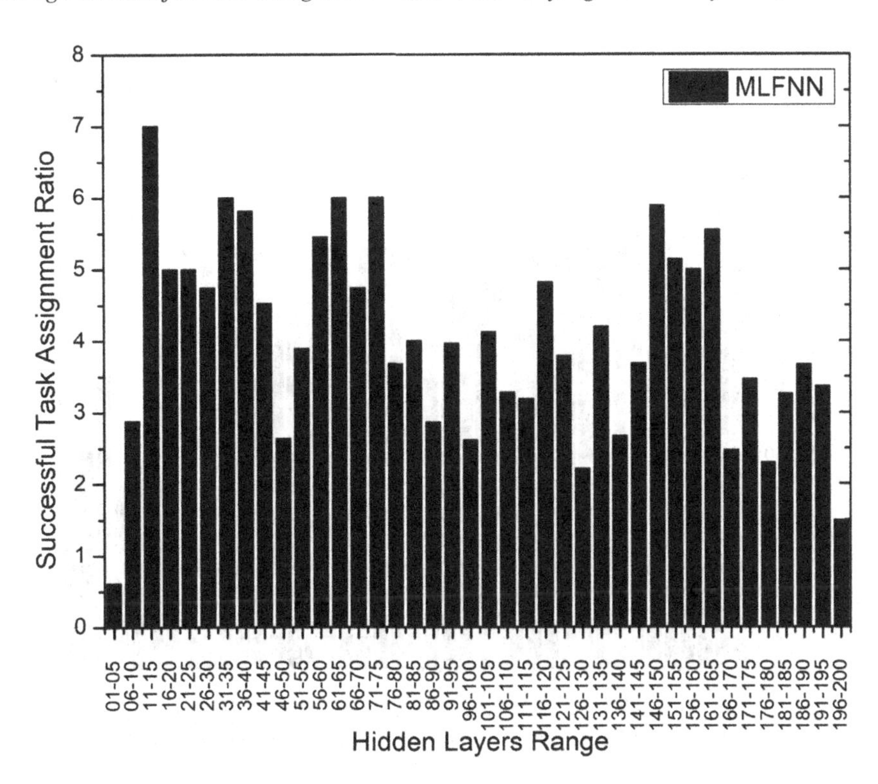

Table 13. Experimentation Results of Successful task assignment ratio based on hidden layers

Hidden Layer Size	Max Value	Avg Value	Hidden Layer Size	Max Value	Avg Value
00-05	80.00	0.61	101-105	81.00	4.13
06-10	85.00	2.88	106-110	81.00	3.28
11-15	85.00	7.01	111-115	84.00	3.19
16-20	80.00	5.00	116-120	85.00	4.82
21-25	81.00	5.00	121-125	85.00	3.79
26-30	85.00	4.75	126-130	80.00	2.21
31-35	93.00	6.00	131-135	85.00	4.20
36-40	84.00	5.81	136-140	83.00	2.67
41-45	82.00	4.53	141-145	80.00	3.69
46-50	84.00	2.64	146-150	83.00	5.89
51-55	81.00	3.89	151-155	83.00	5.14
56-60	84.00	5.45	156-160	83.00	5.00
61-65	85.00	5.99	161-165	84.00	5.55
66-70	83.00	4.74	166-170	80.00	2.47
71-75	81.00	6.00	171-175	83.00	3.46
76-80	83.00	3.68	176-180	80.00	2.29
81-85	85.00	4.00	181-185	84.00	3.27
86-90	85.00	2.87	186-190	83.00	3.67
91-95	83.00	3.97	191-195	81.00	3.37
96-100	82.00	2.62	196-200	80.00	1.49

Figure 11. Maximum successful task assignment ratio with varying training set size

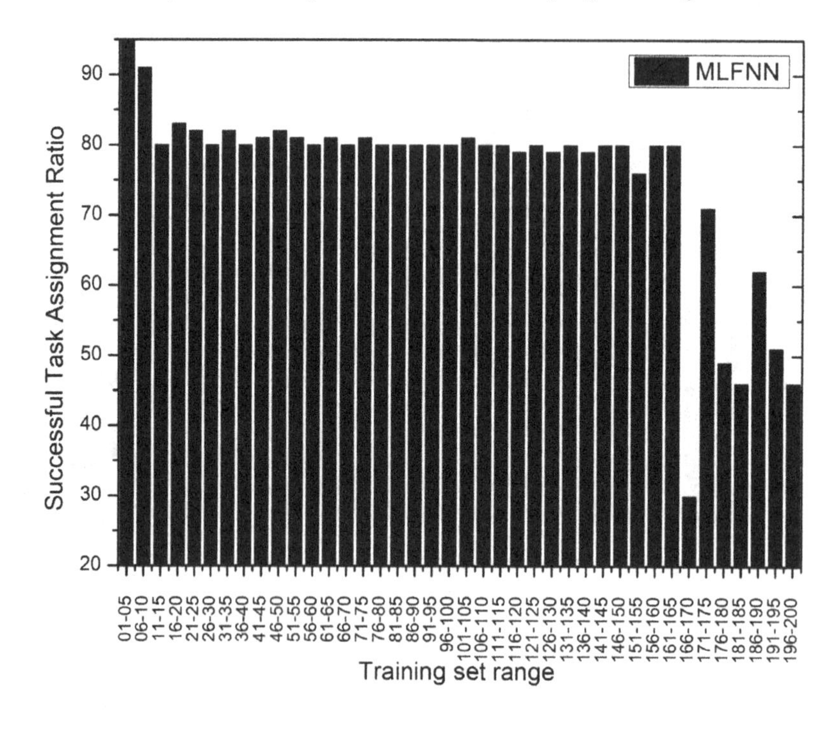

Figure 12. Average successful task assignment ratio with varying training set size

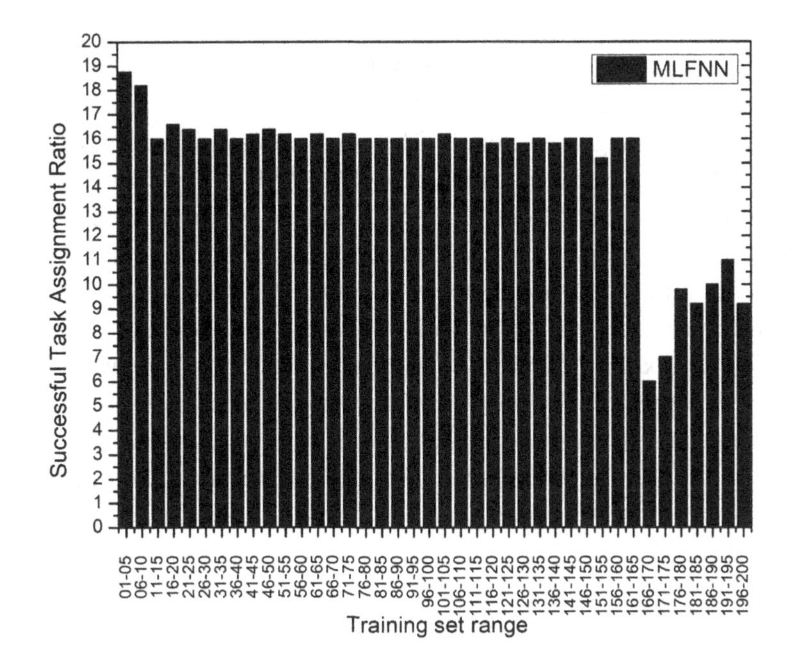

Table 14. Experimentation results of successful task assignment ratio based on training set size

Training Set Size	Max Value	Avg Value	Training Set Size	Max Value	Avg Value
00-05	93.90	18.78	101-105	81.00	16.20
06-10	91.00	18.20	106-110	80.00	16.00
11-15	80.00	16.00	111-115	80.00	16.00
16-20	83.00	16.60	116-120	79.00	15.80
21-25	82.00	16.40	121-125	80.00	16.00
26-30	80.00	16.00	126-130	79.00	15.80
31-35	82.00	16.40	131-135	80.00	16.00
36-40	80.00	16.00	136-140	79.00	15.80
41-45	81.00	16.20	141-145	80.00	16.00
46-50	82.00	16.40	146-150	80.00	16.00
51-55	81.00	16.20	151-155	76.00	15.20
56-60	80.00	16.00	156-160	80.00	16.00
61-65	81.00	16.20	161-165	80.00	16.00
66-70	80.00	16.00	166-170	30.00	6.00
71-75	81.00	16.20	171-175	71.00	7.00
76-80	80.00	16.00	176-180	49.00	9.80
81-85	80.00	16.00	181-185	46.00	9.20
86-90	80.00	16.00	186-190	62.00	10.00
91-95	80.00	16.00	191-195	51.00	11.00
96-100	80.00	16.00	196-200	46.00	9.20

CONCLUSION AND FUTURE RESEARCH DIRECTIONS

The future perspective of making the system to have less human interaction led to the investigation of task assignment algorithms based on deep learning methods. The chapter has presented the inception and genesis of various computing technologies. The contemporary Wireless Sensor Network (WSN) application requirements led to the investigation of various WSN task assignment algorithms. Starting from the genesis of Wireless Sensor Network (WSN) till the current technological advancements in WSN have been investigated and presented. The major factor influencing the task assignment in both homogeneous and heterogeneous environments was identified. To the best of the authors' knowledge, the proposed Particle Swarm Optimization trained Multilayer Feed Forward Network is one of the fore-runners in the deep learning based WSN task assignment. A feasibility study was carried out by varying the hidden layers and data set size. The maximum and average successful task assignment ratio under various hidden layer and data set size was studied considering the future perspective of the WSN. The chapter would serve as an ideal benchmark model for the researchers in the field of deep learning and wireless network.

REFERENCES

Abacus. (n.d.). Retrieved May 5, 2020, from https://en.wikipedia.org/wiki/Abacus

Akhand, M. A. H., Ayon, S. I., Shahriyar, S. A., Siddique, N., & Adeli, H. (2020). Discrete Spider Monkey Optimization for Travelling Salesman Problem. *Applied Soft Computing*, *86*, 105887. doi:10.1016/j.asoc.2019.105887

Bansal, J. C. (2019). Evolutionary and Swarm Intelligence Algorithms. In J. C. Bansal, P. K. Singh, & N. R. Pal (Eds.), *Evolutionary and Swarm Intelligence Algorithms*. doi:10.1007/978-3-319-91341-4

Bhushan, B., & Sahoo, G. (2018). Recent advances in attacks, technical challenges, vulnerabilities and their countermeasures in wireless sensor networks. *Wireless Personal Communications*, *98*(2), 2037–2077. doi:10.100711277-017-4962-0

Bhushan, B., & Sahoo, G. (2019a). E2SR2 : An acknowledgement-based mobile sink routing protocol with rechargeable sensors for wireless sensor networks. *Wireless Networks*, *25*(5), 2697–2721. doi:10.100711276-019-01988-7

Bhushan, B., & Sahoo, G. (2019b). ISFC-BLS (Intelligent and Secured Fuzzy Clustering Algorithm Using Balanced Load Sub-Cluster Formation) in WSN Environment. *Wireless Personal Communications*, *111*(3), 1667–1694. doi:10.100711277-019-06948-0

Bhushan, B., & Sahoo, G. (2019c). Routing Protocols in Wireless Sensor Networks. In B. Mishra, S. Dehuri, B. Panigrahi, A. Nayak, & B. Mishra (Eds.), *Studies in Computational Intelligence* (Vol. 776, pp. 215–248). doi:10.1007/978-3-662-57277-1_10

Bhushan, B., & Sahoo, G. (2020). Requirements, Protocols, and Security Challenges in Wireless Sensor Networks: An Industrial Perspective. In Handbook of Computer Networks and Cyber Security (pp. 683–713). doi:10.1007/978-3-030-22277-2_27

Brown, J. J. C. J. M. (2020). *Sensor Terminology*. Retrieved February 11, 2020, from https://www.ni.com/en-in/innovations/white-papers/13/sensor-terminology.html

Bult, K., Burstein, A., Chang, D., Dong, M., Fielding, M., & Kruglick, E., … Pottie, G. (1996). Low Power Systems for Wireless Microsensors. *International Symposium on Low Power Electronics and Design*. 10.1109/LPE.1996.542724

Carr, J. J., & John, M. B. (2002). *Introduction to Biomedical Equipment Technology* (4th ed.). Pearson.

Deep Learning. (2020). Retrieved April 15, 2020, from https://en.wikipedia.org/wiki/Deep_learning

ENIAC. (n.d.). Retrieved May 4, 2020, from https://en.wikipedia.org/wiki/ENIAC

Ferjani, A. A., Liouane, N., & Kacem, I. (2016). Task allocation for wireless sensor network using logic gate-based evolutionary algorithm. *2016 International Conference on Control, Decision and Information Technologies (CoDIT)*, 654–658. 10.1109/CoDIT.2016.7593640

Friday, H., Wah, Y., Al-garadi, M. A., & Rita, U. (2018). Deep learning algorithms for human activity recognition using mobile and wearable sensor networks : State of the art and research challenges. *Expert Systems with Applications*, *105*, 233–261. doi:10.1016/j.eswa.2018.03.056

Ganeriwal, S., Balzano, L. K., & Srivastava, M. B. (2008). Reputation-based framework for high integrity sensor networks. *ACM Transactions on Sensor Networks*, *4*(3), 1–37. doi:10.1145/1362542.1362546

Guo, Y., Zhang, Y., Mi, Z., Yang, Y., & Obaidat, M. S. (2019). Distributed task allocation algorithm based on connected dominating set for WSANs. *Ad Hoc Networks*, *89*, 107–118. doi:10.1016/j.adhoc.2019.03.006

Hamouda, Y. E. M. (2019). Modified random bit climbing (λ-MRBC) for task mapping and scheduling in wireless sensor networks. *Jordanian Journal of Computers and Information Technology*, *5*(1), 17–33. doi:10.5455/jjcit.71-1541688581

Hecht-Nielsen, R. (1992). Theory of the Backpropagation Neural Network. In Neural Networks for Perception (pp. 65–93). doi:10.1016/b978-0-12-741252-8.50010-8

Huang, H., Savkin, A. V., Ding, M., & Huang, C. (2019). Mobile robots in wireless sensor networks: A survey on tasks. *Computer Networks*, *148*, 1–19. doi:10.1016/j.comnet.2018.10.018

Industrial Revolution. (n.d.). Retrieved May 4, 2020, from https://en.wikipedia.org/wiki/Industrial_Revolution

Information Age. (n.d.). Retrieved May 4, 2020, from https://en.wikipedia.org/wiki/Information_Age

Issac, T., Silas, S., & Rajsingh, E. B. (2019a). Luminaire aware centralized outdoor illumination role assignment scheme: A smart city perspective. In J. D. Peter, A. H. Alavi, & B. Javadi (Eds.), *Advances in Big Data and Cloud Computing* (pp. 443–456). doi:10.1007/978-981-13-1882-5_38

Issac, T., Silas, S., & Rajsingh, E. B. (2019b). Role assignment in lighting application for a greener and safer smart city. *2019 3rd International Conference on Computing and Communications Technologies (ICCCT)*, 1–8. 10.1109/ICCCT2.2019.8824914

Issac, T., Silas, S., & Rajsingh, E. B. (2020a). Dynamic and static system modelling with simulations of an eco-friendly smart lighting system. In D. Peter & S. L. Fernandes (Eds.), *Systems Simulation and Modeling for Cloud Computing and Big Data Applications* (pp. 81–97). doi:10.1016/B978-0-12-819779-0.00005-8

Issac, T., Silas, S., & Rajsingh, E. B. (2020b). Investigations on PSO based task assignment algorithms for heterogeneous wireless sensor network. *2nd International Conference on Signal Processing and Communication (ICSPC)*, 89–93. 10.1109/ICSPC46172.2019.8976850

Jin, Y., Jin, J., Gluhak, A., Moessner, K., & Palaniswami, M. (2012). An Intelligent Task Allocation Scheme for Multihop Wireless Networks. *IEEE Transactions on Parallel and Distributed Systems*, *23*(3), 444–451. doi:10.1109/TPDS.2011.172

Lalwani, S., Sharma, H., Chandra, S., Kusum, S., Jagdish, D., & Bansal, C. (2019). A Survey on Parallel Particle Swarm Optimization Algorithms. *Arabian Journal for Science and Engineering*, *44*(4), 2899–2923. doi:10.100713369-018-03713-6

Misra, S., & Vaish, A. (2011). Reputation-based role assignment for role-based access control in wireless sensor networks. *Computer Communications*, *34*(3), 281–294. Advance online publication. doi:10.1016/j.comcom.2010.02.013

Munir, A., Gordon-Ross, A., & Ranka, S. (2014). Multi-core embedded wireless sensor networks: Architecture and applications. *IEEE Transactions on Parallel and Distributed Systems*, *25*(6), 1553–1562. doi:10.1109/TPDS.2013.219

Pant, M., Deep, K., Bansal, J. C., Das, K. N., & Nagar, A. K. (2018). Soft computing for problem solving (SocProS 2015). *International Journal of System Assurance Engineering and Management*, *9*(1), 1–1. doi:10.100713198-018-0713-1

Păun, G. (2000). Computing with Membranes. *Journal of Computer and System Sciences*, *61*(1), 108–143. doi:10.1006/jcss.1999.1693

Pilloni, V., Navaratnam, P., Vural, S., Atzori, L., & Tafazolli, R. (2014). TAN: A distributed algorithm for dynamic task assignment in WSNs. *IEEE Sensors Journal*, *14*(4), 1266–1279. doi:10.1109/JSEN.2013.2294540

Römer, K., Frank, C., Marrón, P. J., & Becker, C. (2004). Generic role assignment for wireless sensor networks. *Proceedings of the 11th Workshop on ACM SIGOPS European Workshop: Beyond the PC - EW11*, 2. 10.1145/1133572.1133588

Roozeboom, C. L., Hopcroft, M. A., Smith, W. S., Sim, J. Y., Wickeraad, D. A., Hartwell, P. G., & Pruitt, B. L. (2013). Integrated Multifunctional Environmental Sensors. *Journal of Microelectromechanical Systems*, *22*(3), 779–793. doi:10.1109/JMEMS.2013.2245400

Second Industrial Revolution. (n.d.). Retrieved from https://en.wikipedia.org/wiki/Second_Industrial_Revolution

Sohraby, K., Minoli, D., & Znati, T. (2007). *Wireless Sensor Network Technology, Protocols, and Applications*. John Wiley & Sons. doi:10.1002/047011276X

Svozil, D., Kvasnicka, V., & Pospichal, J. (1997). Introduction to multi-layer feed-forward neural networks. *Chemometrics and Intelligent Laboratory Systems*, *39*(1), 43–62. doi:10.1016/S0169-7439(97)00061-0

Titus, I., Rajsingh, E. B., & Silas, S. (2015). E-Health Navigator : A Tool for Selection of Desired Health Care Services. *Advances in Computer Science and Information Technology (ACSIT)*, *2*(14), 54–58. Retrieved from https://www.krishisanskriti.org/Publication.html

Titus, I., Silas, S., & Rajsingh, E. B. (2016). Investigations on task and role assignment protocols in Wireless Sensor Network. *Journal of Theoretical and Applied Information Technology*, *89*(1), 209–219. http://www.jatit.org/volumes/Vol89No1/21Vol89No1.pdf

Tkach, I., & Edan, Y. (2020). Heterogeneous Distributed Bees Algorithm. In Automation, Collaboration, & E-Services: Vol. 7. Distributed Heterogeneous Multi Sensor Task Allocation Systems (7th ed., pp. 34–27). doi:10.1007/978-3-030-34735-2

Waspmote Technical guide. (2018). Retrieved April 24, 2018, from Libelium Comunicaciones Distribuidas S.L website: http://www.libelium.com/downloads/documentation/waspmote_plug_and_sense_technical_guide.pdf

Yang, J., Zhang, H., Ling, Y., Pan, C., & Sun, W. (2014). Task allocation for wireless sensor network using modified binary particle swarm optimization. *IEEE Sensors Journal*, *14*(3), 882–892. doi:10.1109/JSEN.2013.2290433

Yu, Y., & Prasanna, V. K. (2005). Energy-Balanced Task Allocation for Collaborative Processing in Wireless Sensor Networks. *Mobile Networks and Applications*, *10*(1/2), 115–131. doi:10.1023/B:MONE.0000048550.31717.c5

Chapter 6
The Threat of Intelligent Attackers Using Deep Learning:
The Backoff Attack Case

Juan Parras
Information Processing and Telecommunications Center, Universidad Politécnica de Madrid, Spain

Santiago Zazo
Information Processing and Telecommunications Center, Universidad Politécnica de Madrid, Spain

ABSTRACT

The significant increase in the number of interconnected devices has brought new services and applications, as well as new network vulnerabilities. The increasing hardware capacities of these devices and the developments in the artificial intelligence field mean that new and complex attack methods are being developed. This chapter focuses on the backoff attack in a wireless network using CSMA/CA multiple access, and it shows that an intelligent attacker, making use of control theory, can successfully exploit a sequential probability ratio test-based defense mechanism. Also, recent developments in the deep reinforcement learning field allows that attackers that do not have full knowledge of the defense mechanism are able to successfully learn to attack it. Thus, this chapter illustrates by means of the backoff attack, the possibilities that the recent advances in the artificial intelligence field bring to intelligent attackers, and highlights the importance of researching in intelligent defense methods able to cope with such attackers.

INTRODUCTION

This Chapter studies a threat that the current advances in Deep Learning pose to the security of Wireless Sensor Networks (WSNs). Namely, the Chapter focuses on the fact that Deep Reinforcement Learning (Deep RL) tools can be used to exploit a possibly unknown defense mechanism simply by interacting with it. The remarkable advances and proliferation in wireless networks in the last years have brought a significant interest in the security and threats to WSNs: they can be the target of many attacks due to the limited capabilities of the sensors, as some recent surveys show (Fragkiadakis, Tragos, & Askoxylakis,

DOI: 10.4018/978-1-7998-5068-7.ch006

2013) (Zhang, et al., 2015). One of these attacks is the backoff attack, which affects to the Medium Access Control (MAC) layer when a CSMA/CA (carrier-sense medium access with collision avoidance) scheme is used to regulate the access to the medium. The backoff mechanism minimizes the risk of collision, i.e., that two or more sensors transmit simultaneously, by deferring transmissions during a certain random time period: the backoff window. In a backoff attack, a sensor uses a lower backoff window than the rest of the sensors, thus obtaining a higher throughput at expense of the other sensors (Bayraktaroglu, et al., 2013).

In order to clearly study and explain the intelligent attacks that Deep RL tools bring, this Chapter focuses only in the backoff mechanism, as backoff attacks are a real threat to WSNs. Firstly, because network adapters are highly programmable (Cagalj, Ganeriwal, Aad, & Hubaux, 2005), thus allowing sensors to modify their backoff parameters. And secondly, because many MAC layer protocols proposed for WSNs make use of CSMA as medium access mechanism, for instance, SMAC (Ye, Heidemann, & Estrin, 2004), WiseMAC (Enz, El-Hoiydi, Decotignie, & Peiris, 2004), TMAC (Van Dam & Langendoen, 2003) and DSMAC (Lin, Qiao, & Wang, 2004). Actually, two surveys on MAC layer protocols, (Demirkol, Ersoy, & Alagoz, 2006) and (Yadav, Varma, Malaviya, & others, 2009), show that CSMA is the most common access mechanism in contention based MAC protocols.

Our main contribution consists in highlighting the effect that Deep RL based attackers can pose to current WSN defense mechanisms: such an attacker needs not knowing the defense mechanism used, nor its parameters, as it may learn to exploit it simply by interacting with it. Thus, we show that Deep RL based attackers are flexible due to their learning capabilities and pose a significant threat to many current WSN defense mechanisms.

Hence, the main objectives of the Chapter are: (1) introducing the backoff attack and its effects on the network throughput, (2) show different ways in which a game theory based defense mechanism may cope with such attack, (3) introduce a control theory formulation that allows obtaining an optimal attacker control law to exploit the defense mechanism proposed when the attacker knows all the parameters of the defense mechanism, and (4) show how Deep Reinforcement Learning tools can be used to successfully exploit the defense mechanism when the attacker does not know all the parameters of the defense mechanism.

Regarding the Chapter organization, in the Background Section we introduce the CSMA/CA mechanism, as well as how a backoff attack affects the network throughput, and the main control framework that will be used along this work: Markov Decision Processes. Then, we dedicate a Section to study defense mechanisms against the backoff attack: where we focus on sequential tests as an advanced defense mechanism. The next Section presents an optimal attack against this defense mechanism in the case that the defense mechanism is known, and also presents a Deep RL attack that is successful when the defense mechanism is unknown. The final Section summarizes our results and proposes some future lines of interest.

BACKGROUND

CSMA/CA in IEEE 802.11

Let us start by describing the CSMA/CA mechanism as described in the well-known 802.11 standard. The IEEE 802.11 standard (IEEE, 2016) defines the MAC layer specification for a wireless local area

Figure 1. Network scheme for the case that there are n_1 GSs and n_2 ASs. GSs respect 802.11 binary exponential backoff, whereas ASs can choose to use it or to use a uniform backoff

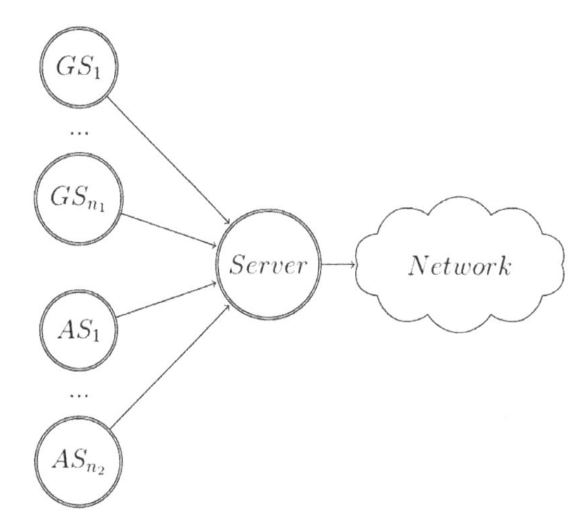

network. The access to the shared medium can be regulated using the Distributed Coordination Function (DCF). In this Chapter, the network topology used is shown in Figure 1, where there are n_1 Good Sensors (GSs) and n_2 Attacking Sensors (ASs) connected to a server using a star topology. The basic mechanism used by the DCF is CSMA/CA to control the medium using two procedures: a carrier sense (CS) which determines whether the channel is busy or idle; and a backoff procedure.

A sensor willing to transmit invokes the CS mechanism to determine the state of the channel. If it is busy, the sensor defers the transmission until the channel is idle and then it starts a counter, called backoff, for an additional deferral time before transmitting: the sensor transmits when its backoff counter reaches 0. This minimizes collisions among multiple sensors that have been deferring to the same event. The backoff follows a uniform random variable in the interval $[0, CW - 1]$, where CW stands for contention window. If a collision is detected, CW is duplicated, which is known as binary exponential backoff, and the backoff procedure starts over. This mechanism is known as Basic Access (BA) and is based on a two-way handshaking.

The standard also defines an alternative procedure, based on a four-way handshaking, called request-to-send/clear-to-send (RTS/CTS). In this case, the transmitter sensor sends an RTS frame to the receiver, using the BA mechanism, to reserve the medium. The receiver reserves the channel sending a CTS frame. When the transmitter receives the CTS frame, starts transmitting its packet. While the channel is reserved, the rest of sensors remain silent. The RTS/CTS procedure helps to ease the hidden node problem (Buehrer, 2006), (Rahman & Gburzynski, 2006) and provides a higher throughput than the BA mechanism when the MAC payload is large (Bianchi, 2000).

Network Throughput Under Backoff Modification

We have described the backoff mechanism that all sensors are supposed to follow. However, ASs may use a different backoff rule in order to obtain a benefit, namely, that they transmit more often: this causes that ASs have a gain in network throughput.

Now, the 802.11 standard does not provide a way to estimate the network throughput. The best-known model to estimate the throughput in a network is Bianchi's model (Bianchi, 2000), which provides expressions both for BA and RTS/CTS mechanisms. The main advantage of this model is that it provides analytical expressions to determine the network throughput. It assumes saturation of the network, that is, that each sensor always has a packet to transmit. This assumption could be relaxed using more complex models (Malone, Duffy, & Leith, 2007).

In (Parras & Zazo, 2018), an extensive study on the effects of the backoff attack in terms of network throughput is carried out. It considers a WSN in which there are a certain number n_2 of Attacking Sensors (ASs) that follow a uniform backoff, i.e., instead of following the binary exponential backoff procedure described above, they select their backoff window W_2 as $W_2 \sim U[0,31]$, where $U[a,b]$ denotes the uniform distribution of integers between a and b, and no change to the backoff interval is done after a collision or successful transmission. As Good Sensors (GSs) follow the IEEE 802.11 standard binary backoff mechanism with $W_1 = CW_{min,1} = 32$ and $CW_{max,1} = 1024$, note that ASs will have smaller backoff window sizes and hence, they will win the contention procedure more often.

After an extensive testing on different settings, the following conclusions are obtained (Parras & Zazo, 2018):

- The throughput of GSs decreases significantly for low values of W_2, regardless of the number of ASs, the mechanism used, i.e., BA or RTS/CTS, and the payload size. Since ASs use lower, they have higher chances to win the contention procedure against GSs. Thus, the throughput is not fairly distributed among sensors. As W_2 increases, the ASs behave more similarly to the GSs and the difference in throughput becomes smaller.
- If there is only one AS, this sensor consumes the major part of the network throughput for low W_2, because it usually wins the contention procedures. However, if there are more than one AS, the total throughput approaches 0 for low W_2 values, because there are several sensors trying to access the network that will collide with a high probability. Also, as n_2 increases, the throughput that an AS obtains decreases: it is better for an AS to be the only AS in the WSN.
- RTS/CTS mechanism provides higher throughput when using larger payloads. The opposite happens when using short payloads.

Hence, if in a WSN using CSMA/CA there is one or more sensors which can modify the binary exponential backoff procedure used by 802.11, the throughput that each sensor gets can be seriously affected.

The MDP Framework

The Markov Decision Process (MDP) framework is a widely used tool to model and solve control problems in dynamical systems (Thrun, Burgard, & Fox, 2005) that we will use to attack the SPRT mechanism. In these problems, an agent repeatedly interacts with an environment which is in a certain state. After each interaction, the agent receives a reward depending on its action and the system moves to a new state. An MDP is defined as:

Definition 1: A Markov Decision Process is a 5-tuple $\langle S,A,P,R,\gamma \rangle$, where:

- S is a finite set of states s.
- A is a finite set of actions a.

- $P_a(s^n, s^{n+1})$ is the probability that action a in state s^n and time n will lead to state s^{n+1} in time $n+1$.
- $R_a(s^n, s^{n+1})$ is the expected immediate reward received after transitioning from state s^n to state s^{n+1} due to action a.
- $\gamma \in (0,1)$ is a discount factor.

A key feature of MDPs is the Markovian property: the probability to reach state s in time n depends only on the previous state in time $n - 1$. In general, MDP can be of finite or infinite horizon, depending on whether the final time N is finite of infinite.

A policy π is a mapping π: $S \rightarrow A$ that associates an action (or a probability distribution over actions) to each state. It is possible to evaluate the performance of a policy π in order to obtain its total cumulative expected reward, i.e., $\sum_i \gamma^i R_a\left(s^i, s^{i+1}\right)$. Also, it is possible to find the optimal policy π^*, which is the policy which provides the highest possible cumulative reward. An MDP is solved when the optimal policy and its cumulative reward have been obtained.

MDP Solving Using Dynamic Programming

An MDP can be solved using the Dynamic Programming (DP) algorithm (Bertsekas, 2005), owed to Bellman. DP is based on optimizing backwards: the optimization starts with the final states and recursively updates a value function V^n that contains the optimal reward for each stage n. Mathematically, for $n=1,..,N$:

$$V^n\left(s^n\right) = \max_a \left\{ R_a\left(s^n, s^{n+1}\right) + \gamma \sum_{s^{n+1}} P_a\left(s^n, s^{n+1}\right) V^{n+1}\left(s^{n+1}\right) \right\} \tag{1}$$

The optimal policy π^* is obtained from (1) as:

$$\pi^*\left(s^n\right) = \arg\max_a \left\{ R_a\left(s^n, s^{n+1}\right) + \gamma \sum_{s^{n+1}} P_a\left(s^n, s^{n+1}\right) V^{n+1}\left(s^{n+1}\right) \right\} \tag{2}$$

Note that (2) can be unfeasible to obtain when N is large or even infinite: for the latter case, known as infinite horizon, there are algorithms that allow obtaining the optimal policy in an iterative way, such as Policy or Value Iteration (Sutton & Barto, 1998). Actually, the efficiency of these iterative methods causes that infinite horizon methods are a reasonable and widely used approximation for finite horizon problems (Bertsekas, 2005).

MDP Solving Using Reinforcement Learning

Reinforcement Learning (RL) is a branch of the Artificial Intelligence field which contains biologically inspired methods to make an agent learn an MDP simply by interacting with a dynamical system, as illustrated in Figure 2. A very complete introduction to the topic is given in (Sutton & Barto, 1998).

Figure 2. Reinforcement learning basic interaction scheme

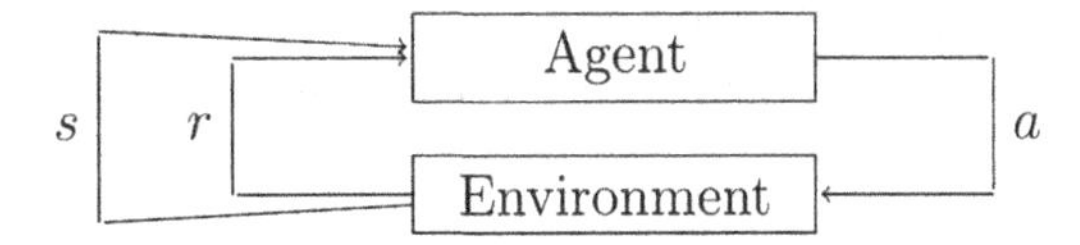

The agent is in state $s \in S$ and chooses an action $a \in A$, then the states transitions to a new state and provides the agent with a certain reward r. The agent learns a policy that maximizes the total reward. Note that RL approximates the optimal policy without knowing the transition probabilities $P_a(s^n, s^{n+1})$.

Q-Learning

There are many different RL algorithms, and one of the best known ones is Q-learning (Sutton & Barto, 1998). Q-learning relies on using the value function $V_\pi(s^n)$ from (1), particularized for a policy π as:

$$V_\pi\left(s^n\right) = R_{a\sim\pi}\left(s^n, s^{n+1}\right) + \gamma \sum_{s^{n+1}} P_{a\sim\pi}\left(s^n, s^{n+1}\right) V_\pi\left(s^{n+1}\right) \quad (3)$$

The state-action value function $Q_\pi(s^n,a)$ is an alternative value function formulation that represents the expected reward that the agent would obtain if it is in state s^n, takes action a and follows policy π afterwards as:

$$Q_\pi\left(s^n, a\right) = R_a\left(s^n, s^{n+1}\right) + \gamma \sum_{s^{n+1}} P_a\left(s^n, s^{n+1}\right) V_\pi\left(s^{n+1}\right) \quad (4)$$

where both value functions are related as: $V_\pi\left(s^n\right) = Q_\pi\left(s^n, \pi\left(s^n\right)\right)$ when π is deterministic.

Q-learning tries to learn a Q-function estimator: in case of discrete actions and spaces, this could be a table. First, the Q-function estimator is initialized as $Q\left(s^n, a\right) = 0, \forall s^n \in S, \forall a \in A$. Then, the agent starts interacting with the dynamical system: it observes the current state s^n, takes action a following an ϵ-greedy policy, in which the agent chooses to take the action that maximizes $Q(s^n,a)$ with probability $1 - \epsilon$ and with probability ϵ, it takes a random action. Low ϵ values may cause that the agent is stuck in poor local maxima of the total reward, as it explores seldom, while high ϵ values facilitate exploration at the cost of obtaining lower total rewards: it is frequent to start with high ϵ values and decrease them during training. To ease the notation, the subscript π from the Q-function is dropped. When the agent takes action a, the environment transitions to a new state s^{n+1} and it returns a reward r to the agent. The Q-function value for state s^n and action a is updated as:

$$Q\left(s^n, a\right) = Q\left(s^n, a\right) + \alpha\left(r + \gamma \max_{a'} Q\left(s^{n+1}, a'\right) - Q\left(s^n, a\right)\right) \quad (5)$$

where $\alpha \in [0,1]$ is a parameter that controls the learning rate: low values of α means a low Q update, but high values of α introduce a high variance on Q values. This algorithm converges in the limit to the actual Q-function under some mild convergence conditions (Sutton & Barto, 1998). A scheme of Q-learning algorithm is:

Initialize $Q(s^n,a)=0$
For n_r repetitions:
    Initialize $n=0$
    Set initial state s^0
    While State s^n is not final:
        Obtain action a using ϵ-greedy policy
        Take action a and obtain s^{n+1} and r
        Update $Q(s^n,a)$ using (5)
        Set $n=n+1$
        Use s^{n+1} for the next iteration
For $s^n \in S$:
    $\pi(s^n)^*=\text{argmax}_a Q(s^n,a)$

Deep Q-Networks

When the $S \times A$ space is large, Q-learning may be unfeasible as too much memory may be required to store the Q-function. In order to overcome this, Deep Q-Networks (DQN) (Mnih, et al., 2013), (Mnih, et al., 2015) approximate the Q-function using Deep Networks.

DQN is based on Q-learning, and the Q-function is replaced by a Deep Neural Network, whose input is the current state s^n and its output is the approximated Q-value function $Q(s^n,a)$ for each action. DQN uses experience replay to learn: the experience vector $e^n=(s^n,a^n,r^n,s^{n+1})$ is stored in a memory E, which is updated with each new interaction. The experience vectors contained in E are used to update the neural network multiple times: this allows a greater data efficiency, allows avoiding the correlation between consecutive samples and it helps avoiding oscillations or divergence in the network.

DQN also uses a target network, which is a clone of the neural network used to update the neural network at each iteration. As target networks are updated only at the end of each episode, they allow that DQN converges, avoiding oscillations in the learning process (Mnih, et al., 2015).

DEFENSE MECHANISMS AGAINST BACKOFF ATTACKS

As the backoff attack is a serious threat, there are many works which propose different defense mechanisms and approaches to counter it. Some works focus on the defense mechanism, such as (Wang, Sun, Li, & Han, 2010) and (Toledo & Wang, 2007). Also, it is possible to use game theory tools in order to study the conflict between agents: this approach is popular on WSN (Akkarajitsakul, Hossain, Niyato, & Kim, 2011), (Ghazvini, Movahedinia, Jamshidi, & Moghim, 2013), and has been used to study backoff attacks in wireless networks (Konorski, 2006), (Cagalj, Ganeriwal, Aad, & Hubaux, 2005), (Parras & Zazo, 2018), (Parras & Zazo, 2019b).

This Chapter follows the model from (Parras & Zazo, 2018) and (Parras & Zazo, 2019b). It models the backoff attack as a game, that is, a conflict between the ASs and a centralized defense mechanism. Each ASs may choose to transmit following the binary backoff procedure prescribed or following a random uniform distribution which provides the ASs with a higher throughput. The defense mechanism may choose to perform a test on the actions of the ASs or not: if the test is performed, the defense mechanism knows whether the ASs has followed the binary backoff procedure or not, but the defense mechanism incurs in a cost associated to the test. The defense mechanism may also choose not to perform this test, which means that it does not detect an AS deviating from the binary backoff procedure. If an AS does not follow the binary backoff procedure and it is detected by the defense mechanism, the packet is dropped, and the AS has to transmit again. Thus, note that the AS may have a gain in terms of throughput by deviating from the binary backoff procedure, but this gains also depends on how often the defense mechanisms performs a test on its actions, as each time that the AS is caught deviating, its throughput decreases. Also, observe that the defense mechanism must reach a compromise between testing sufficiently often as to enforce a fair distribution of the network throughput, while at the same time, minimizing the cost of each test.

Let us focus on the defense mechanism perspective: for each sensor, it observes a stream of binary data x_i^n, where n=0,1,2,... is the time index and i indexes the sensor, and each x_i^n represents the outcome of the test done to sensor i in time step n, where the time step denotes the index of the test. If a sensor follows the binary exponential backoff procedure, then x=0, and x=1 otherwise. In (Parras & Zazo, 2019b), it is shown that a solution to this problem involves reaching an agreement on the frequency with which each AS will deviate from the binary exponential backoff and the frequency with which the defense mechanism will test to detect a deviation. This means that $x_i^n \sim B(\theta_0)$, where $B(\theta_0)$ denotes the Bernoulli distribution of parameter θ_0 and $\theta_0 \in [0,1]$ is the negotiated frequency with which the AS can follow other backoff mechanism than the binary exponential one.

It is important to note that the conditions derived in (Parras & Zazo, 2019b) require detecting whether an AS deviates, i.e., whether $x_i^n \sim B(\theta \neq \theta_0)$. One possible way to address this problem is by using deterministic sequences or a pseudo-random number generator. Even though there are other possible solutions to this problem, as shown in Sections 3.7 and 3.8 in (Mailath & Samuelson, 2006), these two are presented because of their simplicity.

Deterministic Sequences

A possible solution consists is using deterministic sequences of actions that behave as the mixed actions would. This solution was proposed in (Fudenberg & Maskin, 1991). The idea behind this is that all players know what the other player is going to do in each time step, and hence, detecting a deviation is straightforward. One option is simply to generate a sufficiently large vector of actions, that is shared to all players. This allows all players knowing in advance which is going to be their payoffs. This solution can require a very large amount of memory to store the actions vector, thus, a different approach would use periodical sequences in order to require less memory. For instance, in (Fudenberg & Maskin, 1991), the authors derive an algorithm that obtains a periodical vector of actions which allows achieving rational mixed actions.

Thus, this solution requires that the players obtain a deterministic sequence of actions that allows them to detect any deviation instantaneously. This is a very simple and straightforward solution, with two main drawbacks. The first one consists in deciding who is going to obtain the sequences and how: all players must agree on this procedure. The second problem is that this solution requires sharing and storing in memory a potentially very large action vector, since a sequence can be arbitrarily long, even if the sequence is periodical because the periods can be very long. Thus, this simple solution turns out to require a potentially large amount of memory and communications capacity.

PRNG Based Correlator

Another solution consists in using a correlator device. One approach could be using a public randomization device, or equivalently, that all players allowed access to their randomization devices in order that all players could observe a deviation. This was suggested in (Fudenberg & Maskin, 1986). This solution, however, might be hard to implement, requiring either access to the randomizing device of each player or creating a trusted, centralized device that generates the mixed action for each player.

It is possible to overcome these problems by using a Pseudo-Random Number Generator (PRNG). A PRNG is an algorithm that allows obtaining a sequence of numbers that has similar properties to a sequence of random numbers. But the sequence can be completely determined by an initial value, called seed. If the seed is known, the whole sequence can be obtained. A good choice are Linear Congruential Generators (LCG), due to their simplicity. An LCG generates the sequence of numbers x^n using the following recursive formula (Hull & Dobell, 1962):

$$x^{n+1} = \left(bx^n + c\right) \mathrm{mod} m, \tag{6}$$

where b, c and m are the parameters of the LCG, and mod denotes the module operation. The value x^0 is the seed, which can take any value. In order to obtain $x \in [0,1)$, each value x^n is divided by m: by doing so, the values x/m approximately follow a uniform distribution in the interval [0,1). As it is shown in (Hull & Dobell, 1962), the sequence generated using (6) is periodical, with at most period m; and the sequence has period m for all seed values if and only if:

- m and c are relatively prime.
- $b - 1$ is divisible by all prime factors of m.
- $b - 1$ is divisible by 4 if m is divisible for 4.

The expression in (6) is fast to compute, and this is one of the main advantages of LCG, together with their low memory consumption. However, LCG also presents some known problems (Entacher, 1998), (L'Ecuyer & Simard, 1999), that make them not suitable for applications where high randomness quality is required, which is not the backoff attack case.

It is also possible to use of the Inverse Transform method, which allows transforming a uniform random variable $U \sim U[0,1]$ in a random variable following another distribution. Thus, a Bernoulli random variable X of parameter θ can be obtained from U as follows: $x=0$ if $U \leq 1 - \theta$ and $x=1$ if $U > 1 - \theta$. This procedure can be generalized to multinomial random variables. Observe that the mixed actions of a game are the outcomes of a binomial distribution when a player has two actions, and a multinomial in case that there she has three or more actions.

This procedure is easy to implement, computationally fast and does not require a high amount of memory. Note that all players need to use the same LCG parameters, which could be fixed beforehand. However, an important drawback comes from finite precision effects: if two players operate using different precisions, eventually, they will obtain different pseudorandom numbers x/m and this may lead to error. This problem could be alleviated by using, for instance, a periodical reset to the seed or sharing the x value periodically.

Sequential Tests

The two methods already presented allows detecting whether a sensor deviates from a prescribed θ_0. However, they imply modifying the network procedures, and hence, may not be an option in many realistic problems. A tool that the defense mechanism may use to detect deviations is a Hypothesis Test (HT): the defense mechanism has access to the stream of data x_i^t, and it must decide whether the behavior exhibited by the data stream corresponds to an expected behavior or to a behavior under attack. This is a problem that has been widely studied in the signal theory field (Kay, 1993). One of many possible taxonomies to classify HTs is based on the sample size: whether the needed number of samples to make a decision is fixed in advance or not. The former case is the most usual: the well-known Neyman-Pearson HT belongs to this kind (Neyman & Pearson, 1933), (Kay, 1993), among many others, such as Rao or Wald tests (Ciuonzo, De Maio, & Rossi, 2015). The latter case is known as sequential hypothesis test, and traces back to the work of Wald on the Sequential Probability Ratio Test (SPRT) (Wald, 1945), (Wald, 1973). In an SPRT, a sample from the signal of interest is collected at each time step n and used to update a statistic. The updated statistic may be used either to make a decision if there is enough information or to collect another sample. Note that SPRT is very attractive in many scenarios, as in WSN (Chen, Park, & Bian, 2008), (Shei & Su, 2008), in which the ability to make a decision requiring fewer communications among sensors means a lower battery and bandwidth consumption. SPRT also allows making a decision as soon as possible and adapts easily to working with online data.

Thus, it is not a surprise that many WSN mechanisms make use of SPRT when an HT is needed. It has been used for Cooperative Spectrum Sensing (CSS), in which several sensors send to a central entity their local spectrum sensing report and SPRT mechanisms are used to implement the information fusion (Shei & Su, 2008), (Chen, Park, & Bian, 2008), (Wu, Yu, Song, & Hu, 2018). In (Ho, Wright, & Das, 2011), (Das & Ho, 2011) and (Vamsi & Kant, 2014), SPRT is used for detecting sensors that have been compromised and replicated. In (Li, Liu, & Wang, 2010) and (Gara, Saad, & Ayed, 2017), SPRT is used to detect a selective forwarding attack, in which a compromised sensor drops packets. SPRT can also be used for Distributed Denial of Service (DDoS) attack detection (Dong, Du, Zhang, & Xu, 2016) and spam detection (Bhadre & Gothawal, 2014). Thus, SPRT finds many applications currently in WSN, especially when trying to detect a malicious behavior, and this justifies choosing it for the backoff problem.

Let us now describe an SPRT based detection mechanism which does not only apply to the backoff attack problem, but it is also very similar to the tests used in current literature, such as (Bhadre & Gothawal, 2014), (Ho, Wright, & Das, 2011), (Chen, Park, & Bian, 2008), (Vamsi & Kant, 2014), (Li, Liu, & Wang, 2010), (Dong, Du, Zhang, & Xu, 2016) or (Das & Ho, 2011). Even though concrete details differ between these works, the main lines of the detection mechanisms are similar to the mechanism introduced in this Chapter.

The rest of the Chapter focuses on Bernoulli distributions, which have many applications in the signal processing field. For instance, it appears in radar applications (Shnidman, 1998), pattern identification (Fernandez & Aridgides, 2003) and fusion in sensor networks (Niu & Varshney, 2005), (Niu & Varshney, 2008), (Ciuonzo & Rossi, 2014), (Rahaman & Khan, 2018). For the concrete case of Bernoulli distributions, the Neyman Pearson test reduces to the Counting Rule for equal confidences. Other tests which are also used are the Rao and Wald tests (Ciuonzo, De Maio, & Rossi, 2015). While all these works use fixed sample tests, this Chapter focuses on sequential tests.

The Detection Problem

Let us assume a discrete time signal x^n, where the subindex i is dropped for clarity of notation. The detection problem consists in deciding which of two hypotheses H_0 and H_1 best corresponds to the data stream. If each x^n follows a Bernoulli distribution of parameters θ_0 under H_0 and θ_1 under H_1, where θ_0 and θ_1 are known, the problem can be formulated as:

$$\begin{cases} H_0 : x^n \sim B(\theta_0), n = 0,1,2,... \\ H_1 : x^n \sim B(\theta_1), n = 0,1,2,... \end{cases} \tag{7}$$

In the backoff problem, note that a modification of the frequency with which the AS does not follow the backoff procedure provides a better payoff to it, at the expense of the network fair distribution of resources. Note that $\theta_1 > \theta_0$ means that the AS deviates more often from the binary exponential backoff mechanism, which causes it to have a gain in terms of throughput.

It is also important noting that having a Bernoulli test in which the malicious behavior corresponds to $\theta_1 > \theta_0$ happens often in the literature, for instance, in (Das & Ho, 2011), (Vamsi & Kant, 2014), (Li, Liu, & Wang, 2010) and (Bhadre & Gothawal, 2014). Thus, the attack strategies in this Chapter could be extended to many other attack situations in WSNs.

In an HT, there are two different errors: the type I error or false alarm probability, denoted by α, is the probability that H_0 is rejected, provided that H_0 is true; and the type II error, denoted by β, is the probability of accepting H_0, provided that H_1 is true. There is always a tradeoff between having a low false alarm probability and a high probability of correctly rejecting the null hypothesis.

The Counting Rule

A simple an efficient decision rule used in WSN is the counting rule: it is able to outperform more complex mechanisms (Ciuonzo & Rossi, 2014), because it is the universally most powerful fixed length test (Ciuonzo, De Maio, & Rossi, 2015). Its test statistic is $s^n = \sum_{i=0}^{n} x^i$ for a fixed value of n, and decides H_1 if $s^n \geq \delta$, where $P(s^n \geq \delta | H_0) \leq \alpha$ allows fixing the decision threshold δ as a function of α. A large n allows rejecting the null hypothesis with high accuracy, at the cost of a longer delay to decide and a larger number of communications required among sensors.

Sequential Probability Ratio Test

The SPRT for the Bernoulli distribution presents the following test statistic for the sample n:

$$LR^n = \frac{\theta_1^{s^n}(1-\theta_1)^{n+1-s^n}}{\theta_0^{s^n}(1-\theta_0)^{n+1-s^n}}, \tag{8}$$

where $s_n = \sum_{i=0}^{n} x^i$, $s^n \in [0,n+1]$ and $n \in \{0,1,\ldots\}$. By taking logarithms, it is possible to obtain the log-likelihood ratio:

$$LLR^n = s^n \log\left(\frac{\theta_1}{\theta_0}\right) + \left(n+1-s^n\right)\log\left(\frac{1-\theta_1}{1-\theta_0}\right), \tag{9}$$

and the decision rules of (9) are (Wald, 1973):

$$\begin{cases} \text{Reject } H_0 & \text{if} \quad LLR^n \geq h \\ \text{Accept } H_0 & \text{if} \quad LLR^n \leq l \\ \text{Take another sample} & \text{if} \quad \text{otherwise} \end{cases}, \tag{10}$$

where h and l are:

$$h = \log\left(\frac{1-\beta}{\alpha}\right), l = \log\left(\frac{\beta}{1-\alpha}\right). \tag{11}$$

The equation (10) shows that SPRT collects new samples until it has enough information to make a decision, which mathematically means that LLR^n surpasses a certain threshold: this is illustrated in Figure 3. Also, (9) can be rewritten as:

$$LLR^n = s^n \log\left(\frac{\theta_1(1-\theta_0))}{\theta_0(1-\theta_1)}\right) + (n+1)\log\left(\frac{1-\theta_1}{1-\theta_0}\right), \tag{12}$$

and by defining:

$$A = \log\left(\frac{\theta_1(1-\theta_0))}{\theta_0(1-\theta_1)}\right), B = \log\left(\frac{1-\theta_1}{1-\theta_0}\right), \tag{13}$$

it is possible to rewrite (11) as:

Figure 3. Illustration of an SPRT. The upper dotted line is h, the lower dotted line is l. The black continuous line is the LLR^n, the test statistic of the SPRT. The dashed line indicates N -1, the time in which a decision is made by the SPRT. In this case, since $LLR^n \geq h$, H_0 the test decision is to reject H_0. Note that in samples $n \leq 7$, SPRT does not have enough information to make a decision and hence, another sample is collected.

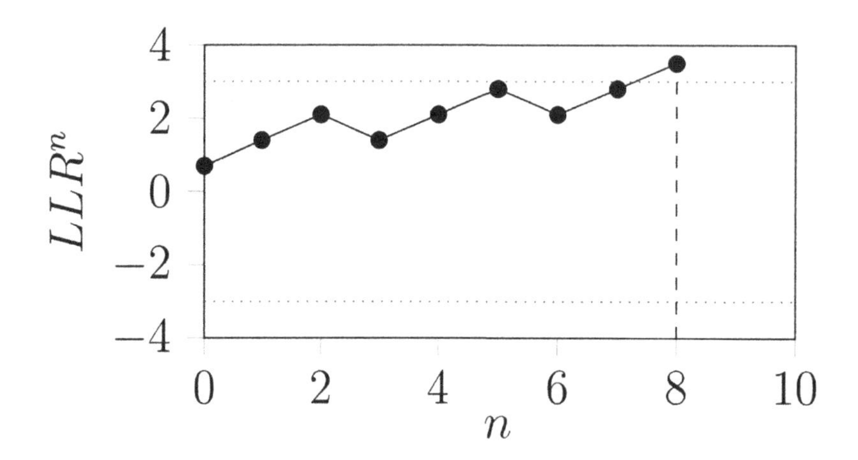

$$LLR^n = As^n + B(n+1). \tag{14}$$

Also, a sequential formulation for (9) can be obtained by noting that:

$$LLR^n = \begin{cases} LLR^{n-1} + A + B & \text{if} \quad x^n = 1 \\ LLR^{n-1} + B & \text{if} \quad x^n = 0 \end{cases}, \tag{15}$$

or even more compactly as:

$$LLR^n = LLR^{n-1} + B + Ax^n, LLR^0 = B. \tag{16}$$

Note that (16) facilitates a sequential implementation of the SPRT: in case of receiving $x^n=1$, the LLR^n adds up $A+B$, and in case that $x^n=0$, only B is added.

SPRT can be extended to deal with composite hypothesis (Lai, 2001), (Lai, 1988). For simplicity, this Chapter limits to the simple hypothesis case.

Overview of Attacks Against Sequential Probability Ratio Test

Although SPRT is widely used in defense mechanisms, it has a very dangerous underlying assumption because it considers that the statistical behavior of the signal x^n does not change with n. If the attacker can change x^n dynamically, it would be possible to exploit an SPRT based defense mechanism. However, in many works this limitation of SPRT is not considered. For instance, in (Ho, Wright, & Das, 2011), (Chen, Park, & Bian, 2008), (Vamsi & Kant, 2014), (Li, Liu, & Wang, 2010) or (Dong, Du, Zhang, &

Xu, 2016), SPRT is used in environments in which the attackers may use a dynamic attack strategy and hence, compromise the defense mechanism. As the next Sections show, this is a dangerous assumption that allows intelligent attackers to exploit the defense mechanism even if they do not have full information of it.

ATTACKS AGAINST SPRT

Now, let us turn to derive attack rules against SPRT. The incoming Sections focus on the interaction between a single AS and the defense mechanism for simplicity. As a malicious behavior means that the AS uses $x^n=1$ as often as possible, the AS receives an instantaneous reward of +1 each time that $x^n=1$: note that the results hold for any positive reward. The AS tries to maximize its total discounted reward R, defined as:

$$R\left(x^n\right)=\sum_{n=0}^{\infty}\gamma^n x^n \tag{17}$$

where $\gamma\in(0,1)$ is the discount factor that gives more weight to the rewards obtained in closer time steps than in the future. The use of γ fits volatile environments such as WSN, as an AS does not know how long it will be able to attack and thus, it cannot be infinitely patient. Also, γ allows that the total reward remains finite, as the minimum value for R using (17) is $R=0$ for $x^n=0$, $\forall n$, and the maximum value is $R=(1-\gamma)^{-1}$, for $x^n=1$, $\forall n$, where these values are obtained by using:

$$\sum_{n=n_0}^{n_1}\gamma^n=\frac{\gamma^{n_0}-\gamma^{n_1+1}}{1-\gamma},\gamma\neq 1. \tag{18}$$

Optimal Attack Strategy

It is possible to obtain the optimal control that the AS should use in order to maximize its reward when facing an SPRT detection mechanism by assuming that the AS knows all the parameters of the defense mechanism. It is possible to obtain the optimal attack by solving the following control problem, where (18) is used and $N-1$ is the time in which the SPRT makes a decision:

$$\begin{aligned}
\max_{x^n}\quad & \sum_{n=0}^{\infty}\gamma^n x^n=\sum_{n=0}^{N-1}\gamma^n x^n+\frac{\gamma^N}{1-\gamma}\\
\text{s.t.}\quad & x^n\in\{0,1\},s^n=\sum_{i=0}^{n}x^i\\
& LLR^n<h,\forall n\leq N-1\\
& LLR^{N-1}\leq l<h
\end{aligned} \tag{19}$$

Note that in (19):

- The function that the AS needs to maximize is split in two terms. The first term refers to the timesteps in which the SPRT detection mechanism is active, in which the AS needs to find an optimal control law for x^n such that it is not discovered. The second term includes the timesteps after a decision is made: at these timesteps the SPRT mechanism is not active, and hence, the AS can always use $x^n=1$.
- The constraint $LLR^n<h$ enforces that the AS is not discovered by the SPRT mechanism.
- The constraint LLR^{N-1} simply indicates that, at timestep $N-1$, in which a decision is made, H_0 is accepted.
- It is necessary to assume that $l<LLR^0<h$, that is, that the LLR initial value does not allow SPRT to make a decision. This condition is usually satisfied by most standard SPRT parameters; using (13) and (14), this condition turns out to be:

$$\frac{\beta}{1-\alpha}<\frac{1-\theta_1}{1-\theta_0}<\frac{1-\beta}{\alpha}. \tag{20}$$

Note that this means also that $l<h$ in the last constraint.

It is important noting that the problem formulated in (19) can be modeled using the MDP framework already introduced, where the states are $s^n=LLR^n$, the actions are $a^n=x^n$, the reward function is $r(s^n,a^n)=a^n$ and the transition function is deterministic and can be obtained using (11). Thus, it is possible to apply Dynamic Programming methods to obtain the optimal policy. In other words, SPRT can be modeled using MDP tools: the target of the AS is finding a policy such that it can attack and camouflage while the SPRT mechanism is running. When the SPRT mechanism is not running, i.e., $n\geq N$, the AS can attack without needing to camouflage: under the MDP notation, when the final state LLR^{N-1} is reached, the AS always uses the attack action. Note that the actions of the AS will determine the final time N, which means that N is another value to optimize.

An important result is that SPRT can be successfully exploited by a dynamic attacker, as the next Theorem, whose proof in in (Parras & Zazo, 2019c), shows:

Theorem 1: Consider the discrete time control problem described in (19), in which the controller chooses x^n and $N-1$. In this problem, the optimal control for $n\in[0,N-1]$ is:

$$\begin{cases} x^n=1 & \text{if} \quad LLR^{n-1}+A+B<h \\ x^n=0 & \text{if} \quad \text{otherwise} \end{cases}$$

By reminding Figure 3, the optimal control from Theorem 1 makes the AS use $x^n=1$ as much as possible while never surpassing the upper threshold, so that the AS is never detected.

Learning an Attack Strategy

The previous Section assumed that the AS knew all the parameters of the defense mechanism, i.e., α, β, θ_1 and N, and hence, it is possible to derive an optimal attack law using Theorem 1. However, in many realistic problems, the AS may not know these parameters. However, Deep RL allows an AS to learn to attack the defense mechanism simply by interacting with it: thus, the AS does not need to know the parameters of the defense mechanism.

As mentioned before, the control problem (19) can be modeled using the MDP framework, where:

- The states are $s^n=LLR^n$, which are continuous.
- The actions are $a^n=x^n$, and note that $A=\{0,1\}$, i.e., the actions are binary variables.
- The rewards are $r(s^n,a^n)=x^n$, i.e., the AS is rewarded when using $x^n=1$. In the backoff problem, this means that the AS is rewarded each time that deviates from the binary backoff procedure.

As the problem has continuous states and discrete actions, it is not possible using Q-learning as presented in this Chapter, but it is possible to use the DQN algorithm to approximately solve the control problem. It is important to emphasize that the AS does not know the concrete parameters of the defense mechanism: it only observes the state and its reward.

Empirical Results: Intelligent Attacks Against the Backoff Defense Mechanism

It is time to evaluate the impact that the attacks described would have on the backoff defense mechanism using SPRT. Let us assume that there is a single AS in the WSN, whose actions are a binary variable that reflect whether the AS follows the binary exponential backoff or not. In all simulations, $\alpha=\beta=0.05$, and for simplicity, $\theta_0=0.2$ and $\theta_1=\theta_0+\{0.05,0.1,0.2\}$ in order to test several cases.

There are several attack situations possible:

- The case in which there is no attack, i.e., the AS behaves as a GS, which means that $x^n\sim B(\theta_0)$.
- The case in which there is a naive attack, i.e., the AS uses $x^n\sim B(\theta_1)$.
- The optimal attacker, i.e., the AS follows the optimal attack using Theorem 1.
- The Deep RL based attacker which uses DQN.

For practical reasons, all simulations are interrupted after $N=200$ SPRT stages; note that the test might not have decided by that time.

In DQN, the Q-function estimator is implemented using a three layers neural network, where each of the three layers is fully connected to its neighbors. The first layer has as input size the state size, i.e., 1, it has 24 neurons and uses rectifier linear units for activation, i.e., they follow the function $f(x)=\max(0,x)$. The second layer has also 24 neurons and uses rectifier linear units for activation. The final layer has 24 neurons, an output size equal to the possible number of actions, i.e., 2, and it uses linear units for activation, i.e., $f(x)=x$. The loss function used is the mean squared error, and adam is chosen as optimizer (Kingma & Ba, 2015). The maximum number of experiences stored, i.e., the dimension of E, is of 2000 experience vectors. After a new experience e^n has been obtained and added to the set E, a batch of 64 experience vectors is randomly picked and used to train the neural network. The policy used is ϵ-greedy with variable ϵ: the initial value is $\epsilon=1$ and after each step n, a new epsilon is obtained as $\epsilon_{n+1}=0.999\epsilon_n$, with a minimum value ϵ=0.01. DQN is trained using 300 episodes, where each episode consists on an SPRT test that either makes a decision before reaching 200 samples or ends in the sample 200. The whole DQN training is repeated using 10 different seeds in order to average the results. With these parameters, DQN algorithm converged successfully as shown in Figure 4.

Figure 4. DQN training curves example in seed 0 for the case in which θ_1=0.25. The solid line corresponds to γ=0.9, the dashed line to γ=0.7 and the dotted line to γ=0.5. DQN converges in all cases, as the rewards become stable with the number of episodes. Note that low γ values lead to smaller total rewards because the agent optimizes myopically: as γ increases, the agent maximizes on the long run, and hence, better total rewards are obtained.

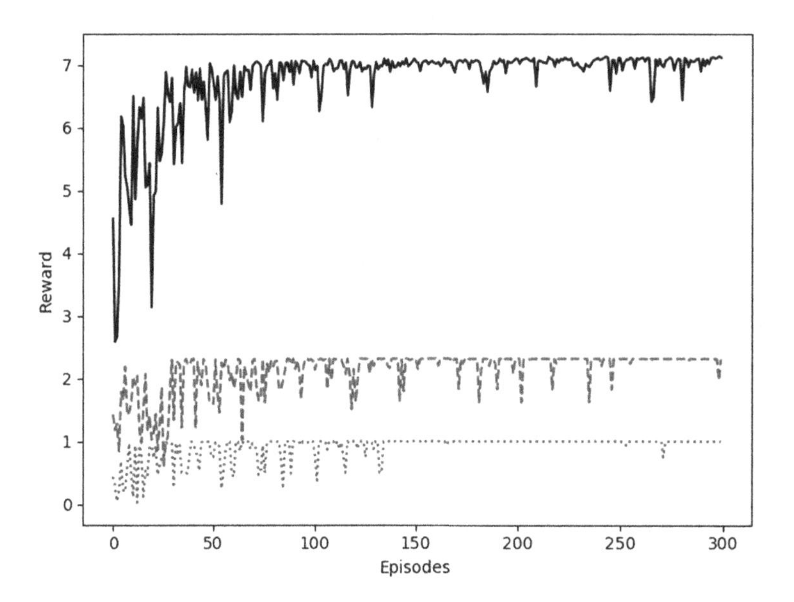

For each attack, 500 runs of each test are averaged and the results are in Table 1, where it is possible to observe that:

- In case that there is no attack, the result of the SPRT correctly detects that $x^n \sim H_0$. However, note that as $\theta_1 - \theta_0$ decreases, the number of samples required to make a decision increases and hence, the number of tests which have not made a decision after 200 samples increases. This is an expected result: when both θ values are close, the decision is harder to make as the distributions under both hypotheses overlap significantly.
- The naive attack is easily detected in most cases, especially as $\theta_1 - \theta_0$ increases. Note that, again, as $\theta_1 - \theta_0$ increases, the number of samples required to make a decision decreases. Observe that in all cases, the reward obtained by the naive attack is larger than the reward obtained if there is no attack: hence, even a naive attack is an option for an attacker against an SPRT.
- The optimal attack is able to successfully overcome the defense mechanism: it is never detected and its reward is significantly higher than in the other two cases, for all γ and θ_1 values.

Table 1. Results for the backoff attack detection problem using SPRT, when θ_0=0.2 and θ_1=θ_0+{0.05,0.1,0.2}. H_0, H_1 and ND are the probabilities that the SPRT decides H_0, rejects H_0 and does not reach a decision respectively. Length is the average samples needed to make a decision: in the DQN case, the test lengths for each γ value are shown, as the differences are significant. R is the total reward (17), for different values of γ. Note that the optimal attack and DQN are able to successfully overcome the SPRT based defense mechanism, providing larger rewards than the naive attack.

θ1	Attack	H_0	H_1	ND	Test length	R, γ=0.5	R, γ=0.7	R, γ=0.9
0.25	No attack	0.256	0.01	0.734	185.42	0.39	0.66	1.98
	Naive	0.016	0.286	0.698	182.01	0.48	0.82	2.48
	Optimal	0	0	1	200.0	2.00	3.31	7.94
	DQN	0	0.83	0.17	[13,13,127]	1.00	2.32	7.14
0.3	No attack	0.854	0.042	0.104	99.02	0.36	0.61	1.95
	Naive	0.034	0.854	0.112	93.83	0.62	1.02	3.00
	Optimal	0	0	1	200.0	1.99	3.10	6.23
	DQN	0	0.83	0.17	[7,7,121.5]	1.00	2.20	5.61
0.4	No attack	0.964	0.036	0.0	30.33	0.44	0.70	2.00
	Naive	0.044	0.956	0.0	26.88	0.81	1.32	3.44
	Optimal	0	0	1	200.0	1.89	2.69	5.15
	DQN	0.03	0.7	0.27	[4,4,168.7]	0.97	1.94	4.63

- Regarding DQN, its behavior strongly depends on the value of the discount factor γ. For low γ value, it is not able to learn a satisfactory strategy, as it focuses on maximizing the return myopically rather than on the long run. However, for γ=0.9 and as θ_1 increases, DQN is able to obtain an attack which is significantly closer to the optimal value.

Hence, SPRT is vulnerable to intelligent attackers, that may know or not all the system parameters. Note that all attacks proposed are successful in that the reward obtained is higher than in the no attack case, and thus, an SPRT mechanism does motivate an intelligent AS to deviate.

In current literature, there are several tools designed for dealing with changes in the statistical behavior of a signal, such as quickest detection tools (Poor & Hadjiliadis, 2009), or, for discrete time signals, repeated hypothesis tests (Basseville, Nikiforov, & others, 1993). The optimal attacker proposed could be detected using misbehavior tools, such as One Class Supporting Vector Machines, as in (Parras & Zazo, 2019c). However, even though this method may detect the optimal attacker, it may not deal with a Deep RL based attacker, as they adapt to the defense mechanism simply by interacting with it, as the simulations show.

CONCLUSION AND FUTURE WORK

In this Chapter, a CSMA/CA based WSN under a backoff attack is studied: some sensors deviate from the defined contention mechanism and this causes the network throughput not to be fairly distributed.

This impact is assessed using Bianchi's model, where it is possible to observe that having ASs do affect having a fair distribution of the network throughput.

Then, a defense mechanism against the backoff attack is proposed, which relies on detecting deviations in the frequency with which an AS deviates from the binary backoff procedure. The key methods studied are sequential tests, in which a decision is made on whether a data stream follows a certain distribution or not. One popular sequential test is SPRT, which is the base for many current WSN defense mechanisms.

However, as noted, SPRT has a dangerous underlying assumption, which is that the distribution of the data stream does not change with time. A dynamic, intelligent attacker as the ones proposed in this Chapter is able to successfully exploit an SPRT procedure, both when the defense parameters are known and unknown to the attacker: in the first case, optimal control tools are used, and in the second, Deep RL tools are employed to obtain an attack policy. Namely, DQN algorithm is used on the backoff attack setup and it is shown that an attacker using these tools is able to obtain good attack policies against the SPRT defense mechanism. It is important to remark that the main threat of Deep RL tools is that they learn simply by interacting with the defense mechanism, and hence, they may adapt potentially to many defense mechanisms: another work which highlights this fact is (Parras & Zazo, 2019a).

Thus, recent advances in Deep Learning, and more concretely, in Deep RL, have brought new threats to WSNs. As indicated, a lot of effort is devoted nowadays to WSN research (Yang, 2014), (Rawat, Singh, Chaouchi, & Bonnin, 2014), (Ndiaye, Hancke, & Abu-Mahfouz, 2017), where security is one of the key challenges addressed. On one side, the communication protocols and standards used in WSN include security solutions, but most of them are still at a proof-of-concept level according to (Tomić & McCann, 2017). On the other side, the existing defense mechanisms are addressed to concrete attacks in concrete setups, and hence, an attacker based on Deep RL is flexible enough as to be a potential threat to these defense mechanisms.

Hence, the attacking approach that proposed in this Chapter presents strong challenges to current WSN defense mechanisms. First, because of the growing computational capabilities of current hardware, there could be soon, if not already, sensors with enough computational resources to implement such an attacker (Payal, Rai, & Reddy, 2015). Second, because using Deep RL tools is an approach that is adaptive and flexible, not requiring an a priori modeling of the defense mechanism nor knowledge of their parameters, and hence, it can learn to exploit a wide range of defense mechanisms. Thus, it is of capital importance researching on defense mechanisms against such attack mechanisms, in order to minimize the threat they pose. A promising defense mechanism could be one in which the defense mechanism also learns how to defend, which means entering the field of Multi Agent Competitive Learning (Hernandez-Leal, Kaisers, Baarslag, & de Cote, 2017), which until today poses strong challenges.

REFERENCES

Akkarajitsakul, K., Hossain, E., Niyato, D., & Kim, D. I. (2011). Game theoretic approaches for multiple access in wireless networks: A survey. *IEEE Communications Surveys and Tutorials*, *13*(3), 372–395. doi:10.1109/SURV.2011.122310.000119

Basseville, M., Nikiforov, I. V., & ... (1993). *Detection of abrupt changes: theory and application* (Vol. 104). Prentice Hall Englewood Cliffs.

Bayraktaroglu, E., King, C., Liu, X., Noubir, G., Rajaraman, R., & Thapa, B. (2013). Performance of IEEE 802.11 under jamming. *Mobile Networks and Applications*, *18*(5), 678–696. doi:10.100711036-011-0340-4

Bertsekas, D. P. (2005). *Dynamic programming and optimal control* (Vol. 1). Athena Scientific.

Bhadre, P., & Gothawal, D. (2014). Detection and blocking of spammers using SPOT detection algorithm. *Networks & Soft Computing (ICNSC), 2014 First International Conference on*, 97-101.

Bianchi, G. (2000). Performance analysis of the IEEE 802.11 distributed coordination function. *IEEE Journal on Selected Areas in Communications*, *18*(3), 535–547. doi:10.1109/49.840210

Buehrer, R. M. (2006). *Synthesis Lectures on Communications*. Morgan & Claypool Publishers.

Cagalj, M., Ganeriwal, S., Aad, I., & Hubaux, J.-P. (2005). On selfish behavior in CSMA/CA networks. *Proceedings IEEE 24th Annual Joint Conference of the IEEE Computer and Communications Societies*, *4*, 2513-2524. 10.1109/INFCOM.2005.1498536

Chen, R., Park, J.-M., & Bian, K. (2008). Robust distributed spectrum sensing in cognitive radio networks. *INFOCOM 2008. The 27th Conference on Computer Communications,* 1876-1884.

Ciuonzo, D., De Maio, A., & Rossi, P. S. (2015). A systematic framework for composite hypothesis testing of independent Bernoulli trials. *IEEE Signal Processing Letters*, *22*(9), 1249–1253. doi:10.1109/LSP.2015.2395811

Ciuonzo, D., & Rossi, P. S. (2014). Decision fusion with unknown sensor detection probability. *IEEE Signal Processing Letters*, *21*(2), 208–212. doi:10.1109/LSP.2013.2295054

Das, S. K., & Ho, J.-W. (2011). A synopsis on node compromise detection in wireless sensor networks using sequential analysis (Invited Review Article). *Computer Communications*, *34*(17), 2003–2012. doi:10.1016/j.comcom.2011.07.004

Demirkol, I., Ersoy, C., & Alagoz, F. (2006). MAC protocols for wireless sensor networks: A survey. *IEEE Communications Magazine*, *44*(4), 115–121. doi:10.1109/MCOM.2006.1632658

Dong, P., Du, X., Zhang, H., & Xu, T. (2016). A detection method for a novel DDoS attack against SDN controllers by vast new low-traffic flows. *Communications (ICC), 2016 IEEE International Conference on*, 1-6.

Entacher, K. (1998). Bad subsequences of well-known linear congruential pseudorandom number generators. *ACM Transactions on Modeling and Computer Simulation*, *8*(1), 61–70. doi:10.1145/272991.273009

Enz, C. C., El-Hoiydi, A., Decotignie, J.-D., & Peiris, V. (2004). WiseNET: An ultralow-power wireless sensor network solution. *Computer*, *37*(8), 62–70. doi:10.1109/MC.2004.109

Fernandez, M. F., & Aridgides, T. (2003). Measures for evaluating sea mine identification processing performance and the enhancements provided by fusing multisensor/multiprocess data via an M-out-of-N voting scheme. *Detection and Remediation Technologies for Mines and Minelike Targets VIII, 5089*, 425-437.

Fragkiadakis, A. G., Tragos, E. Z., & Askoxylakis, I. G. (2013). A survey on security threats and detection techniques in cognitive radio networks. *IEEE Communications Surveys and Tutorials*, *15*(1), 428–445. doi:10.1109/SURV.2011.122211.00162

Fudenberg, D., & Maskin, E. (1986). The folk theorem in repeated games with discounting or with incomplete information. *Econometrica*, *54*(3), 533–554. doi:10.2307/1911307

Fudenberg, D., & Maskin, E. (1991). On the dispensability of public randomization in discounted repeated games. *Journal of Economic Theory*, *53*(2), 428–438. doi:10.1016/0022-0531(91)90163-X

Gara, F., Saad, L. B., & Ayed, R. B. (2017). An intrusion detection system for selective forwarding attack in IPv6-based mobile WSNs. *2017 13th International Wireless Communications and Mobile Computing Conference (IWCMC)*, 276-281.

Ghazvini, M., Movahedinia, N., Jamshidi, K., & Moghim, N. (2013). Game theory applications in CSMA methods. *IEEE Communications Surveys and Tutorials*, *15*(3), 1062–1087. doi:10.1109/SURV.2012.111412.00167

Hernandez-Leal, P., Kaisers, M., Baarslag, T., & de Cote, E. M. (2017). *A Survey of Learning in Multiagent Environments: Dealing with Non-Stationarity.* arXiv preprint arXiv:1707.09183

Ho, J.-W., Wright, M., & Das, S. K. (2011). Fast detection of mobile replica node attacks in wireless sensor networks using sequential hypothesis testing. *IEEE Transactions on Mobile Computing*, *10*(6), 767–782. doi:10.1109/TMC.2010.213

Hull, T. E., & Dobell, A. R. (1962). Random number generators. *SIAM Review*, *4*(3), 230–254. doi:10.1137/1004061

IEEE. (2016). *IEEE Standard for Information technology--Telecommunications and information exchange between systems Local and metropolitan area networks--Specific requirements - Part 11: Wireless LAN Medium Access Control (MAC) and Physical Layer (PHY) Specifications.*

Kay, S. M. (1993). *Fundamentals of Statistical Signal Processing: Estimation Theory*. Prentice-Hall, Inc.

Kingma, D. P., & Ba, J. (2015). Adam: A Method for Stochastic Optimization. *3rd International Conference on Learning Representations, ICLR.*

Konorski, J. (2006). A game-theoretic study of CSMA/CA under a backoff attack. *IEEE/ACM Transactions on Networking*, *14*(6), 1167–1178. doi:10.1109/TNET.2006.886298

L'Ecuyer, P., & Simard, R. (1999). Beware of linear congruential generators with multipliers of the form a=±2 q±2 r. *ACM Transactions on Mathematical Software*, *25*(3), 367–374. doi:10.1145/326147.326156

Lai, T. L. (1988). Nearly Optimal Sequential Tests of Composite Hypotheses. *Annals of Statistics*, *16*(2), 856–886. doi:10.1214/aos/1176350840

Lai, T. L. (2001). Sequential analysis: Some classical problems and new challenges. *Statistica Sinica*, *11*, 303–351.

Li, G., Liu, X., & Wang, C. (2010). A sequential mesh test based selective forwarding attack detection scheme in wireless sensor networks. *Networking, Sensing and Control (ICNSC), 2010 International Conference on*, 554-558.

Lin, P., Qiao, C., & Wang, X. (2004). Medium access control with a dynamic duty cycle for sensor networks. *Wireless Communications and Networking Conference, 2004. WCNC. 2004 IEEE, 3*, 1534-1539. 10.1109/WCNC.2004.1311671

Mailath, G. J., & Samuelson, L. (2006). *Repeated games and reputations: long-run relationships*. Oxford university press. doi:10.1093/acprof:oso/9780195300796.001.0001

Malone, D., Duffy, K., & Leith, D. (2007). Modeling the 802.11 distributed coordination function in nonsaturated heterogeneous conditions. *IEEE/ACM Transactions on Networking*, *15*(1), 159–172. doi:10.1109/TNET.2006.890136

Mnih, V., Kavukcuoglu, K., Silver, D., Graves, A., Antonoglou, I., Wierstra, D., & Riedmiller, M. (2013). *Playing atari with deep reinforcement learning*. arXiv preprint arXiv:1312.5602

Mnih, V., Kavukcuoglu, K., Silver, D., Rusu, A. A., Veness, J., Bellemare, M. G., Graves, A., Riedmiller, M., Fidjeland, A. K., Ostrovski, G., Petersen, S., Beattie, C., Sadik, A., Antonoglou, I., King, H., Kumaran, D., Wierstra, D., Legg, S., & Hassabis, D. (2015). Human-level control through deep reinforcement learning. *Nature*, *518*(7540), 529–533. doi:10.1038/nature14236 PMID:25719670

Ndiaye, M., Hancke, G. P., & Abu-Mahfouz, A. M. (2017). Software defined networking for improved wireless sensor network management: A survey. *Sensors (Basel)*, *17*(5), 1031. doi:10.339017051031 PMID:28471390

Neyman, J., & Pearson, E. S. (1933). IX. On the problem of the most efficient tests of statistical hypotheses. *Philosophical Transactions of the Royal Society of London. Series A, Mathematical and Physical Sciences*, *231*, 289–337.

Niu, R., & Varshney, P. K. (2005). Decision fusion in a wireless sensor network with a random number of sensors. *Acoustics, Speech, and Signal Processing, 2005. Proceedings.(ICASSP'05). IEEE International Conference on, 4*.

Niu, R., & Varshney, P. K. (2008). Performance analysis of distributed detection in a random sensor field. *IEEE Transactions on Signal Processing*, *56*(1), 339–349. doi:10.1109/TSP.2007.906770

Parras, J., & Zazo, S. (2018). Wireless Networks under a Backoff Attack: A Game Theoretical Perspective. *Sensors (Basel)*, *18*(2), 404. doi:10.339018020404 PMID:29385752

Parras, J., & Zazo, S. (2019a). Learning attack mechanisms in Wireless Sensor Networks using Markov Decision Processes. *Expert Systems with Applications*, *122*, 376–387. doi:10.1016/j.eswa.2019.01.023

Parras, J., & Zazo, S. (2019b). Repeated Game Analysis of a CSMA/CA Network under a Backoff Attack. *Sensors (Basel)*, *19*(24), 5393. doi:10.339019245393 PMID:31817778

Parras, J., & Zazo, S. (2019c). Using one class SVM to counter intelligent attacks against an SPRT defense mechanism. *Ad Hoc Networks*, *94*, 101946. doi:10.1016/j.adhoc.2019.101946

Payal, A., Rai, C. S., & Reddy, B. R. (2015). Analysis of some feedforward artificial neural network training algorithms for developing localization framework in wireless sensor networks. *Wireless Personal Communications*, *82*(4), 2519–2536. doi:10.100711277-015-2362-x

Poor, H. V., & Hadjiliadis, O. (2009). *Quickest detection* (Vol. 40). Cambridge University Press Cambridge.

Rahaman, M. F., & Khan, M. Z. (2018). Low-Complexity Optimal Hard Decision Fusion Under the Neyman—Pearson Criterion. *IEEE Signal Processing Letters*, *25*(3), 353–357. doi:10.1109/LSP.2017.2766245

Rahman, A., & Gburzynski, P. (2006). Hidden problems with the hidden node problem. *23rd Biennial Symposium on Communications*, 270-273. 10.1109/BSC.2006.1644620

Rawat, P., Singh, K. D., Chaouchi, H., & Bonnin, J. M. (2014). Wireless sensor networks: A survey on recent developments and potential synergies. *The Journal of Supercomputing*, *68*(1), 1–48. doi:10.100711227-013-1021-9

Shei, Y., & Su, Y. T. (2008). A sequential test based cooperative spectrum sensing scheme for cognitive radios. *Personal, Indoor and Mobile Radio Communications, 2008. PIMRC 2008. IEEE 19th International Symposium on*, 1-5.

Shnidman, D. A. (1998). Binary integration for Swerling target fluctuations. *IEEE Transactions on Aerospace and Electronic Systems*, *34*(3), 1043–1053. doi:10.1109/7.705926

Sutton, R. S., & Barto, A. G. (1998). *Reinforcement learning: An introduction.* MIT Press.

Thrun, S., Burgard, W., & Fox, D. (2005). *Probabilistic robotics*. MIT Press.

Toledo, A. L., & Wang, X. (2007). Robust detection of selfish misbehavior in wireless networks. *IEEE Journal on Selected Areas in Communications*, *25*(6), 1124–1134. doi:10.1109/JSAC.2007.070807

Tomić, I., & McCann, J. A. (2017). A Survey of Potential Security Issues in Existing Wireless Sensor Network Protocols. *IEEE Internet of Things Journal*, *4*(6), 1910–1923. doi:10.1109/JIOT.2017.2749883

Vamsi, P. R., & Kant, K. (2014). A lightweight sybil attack detection framework for wireless sensor networks. *Contemporary computing (IC3), 2014 Seventh International conference on*, 387-393.

Van Dam, T., & Langendoen, K. (2003). An adaptive energy-efficient MAC protocol for wireless sensor networks. *Proceedings of the 1st international conference on Embedded networked sensor systems*, 171-180. 10.1145/958491.958512

Wald, A. (1945). Statistical decision functions which minimize the maximum risk. *Annals of Mathematics*, *46*(2), 265–280. doi:10.2307/1969022

Wald, A. (1973). *Sequential analysis*. Courier Corporation.

Wang, W., Sun, Y., Li, H., & Han, Z. (2010). Cross-layer attack and defense in cognitive radio networks. In *Global Telecommunications Conference (GLOBECOM 2010), 2010 IEEE* (pp. 1-6). IEEE. 10.1109/GLOCOM.2010.5684069

Wu, J., Yu, Y., Song, T., & Hu, J. (2018). Sequential 0/1 for Cooperative Spectrum Sensing in the Presence of Strategic Byzantine Attack. *IEEE Wireless Communications Letters*.

Yadav, R., Varma, S., Malaviya, N., & others. (2009). A survey of MAC protocols for wireless sensor networks. *UbiCC Journal, 4*, 827-833.

Yang, K. (2014). *Wireless sensor networks*. Springer. doi:10.1007/978-1-4471-5505-8

Ye, W., Heidemann, J., & Estrin, D. (2004). Medium access control with coordinated adaptive sleeping for wireless sensor networks. *IEEE/ACM Transactions on Networking, 12*(3), 493–506. doi:10.1109/TNET.2004.828953

Zhang, L., Ding, G., Wu, Q., Zou, Y., Han, Z., & Wang, J. (2015). Byzantine attack and defense in cognitive radio networks: A survey. *IEEE Communications Surveys and Tutorials, 17*(3), 1342–1363. doi:10.1109/COMST.2015.2422735

KEY TERMS AND DEFINITIONS

Attacking Sensor: Sensor in a CSMA/CA WSN that may choose between respecting the binary exponential backoff procedure or using a uniform backoff procedure that provides the sensor with an advantage in terms of throughput.

Deep Neural Network: Graph structure which provides as output a nonlinear combination of its inputs. It has been proven to be a universal function approximator.

Good Sensor: Sensor in a CSMA/CA WSN that always respects the binary exponential backoff procedure.

Hypothesis Test: Mathematical tool used to decide whether a certain data follows a certain distribution or not.

Markov Decision Process: Mathematical framework used to model dynamical systems and obtain their optimal control policies.

Reinforcement Learning: Brach of the Artificial Intelligence field devoted to obtaining optimal control sequences for agents only by interacting with a concrete dynamical system.

Sequential Probability Ratio Test: Hypothesis test of variable sample size in which the test statistic is updated as new samples arrive and makes a decision once it has collected enough information.

Chapter 7
Attacks, Vulnerabilities, and Their Countermeasures in Wireless Sensor Networks

G. Jaspher Willsie Kathrine
https://orcid.org/0000-0003-3055-0356
Karunya Institute of Technology and Sciences, India

C. Willson Joseph
Sahrdaya College of Engineering and Technology, India

ABSTRACT

Wireless sensor network (WSN) comprises sensor nodes that have the capability to sense and compute. Due to their availability and minimal cost compared to traditional networks, WSN is used broadly. The need for sensor networks increases quickly as they are more likely to experience security attacks. There are many attacks and vulnerabilities in WSN. The sensor nodes have issues like limited resources of memory and power and undependable communication medium, which is further complicated in unattended environments, secure communication, and data transmission issues. Due to the complexity in establishing and maintaining the wireless sensor networks, the traditional security solutions if implemented will prove to be inefficient for the dynamic nature of the wireless sensor networks. Since recent times, the advance of smart cities and everything smart, wireless sensor nodes have become an integral part of the internet of things and their related paradigms. This chapter discusses the known attacks, vulnerabilities, and countermeasures existing in wireless sensor networks.

DOI: 10.4018/978-1-7998-5068-7.ch007

INTRODUCTION

Wireless Sensor Networks

WSN are a type of wireless network which consist of small, battery powered sensor nodes. These are used in unmanned places where there exists a need for data collection without any kind of human intervention. WSN gathers and communicates the data through wireless communication as shown in figure 1. The sensor nodes are very small but contain power, storage and communication facilities which are in turn limited as discussed by Akyildiz I.F., Su W., Sankarasubramaniam Y. and Cayirci E (2002).

Further advanced development in wireless communication has empowered the development of low cost, minimal power and multifunctional sensor nodes. Each sensor node has sensing capability, data processing unit and communicating components. The Sensor network uses protocols and algorithms that possess self- organizing capabilities.

Figure 1. Wireless Sensor Network

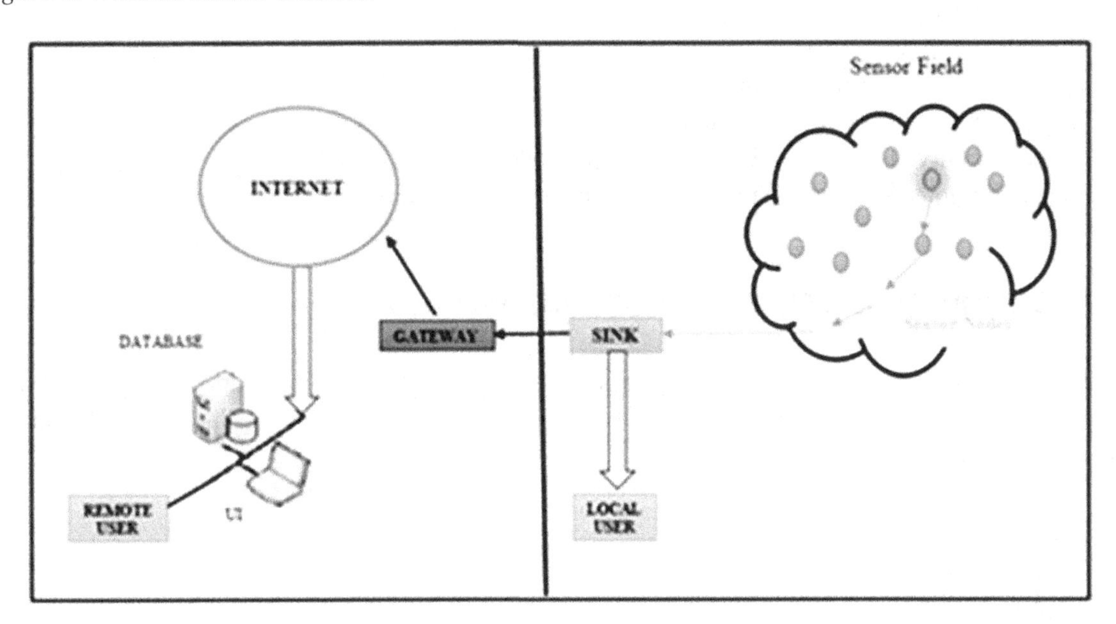

The added feature of sensor network is the cooperative process of sensor nodes. The sensor nodes send the processed data instead of sending the raw data which is sensed and transmit the required data to the network. The important constraints on sensor nodes are the requirement of low power consumption for the purpose of computation and communication. Sensor nodes use the inbuilt trade off mechanism for long network lifetime with low cost power system as discussed by Ian F. Akyildiz, Weilian Su, Yogesh Sankarasubramaniam, and Erdal Cayirci (2002) Since the sensor nodes of WSN are capable of gathering and sending data, they can be deployed in any distributed environment as discussed by Desai Shivani, Nair Tarjni Vyas Anuja R., Jain Hunar (2017) and are used in many applications as discussed by Akyildiz I.F., Melodia T., and Chowdhury K.R.(2007) like smart buildings, medical care applications, industry, defense and internet of things applications as discussed in Ayoub Benayache, Azeddine

Bilami Sami Barkat and Pascal Lorenz Hafnaoui Taleb (2019) and C. Thilagavathi, Dr.P. Saveetha, C. Willson Joseph, K. Anusree, R. Roshni Menon (2018). This chapter aims to discuss and analyze the various security requirements, vulnerabilities possible in wireless sensor networks. The attacks that can occur and its counter measures are also discussed. Finally a framework to improve security in wireless sensor networks is given which has been designed keeping in mind the integration of WSN in modern technologies like internet of things, internet of vehicles and various artificially intelligent automations.

CHARACTERISTICS OF WIRELESS SENSOR NETWORKS

The important characteristics of wireless sensor nodes as discussed in Ian F. Akyildiz, Weilian Su, Yogesh Sankarasubramaniam, and Erdal Cayirci (2002) are

- Consumption of low power and long lasting inbuilt battery life.
- Ability to handle failures in network.
- Simple to use.
- It uses Cross-Layer design.
- Adaptability to enormous scale of distribution.
- Versatility to huge scale of distribution.
- Some mobility of nodes and Heterogeneity of nodes.
- Scalability to enormous Capability to guarantee strict environmental conditions.

APPLICATION OF WIRELESS SENSOR NETWORKS

Wireless sensor networks have high flexibility in solving different problems in application domains. Some of the major applications of WSN as discussed by Ayoub Benayache, Azeddine Bilami Sami Barkat and Pascal Lorenz Hafnaoui Taleb (2019) are

- Military application
- Area Monitoring
- Transportation
- Health Application
- Agriculture Application
- Environment sensing

Military Applications

In the military, wireless sensors work in the areas of battlefield surveillance, command passing, control mechanism, communicating media, computing, intelligence management and targeting system controllers. Overall, WSN is considered as one of the internal parts of military operations as discussed by Akyildiz I.F., Su W., Sankarasubramaniam Y., Cayirci E. (2002).

Area Monitoring

For monitoring certain areas where some events need to be observed, the sensor nodes can be deployed in the area. The sensor nodes observe the events like heat, pressure etc. The observed data is processed and sent to the base station. The appropriate action will be taken from the base station. For example in monitoring the forest fires, in case of fire indication, the sensor collects the data and sends it to the base station. The base station checks the pressure of the fire extinguisher containers and activates the fire extinguishing operation. In the case of fire monitoring and home automation systems, a distributed one class classification technique for detection of faults in water pipelines based on WSNs is proposed by Aya Ayadi, Oussama Ghorbel, M.S.Ben Salah and Mohamed Abid (2020). Agriculture based area monitoring is proposed by A. Maurya, V.K. Jain (2016).

Transportation

WSN is used to collect the real time traffic information and the data is passed to application models. Users will get the updated traffic congestion details from the application domain which can be integrated into vehicles to provide Internet of Vehicles (IoV).

Health Applications

In the medical health sector, sensor networks are used as a patient monitoring system, a supporting interface for disabled patients, for the purpose of medical diagnostics and for drug intake monitoring for disabled patients. It is also used for analyzing and monitoring human physiological activities and to track, monitor the patient behavioral changes as discussed by Akyildiz I.F., Su W., Sankarasubramaniam Y., Cayirci E. (2002).

Agriculture Application

Wireless Sensor Networks are more recently being deployed for analysis and deciding of the crops for agriculture as discussed in T.Sun, G. Hu, G. Yang and J. Jia (2015); H. Liu, Z. Meng, H. Wang and M. Xu (2016) and K. Wang, W.T. Li, L. Deng, Q. Lyu, Y.Q. Zheng, S.L. Yi (2018).

A combination of wireless sensor networks and unmanned aerial vehicles (UAV) has proven to be more effective for identifying the crops effectively and increasing the yield of the crop as discussed by Fan Ouyang, Hui Cheng, Yubin Lan, Yali Zhang, Xuanchun Yin, Jie Hu, Xiaodong Peng, Guobin Wang and Shengde Chen (2019).

Environmental Sensing

The sensor network has vast applications in the environment sector for research and monitoring. Some environments includes volcanoes, ocean, glaciers, landscaping, agriculture, forest etc. the most important environmental sensing areas as discussed by Akyildiz I.F., Su W., Sankarasubramaniam Y., Cayirci E. (2002) are

- Air Pollution Monitoring
- Fire detection in Forest
- GreenHouse monitoring
- Landslide Monitoring
- Volcanoes Monitoring
- Ocean pollution detection
- Agriculture area Monitoring and control.

SECURITY IN WIRELESS SENSOR NETWORKS

The sensor network operates in an ad hoc manner, which covers the security goals of both traditional networks and ad hoc sensor networks security mechanisms. The security mechanism is classified into primary and secondary.

The primary need is the standard security goal includes Confidentiality, Integrity, Authentication and Availability (CIAA) as discussed by Bhushan, B., & Sahoo, G. (2017). The secondary are Data Freshness, Self-Organization, Time Synchronization and Secure Localization as discussed by Pooja and Dr. R. K. Chauhan (2017).

SECURITY REQUIREMENTS

The security requirements of wireless sensor network are data confidentiality, authentication, data integrity, data freshness, availability, self-organization, auditing, time synchronization, secure localization, non-repudiation, forward secrecy and backward secrecy.

- **Data confidentiality:** The sensor node records the data that should not be leaked to unauthorized neighbour nodes because the data have high sensitive value. To enable the confidentiality the public key encryption is induced in the communication. For secure communication, the extremely important thing is key distribution.
- **Authentication:** Data authentication is done by the receiver node and based on the sender original data. Decision making process is carried on. Authentication process guarantees the receiver that data is transferred from the authentic sender.
- **Data integrity:** Data integrity guarantees that the message is never altered or corrupted by an external adversary during data transfer. A malignant element can likewise change the whole data packet by infusing extra packets.
- **Data freshness:** Data freshness characterizes that no old messages have been replayed and that the information is latest. Particularly, freshness of data is significant for implementing shared key schemes.
- **Availability:** Availability is the condition that at the time of requirement, all the required resources and the data should be accessed by all the nodes in the network. A DoS attack influences the availability at any layer of sensor networks.

- **Self-organization:** This guarantees that every node in the WSNs should be self-mending and self-organization. In addition self-organization is important to help multi-hop routing, and to lead key management and building trust relations as discussed by A. Perrig, J. Stankovic and D. Wagner (2004).
- **Auditing:** The nodes of a sensor network are capable of storing particular events that happen inside the network. Through the base station, the sensor nodes are operated and users are not directly operating the sensor node. Users do not know about the presence of specific events unless the node records it. Auditing data is utilized to examine the behaviours of the system in case of failure occurrence.
- **Time Synchronization:** To enhance the battery power consumption, every individual sensor's radio is turned off for specified time intervals. The end to end delay of packets is computed by the sensors by including the time the sensors are turned off, when it travels between two pair-wise sensors. Live tracking of applications like real-time healthcare monitoring, defence applications, weather monitoring, etc., always requires group synchronization.
- **Secure localization:** One of the important requirements of sensor networks is the ability to automatically and correctly locate each sensor in the network. In case any faults occur in a node, it should pinpoint the location of the failure node.
- **Non-repudiation:** It cannot be denied a transferred message which has been sent or received by a real or legitimate entity.
- **Forward secrecy:** Once the node leaves the network, it is unable to read any message forwarded in the network.
- **Backward secrecy:** Once the transmission is over, the joint sensor node is unable to read any previous messages.

WSN SECURITY VULNERABILITIES

Regardless of the way that the WSN offers a great deal, the security challenges must be observed and handled as needs be. Failure to do this may render it not quite useful to say the least just like any kind of network, security of WSN needs to achieve the following,

1. Confidentiality- hiding message from unauthorized nodes.
2. Integrity- guaranteeing that message is not modified over the network.
3. Authenticity- guaranteeing the other party is who it professes to be
4. Availability- capacity to utilize the network resource

WSN has its own eccentricities, so it is noted that security should take more care than the other kinds of networks. The existing security mechanism is not up to the standard, new techniques to be adopted because of the following reasons,

- Energy Limitation
- Nodes are more open to physical attack
- Very close and frequent interaction with other environments and users.

WSN must be secured because of its unique nature other than the traditional network security techniques as discussed by Pooja and Dr. R. K. Chauhan (2017).

ATTACKS IN SENSOR NETWORKS

Wireless networks are increasingly inclined to security attacks because of their communication nature of the transmission medium when contrasted with wired networks. In large networks it is very difficult to protect and monitor each node from physical and logical attacks. Attackers can send different kinds of security attacks to block the security of WSNs.

In each layer such as physical, link, network, transport, and application the attack can occur in different forms. The routing protocol does not provide high security mechanisms and it allows the attacker to break the security. For example, in the physical layer, attack can be done by jamming the radio signal and also by physical device tampering.

The attacks that occur in Wireless Sensor Networks can be classified as Physical layer attacks, Link layer attacks, Network layer attacks, Transport layer attacks and Application layer attacks as shown in figure 2.

Figure 2. Types of Attacks in WSN

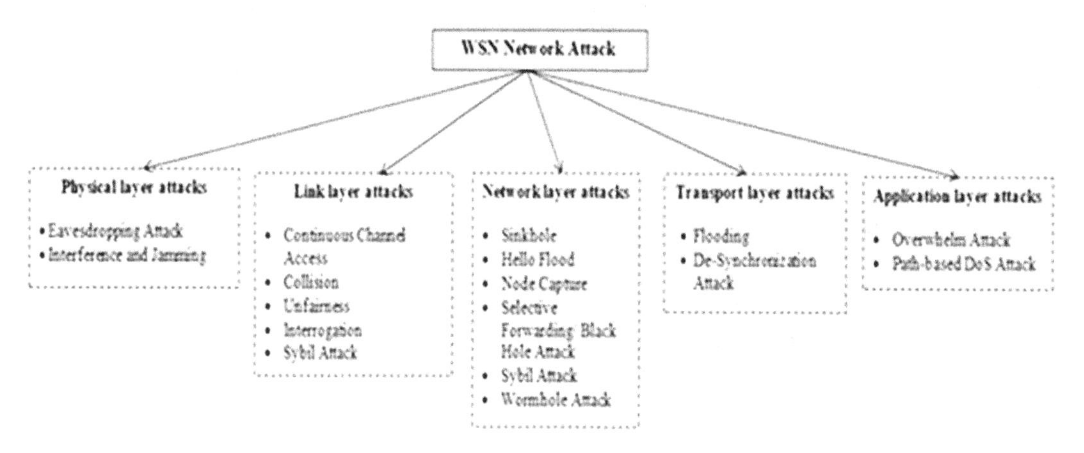

Physical Layer Attacks

Wireless communication is communicated naturally. It is easy to jam or intercept the normal radio signal. The service of wireless network can be attacked by the attacker physically such as overheard or disrupt the service.

Eavesdropping

Eavesdropping is the capturing and reading of messages and discussions by unintended receivers. In mobile ad hoc networks, mobile hosts share the wireless medium. The RF spectrum is mainly used by wireless communication and it broadcast by nature. If the receivers tuned up to proper frequency, it can easily intercept the signals broadcast over airwaves as discussed by A. Perrig, J. Stankovic and D. Wagner (2004). Thus, the transmitted message can be interrupted, overheard and false message can be inserted to the network.

Interference and Jamming

The message will be corrupted or lost when the radio signals are jammed or interfered. To disrupt the communication, if the attacker has a powerful transmitter, the signal produced is strong enough to overwhelm the targeted signals. The random noise and pulse are the common type of signal jamming. Jamming equipment is easily available. Added advantage is jamming attacks can be done from remote locations to the target networks as discussed in A. Mpitziopoulos, Damianos Gavalas, Charalampos Konstantopoulos and Grammati Pantziou (2009).

Link Layer Attacks

Continuous Channel Access (Exhaustion)

Media Access Control protocol gets disrupted by malicious node, by continuous transmitting or requesting over the network. Thus lead to starvation among the other nodes to access the channel in the network. Two main countermeasures, Rate Limiting and Time Division Multiplexing techniques are used to prevent the network.

Rate limiting is applied on MAC admission control and it can neglect the excessive request and prevent the energy drain from repeated transmission. In the Time Division multiplexing technique (TDM), time slot is allotted for each node to transmit.

Collision

It is similar to continuous channel attack. When two nodes try to transmit on the same frequency simultaneously, the collision occurs. When the packets collide, there will be a change in the data portion which results in checksum mismatch at the receiving end. So the data will be considered invalid and it will be discarded. Collision can be prevented by using the technique called error-correcting codes (ECC).

Unfairness

Due the collision on MAC layer attacks or improper use of MAC layer the priority mechanisms lead to unfairness in allocating resources for communication. It is a kind of partial DOS attack and results in marginal performance degradation. To overcome the unfairness attacks, small frame measures can be used so that the individual node can seize and use the channel for smaller durations only.

Interrogation

The hidden-node problem in MAC protocol is moderated using the two-way request-to send/clear to send (RTS/CTS) handshake. The RTS message is repeatedly sent by the attacker to exhaust a node resource and to elicit CTS responses from the target neighbour node. To overcome this type of attacks a node can set and limit the accepting connection from same identity or Anti replay protection and strong link-layer authentication can be used.

Sybil Attack

Most of this type of attack is carried out in Link Layer. There are two types of Sybil attack, Data Aggregation and Voting. In Data Aggregation a single malicious node will act as different Sybil nodes and make the aggregate message a false one. In Voting, the MAC protocols use the voting technique to find the better link for transmission from the pool of available active links. A ballot box is stuff and the Sybil attack is induced. So the attacker can determine the result of the voting and mislead the transmission link and message as discussed by Preeti Sinha, V. K. Jha, Amit Kumar Rai, Bharat Bhushan (2017).

Network Layer Attacks

Sinkhole

Sinkhole attack depends on the routing algorithm technique, it creates a metaphorical sinkhole with the adversary at the center and it tries to lure all the traffic towards the compromised node. To resist the sinkhole attacks, Geo-routing protocol can be used. It is one kind of protocol in which the topology is constructed using only localized information. And the traffic is naturally routed through the physical location of the sink node and it is difficult to lure and to create a sinkhole.

Hello Flood

In many protocols the Hello packets are used to report nodes to their neighbours, this attack misuses the Hello packets. In a network when a node receives such packets, it assumes that it is in the radio range of the sender. This kind of packet can be sent by laptop class adversary to all the sensor nodes in the network and it creates a belief that compromised nodes belong to their neighbours. It results in a number of nodes sending packets to the imaginary neighbour and creates a disregard. One of the key solutions for this attack is Authentication. It can be avoided by verifying bi-directionality of a link before taking action based on the received information over that link.

Node Capture

Even a single node capture is more sufficient for an attacker to take control and corrupt the entire network. Well defined ground breaking work in WSN will be a good solution to over this attack.

Selective Forwarding/ Black Hole Attack (Neglect and Greed)

Multi-hop networks will be usually in WSNs and due to that assumption all participating nodes will forward the message in that trust. The attacking or malicious node can drop and refuse to route certain messages. Black Hole Attack is to drop all the packets which pass through them. In some cases it selectively forwards some packets, and then it is called selective forwarding. To resolve and overcome the issue multi path routing with combination of random selection of paths to destination can be used.

Another method is braided paths which represent paths which have no common link or which do not have two consecutive common nodes, or use implicit acknowledgments, which ensure that packets are forwarded as they were sent.

Sybil Attack

In this attack, a single node presents multiple identities to all other nodes in the WSN. This may mislead other nodes, and hence routes believed to be disjoint with respect to nodes can have the same adversary node. A countermeasure to Sybil Attack is by using a unique shared symmetric key for each node with the base station.

Wormhole Attack

An adversary can tunnel messages received in one part of the network over a low latency link and replay them in another part of the network. This is usually done with the coordination of two adversary nodes, where the nodes try to understate their distance from each other, by broadcasting packets along an out-of-bound channel available only to the attacker. To overcome this, the traffic is routed to the base station along a path, which is always geographically shortest or uses very tight time synchronization among the nodes, which is infeasible in practical environments which is discussed in A. Perrig, J. Stankovic and D. Wagner (2004).

Transport Layer Attacks

Flooding

In this new connection requests are repeatedly made by the attacker until the maximum limit reaches or each connection is exhausted due to many resource requests. So it results in the resource constraints for legitimate nodes. Countermeasures for this attack are, limits can be assigned on the number of connections from the particular node or each connecting client should authenticate commitment to the connection by puzzle solving.

De-Synchronization Attacks

The attacker requests transmission of missed frames by repeatedly frame messages to one or both end-points. Then the message is transferred again and the attacker maintains the proper timing and can avoid the end points from exchanging any useful information. This will cause an extensive waste of energy of genuine nodes in the network in an endless synchronization- recovery protocol. The countermeasure for this attack is, making authentication of all packets with control fields communicated between hosts. Full packet or header authentication resolves the desynchronization attacks which is discussed in Preeti Sinha, V. K. Jha, Amit Kumar Rai and Bharat Bhushan (2017)

Application Layer Attacks

Overwhelm Attack

In this attack the sensor stimuli is used to overwhelm the network nodes and it results in a network to forward huge volumes of traffic to base station. It drains node energy and consumes network bandwidth. To prevent the attack, tune carefully the sensors to specific desired stimulus, like vehicular movement, so it will not respond to other movement which triggers them. Rate-limiting and efficient data-aggregation algorithms can also prevent the attack.

Path-Based DoS Attack

It includes infusing false or replayed packets into the network at leaf nodes. This assault can keep the network from real traffic, since it expends resources on the path to the base station, in this way keeping different nodes from sending information to the base station. Anti-replay protection and combining packet authentication protects from these attacks as discussed in Preeti Sinha, V. K. Jha, Amit Kumar Rai and Bharat Bhushan (2017). The table 1 shows the different security attacks that occur in WSN layer and its Countermeasures.

Table 1. Security attacks of WSN and their countermeasures

Network Layer	Security Attack	Countermeasure
Physical layer	Eavesdropping Interference and Jamming	Spread Spectrum, Frequency Hopping
Link layer	Continuous Channel Access Collision Unfairness Interrogation Sybil Attack	Error- correcting codes, Data rate limits, Time division multiplexing, Short frames
Network Layer	Sinkhole Hello Flood Node Capture Black Hole Attack Sybil Attack Wormhole Attacks	Authentication, link-layer encryption, public key cryptography, Multipath routing
Transport layer	Flooding De-synchronization Attacks	Client Puzzles, authenticated broadcast of messages.
Application layer	Overwhelm attack Path-based DOS attack	Authentication Schemes

COUNTER MEASURES AGAINST ATTACKS IN WSN

Conventional Wireless Sensor Networks consists of a network node with an omnidirectional antenna which transmits radio signals in all directions which paves a way to Eavesdropping. Eavesdropping is over hearing something you shouldn't be hearing and in this aspect it denotes the signals transmitted by the sensor nodes. To prevent this type of attack instead of using the omnidirectional antenna, directional antenna can be used which reduces the rate of eavesdropping when compared with omnidirectional antenna is proposed in Dai, Hong-Ning & Wang, Qiu & Dong, Li & Wong, Raymond (2013).

Transmission Outage constrained scheme is proposed to prevent eavesdropping attack in L. Lv, Z. Ding, J. Chen and N. Al-Dhahir, (2019). In Y. Karakurt, H. U. Yıldız and B. Tavlı, (2018) it is proposed that mitigation or reduction of eavesdropping by decreasing the transmission of power levels can lead to decrement of network lifetime of the wireless sensor nodes. So the value of power levels should not be decreased too much which may lead to optimization issues.

Sybil attacks in real-time applications can be seen in social networks when it is used to claim false identities. In S. Singh and H. S. Saini, (2018), the Received Signal Strength Indicator (RSSI) and Random Password Comparison Method (RPC) are combined with Neighboring node information-based Detection System to detect Sybil nodes and thereby preventing a Sybil attack.

Continuous channel access is required in wireless sensor networks when they are integrated with Internet of things or vehicles and for healthcare, military, energy or any other purpose. In Manu Elappila, Suchismita Chinara, Dayal Ramakrushna Parhi (2020) a fuzzy-based mechanism that enables the network components to adapt the preferences to get the channel access has been discussed.

The network nodes in Wireless Sensor Network always communicate between each other and there is a possibility that data collision can occur. Nevertheless, averting collision for every transmission will lead to increase in latency of the broadcast schedules. A Collision-Tolerant Scheduling (CTS) is proposed in Duc Tai Le Thang Le Duc, Vyacheslav V.Zalyubovskiy, Dongsoo S. Kim, Hyunseung Choo (2017) that offers an opportunity to reduce broadcast latency by allowing collisions at non-critical nodes to speed up the broadcast process for critical ones.

A wormhole attack is capable of stealing data and disables the network performance through external channels. In this attack, two malicious nodes in two different network areas and a high speed channel is established to create fake identities and thereby misleads all the network nodes. So that the attacker's node becomes the main source of contact and thereby becomes stronger than other nodes in the network. Trust based security mechanisms can be useful in detecting and prevention of wormhole attacks as proposed by Farjamnia, G., Gasimov, Y. & Kazimov, C. (2019).

There are many network nodes in WSN and not all the nodes are used and the least used sensor is sometimes made to rest which leads to unfairness. Duty MAC cycled protocol Portillo, C.; Martinez-Bauset, J.; Pla, V.; Casares-Giner, V. (2020) combined with TLB routing Dinakaran, K., Adinadh, K.R., Sanjuna, K.R. & P. Valarmathie (2020) can be used to reduce unfairness.

Intrusions in wireless sensor networks can lead to power outages, energy depletion, flooding attacks and complete shutdown of the wireless sensor network. An Integrated IDS scheme (IIS) that integrates the concept of clustering along with and Digital signature has been proposed by Bhushan, B., & Sahoo, G. (2019). In heterogeneous systems, cluster-based protocols can be used since they give best results as they work on divide and conquer strategy. To secure the transmitted data, a secured, lightweight digital signature scheme that employs symmetric cryptography has been used.

TRUST AND REPUTATION (A WAY TO IMPROVE THE WSNS)

Trust and Reputation is a methodology for preserving the required security among two nodes having transactions and communications in a distributed network. Trust is the subjective probability in which an agent will perform a certain action. Reputation is the expectant behavior of an agent based on the information collected from all of its past behavior as discussed by Zhengwang Ye, Tao Wen, Zhenyu Liu, Xiaoying Song and Chongguo Fu, (2017).

Trust in Wireless Sensor Networks

Trust is a significant factor in the decision making procedures of any network where vulnerability is a factor. That is, the point at which the result of a specific circumstance unable to settled or guaranteed. With no vulnerability, there is no need for a trust management system, if a component of the network knows ahead of time the genuine behavior of their neighbor(for example cooperative, malicious, faulty, etc.), it can settle on an impeccable choice. Subsequently, so as to know whether a trust management system can be applied to a WSN, it is essential first to examine the significance of vulnerability in such condition, discussed in Rodrigo Rom´an, M. Carmen Fern´andez-Gago, and Javier L´opez, (2009). Trust is determined dependent on the past experience of a specific node and its level of conviction, while reputation is the general impression of one node among other network node.

There are two primary trust parameters: trust qualification and trust. Trust qualification characterizes the various degrees of trust, while trust figuring depicts the methods for estimating the trust value between nodes. It is additionally essential to take note of that Trust and Reputation Management (TRM) is a framework which characterizes the various steps that manage trust. The essential stages in TRM are to gather the realities, update the trust values and actuate the production of decisions, discussed in Saqib Ali, Taiseera Al-Balushi, Zia Nadir, Omar Khadeer Hussain (2018).

Trust and Reputation Management (TRM)

A significant component of any trust management system is the trust entity. This entity is responsible for acquiring, calculating and keeping up reputation and trust values. For sensor networks, it is conceivable to characterize the structure of a generic trust entity, as shown in Figure 3. In the structure, “Information” modules get data about the conduct of the individuals from its neighborhood, either through perception and experience (for example “first-hand information”) or by imparting the watched events to different entities (for example “Second hand information”).

Figure 3. Structure of Trust Entity

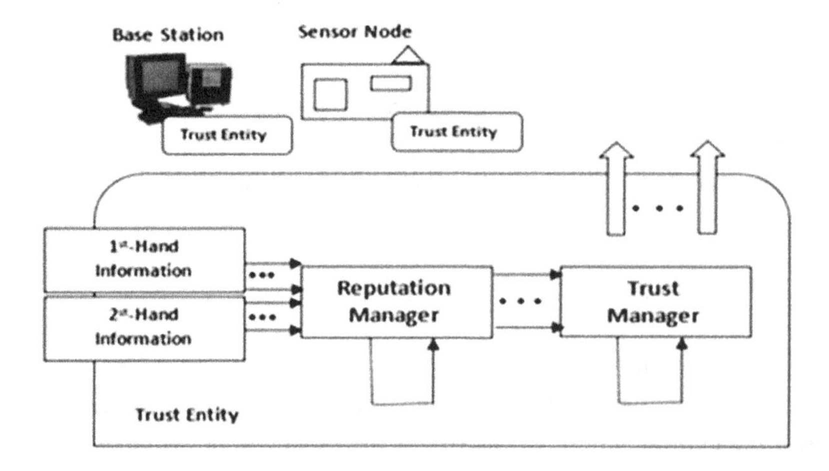

After this procedure, the "reputation manager" module can utilize this rundown of events to gather and store the reputations of the individuals from its neighborhood. Such reputation will be later utilized by the "trust manager" module to get the trust values. They can be utilized to conclude which is the best partner for a specific activity, or find on the off chance that one entity is carrying on malevolently. The two modules need to keep up and update their qualities values during the lifetime of the network.

This structure is plainly appropriate to wireless sensor systems, on the grounds that a sensor node can get information about its environmental factors either directly or indirectly. Furthermore, the sensors have constrained computational capacities. Thusly, by utilizing lightweight algorithm calculations, they can have the option to derive the reputation of its neighbours and choose if they trust them for specific activities. Additionally, this model of a trust entity fits in the plan of the greater part of the existing work on trust for sensor networks, but just a couple of those works consider reputation expressly. All things considered, having both reputation and trust in a similar system is important. By not figuring the trust directly from the conduct of a node, it is conceivable to all the more likely handle perspectives, for example, the development of the node, maturing, and so on.

TRM Techniques

The work discussed in Saqib Ali, Taiseera Al-Balushi, Zia Nadir, Omar Khadeer Hussain (2018), trust and reputation management by adopting the recursive Bayesian methodology. This examination was additionally researched, who found that a mischievous node may even now be viewed as trusted if it behaves properly during communication. This leads the presentation of information trust and communication trust to maintain a strategic distance from any false information infusion into the network. Utilizing tools from various spaces, for example, probability, statistics and mathematical analysis in the trust management structure improves the security actualized in WSNs. Following further examination, the issue that sensor node trust and reputation depends on a measure of certainty in the estimations of that specific node. Then again, stress that WSNs in Ad-hoc and P2P need legitimate trust management solutions.

To give quicker trust evaluation of clustered WSNs of a gathering nature, a lightweight gathering for trust calculation dependent on a trust management scheme was proposed. To conquer the issue of utilizing top end resources in implementing a group trust management scheme, an agent based trust estimation scheme was presented. A way to deal with rearranging the examination among trust and reputation management by building a test system was introduced. An especially scalable cluster based requesting trust management procedures was likewise presented, after analyzing a wide scope of trust and reputation management systems. Many researchers have developed many models based on statistical, logical, nature inspired methods, fuzzy logic, etc. and several orientations like as individual and clustered distributed to obtain trust models and frameworks.

PROPOSED SECURITY MODEL FOR SECURING WSNs

The Framework for securing wireless sensor networks should be secured in all levels as shown in figure 4. The basic security requirements of wireless sensor networks are also discussed in Saqib Ali, Taiseera Al-Balushi, Zia Nadir, Omar Khadeer Hussain (2018). Since WSNs are to be widely used in the field of Internet of Things (IoT) and Artificial Intelligent systems, it should provide a high efficient security system.

Figure 4. Basic security Framework for WSNs

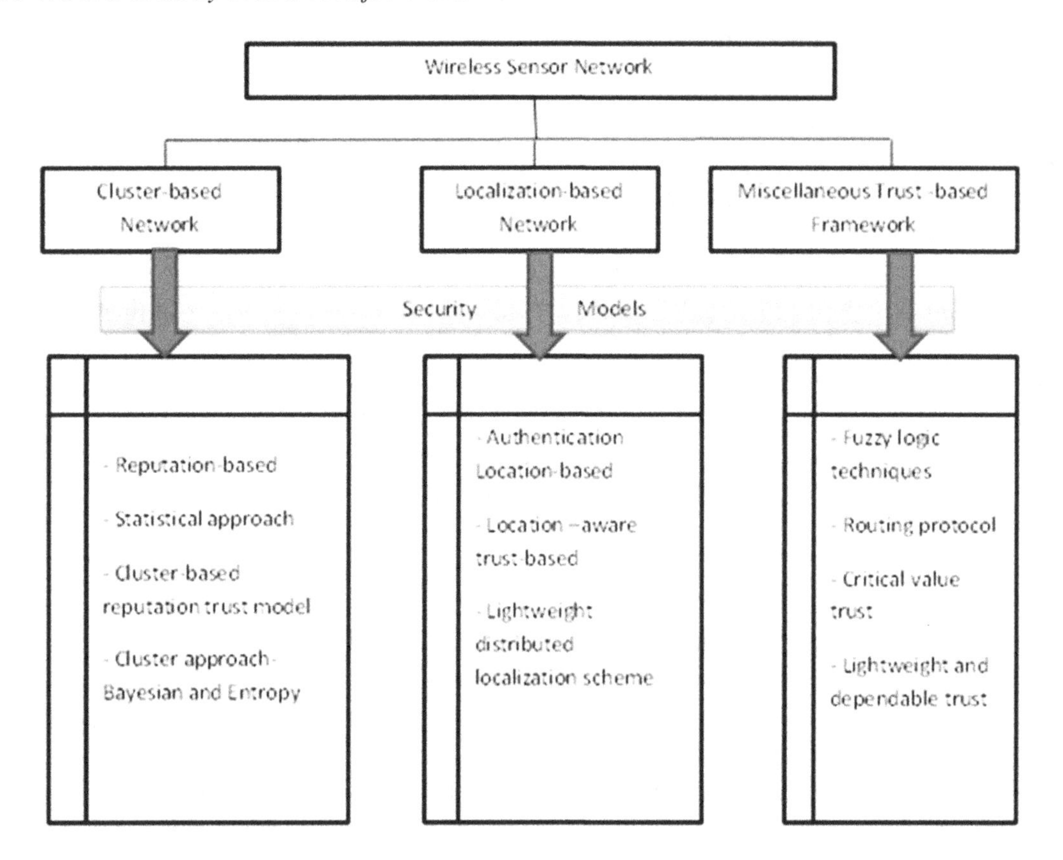

Proposed security mechanisms that are to be integrated in our system are improved authentication, lightweight encryption techniques, Low Energy Consumption Distributed Fault Detection Mechanism and Hamming residue method. Low energy consumption is one of the major requirements of the wireless sensor networks. The reliability of the WSNs depends upon the efficient energy consumption across each individual node in the sensor network.

Secure Communication

Symmetric and asymmetric encryption techniques improve the data being transmitted from a sensor node to another node and also to the base stations. Since WSN are the core devices required for implementing internet of things (IoT) and internet of vehicles (IoV). The Advanced Encryptions Algorithm (AES) is the most secure algorithm for implementing symmetric encryption as proven by Murat Dener (2018). In asymmetric encryptions with two keys RSA of key size 512 bits and Elliptic Curve cryptographic (ECC) algorithm with much smaller key size can also be used. ECC is considered much faster when compared to RSA and is very apt for WSN nodes as discussed by Alakananda Tripathy, Satheesh Kumar Pradhan, Alok Ranjan Tripathy, Ajit Kumar Nayak (2019).

If '*E*' denotes the encryption algorithm, then the encrypted data '*M* is denoted by $E_{nc}(M)=E_k(M)$, where 'k' denotes the key that has been used for encryption. If symmetric encryption is used, the same key can be used for decryption. If asymmetric algorithm is used, then there exists a private and public key pair and the equation becomes, $E_{nc}(M)=E_{pubk}(M)$where 'pubk' denotes the public key of the receiving node. The decryption is as, $M=D_{privk}(E_{nc}(M))$and 'privk' indicates the private key of the receiving node that is known only to that node.

Authentication

To ensure that the server (base station) and the entire wireless sensor network are valid there is a need of a flawless authentication method to confirm the identity of all the entities. The authenticity is to be checked at each level and it should be automatically rechecked if necessary without taking much overhead. This is done to avoid the presence of a malicious node within the network during the course of operation. In the event that there is a malicious node present among the sensors, and the server or cluster head rebuilds a key with the already existing data at regular time intervals and communicate within its network using dynamic key based messages which in turn also performs authentication. If no dynamicity is added then the adversary can enter into a network without much effort and perform passive or active attacks in the network which would be extremely hard to resolve. Authentication scheme using elliptic curve cryptosystem has been proposed in Yoney Kirsal Ever (2020) for an application of IoT. At the point when a cluster head needs to authenticate the sensors under its hierarchy, it calculate a value B_{pq}, where 'p' is the cluster head and 'q' is the sensor node to be verified.

$B_{pq} = H\,(ID_p \parallel T)$ where 'H' - is the Hash Function

Here, T is the current timestamp of sensor 'p' and 'ID_p' is the unique ID of sensor 'p'. The cluster head 'p' sends a message B_{pq} to sensor node 'q'.

After accepting the message AUT, the sensor node 'q' approves the timestamp if $(T' - T) \leq \Delta T$ where T' is the current timestamp of sensor node 'q' and ΔT is the normal time interim of the transmission delay. Since sensor node 'q' is within the node area and in synchronization with the network, the knowledge of the identity of the cluster head and the timestamp of sending and receiving message will be known to a valid node. In the event that the condition is not true, then authentication fails and cluster head 'p' gets to know that 'q' is a malicious sensor node and ends the communication with sensor 'q'.

If the condition is true, at the point sensor node 'q' computes:

$B_{qp} = H\,(ID_p \parallel T)$

Here, T is the calculated timestamp with the help of T' is the current timestamp. Now sensor 'q' compares the value of B_{qp} with the received value of B_{pq}. If $B_{pq} = B_{qp}$ remains true, at that point cluster head 'p' realizes that sensor 'q' has been authenticated effectively and now it is protected to build up a key with sensor 'q'. In a similar way, the server additionally authenticates the cluster heads before the key is shared.

As already discussed since wireless sensors are to be used in various fields of healthcare, agriculture, military, etc. the authentication schemes vary based on the applications. A three-factor authentication scheme has been proposed by Preeti Soni, Arup Kumar and S. K. Hafizul Islam (2019) specifically for WSN in healthcare systems.

Low Energy Consumption Distributed Fault Detection Mechanism

Wireless sensor network is a significant specialized support for pervasive communication. To avoid the network failure due to the uneven energy utilization of sensor nodes, equipment failure and attacker interruption on information transmission, a low energy utilization based fault detection technique is to be used. A time and spatial correlation based faulty node detection scheme is to be used for effective and complete identification of such nodes. This mechanism ensures continuity in data transmission and prevents loss of data in WSN. Such a similar scheme has been identified for WSN in Shuang Jia, Lin Ma and Danyang Qin (2019).

Hamming Residue Method

In general wireless sensor networks comprise of small sensor node with restricted energy. WSNs are self-sufficient and are distributed in nature. Because of the absence of central authority and due to the arbitrary deployment of nodes in the network, WSNs are inclined to security threats. One of the major attacks in WSN is malicious attack to tamper with the data being communicated. Even after encryption and transmission, malicious nodes or attackers can tamper with captured packets. Inclusion of cryptographic methodologies or by time based synchronization analysis does not completely solve the issue due to the completely distributed nature of WSN. A productive methodology called Hamming residue method (HRM) is introduced to prevent the malicious attack as discussed by Majid Alotaibi (2019).

Figure 5. Proposed Security framework for WSN

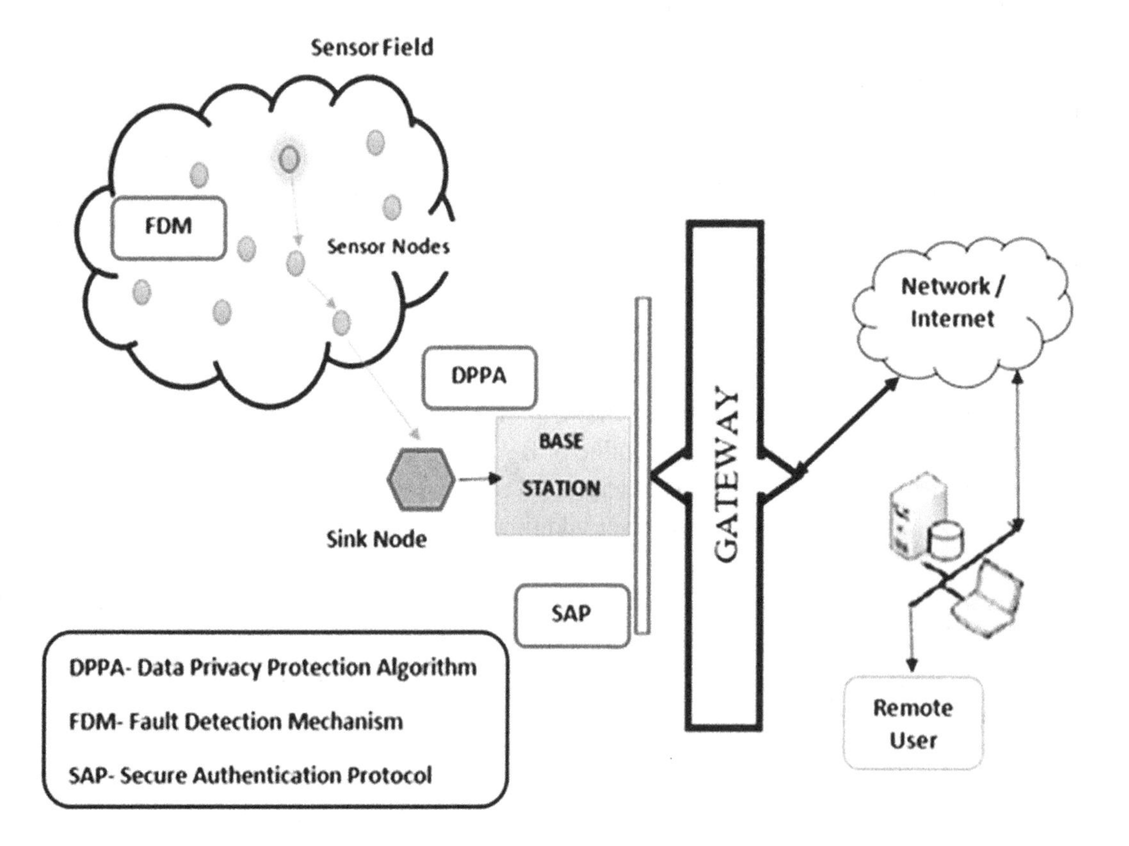

First, a codeword is being created by the utilization of defined initial security bits (user decision) and security check bits (Hamming bits). After which, quadratic buildup technique is being utilized to improve the security at different hops. Utilizing IPV6, the data of HRM is stored in the header and if the code matches with the code created by the intermediate node, then information is transferred else, the node is viewed as an malicious attacker node.

Based on the combined techniques discussed in the proposed methodology, a security framework for wireless sensor networks is given in figure 5.

The data privacy protection algorithm (DPPA) with repetition system is incorporated to guarantee security by implementing privacy. Homomorphic encryption component ensures additional security while conveying concealed information. Routing to find the appropriate path of transmission and communication is proposed by directed Tree Protocol. The data input tree is built up between the destination node and the source node. Then, the destination node quickly sends the packet loss rate information and the encryption key through the data input tree to the source node. Therefore, it improves the dependability and security of information transmission and guarantees the lossless information. The work flow of the proposed work is illustrated figure 6.

Figure 6. Workflow of Data Privacy protection Algorithm

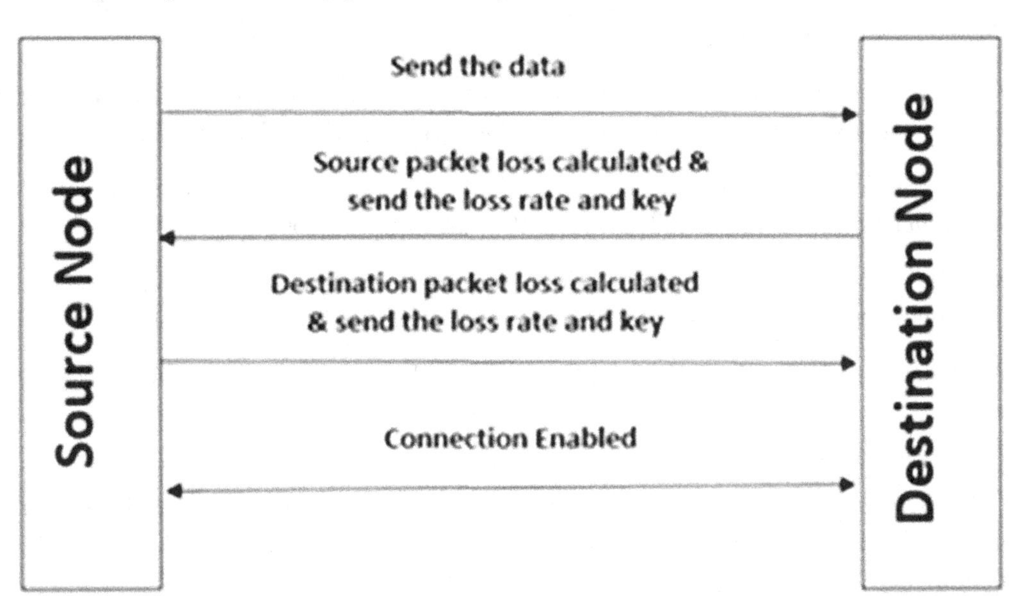

The time correlation information existing among the nodes is utilized to distinguish faulty nodes in fault detection mechanism (FDM). The spatial correlation data is also added to identify all of the faulty nodes in the network in time internals to check the nodes exhaustively thereby improving the efficiency of data transmission. Secure Authentication Protocol (SAP) enables the mutual authentication among the data transmission nodes, base station and gateway. All these processes are repeated dynamically in an unpredictable manner so that an attacker or malicious node is captured at the earliest and loss or manipulation of data is avoided.

CONCLUSION

This chapter examines the various attacks vulnerabilities and security measures related to wireless sensor networks. A large portion of the attacks against security in wireless sensor systems are brought about by the addition of unwanted data by the undermined nodes inside the network. For safeguarding the consideration of false reports by undermined nodes, a method is required for identifying false reports. Many variations of security solutions have been provided but with the involvement of wireless nodes into the internet of things and everything smart, more security vulnerabilities will be unearthed in all areas. So, none can be found to be providing a complete security solution in 360 degrees. There exists a need to devise a novel security component which will make communication inside sensors progressively robust, adaptable, fast and energy efficient. In many cases, implementing such a security system and making it productive will be an extraordinary research challenge. Again, guaranteeing holistic security in wireless sensor networks is a significant research issue.

REFERENCES

Akyildiz, I. F., Melodia, T., & Chowdhury, K. R. (2007). A survey on wireless multimedia sensor networks. *Computer Networks*, *51*(4), 921–960. doi:10.1016/j.comnet.2006.10.002

Akyildiz, I. F., Su, W., Sankarasubramaniam, Y., & Cayirci, E. (2002). A survey on sensor networks. *IEEE Communications Magazine*, *40*(8), 102–114. doi:10.1109/MCOM.2002.1024422

Akyildiz, I. F., Su, W., Sankarasubramaniam, Y., & Cayirci, E. (2002). Wireless Sensor Networks: A Survey. *Computer Networks*, *38*(4), 393–422. doi:10.1016/S1389-1286(01)00302-4

Akyildiz, I. F., Su, W., Sankarasubramaniam, Y., & Cayirci, E. (2002). A Survey on Sensor Networks. *IEEE Communications Magazine*, *40*(8), 102–114. doi:10.1109/MCOM.2002.1024422

Ali, S., Al-Balushi, T., Nadir, Z., & Hussain, O. K. (2018). Improving the Resilience of Wireless Sensor Networks Against Security Threats: A Survey and Open Research Issues. *International Journal of Technology*, *4*(4), 828–839. doi:10.14716/ijtech.v9i4.1526

Alotaibi, M. (2019). Security to wireless sensor networks against malicious attacks using Hamming residue method. *EURASIP Journal on Wireless Communications and Networking*, *2019*(1), 8. doi:10.118613638-018-1337-5

Ayadi, A., Ghorbel, O., BenSalah, M. S., & Abid, M. (2020). Spatio-temporal correlations for damages identification and localization in water pipeline systems based on WSNs. *Computer Networks*, *171*, 107134. doi:10.1016/j.comnet.2020.107134

Benayache, A., Azeddine, B. S. B., & Pascal, L. H. T. (2019). MsM: A microservice middleware for smart WSN-based IoT application. *Journal of Network and Computer Applications*, *144*(15), 138–154. doi:10.1016/j.jnca.2019.06.015

Bhushan, B., & Sahoo, G. (2017). Recent Advances in Attacks, Technical Challenges, Vulnerabilities and Their Countermeasures in Wireless Sensor Networks. *Wireless Personal Communications*, *98*(2), 2037–2077. doi:10.100711277-017-4962-0

Bhushan, B., & Sahoo, G. (2019). A Hybrid Secure and Energy Efficient Cluster Based Intrusion Detection system for Wireless Sensing Environment. *2019 2nd International Conference on Signal Processing and Communication (ICSPC)*. doi: 10.1109/icspc46172.2019.8976509

Dai, H.-N., Wang, Q., Dong, L., & Wong, R. (2013). On Eavesdropping Attacks in Wireless Sensor Networks with Directional Antennas. *International Journal of Distributed Sensor Networks*, *9*(8), 1–13. doi:10.1155/2013/760834

Dener, M. (2018). Comparison of Encryption Algorithms in Wireless Sensor Networks. *ITM Web of Conferences*, *22*, 1-5.

Dinakaran, K., Adinadh, K. R., Sanjuna, K. R., & Valarmathie, P. (2020). Quality of service (QoS) and priority aware models for adaptive efficient image retrieval in WSN using TBL routing with RLBP features. *Journal of Ambient Intelligence and Humanized Computing*. Advance online publication. doi:10.100712652-020-01793-7

Elappila, M., Chinara, S., & Parhi, D. R. (2020). Survivability Aware Channel Allocation in WSN for IoT applications. *Pervasive and Mobile Computing*, *61*, 101107. doi:10.1016/j.pmcj.2019.101107

Ever, Y. K. (2020). A secure authentication scheme framework for mobile sinks used in the Internet of Drones application. *Computer Communications*, *155*, 143–149. doi:10.1016/j.comcom.2020.03.009

Fan, O., Cheng, H., Lan, Y., Zhang, Y., Yin, X., Hu, J., Peng, X., Wang, G., & Chen, S. (2019). Automatic delivery and recovery system of Wireless Sensor Networks (WSN) nodes based on UAV for agricultural applications. *Computers and Electronics in Agriculture*, *162*, 31–43. doi:10.1016/j.compag.2019.03.025

Farjamnia, G., Gasimov, Y., & Kazimov, C. (2019). Review of the Techniques against the Wormhole Attacks on Wireless Sensor Networks. *Wireless Personal Communications*, *105*(4), 1561–1584. doi:10.100711277-019-06160-0

Jia, S., Ma, L., & Qin, D. (2019). Research on Low Energy Consumption Distributed Fault Detection Mechanism in Wireless Sensor Network. *Publisher: IEEE. China Communications*, *16*(3).

Karakurt, Y., Yıldız, H. U., & Tavlı, B. (2018). The impact of mitigation of eavesdropping on wireless sensor networks lifetime. *Proceedings of 26th Signal Processing and Communications Applications Conference (SIU)*, 1-4. 10.1109/SIU.2018.8404252

Le Duc, D. T. L. T., Zalyubovskiy, V. V., Kim, D. S., & Choo, H. (2017). Collision-tolerant broadcast scheduling in duty-cycled wireless sensor networks. *Journal of Parallel and Distributed Computing*, *100*, 42–56. doi:10.1016/j.jpdc.2016.10.006

Liu, H., Meng, Z., Wang, H., & Xu, M. (2016). Spatio-temporal variation analysis of soil temperature based on wireless sensor network. *International Journal of Agricultural and Biological Engineering*, *9*(6), 131–138.

Lv, L., Ding, Z., Chen, J., & Al-Dhahir, N. (2019). Design of Secure NOMA Against Full-Duplex Proactive Eavesdropping. *IEEE Wireless Communications Letters*, *8*(4), 1090–1094. doi:10.1109/LWC.2019.2907852

Maurya, A., & Jain, V. K. (2016). Fuzzy based energy efficient sensor network protocol for precision agriculture. *Computers and Electronics in Agriculture*, *130*, 20–37. doi:10.1016/j.compag.2016.09.016

Mpitziopoulos, A., Gavalas, D., Konstantopoulos, C., & Pantziou, G. (2009). A survey on jamming attacks and countermeasures in WSNs. *IEEE Communications Surveys and Tutorials*, *11*(4), 42–56. doi:10.1109/SURV.2009.090404

Perrig, A., Stankovic, J., & Wagner, D. (2004). Security in Wireless Sensor Networks. *Communications of the ACM*, *47*(6), 53. doi:10.1145/990680.990707

Pooja & Chauhan. (2017). Review on Security attacks and Countermeasures in Wireless Sensor Networks. *International Journal of Advanced Research in Computer Science*, *8*(5), 1275–1283.

Portillo, C., Martinez-Bauset, J., Pla, V., & Casares-Giner, V. (2020). Modeling of Duty-Cycled MAC Protocols for Heterogeneous WSN with Priorities. *Electronics (Basel)*, *9*(3), 467. doi:10.3390/electronics9030467

Rom'an, Fern'andez-Gago, & L'opez. (2009). Trust and Reputation Systems for Wireless Sensor Networks. In Security and Privacy in Mobile and Wireless Networking. Troubador Publishing Ltd.

Shivani, R., & Hunar. (2017). Survey on cluster based data aggregation in wireless sensor network. *International Journal of Advanced Research in Computer Science*, *8*(5), 311–314.

Singh, S., & Saini, H. S. (2018). Security approaches for data aggregation in Wireless Sensor Networks against Sybil Attack. *Proceedings of Second International Conference on Inventive Communication and Computational Technologies (ICICCT)*, 190-193. 10.1109/ICICCT.2018.8473091

Sinha, Jha, Rai, & Bhushan. (2017). Security vulnerabilities, attacks and countermeasures in wireless sensor networks at various layers of OSI reference model: A survey. *Proceedings of IEEE International Conference on Signal Processing and Communication (ICSPC)*, 288-293.

Soni, P., Kumar, A., & Hafizul Islam, S. K. (2019). An improved three-factor authentication scheme for patient monitoring using WSN in remote health-care system. *Computer Methods and Programs in Biomedicine*, *182*, 105054. doi:10.1016/j.cmpb.2019.105054 PMID:31499422

Sun, T., Hu, G., Yang, G., & Jia, J. (2015). Real-time and clock-shared rainfall monitoring with a wireless sensor network. *Computers and Electronics in Agriculture*, *119*, 1–11. doi:10.1016/j.compag.2015.09.023

Thilagavathi, Saveetha, Willson Joseph, Anusree, & Roshni Menon. (2018). Certain Applications of Sensor Network in the Field of IoT, Big Data Analysis and Cloud. *Journal of Advanced Research in Dynamical & Control Systems*, *10*, 2012–2015.

Tripathy, A., Pradhan, S. K., Tripathy, A. R., & Nayak, A. K. (2019). A New Hybrid Cryptography Technique in Wireless Sensor Network. *International Journal of Innovative Technology and Exploring Engineering*, *8*(10), 121–131. doi:10.35940/ijitee.I8736.0881019

Wang, K., Li, W. T., Deng, L., Lyu, Q., Zheng, Y. Q., Yi, S. L., Xie, R., Ma, Y., & He, S. (2018). Rapid detection of chlorophyll content and distribution in citrus orchards based on low-altitude remote sensing and bio-sensors. *International Journal of Agricultural and Biological Engineering*, *11*(2), 164–169. doi:10.25165/j.ijabe.20181102.3189

Ye, Z., Wen, T., Liu, Z., Song, X., & Fu, C. (2017). An Efficient Dynamic Trust Evaluation Model for Wireless Sensor Networks. *Journal of Sensors*, *2017*, 1–16. doi:10.1155/2017/7864671

Chapter 8
Recent Trends in Channel Assignment Techniques in Wireless Mesh Networks

D. Jasmine David
Karunya Institute of Technology and Sciences, India

Jegathesan V.
Karunya Institute of Technology and Sciences, India

T. Jemima Jebaseeli
Karunya Institute of Technology and Sciences, India

Anand Babu Ambrose
Independent Researcher, India

Justin David D.
James College of Engineering and Technology, India

ABSTRACT

Wireless mesh networks have numerous advantages in terms of connectivity as well as reliability. Traditionally, the nodes in wireless mesh networks are equipped with a single radio, but the limitations are lower throughput and limited use of the available wireless channel. To overcome this, the recent advances in wireless mesh networks are based on a multi-channel multi-radio approach. Channel assignment is a technique that selects the best channel for a node or to the entire network just to increase the network capacity. To maximize the throughput and the capacity of the network, multiple channels with multiple radios were introduced in these networks. In this work, algorithms are developed to improve throughput, minimize delay, reduce average energy consumption, and increase the residual energy for multi-radio multi-channel wireless mesh networks.

DOI: 10.4018/978-1-7998-5068-7.ch008

INTRODUCTION

Wireless Mesh Network (WMN) is a fundamental network procedure of the wireless networks. The wireless mesh network is transpired as a capable concept to gather challenges in the subsequent generation networks provided with flexible, adaptive, and self-configurable architecture with low cost. WMN are multi-hop networks and it attracted lots of interest recently. Three different types of architecture in WMN are given by Akyildiz et al. (2005). They are infrastructure or backbone WMN, client WMN, hybrid WMN. WMN enables rapid, flexible and robust network configuration and coverage, without the expensive cabling infrastructure. The characteristics of WMN are self-organizing, self-healing, and self-configuring for high reliability. The capability of WMN can be influenced by issues like the architecture of the network, topology, flow pattern, node concentration, number of channels per node, communication power intensity and mobility of a node. Factors such as inadequacy of protocol, interference from external sources sharing spectrum, and the insufficiency of Electro-Magnetic (EM) spectrum reduces the capacity of single radio WMN. The elements in WMN are mesh client, mesh router or Access Point (AP) to forward in the traffic in network, and gateways connected to the Internet. Generally, Mesh routers are assumed as immobile and they are endowed with one or more than one wireless interface. Clients can be either mobile or immobile. They are linked to the network via routers.

WMN is a modern technology that provides better quality service to users. In wireless internet, most of the nodes are either static or fewer mobility nodes. The nodes automatically create ad hoc networks and maintain their connectivity due their self-organized nature. Due to this reliability is improved and the coverage area is increased and the equipment cost is reduced. Divyansh Puri et al. (2019) introduced the machine learning techniques for WSNs at a very low cost to increase the life span of Wireless Sensor Networks. The competency of the network owing to interference from neighbouring nodes in the network can be affected by this configuration. The challenging approach to improve the capacity of WMN is furnishing all nodes with several radios so that the spectrum can be utilized efficiently and actual available bandwidth in the network can be improved. WMN has grown with time to manage with the user anxieties such as data rate, scalability, reachability, and mobility of the user.

Interference is the main factor that directly impacts the network capacity. A protocol called Priority based Interference Aware Channel Assignment with Bandwidth Reservation is proposed to minimize interference. Though interference is minimized in this algorithm, the performance is degraded due to congestion. To improve the performance of the network Distributed Optimal Congestion Control and Channel Assignment algorithm is proposed. To improve the performance, Traffic Aware Channel Assignment with Node Stability is proposed. To provide secure communication Efficient Key Establishment Protocol is devised. This algorithm automatically removes the misbehaving node in the network and the information is notified to the neighbour nodes. From the comparative study of the performance of the proposed algorithms, it is proven that the performance of traffic aware channel assignment algorithms with node stability is good when compared with the other two modules. Therefore, from all the performance metrics investigated, TACA-NS is identified as the most attractive scheme for channel assignment in wireless mesh networks among the three proposed schemes.

The content of the paper is organized as follows: section 2 is about the literature survey, section 3 is about the design issues in channel assignment, section 4 is about the comparison of channel assignment methods, section 5 is about the priority based interference aware channel assignment with bandwidth reservation, section 6 is about the congestion aware channel assignment, section 7 is about the traffic aware channel assignment with node stability, section 8 is about the efficient key establishment protocol, section 9 is about performance evaluation and the final discussions are concluded in section 10.

CHANNEL ASSIGNMENT MECHANISM

Channel Assignment (CA) is one of the major parameters to improve the efficiency of the network by minimizing the interference and to offer data transmission between multiple radios of the node. Channel assignment is a method which picks the best channels for wireless node individually or the whole network targeting to improve the capacity of the network. In the channel assignment mechanism, the node attempts to identify a possible mapping among channels and radio interfaces at each node. It has been broadly researched for multi-radio WMN, still challenging when it comes to implementation. Routing also plays a vital role by providing end to end route selection. Both channel assignment and routing are interdependent and it affects each other. Channel assignment attempts to reduce network interference by maintaining network connectivity. Routing provides end to end route according to user's requirement. If the nodes operate on the same channel, the performance of the network decreases. If the nodes are endowed with more amounts of radios and channels, the network performance can be improved without any degradation in bandwidth. To attain high speed, the multi-radio nodes operate on the orthogonal channel. The major factor that has a direct impact on the capacity of a network is interference. By using multiple channels, interference is minimized along with the allocation of resources like radios and channels to attain fairness and congestion control. Simultaneous transmission over the same channel may cause interference therefore the reuse of the same channel be avoided. In general, the channel allocation algorithms should satisfy the requirements such as, if there is 'N' number of interfaces available, at a time, only one node can communicate only on 'N' channels. The nodes that are communicating must be destined to a common channel. Also, the available channel count should be fixed and restricted by the user. The condition for solving channel assignment problem is given by Crichigno et al. (2008) that the number of channels allotted to a node must be either equal or little less than the number of radio interfaces it has. Also, neighbouring nodes must have a minimum one radio in the common channel to connect. Channel assignment mechanism can be broadly classified into four types as static channel assignment, dynamic channel assignment, semi-dynamic channel assignment, and hybrid channel assignment. The details are as follows.

Static Channel Assignment

Most of the channel assignment put forward in the literature is a static type in which nodes tuned their radios to particular channels permanently. Static channel allocation methods are more reliable to organize but abortive to adapt to the environmental changes. The static channel assignment aims to maximize the net performance of the networks. Subramanian et al. (2008) and Marina et al. (2010) discussed the static channel allocation schemes. The challenges are such as, without any degradation in the connectivity of the network, the network interference needs to be minimized. But it is not possible in static channel assignment. If an identical channel assignment algorithm is used, it allows the same group of channels to each node. In this approach, the connectivity is not degraded but it has maximum interference due to interfering with each link with the other links on the same channel. On the other hand, if the channel is assigned based on the least used channel in the neighbourhood, the interference can be eliminated but the connectivity of the network is disrupted. Channel assignment and routing are interdependent. The topology of the network and the link bandwidth can affect the traffic routing demand. If traffic routing demand is determined the topology of the network and the required bandwidth of each link can also be determined which in turn affects the allocation of channel. So many studies consider channel assignment and routing together for the effective utilization of channel resources.

Dynamic Channel Assignment

It enforces the nodes to shift their interfaces dynamically from one channel to another channel between successive data transmissions. The dynamic channel allocation is discussed by Gong et al. (2009). They require tight synchronizations among nodes. Dynamic approaches are used only for single radio nodes working over multiple frequencies since they cannot exploit the advantages of multi-radio networks. The challenges are such as the delay while switching the interface from channel to channel called channel switching delay should not be neglected. Inside the same frequency band, channel switching delay is from few hundred microseconds to few milliseconds. If it is switching from one frequency band to another then the delay will be more. Amongst the interfaces, a coordination mechanism is required. If two interfaces need to connect, they need to switch the same channel. A Control mechanism is essential to allow interfaces to exchange.

Semi Dynamic Channel Assignment

Static channel assignments could be easily extended to be semi-dynamic, Mohsenian Rad et al. (2007) if node refreshes the channel assigned to the radios on a consistent period time. In semi-dynamic channel assignment, Alicherry et al. (2006), adapts traffic pattern changes and interference from both internal and external sources. The challenge is that the frequent channel switching takes place among the communicating nodes. Also, higher switching overhead and better coordination mechanism are needed.

Hybrid Channel Assignment

For fixed radio interface semi-dynamic channel assignment and the other interfaces are controlled dynamically. In hybrid channel assignment, wireless nodes will not share a common interface with their neighbours because the dynamic radio switches to the channel of the neighbouring nodes to make the connection. In the fixed allocation scheme the protocol overheads are low and it is based on traffic patterns in the network. Dynamic allocation experiences high channel switching overheads. Hybrid multiple radio multiple channel WMN combines the merits of fixed and dynamic allocation methods. In this dynamic interface will be able to change the channels very often while static interface works on fixed channels. The static allocation intends to maximize the throughput of the network from the end user to the gateway. Dynamic allocation works on-demand basis. This architecture guarantees eminence paths by fixed links and improves the connectivity and adaptivity of the network by dynamic links. Channel switching overhead is reduced considerably.

INTERFERENCE MODELS

The procedure for channel allocation must consider the interference between links internal to the network and from external sources. The interference models are protocol and physical model.

Protocol Model

In the protocol model, the links are assumed to be ideal and the interference depends on the distance separating nodes. The physical interference model is also known as the cumulative interference model. It depends on the distance, Signal to Interference Noise Ratio (SINR) levels, fading, path loss and the other factors that affect the strength of a signal. In the interference model, the links are assumed to be ideal and the interference is determined by the distance separating two nodes. Interference-related constraints are adopted by the IEEE 802.11 standard. It states that any two links that share at least one of their vertices cannot use the same channel at the same time. In the protocol model, each node 'n_i' will have a transmission range 'R' and interference range 'R_0'. Usually, the interference range is larger than the transmission range. In this model if node 'n_i' transmits to node 'n_j', the transmission is successful if the distance between 'n_i' and 'n_j' is within R and no other nodes within 'R_0' of 'n_j' are transmitting at the same time. If 'n_j' needs to send acknowledgment (ACK) packets to 'n_i' for a reliable transmission, the protocol model also requires that no other nodes within 'R_0' of 'n_i' are transmitting simultaneously. This enables transmission in both directions to be successful.

Physical Model

The physical model is more accurate than the protocol model since, in different environments, the attenuation is taken into consideration. The protocol model is simple and easy to derive competent channel assignment procedures. Most of the channel assignment procedures are based on the protocol model. The interference among two links depends on their physical distance and channel separation. If the physical distance between the links is fixed, interference decreases with the increasing channel separation. When the channel separation is fixed, interference decreases with the increasing physical distance. Under fixed channel separation, the interference range is defined as the maximum distance at which the two links do not interfere with each other. The main challenge is the design of the channel allocation algorithm, under an interference model which is more accurate. Most of the works are protocol model, in this case, the interference is not accurate. The protocol model is not suitable for the real-time channel assignment. Multiple channels with multiple radios will overcome interference and the capacity is increased. The study of interference together with carrier sensing can be used in single-channel WMN mainly due to Investigate the behaviour of wireless channels with interference at the first step. The observations can be applied to networks with multiple channels and multiple radios. Interference in WMN can be studied by investigating its relationship with carrier sensing. IEEE 802.11 standard recommends the use of Request to Send and Clear to Send (RTS/CTS) to eliminate the hidden terminal problem. The RTS and CTS procedure is ineffective in eliminating the hidden terminal problem and it fails to increase the multi-hop capacity. The capacity problem in WMN can be eliminated by equipping mesh routs with multiple radios tuned to a non-overlapping channels.

MULTI-RADIO MULTI-CHANNEL WMN

In a single-channel multi-hop network, interference happens among close by flows known as inter-flow interference and among the close by hops in a single flow called intra-flow interference, leading to the degradation of the network performance. Due to this, a single radio single channel approach is not attractive. To completely avail the advantages of multiple channels, the channel must be varied often with the network traffic. Since the single-channel wireless mesh network is not adequate in network capacity and exploitation for, the time being, multi-channel multi-hop wireless networks become the main subject for researchers.

Channel assignment and routing in Multi-Radio Multi-Channel (MRMC) environment is a challenging issue due to the deployment of multiple radios on nodes. To exploit the spectrum for multiple radio multiple channel networks for researches, the non-overlapping channels in the free Industrial Scientific and Medical band (ISM) provides opportunity. The key factors that governing the channel assignment schemes are interference traffic pattern and multipath connectivity converged networks. The wireless mesh network is a fully wireless network that employs multi-hop ad hoc networking techniques to forward traffic to the internet. Mesh network uses dedicated nodes to build wireless backbone to provide multi-hop connectivity between the nodes. The channel assignment and routing problems have been addressed in several studies. But most of the studies single radio is assigned to each node. Equipping the nodes with multiple radios in a non-overlapping channel to utilize the spectrum efficiently is a challenging issue. Routing plays an important role in multi-radio multichannel networks in providing end to end connectivity. The intra flow interference occurs if the nodes are placed in the interference range of each other in the same flow. If multiple flows on the single-channel, then interflow interference will occur. Interference can be classified into two types internal and external. The external interference appears when two or more co-existing wireless networks work in the same frequency channel. It can be eliminated by using different non-overlapping channels. The channel assignment and routing are interdependent. The goal of channel assignment in multiple-radio multiple channel WMN is to minimize the interference as much as possible. Interference cannot be eliminated from the network due to the limited range of available frequency. Both external and internal interference limits system performance.

When the number of concurrent transmissions increases WMN faces interference problems. If any two adjacent nodes (or) links need to communicate on the channel with the same frequency band, the concurrent transmission cannot be successful. Due to interference between the links, the efficiency of the links can be decreased. So, to nullify this problem mesh routers can be furnished with multiple numbers of radios and it can be configured to operate on different channels. With the help of the channel assignment algorithm, radio can be allotted to the channel. Without any interference issues, the transmission and reception of data can be done simultaneously in multiple channel WMN.

The nature of the WMN protocol is self-configuring and self-healing. Due to this nature, the protocol is extremely distributed. So, each node in the networks behaves selfishly to increase its benefit from the resources of the network. Using the channel assignment, the selfishness of nodes can be exploited exclusively for their benefit. Mechanisms are required to address the behaviour of the selfish nodes inside the network for effective functioning. In such a competitive network, the traditional standard protocols need to be revisited to meet these challenges. To reduce the interference, the node should minimize the number of neighbours with which it will share a common channel. Channel assignment is based on the number of channels available, interfaces, and topology of the network and communication request. Due to the limited number of available channels, interference cannot be eliminated. If the channel assignment is not proper then due to hidden terminal problem throughput will be degraded. The key to design an efficient routing protocol is to build up a channel with a proper routing metric. To improve the throughput multi-channel multicast algorithm is proposed to reduce the number of neighbouring nodes and hop count distances between the source and destinations. The multicast protocol for WMN includes two procedures first is to build an effective multicast structure and the second tries to allocate channels for minimizing interference.

APPLICATIONS OF WIRELESS MESH NETWORKS

The applications of Wireless Mesh Networks are as follows:

- Public safety: To have uninterrupted communication even when telephone networks are down, the WMN can be used. In public safety networks, the WMN keeps the emergency networks alive always.
- Health care: To monitor the immovable patient, the optimal method is health care telemedicine. With the help of WMN, patient details can be transferred and secured.
- Disaster recovery: Due to disaster if the network is down the WMN replaces the network as an emergency responder with minimum cost.
- Broadband access network: By using WMN structure, broadband access network services can be provided to the rural area.
- Educational institute: To connect everyone without interruption WMN can be used in universities, colleges, and schools.
- Military requirements: In some countries, the military vehicles are placed with mesh nodes communicate with another vehicle in real-time with secure data transfer.

KEY CONTRIBUTION

The following are the key contributions of the proposed work.

- The channel assignment method ensures minimum interference among the nodes in the network and provides maximum connectivity.
- The link quality is maintained by determining the delay on the links between any two nodes.
- The channel switching delay is considered by routing protocol to determine the end to end connectivity across the WMN.
- The channel assignment scheme quickly responds to the routing load, mobility and node failures by reassigning the channels when and where necessary.

LITERATURE SURVEY

Wireless Mesh Network is a collection of mesh routers that gather and route the traffic flow originated by the mesh clients. Generally, the mesh routers are immobile and are endowed with many radio interfaces.

Clients are mostly movable and through mesh routers the information is routed to the corresponding recipient. In traditional wireless networks, every node will have only one radio and it functions on a single channel. In a network, if the number of nodes increases, the capacity of the network decreases due to interference and collision of packets (Chiu et al. (2009)). In WMN the decrease in network capacity is a serious issue because of the interference of neighbouring nodes on the same path and the neighbouring paths. Channel assignment is an effective mechanism to use multiple orthogonal channels to reduce the number of interfering nodes and to improve the capacity of a network. It seeks to identify a suitable path among channels and radio interfaces at every node to reduce the interference and to improve the capacity of the network. Though channel assignment minimizes the interference it causes some major issues that influence the performance of the network. As in the literature, many CA methods based on different parameters have been proposed. In Skalli et al. (2007), the survey on channel assignment is done based on the type of nodes as static (fixed), dynamic (mobile) and hybrid (both fixed and mobile). Channel assignment is formulated as centralized and distributed CA algorithms (Si et al. (2010)). In Ding et al. (2011), CA is done based on the type of antenna used as an omnidirectional and directional antenna. The channel assignment algorithms are grouped based on joint routing with channel assignment (Qu et al. (2016)). Routing, channel assignment, and topology preservation is jointly addressed (Liu et al. (2012)).

DESIGN ISSUES IN CHANNEL ASSIGNMENT

If a node is enabled to operate on the various channels, the capacity of the node can be improved, but it causes a number of issues that decreases the performance of the network. In this section, the major design issues that needed to be addressed by the CA protocol are discussed.

Connectivity

The channel assignment is an actual mechanism to reduce interference and to maximize the network capacity of multiple radio wireless networks. Some channel assignment techniques may eliminate one or more logical interfaces from the main topology leading network partition (or) path length may increase (Mohsenian Rad et al. (2007)). The CA algorithm aims to minimize interference thereby increasing throughput and capacity. Moreover, while minimizing the interference, the connectivity of the network should be maintained.

Routing

To design a channel assignment algorithm that avoids interference, the knowledge about each link's traffic is required. The routing decision is based on the connectivity of the network. The capacity of each link can be improved by channel assignment. Based on channel assignment the network connectivity can be determined. Channel assignment and routing should be done jointly for multiple radio multiple channel WMN to reduce the interference and balance the network traffic among various channels.

Stability

In distributed CA algorithm, each node self-reliantly selects the channel based on the statistics about the k-hop neighbour. There is a possibility that at the same time, two nodes can choose the single-channel and then shift back due to the overload of the channel. The overload is mainly because the decision taken by the CA scheme is based on the channel interference index. Nodes in the network pick the new links based on the former hop and this condition may re-occur for all the k-hop nodes in the recognized path and it may result in ripple effect (Raniwala et al. (2005)).

Problems in Mesh Client

In WMN, the client may leave or linked to any route with dynamism. When a client leaves from one route and it should join the new router immediately. A new connection should provide a new route with connectivity to receive the data continuously. To reduce handoff, mesh client should identify new network connectivity which has less re-routed traffic.

Fault Tolerance

In WMN, nodes may fail because of hardware and software issues. There may be a chance of link breakage due to co-channel interference or temporary obstacle. If there is a link failure or node failure, the network should not be affected by network partition and it should work by self-healing mechanism. In Multi radio WMN, every node is equipped with a greater number of radios. Therefore, channel assignment combined with fault tolerance is more complex. The channel assignment needs to choose one link among the number of available links to get a better solution.

COMPARISON OF CHANNEL ASSIGNMENT METHODS

Figure 1 shows the categorization of different channel allocation schemes for multiple-radio multiple-channel wireless mesh networks. For each technique, the basic idea, merits, demerits, and comparison are discussed.

Channel Interference Index

There are five types of interference in a multi-hop WMN that influences the performance of a network.

Intra Flow Interference

The interference occurs among two consecutive links of the same flow on the same channel. The simultaneous transmission and reception of data among inter-mediate nodes are not possible in this case.

Inter-flow Interference

The interference arises if two links of different flows are utilising the same channel. In this case, both the flows cannot be able to communicate at the same time over the links.

External Interference

This interference is produced by the devices in the network and it does not have any hold on the network provider but uses the same frequency.

Hidden Terminal Node Interference

Interference occurs due to collisions amongst transmissions by a node that is not able to hear the transmission of its neighbour node.

Rate Interference

It is caused due to the lower rate of a node. Because of the lower rate nodes, the effectiveness of other nodes sending at higher data is affected (Niranjan et al. (2006)).

Figure 1. Taxonomy of channel assignment approaches

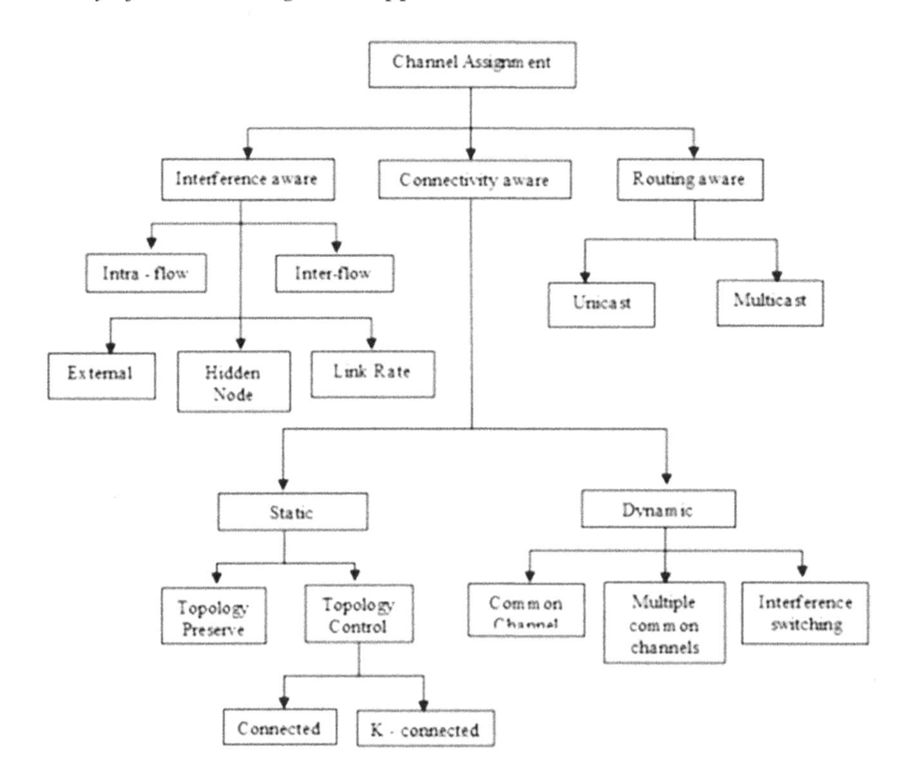

The interference index of a channel can be categorized for channel selection as interference aware channel interference index, traffic aware channel interference index, and throughput aware channel interference index.

Interference Aware Channel Interference Index

The main function of interference aware channel interference index is to calculate the number of nodes per link in every channel. The channel with a minimum number of intermediate nodes can be identified as the suggested channel. The aim is to reduce or to nullify the total interference in the network. The numbers of approaches are available to model the interference according to the protocol model (Mogaibel et al. (2012)) and physical model (Galvez et al. (2013) and (2015)). The number of links is minimized using conflict graph and the same channel is shared in a few interference aware channel assignment approaches (Devare et al. (2014)). Riggio et al. (2011) and Chaudhry et al. (2012), proposed a methodology to reduce the number of interfaces that shares a channel in an interference range of anode or k-hops. The realistic interference index is used in Revathi et al. (2014). In Subramanian et al. (2006) and Jardosh et al. (2005), the external interference of every channel is measured by node through Radio Frequency monitoring periodically. In Reality Check Breadth First Search (RC-BFS) the physical model and the protocol model are merged by reality check mechanism to satisfy the inefficiency of protocol model which is considered in Breadth First Search Channel Assignment (BFS-CA).

Traffic Aware Channel Interference Index

Traffic aware channel assignment allows the channel between the interfaces depends on the current traffic. Sridhar et al. (2009), proposed a channel interference metric that computes the total traffic over all the interfering links in WMN. The channel interference metric is calculated using the following equation.

$$\text{channel interference index} = \sum_{j=1}^{k} Channel\ utilisation\ table(i,j) \tag{1}$$

'k' is the interference range. The channel utilization table can be maintained by every node so that the total of data sent by the node 'j' on the channel can be tracked. The protocol model is enriched by taking an account of traffic carried and path loss. It is proposed by (Chiu et al. (2009)). Using protocol model channel interference index can be computed as,

$$channel\ interference\ index = \sum_{j=1}^{k} \frac{channel\ utilisation\ table(i,j)}{j^{y}} \tag{2}$$

'y' is path loss exponential. Raniwala et al. (2004), proposed Load Aware Channel Assignment (LACA) where the interfaces are arranged in decreasing order based on their traffic load. But Skalli et al. (2007), arranged the interfaces based on their ranks. The formula to find rank includes the traffic load, distance from gateway and number of interfaces of a particular node.

$$Rank\ of\ a\ node = \frac{Traffic\ load\ on\ a\ node}{Distance \times Number\ of\ interfaces} \tag{3}$$

Centralized Rank Based Channel Assignment (CRB-CA) proposed by Sarasvathi et al. (2012), is to improve the Mesh based Traffic and Interference aware Channel Assignment (MesTic) ranking formula by adding link quality with link load, the distance and the number of interfaces.

$$Rank\ of\ a\ node = \frac{WCETT\ of\ a\ node}{Distance \times Number\ of\ Interfaces} \tag{4}$$

WCETT is Weighted Cumulative Expected Transmission Time. Centralized load aware channel assignment proposed by Athanasiou et al. (2011), considers the load in terms of the state of the channel, the number of users involved and the traffic load. It depends on the load in the channel Bakshi et al. (2011), projected traffic aware CA algorithm. In this, a channel with the least traffic is identified to solve congestion in links. Here, a central node will monitor the practicability of the recognized path. If there is any congestion identified in any link or if the path selected is not feasible, the node assigns them with other least traffic channels. In Mohsenian Rad et al. (2009), every node has an eye over the ongoing links traffic and if it identifies any congestion it resolves it by switching the congested interface with the least load channel.

Throughput Aware Channel Interference Index

The total throughput from source to the destination of all the available flows is considered as channel selection index. This is to guarantee fairness between all the existing flows in the network (Shin et al. (2006)). Cross section throughput is proposed by Raniwala et al. (2004), for selecting the channel. Cross section method calculates the total obtainable effective throughput in WMN. Cross section method is proposed to increase the sum of throughput among all pairs of nodes in the network. In the literature, either interflow or intra flow interference is considered to assign channel. It may lead to inaccurate bandwidth estimation. If interference and congestion control mechanism is combined, the network performance can be improved. In the proposed work both interflow and intra flow interference are considered and congestion is also combined. Based on connectivity, due to link failure, there may be performance degradation. The mechanism has to be devised to inform about the link failure immediately to improve the performance of the network. If stability in a node or network is provided, the performance can be further improved. In the proposed work traffic and node, stability is considered to overcome link failure. Though, the number of mechanisms available to assign channel in WMN, interference is the challenging issue in channel assignment. Minimizing the interference and improving throughput can be done using priority based interference aware channel assignment with bandwidth reservation.

PRIORITY BASED INTERFERENCE AWARE CHANNEL ASSIGNMENT WITH BANDWIDTH RESERVATION

There are five phases for the proposed Priority Based Interference Aware Channel Allocation with Bandwidth Reservation Algorithm. They are such as neighbour identification, interference minimization, QoS aware path finding and admission control, Prioritisation, and QoS violation recognition and retrieval.

Neighbour Identification

The node in the network periodically sends HELLO packets to contain its node ID through its all existing radio interfaces. Nodes within its transmission range will receive this and note its neighbour ID, the link with which it received the HELLO packets and the channel which is assigned to that particular interface. With all this information, all nodes in a network must build a neighbour table. Neighbours are connected to the node directly within its communication range. Using neighbour identification phase each node can identify its multiple link neighbours. In the proposed PBIACA-BR approach all the nodes know the reserved bandwidth for every flow that is passing through the node and all the nodes know the link with which the flow is passing. So, using this two information the aggregate flow rate is calculated. This ensures accurate bandwidth estimation. Since the interference region is twice the transmission range, each node sends HELLO packets to cover all the nodes within the transmission region. Each node that receives these HELLO packets from all its neighbours who are in its interference range and their identities are updated in its routing table. The HELLO packet has the details about the outgoing links of a node and the amount of traffic it has reserved for a flow. Using this information local conflict graph is constructed and clique constraints are derived.

QoS Aware Path Finding and Admission Control

QoS aware path finding has four sub-phases. If a node wants to identify a path for an incoming flow which demand for QoS sends route request. During the route discovery phase, bandwidth estimation, admission control, load balancing, and resource reservations are performed.

Prioritisation

The Interference aware channel allocation with bandwidth reservation provides resource reservation for multiple channels multiple radio wireless mesh networks. In this algorithm, every each node is furnished with the number of radios. Every radio in a node is allotted to orthogonal channels to minimize interference. The proposed algorithm develops numerous channel selection principles to exploit all the existing channels of the WMN. By these selection principles, every node in the network can choose the finest channel to satisfy the delay and to minimize the entire interference for all the available multicast sessions. In this method, the priority factor is considered to stop high-priority multicast sessions from experiencing heavy interference than low-priority multicast sessions thereby reducing interference. Based on the interference, each link is assigned with priority. A low interference link is provided with high priority.

QoS Violation Recognition and Retrieval

Once the admission control and resource reservation process are completed, the sender can transmit the data to the destination. If the bandwidth requirements are not met, it is considered as a QoS violation.

CONGESTION AWARE CHANNEL ASSIGNMENT

Multiple radio multiple channel WMN uses IEEE 802.11 radios. It has a cross layer design. Scheduling is done in the physical layer, channel allocation in the data link layer, routing in the network layer and congestion control at the transport layer. To admit a flow, the parameters like the number of packets that travels in a path, the channel used for each link and the capacity of each link have to be determined.

Distributed Optimum Congestion Control Algorithm

The proposed algorithm, distributed optimal congestion control and channel assignment algorithm has six phases. They are as follows: network formation and neighbour identification, price based congestion control, decoupling approach, local channel assignment, distributed channel allocation, and optimum congestion control. The neighbour identification phase is the same as the previous algorithm.

Price Based Congestion Control

In price base congestion control, only the flow is checked. If the available bandwidth is greater than the required bandwidth flow is admitted. Once the flow is admitted, the link price needs to be checked. If the link price is equal to zero at equilibrium, the link is considered as congestion less link and the flow can be transmitted on the same channel. If the link price is not equal to zero the particular link can be identified as a critical link and the flow need to be transmitted on a different link. Therefore, channel reassignment is necessary.

Decoupling Approach

The decoupling approach has five interlinked steps. They are as follows: channel utility, channel rate calculation, the formation of interference matrix, link selection, and channel assignment.

Local Channel Link Assignment

In the local channel link assignment, each node verifies the link price for every flow that passes through the node. If the link price is non zero at equilibrium, the node assumes that the link is critical for the flow and it identifies another link to transmit the flow. To select the channel link, dynamic link quality measurement and link distortion detection schemes are used. To calculate link quality metric, channel idle time, packet delivery ratio and packet lost due to hidden terminal problems are considered.

Distributed Channel Allocation

In the distributed channel allocation phase the channel allocation is considered as binary channel allocation. For any link $l \in L$ and any channel belongs to set of available channels, it can be considered as '1' and channel can be allotted to the link l. Else it is '0'. The transmit power of the sender node at link 'l' is 'p_l'. The path loss (Q_{il}) from the sender node 'i' to the receiver node 'j' can be represented as

$$Q_{il} = \frac{G_t G_r \lambda^2}{(4\pi)^2 d_{il}^2} \tag{5}$$

d_{il}→ Distance among sender node and receiver node
G_t→ Antenna gains of the sender.
G_r→ Antenna gains of the receiver.
λ→ Wavelength of the signal.

The current data rate of the link can be calculated using the equation if the path loss is known.

Optimum Congestion Control

In optimum congestion control, all the possible parameters for the link are estimated. Based on these estimated values, the maximum delay in the link is calculated.

TRAFFIC AWARE CHANNEL ASSIGNMENT WITH NODE STABILITY

The performance of WMN is highly affected by interference and congestion at gateways resulting in remarkable packet losses and delays. To improve the performance of the network directional and smart antenna is used. Routing aims to identify better paths depend on the specified routing metrics. To obtain better network performance, the routing protocol should give a positive impact on the stability of the network, capture the characteristics of the mesh network, compute paths properly and avoid loops. Network stability is one of the most important metrics to consider the performance of the network but it is often neglected by most of the routing protocols. If the rerouting increases, naturally it will affect the stability of the network. This instability is caused due to the quality fluctuations in the link and degradation in the performance. Instability may cause out of order in packet delivery, jitter and packet loss which are not acceptable for efficient data transmission. There are five phases for the proposed traffic aware channel assignment with node stability. They are as follows: network formation and neighbour identification, minimum variation channel allocation, link quality metric, node stability using stability index factor algorithm, and loop-free forwarding.

Minimum Variation Channel Allocation

In MVCA a new set of flow rate f(*l*) and the maximum number of changes allowed to the channel is considered. To identify the radio be allotted to a new channel, collision domain utilization needs to be identified.

Link Quality Metric

Link Quality Metric (LQM) is the characteristics of the link which is stored in the routing table. It is a function of interference ratio and congestion level. Interference Ratio (IR) calculates the interference level, through Signal to Noise Ratio (SNR) and signal to interference plus noise ratio.

Node Stability Using SIF Algorithm

The stability index is calculated using the history of the link quality which includes the current quality of the link also. The link is acceptable if the link quality metric is less than the threshold value. If the link quality metric is greater than the threshold value it is considered as an unacceptable link.

Loop Free Forwarding Algorithm

Due to the presence of a loop, the packets are routed indefinitely in the same path. Loops may cause an increase in network traffic resulting in more packet loss and a decrease in bandwidth availability. Loops may cause unnecessary routes. To prevent loops in the TACA-NS loop-free forwarding algorithm (LFFA) is included.

EFFICIENT KEY ESTABLISHMENT PROTOCOL

WMN is vulnerable to active and passive attacks due to its wireless multi hop nature (Khalili et al. (2003)). Passive attacks lead to confidentiality violation and active attack results in authentication and integrity compromises (Siddiqui et al. (2007)). Therefore, it is necessary to provide the safety mechanisms to guard the transmitted information. To sustain secure communication between the user, and key establishments can be provided. Baseline protocol has two stages namely, distributed key generation stage and authentication stage. Nodes in the network generate the master key collaboratively and create their private keys and then the nodes authenticate each other. Initially, users generate their shares of the master private key collaboratively. The master public key shares are generated by multiplying the elliptic curve group with master private key shares. Elliptic curve cryptography is public encryption based on the elliptic curve theory. Using this faster, a smaller and more efficient cryptographic key can be created. When a user receives these master public key shares, the master public key of the particular network is reconstructed. From then on, all the users send a request for its share of the private keys. The users who received this request can generate its share as a product of the shares of their master private key and the public key of the user who requested the share and transmits. When the requested user receives a sufficient number of shares, the user reconstructs its private key. Though the baseline protocol is distributed, it has two dangerous disadvantages. They are, the transmission delay being very high and these protocols are easily affected by security vulnerabilities. The elliptic curve cryptography employs non-pairing homomorphic encryption for secure WSNs as it is presented by Bharat Bhushan et al. (2019). To secure the data transmission in WSNs, the construction of the master key affects the bandwidth utilization of the network directly. If there are 'n' users in the network involves in the construction of master private key, then a minimum of n(n-1) request messages have to be sent by the users. After receiving this, the user needs to transmit its master public key. So, another 'n' number of packets need to be transmitted. This requires more bandwidth and also delay in packet transmission. Baseline protocols are easily affected by the internal attacker. The internal attackers can get the master private key of the network itself.

Master Private Key Generation

In a master private key generation, all mesh routers generate shares collaboratively. Initially, every mesh router R_i picks 'n' number of shares and $t_{i,y}$ secrets with the polynomial $f_{i,y}(a)$ of degree $(k-1)$ such that $f_{i,y}(0)=t_{i,y}$. Where $1 \leq y \leq n$, i is i^{th} user and y is y^{th} share, k – threshold value. Then all the mesh routers will calculate the sub share $\sigma_{j,i,y}$ for the master private key by estimating the polynomial generated for the mesh router 'i' exchanges information to its components. If the mesh router Ri receives $(r-1)n$ sub shares, it computes its master private key subshare γ_{iy}.

User Private Key Creation

In the user private key creation stage, each node in the network creates its private key in coordination with other nodes in the network. Once the node 'i' completes its share computations for the master private key, the node broadcasts a request packet to construct its private key. When a node 'j' receives this request packet and if it has finished its share computation, it can help in the process of generating user's private key. Else it can save this request and after completing its master private key share it can handle this request. If the node 'j' already estimated the shares of its master's private key, it can transmit a message to the node 'i', requesting to include the number of shares that the node 'j' can contribute.

Key Authentication

The cluster head generates the master private key and checks if the master private key share is more than 1. If the share is more than 1, the cluster head partially shares the private key to the clients. Then the cluster head waits for the response from the clients. Once the response received from the clients, cluster head shares the private key to the clients as discussed by Bhushan et al. (2019). After receiving the private key from the cluster head, mesh client generates the user's private key. Once the user private key is generated, the client gets the public key. As soon as the mesh client receives the public key, it is authenticated so that data can be sent to the clients. Likewise, authentication is provided between the cluster head and clients for secure communication.

PERFORMANCE EVALUATION

The performance of the proposed system is explained as follows.

Figure 2. Comparison of PBIACA-BR performance based on throughput

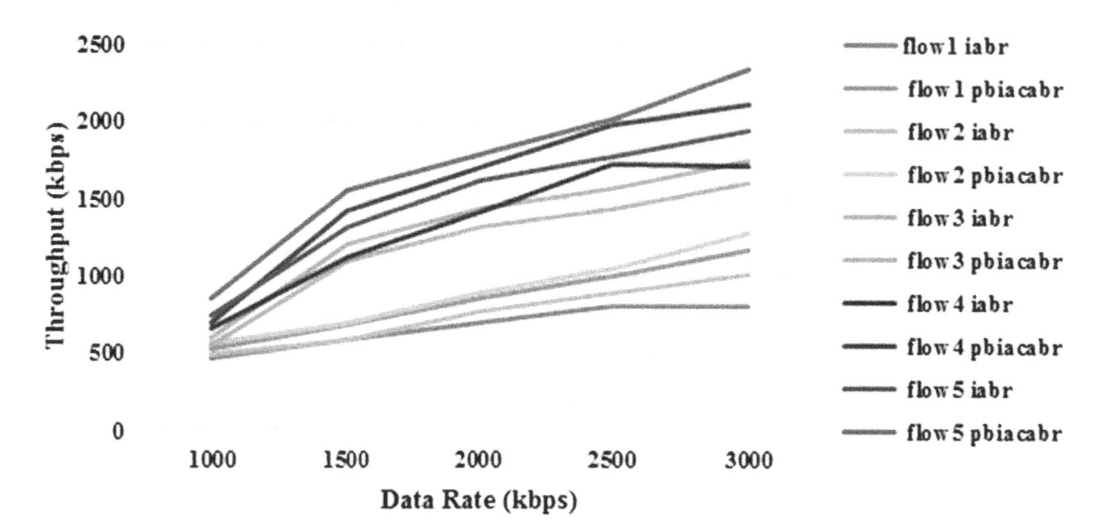

Figure 2 shows the comparison of PBIACA-BR performance based on throughput performance. Bandwidth is properly allocated in PBIACA-BR approach than the IABR approach. In general, as data rate increases throughput also increases but after a certain limit it will be decreased due to insufficient bandwidth but in PBIACA-BR the throughput keeps increasing because the bandwidth is properly allocated.

Figure 3. Comparison of DOCCA performance based on throughput

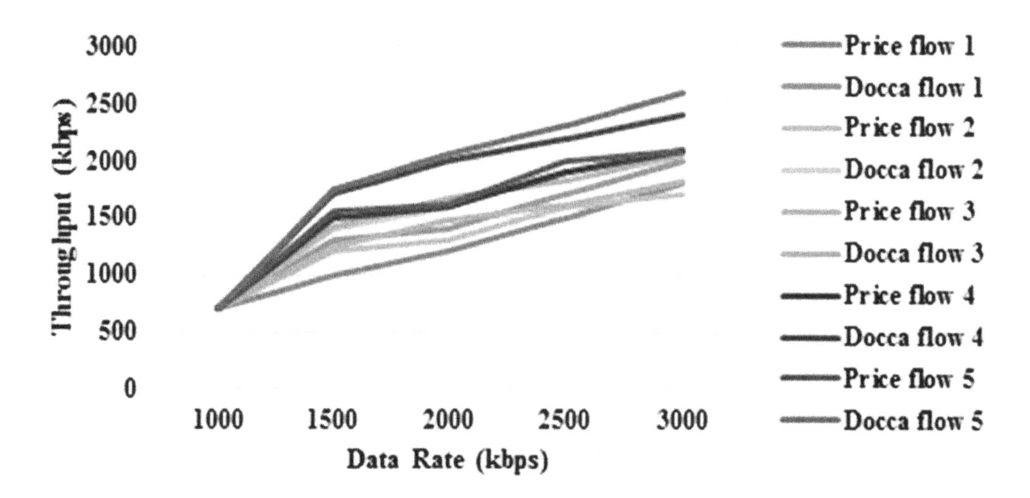

Each node is enriched with four radios. Increasing the number of radios minimizes the number of links sharing a radio. This increases the performance of the network. In the same way, if the number of channels to the neighbouring links can be increased so that the mutual interference and congestion in the network can be reduced and the throughput can be improved. It is shown in Figure 3. The performance of TACA-NS is compared with Traffic aware channel allocation without node stability for increasing data rate. Figure 4 shows that TNCA-NS outperforms TACA.

Figure 4. Comparison of TACA-NS performance based on throughput

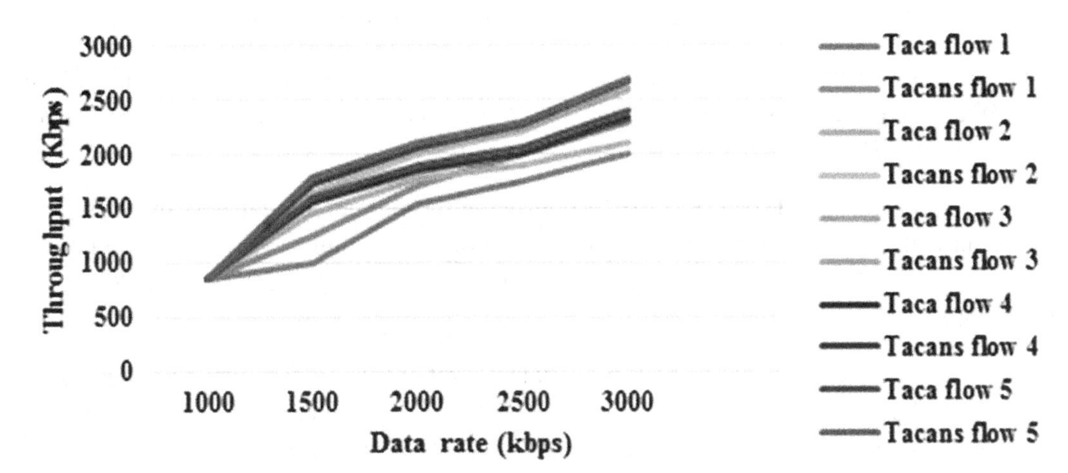

CONCLUSION

The focus of this work is assigning channels to multiple radio and multiple channel wireless mesh networks to improve throughput, delay, average energy, and residual energy. To improve upon these parameters, priority-based interference aware channel assignment with bandwidth reservation, distributed optimal congestion control and channel assignment and traffic aware channel assignment with node stability algorithm is proposed. Comparative analysis was done with all the three algorithms and simulation results of each algorithm were analyzed to highlight the prime aspect of the proposed approaches. When compared with PBIACA-BR, DOCCA achieved to an average of 16% improvement in throughput. This is because that congestion is not considered in PBIACA-BR. Due to congestion packets are dropped there by the effectiveness of the network reduced. With the congestion control method, if traffic is known and node stability is also considered, then the throughput is increased an average of 7% than DOCCA. Traffic aware channel assignment with node stability provides high throughput.

REFERENCES

Akyildiz, I. F., Wang, X., & Wang, W. (2005). Wireless mesh networks- a survey. *Computer Networks*, *47*(4), 445–487. doi:10.1016/j.comnet.2004.12.001

Alicherry, M., Bhatia, R., & Li, E. (2006). Joint channel assignment and routing for throughput optimization in multi radio wireless mesh networks. *IEEE Journal of Selected Areas in Communication*, *24*(11), 1960-1971.

Athanasiou, G., Broustis, I., & Tassiulas, I. (2011). Efficient load-aware channel allocation in wireless mesh networks. *Journal of Computer Networks and Communication*, *31*(7), 1–13. doi:10.1155/2011/972051

Bakshi, B., Khorsandi, S., & Capone, A. (2011). On-line joint QoS routing and channel assignment in multi-channel multi-radio wireless mesh networks. *Computer Communications, 34*(2), 1342–1360. doi:10.1016/j.comcom.2011.02.001

Bhushan, B., & Sahoo, G. (2019). Secure Location-Based Aggregator Node Selection Scheme in Wireless Sensor Networks, *Proceedings of ICETIT 2019, Emerging Trends in Information Technology.*

Bhushan & Sahoo. (2019). A Hybrid Secure and Energy Efficient Cluster Based Intrusion Detection system for Wireless Sensing Environment. *2019 2nd International Conference on Signal Processing and Communication (ICSPC).*

Chaudhry, A., Ahmad, N., & Hafez, R. (2012). Improving throughput and fairness by improved channel assignment using topology control based on power control for multi-radio multi-channel wireless mesh networks. *EURASIP Journal on Wireless Communications and Networking, 2012*(1), 1–25. doi:10.1186/1687-1499-2012-155

Chiu, H. S., Yeung, K., & Lui, K. S. (2009). An efficient joint channel assignment and routing protocol for IEEE 802.11- based multi-channel multi-interface mobile ad hoc networks. *IEEE Transactions on Wireless Communications, 8*(4), 1706–1715. doi:10.1109/TWC.2009.080174

Crichigno, M., Wu, Y., & Shu, W. (2008). Protocols and Architectures for Channel Assignment in Wireless Mesh Networks. *Ad Hoc Networks, 6*(7), 1051–1077. doi:10.1016/j.adhoc.2007.10.002

Devare, M.A.S. (2014). Channel allocation using ARS and BFS-CA analysis in WMN. *International Journal of Advanced Engineering Nano Technology*, 1-6.

Ding, Y., & Xiao, L. (2011). Channel allocation in multi-channel wireless mesh networks. *Computer Communications, 34*(7), 803–815. doi:10.1016/j.comcom.2010.10.011

Gálvez, J. J., & Ruiz, P. M. (2013). Efficient rate allocation, routing and channel assignment in wireless mesh networks supporting dynamic traffic flows. *Ad Hoc Networks, 11*(6), 1765–1781. doi:10.1016/j.adhoc.2013.04.002

Galvez, J. J., & Ruiz, P. M. (2015). Joint link rate allocation, routing and channel assignment in multi-rate multi-channel wireless networks. *Ad Hoc Networks, 29*(c), 78–98. doi:10.1016/j.adhoc.2015.02.002

Gong, M. X., Midkiff, S. F., & Mao, S. (2009). On-demand routing and channel assignment in multi-channel mobile ad hoc networks. *Ad Hoc Networks, 7*(1), 63–78. doi:10.1016/j.adhoc.2007.11.011

Jardosh, A. P., Ramachandran, K. N., Almerith, K. C., & Belding-Royer, E. M. (2005). Understanding congestion in IEEE 802.11b wireless networks. *Proceedings of the 5th ACM SIGCOMM Conference on Internet Measurement*, 279-292. 10.1145/1330107.1330140

Liu, F., & Bai, Y. (2012). An overview of topology control mechanism in multi-radio multi-channel wireless mesh networks. *EURASIP Journal on Wireless Communications and Networking, 2012*(1), 1–12. doi:10.1186/1687-1499-2012-324

Marina, M. K., Das, S. R., & Sunramaniam, A. P. (2010). A topology control approach for utilizing multiple channels in multi-radio wireless mesh networks. *Computer Networks, 54*(2), 241–256. doi:10.1016/j.comnet.2009.05.015

Mogaibel, H. A., Othman, M., Subramaniam, S., & Hamid, N. A. W. A. (2012). On-demand channel reservation scheme for common traffic in wireless mesh networks. *Journal of Network and Computer Applications*, *35*(4), 132–151. doi:10.1016/j.jnca.2012.01.017

Mohsenian Rad, A. H., & Wong, V. W. S. (2007). Joint logical topology design, interface assignment, channel allocation and routing for multichannel wireless mesh networks. *IEEE Transactions on Wireless Communications*, *6*(12), 4432–4440. doi:10.1109/TWC.2007.060312

Niranjan. Pandey, S., & Ganz, A. (2006). Design and evaluation of multichannel multirate wireless networks. *Mobile Networks and Applications*, *11*(5), 697–709. doi:10.100711036-006-7796-7

Puri, D., & Bhushan, B. (2019). Enhancement of security and energy efficiency in WSNs: Machine Learning to the rescue. *2019 International Conference on Computing, Communication, and Intelligent Systems (ICCCIS)*. 10.1109/ICCCIS48478.2019.8974465

Qu, Y., Ng, B., & Seah, W. (2016). A survey of routing and channel assignment in multi-channel multi-radio (WMNs). *Journal of Network and Computer Applications*, *65*(C), 120–130. doi:10.1016/j.jnca.2016.02.017

Raniwala, A., & Chiueh, T. (2005). Architecture and algorithms for an IEEE 802.11-based multi-channel wireless mesh networks. *Proceedings 24th Annual Joint Conference of the IEEE Computer and Communications Societies, 3*, 2223-2234.

Raniwala, A., Gopalan, K., & Chiueh, T. (2004). Centralized channel assignment and routing algorithms for multi-channel wireless mesh networks. *Computer Communications*, *8*, 50–65.

Revathi, M., & Deva Priya, S. (2014). Channel allocation for interference analysis in wireless networks. *Proceedings of ICSEM'14 – 2nd International Conference on Science, Engineering and Management*, 1-7.

Riggio, R., Rasheed, T., Testi, S., Granelli, F., & Chlamtac. (2011). Interfeerence and traffic aware channel assignment in wifi-based wireless mesh networks. *Ad Hoc Networks, 9*(5), 864-875.

Sarasvathi, V., & Iyengar, N. (2012). Centralized rank-based channel assignment for multi-radio multi-channel wireless mesh networks. 2nd *international conference on Computer, Communication, Control and Information Technology*, 182-186. 10.1016/j.protcy.2012.05.027

Shin, M., Lee, S., & Kim, Y. A. (2006). Distributed channel assignment for multi-radio wireless networks. *IEEE International Conference on Mobile Ad Hoc and Sensor Systems*, 417-426. 10.1109/MOBHOC.2006.278582

Si, W., Selvakennedy, S., & Zomaya, A. Y. (2010). An overview of channel assignment methods for multi-radio multi-channel wireless mesh networks. *Journal of Parallel and Distributed Computing*, *70*(5), 505–524. doi:10.1016/j.jpdc.2009.09.011

Siddiqui, M. S., & Hong, C. S. (2007). Security issues in wireless mesh networks. *Proceedings of the International Conference on Multimedia and Ubiquitous Engineering*, 717-722.

Skalli, H., Ghosh, S., Das, S. K., Lenzini, L., & Conti, M. (2007). Channel assignment strategies for multi-radio wireless mesh networks: Issues and solutions. *IEEE Communications Magazine*, *45*(11), 86–95. doi:10.1109/MCOM.2007.4378326

Sridhar, S., Guo, J., & Jha, S. (2009). Channel assignment in multi-radio wireless mesh networks: a graph-theoretic approach. *Proceedings of the First international Conference on Communication Systems and Networks*, 180-189. 10.1109/COMSNETS.2009.4808856

Subramanian, A., Buddhikot, M., & Miller, S. (2006). Interference aware routing in multi-radio wireless mesh networks. *2nd IEEE Workshop on Wireless Mesh Networks*, 55-63. 10.1109/WIMESH.2006.288620

Subramanian, A., Gupta, H., Das, S. R., & Cao, J. (2008). Minimum interference channel assignment in multi radio wireless mesh networks. *IEEE Transactions on Mobile Computing*, *7*(12), 1459–1473. doi:10.1109/TMC.2008.70

Chapter 9
Integrated Intrusion Detection System (IDS) for Security Enhancement in Wireless Sensor Networks

Mini Rani Sharma
Rural Development Department, India

Vikash Kumar Agarwal
RTC Institute of Technology, India

Nitish Kumar
RTC Institute of Technology, India

Santosh Kumar
https://orcid.org/0000-0003-2239-164X
ITER, Siksha 'O' Anusandhan (Deemed), India

ABSTRACT

Rapidly expanding application areas of wireless sensor networks (WSNs) that include critical civilian and military applications intensify the security concerns especially in hostile unattended environments. In order to ensure the dependability and security of continued WSN services, there is a need of alternate defense line called intrusion detection system (IDS). WSNs are comprised of a huge number of energy-constrained nodes whose battery replacement or recharging is a challenging task after their deployment. In this present work, the authors develop a simple integrated IDS scheme (SIIS) that integrates the concept of clustering along with RC4 and digital signature. For heterogeneous systems, cluster-based protocols perform best as they work on divide and conquer strategy. In order to safeguard the transmitted data, a secured, lightweight digital signature scheme that employs symmetric cryptography is used along with RC4, a synchronous stream cipher that satisfies both efficiency and security for lightweight algorithms.

DOI: 10.4018/978-1-7998-5068-7.ch009

INTRODUCTION

Wireless Sensor Networks (WSNs) comprise of several autonomous sensor nodes (SNs) distributed in varied areas of interest for gathering and cooperatively transmitting important data through a wireless medium to the most powerful node called the base station or sink node. WSNs have several applications in the fields of science and technology owing to its cheap and easy deployment process. These are used in gathering information related to human behavior, healthcare monitoring, military surveillance, highway traffic monitoring, monitoring of environmental and physical phenomenon such as earthquake, wildfire, and pollution, monitoring of industrial sites or manufacturing machine performance among others (Bao et al. 2021), (Madsen et al 2005). Security in WSNs is an important concern especially in the case of mission-critical applications, commercial application or for WSNs deployed in hostile environments. A confidential health record of a patient must not be known by the third party in healthcare application or a security gap existence in military application may lead to causalities in the battle field. It is essential to protect WSNs from varied security threats (Sharma et al. 2017), (Diaz and Sanchez 2016).

Adversary can eavesdrop onto the message being passed, compromise a node, inject fake messages, waste network resources and alter the data integrity (Pragya et al. 2017). This objective is very tough to achieve because of the constrained resources in WSNs such as energy, battery power, processing capabilities and memory. Also because of several reasons such as its distributed and open nature, deployment in unattended environment, WSNs are vulnerable to several types of attacks (Pragya et al. 2017), (Raghuram et al. 2018). These limitations of WSNs reduces the impact of traditional security countermeasures such as cryptography and key management. It is virtually impossible to design such a network in which the attackers cannot find any way for entry. Therefore, the network must consider the integration of fault tolerance capabilities and self-awareness. As the compromised node can launch wide range of attacks in a WSN, an alternate line of defense like IDS is required that is capable of detecting and reducing the threat by detecting the misbehaving nodes (Bhusan and Sahoo 2017b), (Butun et al. 2015), (Sagar and Lobiyal, 2015), (Rajeshkumar and Valluvan 2016).

Motivation: Any unauthorized or unwanted activity in the network that can be achieved actively (packet dropping, harmful packet forwarding, hole attacks) or passively (eavesdropping or information gathering) is referred to as intrusions. Intrusion detection system (IDS) detect suspicious activities within the network and triggers an alarm in case of any intrusion (Almomani et al. 2016). It provides information such as location of the intruder (regional node or single), time of intrusion entry (date and time), activity of the intrusion (passive or active) and type of intrusion. This information is helpful in mitigating and remedying the impact of the attack as some very fine and specific information about the intruder is found. Therefore, there is a need of IDS for maintaining the network security (Bhusan and Sahoo 2019) Due to unique and constrained characteristics of WSNs such as low transmission bandwidth, less data storage and memory size, and limited power supply, most of the security schemes including the IDSs designed for wired networks are not applicable directly for wireless environment. Also, there exists no specialized datasets containing the attacks and the normal profiles in WSNs which can be used to detect the signature of the attack. Therefore, it is a big challenge to design an efficient and effective IDS technique which is applicable for WSNs. This motivates us to design a secure and energy-efficient IDS scheme in this paper.

Our contribution: The major contribution of this paper can be summarized as follows.

- In this paper, intrusion detection problem is addressed from a different angle. A specialized attack model is constructed that is comprised of blackhole attack, vampire attack, jelly-fish attack and grey-hole attack. the performance of the normal network, simple IDS scheme and the proposed hybrid IDS scheme under the constructed attack model is analysed. Finally, an efficient, secured and intelligent intrusion detection system that can work even in presence of several attacks is designed.
- Considering both security and the lifetime of the network, we propose a hybrid 3-tier IDS scheme for WSNs that integrates the concept of clustering along with RC4 and Digital signature.
- We compare the performance of simple integrated IDS scheme comprised of clustering concept along with RC4 and digital signature with the normal packet transmission and packet transmission under attack in terms of average delay, energy consumption, PDR and throughput.
- The proposed hybrid 3-tier IDS show consistent delay therefore proves itself suitable for larger networks. Varying the number of nodes also supports our claim that the proposed IDS shows consistent delay.

The remainder of the paper is organized as follows. Attack model under consideration is elaborated in section 2 that comprises of four out of many possible attacks in WSNs: blackhole attack, vampire attack, jelly-fish attack and grey-hole attack. Section 3 describes the developed simple integrated IDS schemes (SIIS) that integrates the concepts of Clustering technique along with RC4 and Digital signatures. Section 4 throws light on the experimental set-up and the considered simulation parameters. The obtained

Table 1. List of Abbreviations

Serial number	Notations or Abbreviations	Description or meaning
1	d	The distance between the receiver and the transmitter
2	E_{wc}	Energy consumed by the wireless circuits for receiving and sending of data
3	$E_T(l,d)$	Energy consumption in transmission of *L bits* data to the receiver
4	E_{pc}	The cost incurred in processing of a bit report to the BS
5	d_{BS}	The distance between the BS and the CH
6	E_{NCH}	Energy dissipated by the non-CH nodes during one round
7	E_{CH}	Energy dissipated by the CH during one round
8	N	Total number of nodes
9	k	Total number of clusters
10	d_{BS}	Distance between CHs and BS
11	d_{CH}	Distance between CHs and their cluster members
12	P_{opt}	The optimal probability of any node to be elected as CH
13	Th_0	Chosen threshold value
14	E_{res}	The nodes current residual energy
15	E_{avg}	Networks average energy
16	E_{elec}	Energy consumption of the transmitters electronic circuit
17	E_{amp}	The amplifiers energy consumption to achieve the required SNR at the receiver end

results are discussed and elaborated in Section 5 followed by conclusion and several open future research issues in the field of WSNs security in section 6.

The table 1 lists the abbreviations used in this paper.

ATTACK MODEL

Security attacks in WSNs can be namely classified into passive and active attacks. Attackers are camouflaged in passive attack and can launch attacks such as eavesdropping, node destruction, node malfunctioning and traffic analysis. Attackers actually affect the network operation in active attacks and can launch attacks such as jamming, DoS, blackhole, sinkhole, flooding and sybil attack. Security solutions like key management, cryptography and authentication enhances the security in WSNs. However, these solutions are not capable of preventing all the possible attacks in WSNs as a compromised node can launch a wide range of attacks in WSNs. In this section, four out of many possible attacks are chosen as the attack model. The chosen attacks include blackhole attack, vampire attack, jelly-fish attack and grey-hole attack. These attack models are described in the section below.

Blackhole Attack

Here, the malicious node advertise itself as the shortest path to destination or as the destination itself. With the propagation of this advertisement message, more traffic gets directed towards this malicious node. This leads to resource contention as handling heavy traffic causes depletion of the resources of the neighboring node. If the malicious node receives data packets from normal nodes, it discards the packets completely or selectively. Blackhole can be launched in two ways: a simple blackhole in which the adversary carries out its dropping activity individually and a cooperative blackhole in which multiple blackholes or the adversary nodes coordinate to manipulate the deployed security mechanisms or the routing protocols specification. Based on their severity, blackholes can be classified into two types: passive blackhole that only drops the packets passing through them without false information injection to the network and active blackhole in which apart from dropping the packets, adversary also injects false information thereby disrupting the normal communication by affecting the network load (Sinha et al. 2017; Tripathi et al. 2013).

A typical blackhole attack involves two routing packets: Route Reply packets (RREP) and Route Request packets (RREQ). The RREQ holds the destination addresses of all network nodes and is broadcasted. Upon receiving the RREQ packet, a node responds directly to the original sender using RREP. In blackhole attack, the malicious adversary upon receiving RREQ replies immediately to the source node using RREP claiming to possess the best route. In this way the adversary attracts higher number of data packets from the normal nodes. Also, the source node receives the blackholes RREP before the original correct RREP packets. Therefore, the source sends data packets by considering the initial RREP packet and discards the subsequently received RREP packets. These packets are then dropped leading to degraded network performance.

Vampire Attack

In several source routing protocols, a malicious packet specifies network paths that are much longer than the normal optimal path thereby leading to energy wastage at intermediate nodes responsible for forwarding the packet on the basis of included source route. These attacks are not protocol specific and are independent of protocol design properties such as beacon and geographic routing, link-state, distance-vector and source routing. The malacious nodes do not overflow the network with huge data volumes but instead pass very small data in order to attain the maximum energy depletion. These attacks use protocol complaint message which makes them tough to get discovered and mitigated. Carousel attack involving a single attacker increases the energy consumption by four times while the stretch attacks magnitude depends on the malicious nodes position. Attacker or adversary can also combine the two attacks and launch a unique type of "stretched cycle attack" [19, 20]. Vampire attack is launched in following two ways.

Carousel attack: In carousel attack, the adversary purposely introduces routing loops in the composed packets. As the packets during this attack are sent in circles therefore this attack is named carousel attack. In this attack, the adversary exploits the limited header verification at the forwarding nodes and affects the source routing protocols by making a packet to traverse repeatedly. As a result of which there is an increase in the O (α), energy usage factor, where α denotes the route length. There is an increase by a factor of 3.96 in the overall energy consumption per message.

Stretch attack: In stretch attack, adversary or an attacker constructs long artificial routes and then traverses potentially each and every node in the network. As the packets path length is increased during this attack causing them to be worked upon by a greater number of nodes therefore this attack is named as stretch attack. As the packets traverse entire network, there is an increase in the O (min (N, α)), energy usage factor, where N denotes the total number of SNs deployed in the sensing region and α represents the maximum allowed path length.

Jelly Fish Attack

It is a type of DoS attack which is a result of TCP vulnerability. Owing to its passive nature, it is difficult to detect as none of the protocol rules are obeyed by the adversary. The main objective is to change the data packets order [Bhushan and Sahoo 2017] The three variations of jelly fish attack are explained below.

Jelly fish reorder attack (JFRA): In this version of jellyfish attack, the adversary sends data in arbitrary manner instead of following FIFO order to forward the packets. Firstly, the packets are placed in a random buffer and the jellyfish node reorders the buffer and sends the packet from the buffer. If the packets are not received in correct order at the destination side, duplicate acknowledgements are sent to the sender. In case, the sender receives three duplicate acknowledgements, they start transmission sometimes without the arrival of timeout and even if the destination receives the packet, the sender assumes the packet to be lost and continues retransmitting the packet.

Jellyfish periodic dropping attack (JPDA): This variation of jellyfish attack is possible due to chosen period by malicious node. The most prominent nodes for this periodic dropping attack are the relay nodes. because of the congestion lossess, the nodes are forced to drop packets periodically thereby reducing the TCP throughput to some very small value or even sometimes zero.

Jellyfish delay variance attack (JDVA): In this variation of the jellyfish attack the mischievous node introduce random packet delays without bringing any change in the packets ordering. The jellyfish adversary initially gains access to routing paths. Once they get the access, they delay almost all packets they receive before forwarding them. The malicious nodes in this type of attack attract data packets and delays them. Once the delay time is over, the packets are resolved thereby significantly affecting the performance of the network. This leads to low throughput and high delay.

A comparison of the three types of jelly fish attack in terms of their insight, cause and effect is given in Table 2.

Table 2. comparison of JFRA, JPDA and JDVA attack

Attack	Insight	Cause	Effect
Jelly Fish reorder attack	Packets are reordered	TCP vulnerability	Retransmissions and degraded throughput
Jelly fish periodic dropping attack	Packets are dropped in periodic manner	Due to malicious period chosen by adversary	Near zero throughput
Jelly fish delay variance attack	Packets are delayed arbitrarily	Exploiting TCP vulnerability	Congestion Interference

Grey Hole Attack

An extension of black hole attack that bluffs the source and uses partial forwarding for system monitoring is called a grey-hole attack. Adversaries employ dropping method only for some selected data packets and rest of the time behaves like a genuine node and participates in the normal data communication. These malicious nodes update the routing path or the source route cache as the shortest path by participating in the process of route discovery. This bluffs the source node to consider the adversary as the next hop node and thereby forwarding the packet to them. And now when the malicious node or adversary receives the packets, they drop them selectively on the random basis. The prevention or even the detection of such attack is tough because the partial dropping of packets can also be because of several other reasons such as selfish nature, overload, or congestion. Routing table information such as final packet destination or the next hop information can be utilized for determining the best route [Bhusan and Sahoo, 2017; Schweitzer et al. 2017).

Model: Typically, a grey-hole attack involves two distinct stages. The vulnerabilities of the routing protocol are exploited by the adversary to update the routing table of the source as the shortest path in the first stage. This is intended to divert all the traffic towards the malicious node instead of the legitimate or genuine route. In the second stage, the malicious nodes drop packets on the basis of certain probability. The malicious node rapidly changes its behavior thereby enabling them to sometimes transfer packets and sometimes drop them. Table IV below presents the attack model for grey-hole attack.

PROPOSED INTEGRATED IDS SCHEME

In this section simple existing intrusion detection schemes are used for the network's performance enhancement. This integrates the concept of Clustering technique along with RC4 algorithm and Digital signature. Clustering technique is employed which works in rounds and every round is performed in three phases: *Optimal clustering phase, CH selection phase,* and *The Steady state phase.* In order to aggregate data from multiple nodes, the concept of data aggregation is used in order for reducing the redundant data transmission in clustering approach where CHs aggregate the sensed data and transmits only the aggregated value to the BS. It is used in integration with RC4 algorithm, a type of symmetric, synchronous stream cipher that satisfies both efficiency and security for lightweight algorithms. In order to maintain the accuracy of the received data, digital signature is used to ensure trust, non-repudiation and integrity of the data in the SIIS scheme. The used digital signature cryptosystem uses *Two-fish algorithm* in which the *query packet* is encrypted at the source node before it is forwarded to their neighbors. The used three concepts are elaborated in the below section.

Assumptions: During the experiment and analysis of various schemes, several assumptions taken are listed in the section below.

- All the SNs are dispersed uniformly with the sensing region assumed to be a square field.
- Base station and the SNs are left unattended after the initial deployment phase complete.
- Communication between every member node and the CHs are done using single-hop approach and the communication of inter-cluster type is done using multi-hop approach.
- The considered WSNs model is heterogeneous type in terms of the energy content of the nodes.
- There is no modification or altering of the payload of the packet.
- Packets retransmission must occur within some specified time threshold.
- There is a limit on the packet sending rate.
- There can be resubmission of the same packet for only a limited number of times.

Clustering

The employed clustering technique in SIIS makes an improvement in the CH selection phase of LEACH protocol. The scheme uses enhanced method of threshold value selection and use nodes residual energy as a metric for calculation. The employed clustering technique operates in rounds where every round is performed in three phases described in this section below: Optimal clustering phase, CH selection phase, and Steady state phase. Before first round commences, a neighbor discovery phase is executed (Anguraj and Smys 2018; Bhusan and Sahoo 2019; Mehmood et al. 2015). In the following section, the phases involved in the clustering approach is presented.

Neighbor Discovery Phase

The sink initiates this phase by transmitting a Hello packet that holds the sender id, euclidean distance and hop counts to reach the sink. The two metrics euclidean distance and hop count measures the sink distance. The nodes that receive the Hello packet records the data of the packet and add the sender as its neighbor and replies to the sender with a Hello reply. Also, every node forwards the received Hello packet by changing the packet content in terms of its own sender id, euclidean distance and hop counts to reach the sink. In case, a node has its energy value lesser than the threshold value, it sends Dead message and broadcasts itself as dead node. The nodes that receive this dead message updates their routing table.

Optimal Clustering Phase

A square area of $M \times N$ metrics is assumed with N nodes distributed uniformly over the entire area. For simplicity purpose, the location of the BS is chosen to be the center of the network. A CH is elected based on optimal value discussed in the section below and every non-CHs transmit data (L bits) to the CHs. In order to estimate the energy dissipation, a radio model is used.

Suppose d denotes the distance between the receiver and the transmitter, E_{wc} denotes the energy incurred by the wireless circuits for receiving and sending of data, then the energy consumption in transmission of L bits is estimated by the equation below.

$$E_T(L,d) = \begin{Bmatrix} L * E_{wc} + L * \varepsilon_{fs} * d^2, if\ d \leq d_0 \\ L * E_{wc} + L * \varepsilon_{mp} * d^4, if\ d \geq d_0 \end{Bmatrix} \quad (1)$$

Theorem 1: The total energy consumed in one round for a clustered wireless sensor network can be estimated as

$$E_{Total} = \left\{ 2*N*E_{wc} + N*E_{pc} + \varepsilon_{fs} \times \left(k*d_{BS}^2 + N*d_{CH}^2 \right) \right\} \times L \quad (2)$$

Proof: Let k represent the total number of clusters formed, E_{pc} represent the cost incurred in processing of a bit report, d_{BS} represent the distance between the BS and the CH. The energy dissipated by the CH during one round is estimated as E_{CH} is given by equation (3) (Kumar et al 2009)

$$E_{CH} = \left(\frac{N}{k} - 1 \right) \times L * E_{wc} + \left(\frac{N}{k} \right) \times L * E_{pc} + L \times E_{wc} + L \times \varepsilon_{fs} * d_{BS}^2 \quad (3)$$

The energy dissipated by the non-CH nodes during one round is estimated as E_{NCH} is given by the equation (4).

$$E_{NCH} = L \times E_{wc} + L \times \varepsilon_{fs} * d_{BS}^2 \quad (4)$$

In majority there are two families of nodes in a network: CHs and non-CH nodes. Therefore, combining the energy dissipated by the CHs and the non-CH nodes, the overall total energy dissipated per round can be obtained as given in the equation (1).

Theorem 2: The optimal probability of any node to be elected as CH can be estimated as

$$p_{opt} = \frac{1}{0.765} \times \sqrt{\frac{2}{N\pi}} \times \sqrt{\frac{\varepsilon_{fs}}{\varepsilon_{mp}}} \quad (5)$$

Proof: Let d_{CH} represent the distance between the CHs and their cluster members which is given by the following equation (6).

$$d_{CH}^2 = \frac{M^2}{2\pi k} \tag{6}$$

If for several number of nodes, the distance to the base station is more than d_0 then, d_{BS} can be obtained by the following equation (7).

$$d_{BS} = 0.765 * \frac{M}{2} \tag{7}$$

From the above two equations, (6) and (7), the optimal probability of any node to be elected as CH can be estimated as equation (5) [Kumar et al. 2009; Smaragdakis et al. 2004).

$$\rho_{opt} = \frac{1}{0.765} \times \sqrt{\frac{2}{N\pi}} \times \sqrt{\frac{\varepsilon_{fs}}{\varepsilon_{mp}}} \tag{5}$$

This is of utmost importance because if there is no optimal construction of clusters, it results in exponential increase in the total energy consumption in any round which is always constrained in the case of WSNs.

CH Selection Phase

In LEACH, a number is generated between 0 and 1 randomly by every node and if this is lesser than the chosen threshold value Th_0, then it serves as the CH for that round. For balancing the energy consumption and distributing the load efficiently among the entire network, in every round, new CHs are elected. The value of Th_0 is given by (8).

$$Th_o = \begin{Bmatrix} \dfrac{\rho_{opt}}{1 - \rho_{opt} \times \left(r \bmod \left(\dfrac{1}{\rho_{opt}} \right) \right)}, & if\ s \in G \\ 0 & otherwise \end{Bmatrix} \tag{8}$$

where r represents the current round count and G represents the SNs that has not become the CH in last $\frac{1}{p_{opt}}$ rounds (Xiangling & Yulin, 2007).

In the used SIIS, we propose an improvement in choosing of the threshold by including E_{res} as a metric for calculation. The threshold is reduced relative to the ratio between the residual energy of nodes and networks average energy and is calculated as (9) as given in [49].

$$Th_{improved} = \begin{cases} \dfrac{p}{1 - p_i \times \left(r \bmod \left(\dfrac{1}{p_{opt}} \right) \right)} \times \left(\dfrac{E_{res}}{E_{avg}} \right), & if\ s_i \in G \\ 0, & otherwise \end{cases} \tag{9}$$

where E_{res} represents the current status of the residual energy of the nodes and E_{avg} represents the networks average energy.

Steady State Phase

In this phase, TDMA schedule is fixed once the clusters are formed and then the data communication starts. The active nodes senses data periodically and sends it to the CHs. TDMA schedule is used as it avoids data collision. In this phase, the nodes remain in sleep mode except the CHs. Only for a short duration of time, the member nodes remain in transmit state therefore the energy of the member nodes is saved. Once the CHs receive all the data, it aggregates and routes them to the BS via multi-hop transmission. Two commonly used communication patterns are multiple-hop and single-hop transmissions but in case of single-hop transmission, the CHs located at a distance deplete their energy faster in comparison to normal nodes as the data packets are directly sent to BS in such transmission strategy without any relay. This is because nodes located at distant places incurs higher energy burden due to its larger transmission range and therefore they die out early. In order to overcome this issue, optimal path selection is employed to provide the most optimal route to the BS [Bhusan and Sahoo 2018].

RC4

RC4, a type of symmetric, synchronous stream cipher that satisfies both efficiency and security for lightweight algorithms. It was proposed by R. Rivest in 1987 for the purpose of RSA Security. It makes use of S-box and an N length array where 1 byte is stored at each location k. This permutation is scrambled using a l byte secret key, key. Array K holds the major key and the secret key (key) is replicated as $K[y]= K[y \bmod l]$ for $(0 \leq y \leq N - 1)$. RC4 is an integration of two algorithms, namely PRGA (Pseudo-Random Generation) and KSA (Key Scheduling Algorithm). In the former step that is in KSA, a pseudorandom permutation is generated using the key K. In the second step that is the PRGA, the pseudorandom permutation is used to generate pseudorandom keystream bytes. Table 3 below presents the KSA and PRGA algorithm [Gupta et al. 2013; Khalid et al. 2017; Sarkar 2017].

Table 3. pseudocode for KSA and PRGA algorithm

Algorithm: 1 (KSA)	**Algorithm:2 (PRGA)**
1. **procedure** KSA (Key, *K*)	**1. procedure** PRGA (S-box, K)
2. Initialize $S \leftarrow \{0 \text{ to } N-1\}$ and $j \leftarrow 0$	**2.** Initialize $i \leftarrow 0$ and $j \leftarrow 0$
3. **for** i={0 to $N-1$} do	**3. while** TRUE do
4. $j=j+K[i] + key[i]$ // Increment	**4.** $j=j+K[i]$, $i \leftarrow i+1$ // Increment
5. $K[i] \leftrightarrow K[j]$ // Swap	**5.** $K[i] \leftrightarrow K[j]$ // Swap
6. end for	**6. Output** $Z \leftarrow K[K[i] \leftrightarrow K[j]]$
7. return K	**7. end while**
8. end	**8. end**

Efficiency of RC4 is dependent on the two algorithms: KSA and PRGA. Initial pseudorandom state generation during the KSA invokes a fixed cost whereas a variable cost is incurred in PRGA for keystream byte generation. RC4 is efficiently implemented to minimize the cost for every round and therefore leading to enhanced throughput. This stream cipher is used for encryption and decryption of data to secure long-term communication. RC4 is a binary additive cipher that operates in output feedback (OFB) mode. The above discussed clustering approach is used in integration with RC4 algorithm. RC4 has huge number of network applications including secure socket layer that secures the electronic commerce of the world over World Wide Web. These are especially useful for applications that operate in resource constrained environment. Therefore, it is chosen to be used in SIIS for WSNs as WSNs are also resource constrained. These are advantageous for encryption of larger packets. This is because the process of key stream generation converts the cipher text within negligible time once the initialization process is completed. Table 4 below shows the comparison of RC4 with other security algorithms in terms of their operation, speed and cost.

Table 4. comparison of RC4 with other security schemes

Parameters	Other security algorithms	RC4
Operation	Complex	Simple
Speed	>100 Mbps	22 Mbps
Cost	High	Low

Why RC4? One of the major significances of RC4 is that it employs a variable length key that varies from 0 to 255 bytes. The proposed SIIS shows enhanced performance and runs efficiently because of the benefits of RC4 algorithm as RC4 is fast, simple and have remarkably low overhead.

Digital Signature

Why Digital Signature? The accuracy of the received data is of utmost importance in wireless sensor networks therefore the concept of digital signature is used to ensure trust, non-repudiation and integrity of the data. SNs operate in resource constrained hardware scenario. The use of digital signature improves the security aspect of the protocol, increases the computation speed and lowers the overall energy consumption. Therefore, digital signature is chosen as a technique that can be integrated with the clustering approach along with RC4 [Yaruv et al. 2017; Mughal et al. 2018].

In the process of digital signature, a signature key is used to encrypt the data and generate the digital signature. SIIS uses digital signature cryptosystem that employs *Two-fish algorithm* in which the *query packet* is encrypted before it is forwarded to their neighbors. The nodes that are aware of keys are only able to decrypt the packet correctly. It generates *update packets* and sends it to the source that verifies the validity of the sender by calculating the trust index of the link.

Two-fish is a block cipher of 128-bits that can accept a key of variable lengths and up to 256 bits. The Two-fish algorithm involves two different operations namely *byte-oriented* and *non-byte oriented.* Byte-oriented operations include *F-function* and the *whitening operation.* Whitening operation is the XORing of data with 8 sub-keys. If this operation is performed before the round 1 at the input level, it is called input whitening and if it is performed after the round 16th at the output level, it is called as output whitening. This process of whitening makes it difficult for the adversaries to get an access to the key by hiding the inputs to last and the first round. Non-byte-oriented operations include S-boxes and Pseudo-Hadamard Transformations (PHT). Two-fish uses 4 separate S-boxes together which forms *h-function* with the MDS matrix (Maximum Distance Separable). Every S-box consists of three fixed permutations of 8-by-8 bit. The sub-keys are XORed between the three permutations. These are generated once and acts as global key which remains fixed during the process of encryption and decryption. The plaintext is divided into 32-bit words which are XORed with four key words followed by 16 rounds. In each of the 16 rounds, two h-functions are computed and their results are compared using PHT. Two-fish use 32-bit PHT, a simple and efficient mixing operation for mixing the outputs of two parallel h-functions of 32-bit.

$$a' = a + b mod 2^{32}$$

$$b' = a + 2b mod 2^{32}$$

Why Two-fish? Two-fish algorithm is selected because of several reasons. First and the foremost is that it works on high-end CPUs and is flexible to be used with low overhead. Every encryption algorithm possesses key-setup routine in which they take up keys and find round subkeys but Two-fish algorithm takes up the keys, makes S-box which is dependent on the keys and finds the round subkeys. The key-setup phase of Two-fish algorithm is much faster as compared to other encryption algorithms. Also, if large amount of plaintext is to be encrypted using the same key, Two-fish is an efficient encryption technique as it involves faster encryption. Therefore, because of its unique combination of flexibility and speed, Two-fish is the best choice that can be used for security enhancement.

EXPERIMENTAL SETUP

In this section, numerical experiments are conducted in order to verify the efficiency of our proposed three tier hybrid IDS technique. Our simulation is conducted by NS2, a network simulation platform useful for both small and large networks. Here we considered a rectangular WSN region of 1300×1000 m, consisting of 50 to 60 sensors evenly distributed in the network. We considered flat grid topology, two ray ground propagation model, 802.15.4 MAC layer and the traffic is evenly and randomly generated. The performance of the proposed hybrid 3-tier IDS that uses secure energy efficient disjoint routing scheme in integration with the watchdog and stackelberg game theory approach is compared with three other techniques: Normal network topology without attack, Normal network topology under the considered attack model (that includes blackhole attack, vampire attack, jelly-fish attack and grey-hole attack) and simple integrated intrusion detection scheme that integrates clustering approach along with RC4 and digital signatures. Tests are conducted using plane coordinates and mobile as well as static nodes. Initially all the four above mentioned topology is constructed with 10 nodes and comparative performance evaluation of these topologies are done in terms of their average delay, energy consumption, packet delivery ratio and throughput. In order to support the claim of this research, the number of nodes is varied from 10 nodes to 30 nodes and later 50 nodes keeping the topology same. Table 5 shows the details of the simulation parameters considered.

Table 5. simulation parameters considered for the experiment

Simulation Parameters	Considered values
Simulation Period	10ms
Coverage Area	1300 X 1000
Antenna type	Omni-antenna
Propagation model	Two-ray ground propagation
Routing	Ad-hoc
MAC Type	802.11
Node placement	Uniform
No of Nodes	10 to 55
No. of BS	1
Length of every data packet	1024 bits
Attacks	Black hole, Vampire, Jelly fish, Grey hole
No of malicious nodes	4 – 8
Malicious node percentage	10 – 15%
Traffic Type	CBR
Agent Type	UDP
Routing Protocol	AODV
Initial power	1000 J
Transmission Power	4.6 J
Receiving Power	5.0 J
Queue Type	Drop-Tail

RESULTS AND DISCUSSIONS

Performance of the proposed hybrid IDS scheme is compared with normal packet transmission, normal transmission under the considered attack model and the developed simple IDS scheme in terms of various parameters as detailed below.

Average Delay

Average E2E delay includes all the types of possible delays that might occur in a wireless environment such as delays due to buffering during the process of route discovery, queuing delay at the interface queue, propagation delay during packets transfer or the retransmission delay at the MAC. This problem is even more critical for several real-time transmission tasks. Figure 1, shows the variation of average delay with time for a network of 10 nodes. The average delay is maximum for normal packet transmission under attack due to its reactive nature and also because the path selection to the destination may not be of minimum or optimal number of hops. Figure 2 depicts the variation of average delay with time for 50 nodes. In all the three cases, normal packet transmission under attack shows maximum average delay followed by normal packet transmission. The average delay is considerably reduced by the introduction of simple

Figure 1. Variation of average delay with time for 10 nodes

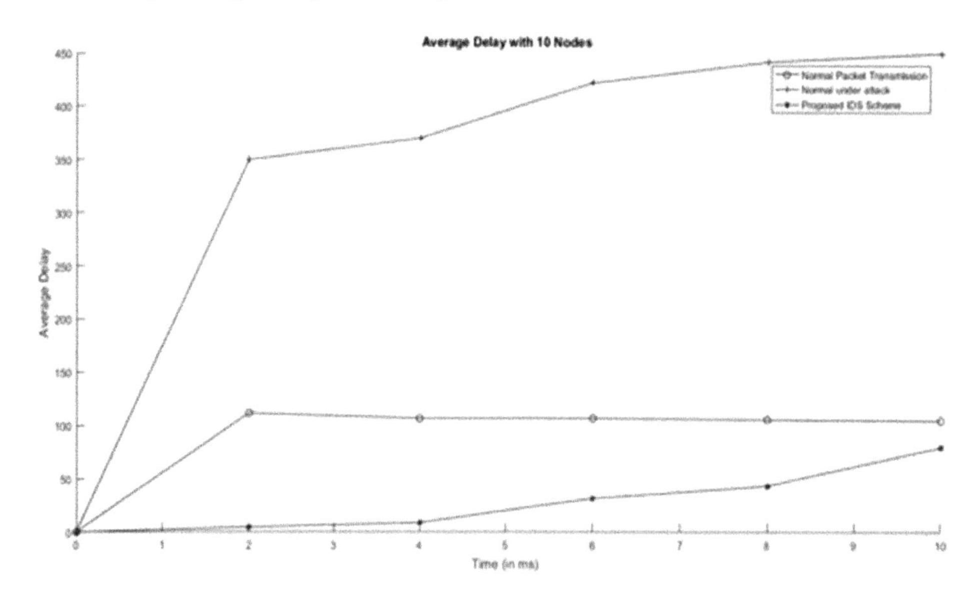

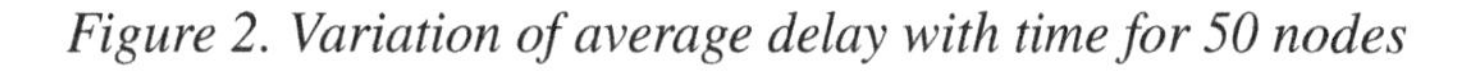

Figure 2. Variation of average delay with time for 50 nodes

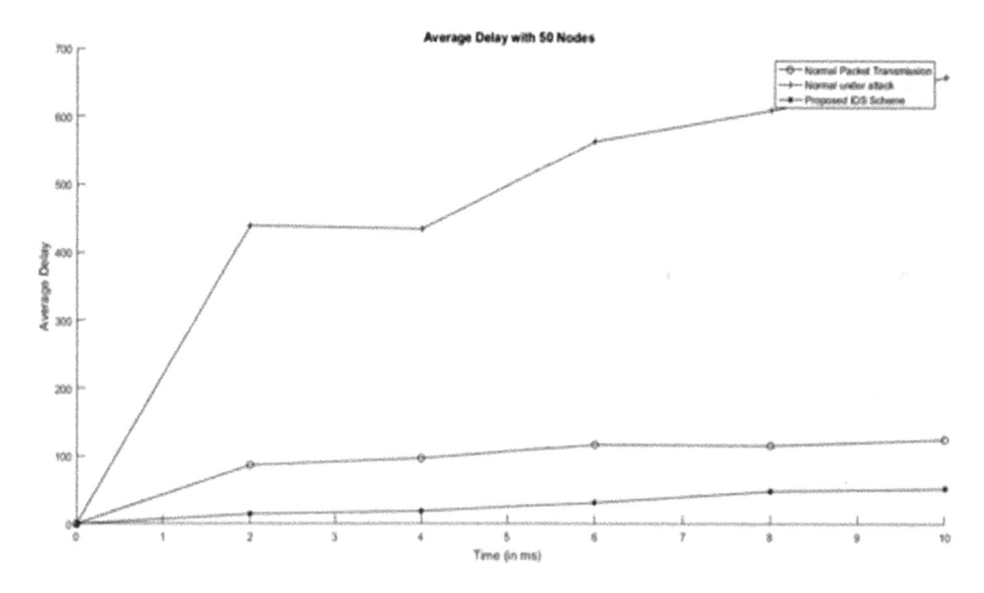

integrated IDS scheme. This is because of the use of clustering approach that reduces the transmission of redundant data due to data aggregation where the CHs transmits only the aggregated value to the BS.

Energy Consumption

Energy expenditure of any given node is directly proportional to the transmission distance involved. Average energy consumed by the SNs in the network helps in evaluation of the proposed techniques energy efficiency. Lifetime of the network depends on the lifetime of individual nodes that constitute the heterogeneous system therefore energy consumed in every round and the residual energy for use is the major factor on which the networks lifetime depends. It is of utmost importance for all the SNs in the network to stay alive and there is a degrading of networks performance once even a single node dies out. Figure 3, shows the variation of energy consumption with time for a network of 10 nodes. The normal packet transmission under the considered attack model shows maximum energy consumption followed by the normal packet transmission without attacks. Compared to these, the Simple integrated

Figure 3. Variation of energy consumption with time for 10 nodes

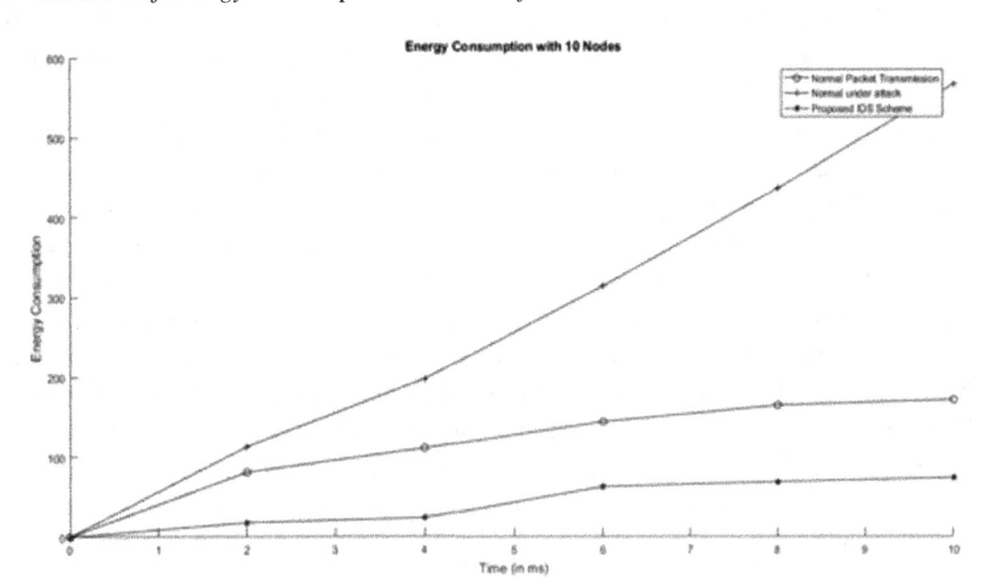

Figure 4. Variation of energy consumption with time for 50 nodes

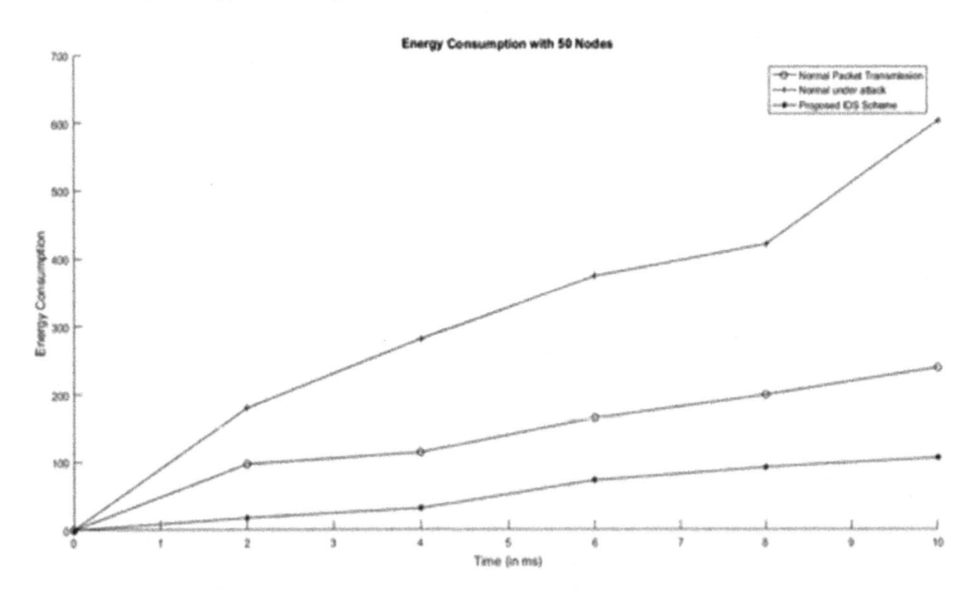

IDS scheme shows enhanced performance of networks in terms of energy consumption. In the steady state phase of the clustering approach, the normal nodes remain in the state of *sleep mode* and data transmission between the CHs and the normal node is done using the TDMA schedule. This avoids data collision and every member remains in *transmit mode* only for a small duration thereby conserving the energy of the member nodes. Therefore, SIIS scheme shows less energy consumption as compared to normal packet transmission and normal transmission under attack. In order to support the claim of the research, we vary the number of nodes in the network to 50 nodes as depicted in Figure 4.

Packet Delivery Ratio

It accounts for the ratio of the total number of delivered data packets to those generated by the CBR sources (Constant Bit Rate). Figure 5 shows the variation of PDR with time for a network consisting of 10 nodes. In normal data transmission scenario, data packets are transmitted to the sink via randomly selected multiple paths. This random path selection makes the network more vulnerable thereby leading to increased number of packets drops. Therefore, the normal transmission yields less PDR. The networks packet delivery ratio falls further when there is a launch of attacks in the network as this leads to increased packet loss. Therefore, the normal packet transmission under attack yields the least PDR in comparison to all the other three routing schemes. The PDR of the data transmission increases with the use of simple integrated IDS scheme. The SIIS uses clustering with RC4 and digital signatures. RC4 facilitates encryption and decryption of data to secure long-term communications as it employs a variable length key that varies from 0 to 255 bytes. The use of digital signature in SIIS also ensures the trust, non-repudiation and integrity of data. These security schemes nullified the impact of attacks to some extent and thereby enhances the PDR of the network. In order to support the claim of this research, the PDR of the network is also calculated by increasing the number of nodes in the network to 50 nodes as depicted in the Figure 6.

Figure 5. Variation of PDR with time for 10 nodes

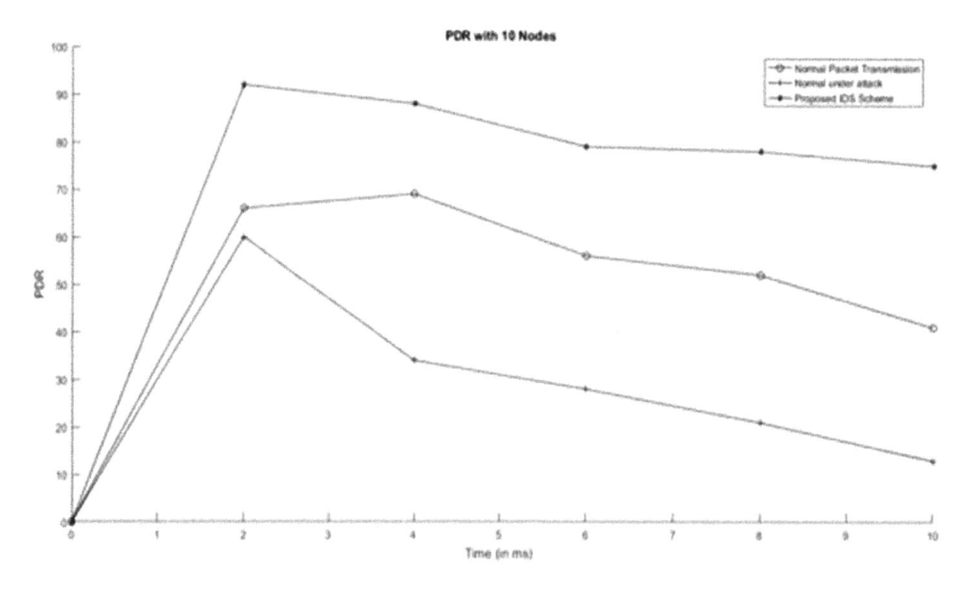

Figure 6. Variation of PDR with time for 50 nodes

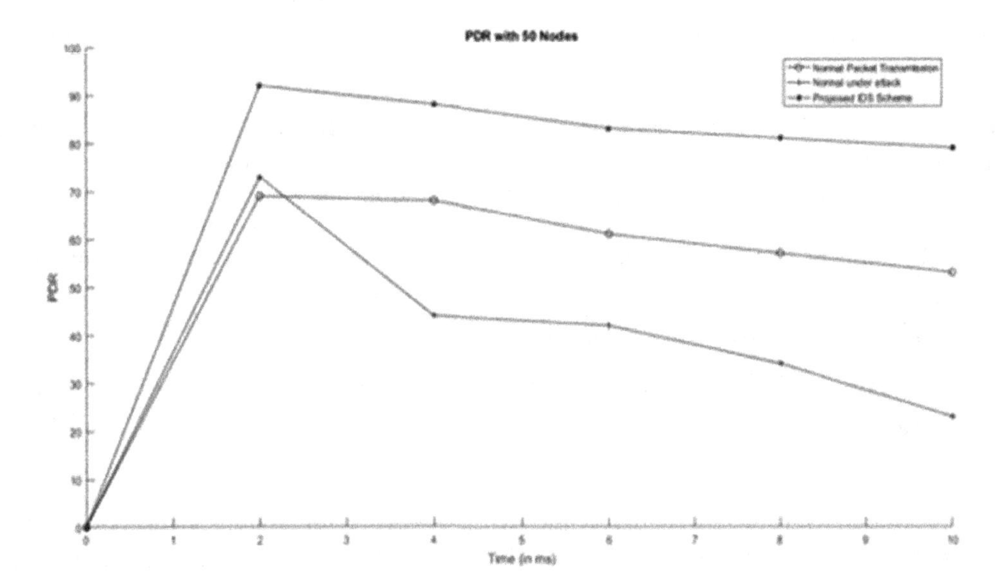

CONCLUSION

The aim of this work is to design an efficient, secured and intelligent intrusion detection system that can even work in the presence of several attacks. In order to achieve this goal, a specialized attack model is constructed here that comprise of blackhole attack, vampire attack, jelly-fish attack and grey-hole attack. We have proposed a 3-tier hybrid IDS scheme that integrates the concept of clustering along with RC4 and Digital signature. Our proposed scheme is applicable to the WSNs for the advantage of low energy consumption, fast speed and secured wireless communication. In this paper, the proposed IDS scheme is not suitable for multimedia data routing and is limited to only physical data routing. this issue is to be solved in our future works making it suitable for all types of data. A secured and energy-efficient routing protocol needs to be developed that considers jointly the delay and the packet loss due to the fading channel. Also, in future work, we will try to design a new metric that integrates both QoS and energy with link reliability.

REFERENCES

Almomani, I., Al-Kasasbeh, B., & Al-Akhras, M. (2016, August). WSN-DS: A dataset for intrusion detection systems in wireless sensor networks. *Journal of Sensors, 2016*, 1–16. doi:10.1155/2016/4731953

Anguraj, D. K., & Smys, S. (2018). Trust-Based Intrusion Detection and Clustering Approach for Wireless Body Area Networks. *Wireless Personal Communications*. Advance online publication. doi:10.100711277-018-6005-x

Bao, F., Chen, I.-R., Chang, M., & Cho, J.-H. (2012, June). Hierarchical trust management for wireless sensor networks and its applications to trust-based routing and intrusion detection. *IEEE eTransactions on Network and Service Management*, *9*(2), 169–183. doi:10.1109/TCOMM.2012.031912.110179

Bhushan, B., & Sahoo, G. (2017). Recent Advances in Attacks, Technical Challenges, Vulnerabilities and Their Countermeasures in Wireless Sensor Networks. *Wireless Personal Communications*, *98*(2), 2037–2077. doi:10.100711277-017-4962-0

Bhushan, B., & Sahoo, G. (2017). A comprehensive survey of secure and energy efficient routing protocols and data collection approaches in wireless sensor networks. *2017 International Conference on Signal Processing and Communication (ICSPC)*. 10.1109/CSPC.2017.8305856

Bhushan, B., & Sahoo, G. (2017). Detection and defense mechanisms against wormhole attacks in wireless sensor networks. *2017 3rd International Conference on Advances in Computing, Communication & Automation (ICACCA)*. DOI: 10.1109/icaccaf.2017.8344730

Bhushan, B., & Sahoo, G. (2018). Routing Protocols in Wireless Sensor Networks. *Computational Intelligence in Sensor Networks Studies in Computational Intelligence,* 215-248. DOI: doi:10.1007/978-3-662-57277-1_10

Bhushan, B., & Sahoo, G. (2019). Secure Location-Based Aggregator Node Selection Scheme in Wireless Sensor Networks. *Proceedings of ICETIT 2019 Lecture Notes in Electrical Engineering*, 21–35. DOI: 10.1007/978-3-030-30577-2_2

Bhushan, B., & Sahoo, G. (2019). ISFC-BLS (Intelligent and Secured Fuzzy Clustering Algorithm Using Balanced Load Sub-Cluster Formation) in WSN Environment. *Wireless Personal Communications*. Advance online publication. doi:10.100711277-019-06948-0

Bhushan, B., & Sahoo, G. (2019). $$E^{2} SR^{2}$$ E 2 S R 2: An acknowledgement-based mobile sink routing protocol with rechargeable sensors for wireless sensor networks. *Wireless Networks*, *25*(5), 2697–2721. doi:10.100711276-019-01988-7

Bhushan, B., Sahoo, G., & Rai, A. K. (2017). Man-in-the-middle attack in wireless and computer networking — A review. *2017 3rd International Conference on Advances in Computing, Communication & Automation (ICACCA)*. DOI: 10.1109/icaccaf.2017.8344724

Butun, I., Morgera, S. D., & Sankar, R. (2014). A survey of intrusion detection systems in wireless sensor networks. IEEE Commun. Surveys Tuts., 16(1), 234-241. doi:10.1109/SURV.2013.050113.00191

Chaudhary, A., Tiwari, V., & Kumar, A. (2014). A novel intrusion detection system for ad hoc flooding attack using fuzzy logic in mobile ad hoc networks. *Proc. IEEE Recent Adv. Innov. Eng. (ICRAIE),* 1-4. 10.1109/ICRAIE.2014.6909148

Diaz, P. S., & Sanchez, P. (2016). Simulation of attacks for security in wireless sensor network. *Sensors (Basel)*, *16*(11), 1932. doi:10.339016111932 PMID:27869710

Diddigi, R. B., Prabuchandran, K. J., & Bhatnagar, S. (2018). Prabuchandran K.J., Shalabh Bhatnagar, "Novel Sensor Scheduling Scheme for Intruder Tracking in Energy Efficient Sensor Networks. *Wireless Communications Letters IEEE*, *7*(5), 712–715. doi:10.1109/LWC.2018.2814576

Gupta, S. S., Chattopadhyay, A., Sinha, K., Maitra, S., & Sinha, B. P. (2013). High-Performance Hardware Implementation for RC4 Stream Cipher. *IEEE Transactions on Computers*, *62*(4), 730–743. doi:10.1109/TC.2012.19

Jaitly, S., Malhotra, H., & Bhushan, B. (2017). Security vulnerabilities and countermeasures against jamming attacks in Wireless Sensor Networks: A survey. *2017 International Conference on Computer, Communications and Electronics (Comptelix)*. 10.1109/COMPTELIX.2017.8004033

Kaur, M., Sarangal, M., & Nayyar, A. (2014). *Simulation of Jelly Fish Periodic Attack in Mobile Ad hoc Networks. International Journal of Computer Trends and Technology* , 15.

Khalid, A., Paul, G., & Chattopadhyay, A. (2017). RC4-AccSuite: A Hardware Acceleration Suite for RC4-Like Stream Ciphers. IEEE Transactions on Very Large Scale Integration (VLSI). *Systems*, *25*(3), 1072–1084. doi:10.1109/tvlsi.2016.2606554

Kulkarni, N., Prasad, N. R., & Prasad, R. (2017). Q-MOHRA: QoS Assured Multi-objective Hybrid Routing Algorithm for Heterogeneous WSN. *Wireless Personal Communications*, *100*(2), 255–266. doi:10.100711277-017-5064-8

Kumar, D., Aseri, T. C., & Patel, R. B. (2009). EEHC: Energy efficient heterogeneous clustered scheme for wireless sensor networks. *Elsevier Comput. Commun.*, *32*(4), 662–667. doi:10.1016/j.comcom.2008.11.025

Lina, Deshmukh, & Potgantwar. (2015). Ensuring an early recognition and avoidance of the vampire attacks in WSN using routing loops. *Advance Computing Conference (IACC) 2015 IEEE International*, 61-66.

Madsen, T. K., Fitzek, F. H., Prasad, R., & Schulte, G. (2005). Connectivity Probability of Wireless Ad Hoc Networks: Definition, Evaluation, Comparison. *Wireless Personal Communications*, *35*(1-2), 135–151. doi:10.100711277-005-8745-7

Mehmood, A., Khan, S., Shams, B., & Lloret, J. (2015). Energy-efficient multi-level and distance-aware clustering mechanism for WSNs. *International Journal of Communication Systems*, *28*(5), 972–989. doi:10.1002/dac.2720

Mughal, M. A., Luo, X., Ullah, A., Ullah, S., & Mahmood, Z. (2018). A Lightweight Digital Signature Based Security Scheme for Human-Centered Internet of Things. *IEEE Access: Practical Innovations, Open Solutions*, *6*, 31630–31643. doi:10.1109/ACCESS.2018.2844406

Patel, H. P., & Chaudhari, M. B. (2013). A Time Space Cryptography Hashing Solution for Prevention Jellyfish Reordering Attack in Wireless Adhoc Network. *International Conference on Computing Communication and Networking Technologies (ICCCNT)*. 10.1109/ICCCNT.2013.6726689

Pragya, M., Arya, K. V., & Pal, S. H. (2017). Intrusion Detection System Against Colluding Misbehavior in MANETs. *Wireless Personal Communications*, *100*(2), 491–503. doi:10.100711277-017-5094-2

Rajeshkumar, G., & Valluvan, K. R. (2016). An Energy Aware Trust Based Intrusion Detection System with Adaptive Acknowledgement for Wireless Sensor Network. *Wireless Personal Communications*, *94*(4), 1993–2007. doi:10.100711277-016-3349-y

Sagar, A. K., & Lobiyal, D. K. (2015). Probabilistic Intrusion Detection in Randomly Deployed Wireless Sensor Networks. *Wireless Personal Communications*, *84*(2), 1017–1037. doi:10.100711277-015-2673-y

Sarkar, S. (2017). Results on significant anomalies of state values after key scheduling algorithm in RC4. *IET Information Security*, *11*(5), 267–272. doi:10.1049/iet-ifs.2016.0451

Schweitzer, N., Stulman, A., Margalit, R. D., & Shabtai, A. (2017). Contradiction Based Gray-Hole Attack Minimization for Ad-Hoc Networks. *IEEE Transactions on Mobile Computing*, *16*(8), 2174–2183. doi:10.1109/TMC.2016.2622707

Sharma, M., Tandon, A., Narayan, S., & Bhushan, B. (2017). Classification and analysis of security attacks in WSNs and IEEE 802.15.4 standards: A survey. *2017 3rd International Conference on Advances in Computing,Communication & Automation (ICACCA)*. DOI: 10.1109/icaccaf.2017.8344727

Sinha, P., Jha, V. K., Rai, A. K., & Bhushan, B. (2017). Security vulnerabilities, attacks and countermeasures in wireless sensor networks at various layers of OSI reference model: A survey. *2017 International Conference on Signal Processing and Communication (ICSPC)*. 10.1109/CSPC.2017.8305855

Smaragdakis, G., Matta, I., & Bestavros, A. (2004). SEP: a stable election protocol for clustered heterogeneous wireless sensor networks. *Proc. Int. Workshop Sensor and Actor Network Protocols and Applications*, 251-261.

Tripathi, M., Gaur, M. S., & Laxmi, V. (2013). Comparing the impact of black hole and gray hole attack on LEACH in WSN. *Procedia Computer Science*, *19*, 1101–1107. doi:10.1016/j.procs.2013.06.155

Xianging, F., & Yulin, S. (2007). Improvement on LEACH protocol of wireless sensor networks. *Proc. Int. Conf. Sensor Technologies and Applications*, 260-264.

Yavuz, A. A., Mudgerikar, A., Singla, A., Papapanagiotou, I., & Bertino, E. (2017). Real-Time Digital Signatures for Time-Critical Networks. *IEEE Transactions on Information Forensics and Security*, *12*(11), 2627–2639. doi:10.1109/TIFS.2017.2716911

Chapter 10
A Speed Control-Based Big Data Collection Algorithm (SCBDCA) Using Clusters and Portable Sink WSNs

Rajkumar Krishnan
PSNA College of Engineering and Technology, India

Jeyalakshmi V.
College of Engineering, Guindy, India

V. Ebenezer
Karunya Institute of Technology and Sciences, India

Ramesh G.
KLN College of Engineering, India

ABSTRACT

WSNs collects a huge amount of heterogeneous information and have been widely applied in various applications. Most of the information collecting techniques for WSNs can't keep away from the hotspot trouble. This problem affects the connectivity of the community and reduces the life of the entire community. Hence, an efficient speed control based data collection algorithm (SCBDCA) using mobile sink in WSNs is proposed. A tree construction technique is introduced based on weight such that the origin nodes of the built timber are described as rendezvous points (RPs). Also, other unique type of nodes named sub or additional rendezvous points (ARPs or SRPs) is chosen in accordance to their traffic consignment and hops to origin nodes in the network. Mobile sink gathers the information from the cluster head (CH) and is controlled by speed control mechanism. Simulation results among different existing procedures exhibit that the SCBDCA can extensively stabilize the consignment of the communication network, decrease the energy consumption, and extend the lifetime in the communication network.

DOI: 10.4018/978-1-7998-5068-7.ch010

INTRODUCTION

WSN contains numerous minute and minimum energy sensors that utilize radio frequencies to function allotted sensing jobs. WSNs locate their purposes in various areas that encompass affected person monitoring and sickness investigation, pollution supervising and source detection, equipment supervising and fault prediction, tide monitoring and sea searching. Large range of sensors is arranged in a Field of Interest (FoI) in speculator manner in such WSNs. The sensors nodes are frequently placed in huge amount randomly to assure trustworthiness in stochastic deployment. In data gathering process, a sensor node consumes plenty of energy in communication phase, particularly for a huge quantity of diverse information. Furthermore, in multi-hop transmission, sensors close to the sink devour more power as they are accountable for accepting and transferring information from the entire network. It gives the problem of uneven power utilization among sensors in the network, (i.e.) the hotspot problem. To solve this difficulty and stable the power utilization in the network, some Mobile Elements (MEs) like sink or information collector have been taken for information aggregation. In many information aggregation procedures, the latency is predictable if the MEs function for gathering information through shortest path in the network. Gathering of data in WSN in depicted in figure 1.

Figure 1. Data gathering in WSN

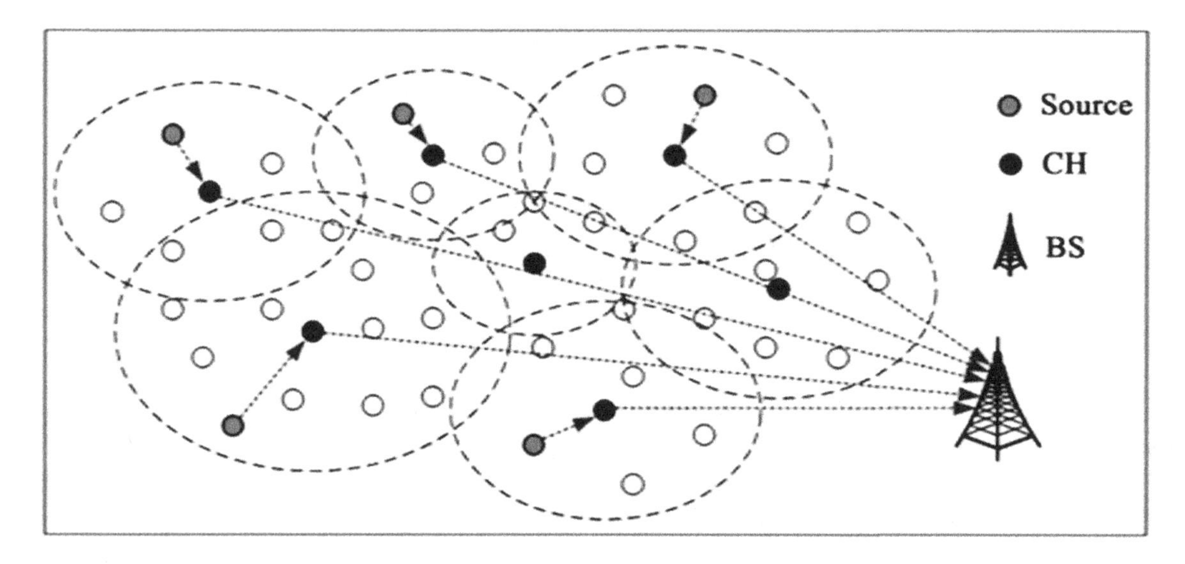

Mobile information collecting procedures will be separated into two types: homogeneous (uniform) and heterogeneous sensor network. In uniform sensor arrangement, the sensors with minimum distance to the BS or sink path are fixed as polling spot for the MEs to see in the communication network. The different sensor network made up of fixed sensors, super nodes, MEs and a fixed BS. The super nodes usually have greater assets to perform as Rendezvous Points (RPs) and Sub Rendezvous Points (SRPs) for information collection. The range and place of these super nodes ought to be restricted to reduce the MEs tour size which shortens the information collecting delay in the communication network.

The main contributions of the paper include the following key points: Clustering and Cluster Head (CH) selection by taking remaining power and the utilization of the nodes to its nearby nodes.

- Both RP and SRP selection after positive collection series and balancing the energy utilization.
- Controlling the speed of the mobile sink using speed control mechanism.

The remaining part of this paper is prepared as follows: part 2 describes the literature of data gathering algorithms. In part 3, the network model and the details of proposed scheme, Speed Control based Data Collection Algorithm (SCBDCA) using mobile sink in WSNs is given. The part 4 illustrates the simulation outcomes and performance investigation. As a final point, the conclusion and future work are introduced in part 5.

RELATED WORKS

Some related papers on mobile information collecting in WSNs are discussed and grouped into two classes in this section: one-hop communication gathering where one or many MEs sees every sensors and accumulate its information solely by using one-hop path and multi-hop communication gathering where the MEs solely go to a division of sensors or some specific places and other sensors have to transmit their information through multi-hop route. Multi-hop communication gathering again classified into two types based totally on the quantity of information required to be gathered: neighboring information gathering where only collects information from origin nodes required to be gathered and complete information gathering where information of the entire arrangement required to be gathered.

An information mule means a mobile gadget that gathers information in a sensor node by seeing the places in a sensor group in physical manner (Sugihara & Gupta., 2010). The information mule gathers information when the node is in the closeness of a sensor node. Data mule will be considered as a substitute to multi-hop transmission of information while mobility of node is utilized in a sensor group. A information mule strategy requires to reduce the delay of information delivery. The data or information Mule Scheduling (DMS) is a forecast hassle that has limitation of both place and time in the network. A proficient heuristic procedures was designed for 1D problem that contains limitations on the information mule movement dynamics.

To achieve higher efficiency in the network mobile sink is taken in WSNs and Wireless Sensor and Actor Network (WSAN) (Zhang et al., 2013). Rechargeable sensor node is introduced (Bhushan B., & Sahoo G., 2019). Using mobile sink, sensors ought to obtain a longer life span. A short and capable information collection method was suggested such that the considered place will be separated into numerous zones with fair information collection delay in the communication network. An algorithm is designed to balance the information collection delay by designing the dividing problem like Traveling Salesman Problem (TSP) among all the zones.

Most of the distribution algorithms on mobility assist in WSNs have been modeled depends on the guess that the progress monitor of sink is irregular (Lee et al.,2008). Though, the movements of the sinks in most of the appliances, considering an illustration, the progress monitor on operation of a warrior in a fight ground or firelight on salvage action in the calamity region will be decided beforehand because the soldiers go in conformity with the planed operation or the training rules to function their work. Probable Movement depends on Data or information Distribution Protocol (PMDD) permits information to straight path from origin to dynamic sinks by considering the predictable activities.

SinkTrail and SinkTrail-S are two energy efficient proactive information exposure algorithms proposed for mobile sink depends on information gathering to stay away from steady sink place's traffic updates when a sink's forthcoming places can't be planned well in advance (Liu et al., 2013) which features minimum complication and decreased control overheads. The proposed approach is distinguished from existing system based on two unique aspects: Required flexibility is allowed in the motion of mobile sinks to animatedly comply with different global changes in the network. SinkTrail inaugurates a rational coordinate scheme for steering and transmitting information packs barring necessities of GPS gadgets or well planned milestones applicable for various appliance scenarios.

A voting based mobile information collection method was introduced and formulated into a making a best solution called Bounded Relay Hop Mobile Data Gathering (BRH-MDG) (Zhao et al., 2013). A division of sensors will be chosen as voting stations that store itself accumulated information and add it to the mobile gathering point in the network. When nodes are associated with these voting stations, it is confirmed that any packet transition is circumscribed within a specified quantity of hops in the network. Two well-organized procedures for choosing voting stations surrounded by sensors are identified.

Towards enhanced power delay back-and-forth, an inclined, adjustable sink movement system was proposed that adapts to neighborhood network situation, like the adjoining solidity, available power and the range of previous look up in every communication area (Kinalis et al., 2014). The sink goes probabilistically considering minimum noticed locations to wrap the communication location quicker in the communication network. This scheme attains appreciably decreased delay in contrast to recognized blind unsystematic and non-flexible methods particularly in networks with heterogeneous sensor allocation except negotiating the power effectively and dispatch success.

The simplest way was to construct a tree from every sensor hubs to the sink hub, however this increases the hassle of power distort due to the fact that the sensors nearer to the sink hub would have a lot greater outstanding tasks from transferring information (Chang et al., 2012) Energy Balanced Data Collection scheme (EBDC) was proposed which decides the direction of a mobile information authority to such an extent that the information transferring tasks of every sensors will be completely adjusted.

In Clustering based Routing Algorithm (CRA), sink node moves in a circle and the globular region close to this direction is fixed as storage place to which the hubs are linked as spine data link in the network (Wang et al., 2014). Information is sent to the storage place and subsequently the data reaches the sink via the spine data link. The most reliable region and volume of the storage place is figured so as to reduce the normal load of the hubs in the region. This article also introduces CRA-1, which improves the calculation by using solidity controlling for the motive of stabilizing the power utilization of the entire system.

A mobile information gatherer (M-collector) can be a portable artificial machine or a vehicle furnished with an amazing handset and battery functioning similar to a portable Base Station (BS) and collecting information while travelling through the area (Ma et al., 2013). A M-collector begins the information collection visit at times from the fixed information sink and surveys every sensor during the navigation within its communication limit. Afterward data is collected legitimately from the sensor in one-bounce communication and lastly moves the information to the fixed sink. The duration of sensors is predictable to be protracted because information packs are straightforwardly collected exclusive of transfers and crashes. This paper focuses on the hassles of reducing the distance of every information collection visit and is referred as Single or one Hop Data Gathering Problem (SHDGP). It was formalized into combined numeral software and a heuristic tour planning software was then presented for the circumstances the place a solo M-collector is utilized.

Maximizing network lifetime was focused utilizing portable Information Gatherer (IGs) besides negotiating on the reliability necessities (Vupputuri et al., 2010). A dissimilar WSN was first considered which includes a huge quantity of sensor hubs, a couple of IGs and a fixed BS. The sensor hubs are fixed one and built in the territory uniformly. The IGs have moving skills and their motion will be managed in the network. Every sensor hub intermittently transmits detected occasion packets to its closest IG in the communication network. The IGs combine the occasion packets obtained from the sensor hubs and transmit these combined occasion packets to the fixed BS. The IGs should send the combined occasion packets to the BS by means of a agreed reliability whilst averting the hotspot areas such that the system life span is increased. Reliability is attained by transmitting every combined occasion packet through several routes to the BS in the communication system. The network life span is thus increased by shifting the IGs as it would be the transmitting work is scattered amongst the sensor hubs in the network. The H2B2H protocol was introduced for the efficient transmission of data even after the links between any cluster heads fail (Krishnan & Perumal., 2019). Family based algorithm is proposed for fruitful data transmission even if the node fails (Krishnan & Perumal., 2018). The secure data aggregation was done with reduced power consumption without decrypting the data at the receiver side (Bhushan B., & Sahoo G., 2019). Energy efficiency is achieved by using small and static number of nodes (Bhushan B., & Sahoo G., 2018).

SPEED CONTROL BASED DATA COLLECTION ALGORITHM (SCBDCA)

Network Model

The model consists of static and homogeneous sensor hubs built approximately. There are n hubs built in a system. A sink is situated at the middle of the system. Every hub is not now solely an origin of information but additionally transmit to another origin to arrive the sink. The system deployed is hierarchical where groups are created. In hierarchical clustering, groups are created so sources inside a group transmit their information through a one-hop or many-hop based on the length of the group to a selected hub called as the Cluster Head (CH).

The CH will probably achieve information collection and handling and then transmit information to the sink. For this situation, it is assumed that the CHs function as easy relays and no information accumulation is done. It is additionally assumed that the conversation among hubs and CHs and conversation among CHs and the sink are on discrete frequency funnel with the goal that the two layers don't meddle. In common, groups will be created by choosing a small number of nodes in the system to function as CHs or with the aid of including unique CH to the system.

Speed Control Mechanism

The existing algorithm clearly provided RP's and SRP's to get data from the CH's. However, there is no speed control of the mobile sink present in the networks while collecting information from the CH's while taking the location of the RP's and SRP's. As an extension, a new algorithm is designed to facilitate the collection of packets by the mobile sink by controlling the speed. If low speed is required then the node will move slowly and if more speed is required the speed will be changing too fast.

The primary goal of the information gathering procedure is to decrease the delay and improve the system's life span. The information gathering algorithm is utilized to acquire the cumulative information from the sensor hub to the sink. There are a number of methods used to gather the information from origin node to the sink. Every node is initially fixed and afterward the system is regarded as fixed system. The fixed sensor hub transmits the information to the sink by any number of bounce. Therefore the sensor situated closer to the sink is exhausted quickly. The hubs will be divided as bottom and top layer. The hubs in the bottom layers are similar sensor hubs whereas the hubs in the top layer are greater effective than the hubs in the bottom layer. The top layer hubs are termed as CHs. The mobile sink will gather the information periodically and dynamically. The hubs that are situated nearer to the information viewer will add the information straightly. The hubs that are situated distant from the viewer will transmit the information by relaying.

In Speed Control based Data Collection Algorithm (SCBDCA), a portable sink is introduced which collects the data trip at times from the Base Station (BS), breaks at every RP, gathers information from the single bounce hubs straightly and at last return again to the BS for single round. The proposed procedure includes three phases: clustering phase, RPs and SRPs choice phase and the information gathering phase which are controlled by speed control mechanism.

In clustering phase, the parameters: residual energy, distance from the node to the BS, the average residual energy of its one-bounce nearest hubs and the utility of the node to its neighbors are calculated for every sensor hub in the communication system. After performing single bounce communication, the nearest single bounce hub with highest energy will be selected as CH. The information collection group is thus constructed along with the CH. The standard remaining power of single-bounce neighbor is obtained using equation 1.

$$\mathrm{E}^{1\text{-hop}} = \frac{\sum_{1}^{N_i^{1-hop}} \mathrm{E}_{\mathrm{r}}}{N_i^{1-hop}} \tag{1}$$

In RPs and SRPs determination phase, each CH is considered as a RP. And each cluster is break down to a lot of small clusters as indicated by its traffic load. The places of RPs and SRPs are viewed as the breaking spots of the portable sink. All sensor hubs can transmit their information to the anticipated single bounce closest hubs of its related RP or SRP. Finally it will buffer and add to the portable sink while it appears.

Figure 2. Network model for SCBDCA algorithm

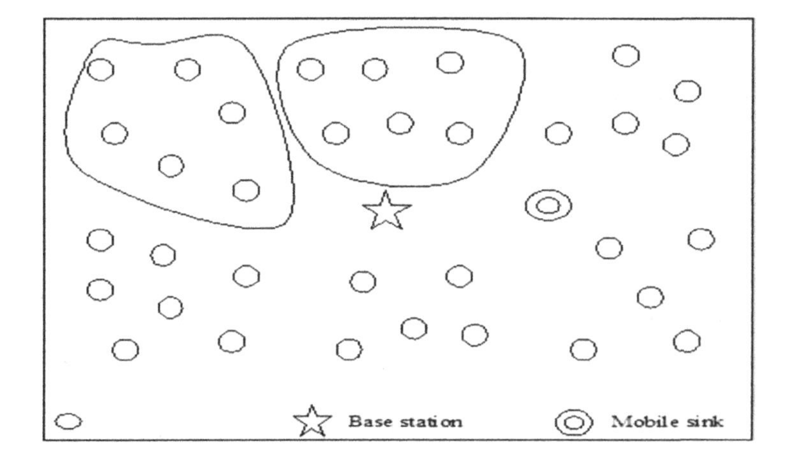

In information gathering phase, when the portable sink reaches every RP/SRP each time, chooses whether to request for tree-rebuilt based on the remaining power proportion of its single-bounce closest hub. If it meets the constraints, the intended RP/SRP make the request to the portable sink after it comes. At last, the sink determines about tree rebuilt according to the based on the quantity of requests to the number of RPs and SRPs.

Figure 3. Flowchart of SCBDCA scheme

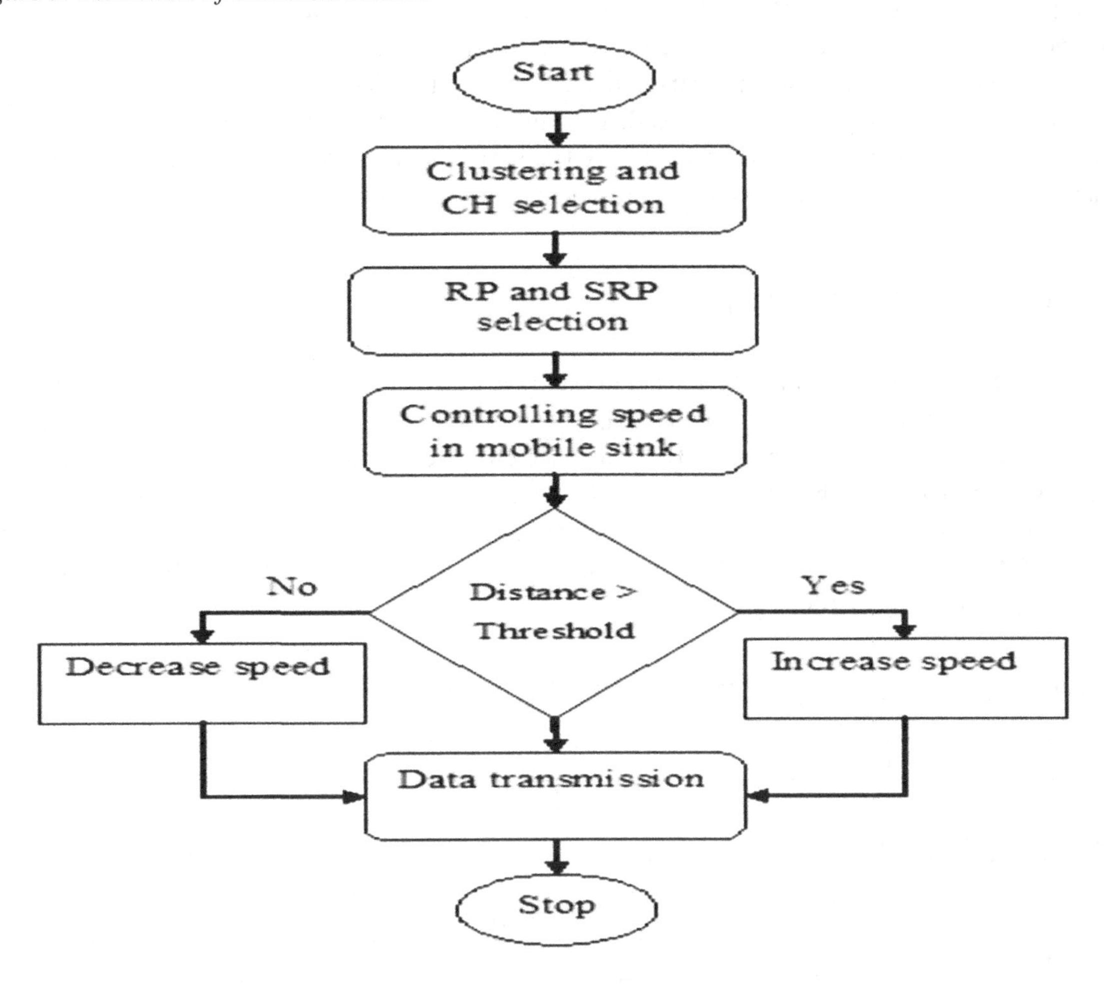

The flowchart of SCBDCA algorithm is shown in figure 3. Initially consider a homogeneous network with 50 nodes. The nodes are clustered depending upon the geographic location of the nodes. Once the nodes are clustered, CH is selected based on the remaining power and the utilization of the node to its nearby hubs. To balance the data gathering latency, a mobile sink is introduced in each zone to see all CH of that zone for data collection. The new algorithm introduced in this paper is the speed control mechanism which is mainly used to control the speed of the portable sink. The remoteness among the portable sink and CHs are calculated and the average distance is taken as the threshold. If the distance is found to be exceeding the fixed limit, increase the speed of the mobile sink else the speed is decreased or adjusted accordingly depending upon the distance.

Algorithm of SCBDCA Scheme:

```
For the mobile sink (MS)
if distance > threshold
increase speed
else
decrease speed
```

A special speed control mechanism estimates the distance between the present point of the portable sink and the next point of the portable sink and adjusts the speed of the portable sink accordingly. Sometimes the portable sink needs to gather the information from the CH that is closer to the sink and also there is a chance for the portable sink to gather the information from the CH that is far away from the sink. Depending upon the remoteness amid the CH and the mobile sink, the speed control mechanism controls the speed of the portable sink in a proficient approach.

PERFORMANCE EVALUATION

The achievement of SCBDCA is investigated through the usage of Network simulator version-2 (NS2). NS2 is a freely available programming language penned both in C++ which is used in back end and OTCL (Object Oriented Tool Command Language) which is used in front end. NS2 is a separate occasion time driven simulator which is primarily utilized to model the protocols. The hubs are scattered in the simulation atmosphere in the communication network. The measuring factors taken for the simulation of SCBDCA are shown in Table 1. The simulation of the projected SCBDCA has 50 hubs built in the simulation circumstances 1000×1000m.

Table 1. Simulation parameters of SCBDCA

Parameter	Value
Channel Type	Wireless Channel
Simulation Time (in seconds)	50
Number of nodes	50
MAC type	802.11
Traffic model	CBR
Simulation Area	1000×1000
Transmission range	250m
Network interface Type	WirelessPhy
Mobility Model	Random Way Point

The hubs are contacted with every other hub through the use of communication protocol User Datagram Protocol (UDP). The nodes are moved randomly inside the simulation region through the use of portability scheme Random Way Point (RWP). The radio signal is propagated with the aid of double ray floor proliferation schemes. The passages in the set of connection are handled using the traffic model

Constant Bit Rate (CBR). Every hub gets the indication from every track through the Omni track aerial. The presentation of the SCBDCA is assessed with the help of parameters packet delivery rate, loss rate, normal impediment, throughput, remaining power and network life span.

Packet Delivery Rate

Packet Delivery Rate (PDR) is the ratio of amount of data parcels given to every recipient to the quantity of packets transmitted from the origin hub. The PDR is obtained through the equation 2.

$$PDR = \frac{Total\ Packets\ Received}{Total\ Packets\ Send} \tag{2}$$

Figure 4. Packet Delivery Rate

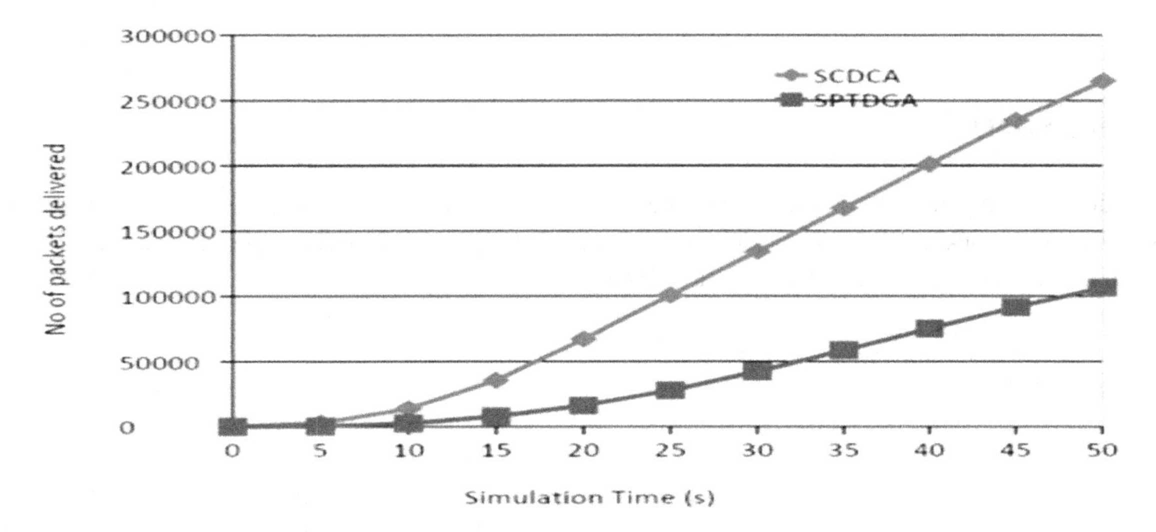

The diagram 4 gives the PDR of the new method SCBDCA is greater than the PDR of already available approach SPTDGA. The more noteworthy estimation of PDR implies the better execution of the convention.

Packet Loss Rate

The Packet Loss Rate (PLR) is the proportion of the quantity of packets discarded to the range of packets transmitted. The method to compute the PLR is specified in formula 3.

$$PLR = \frac{Total\ Packets\ Dropped}{Total\ Packets\ Send} \tag{3}$$

The PLR of the new method SCBDCA is lesser than the present scheme SPTDGA in Figure 5. Lesser the PLR shows the greater performance of the set of connections.

Figure 5. Packet Loss Rate

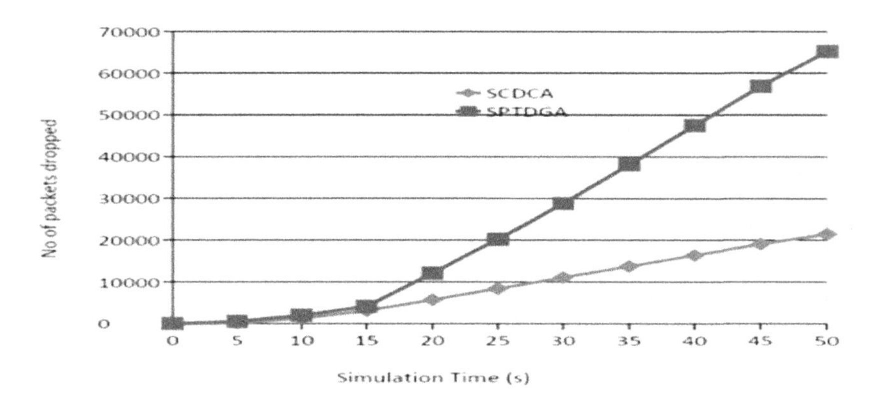

Average Delay

The common impediment is described as the dissimilarity amid the existing packets arrived instance and the earlier packet arrived time. The impediment in the arrangement reduces the effectiveness of the arrangement. The average delay is measured by equation 4.

$$Delay = \frac{\sum_{0}^{n} Pkt\ Send\ Time - Pkt\ Recvd\ Time}{Time} \tag{4}$$

Diagram 6 gives the details of impediment assessment. It is low for the new method SCBDCA than the present method SPTDGA. The least amount of assessment of impediment is the greater assessment of the throughput of the arrangement.

Figure 6. Average delay

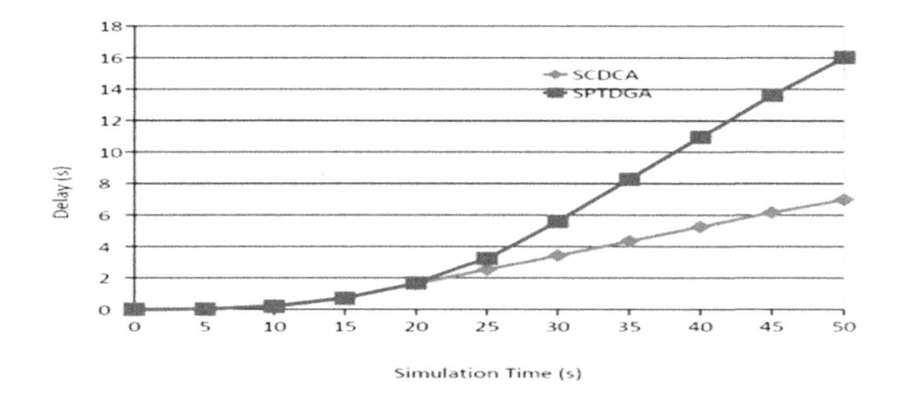

Throughput

It is the normal of fruitful messages conveyed to the goal. The normal throughput is evaluated through the given formula 5.

$$Throughput = \frac{\sum_{0}^{n} Pkts\ Received(n) * Pkt\ Size}{1000} \quad (5)$$

Diagram 7 presents the new method SCBDCA. It has higher normal throughput while analyzing with the current method SPTDGA.

Figure 7. Throughput

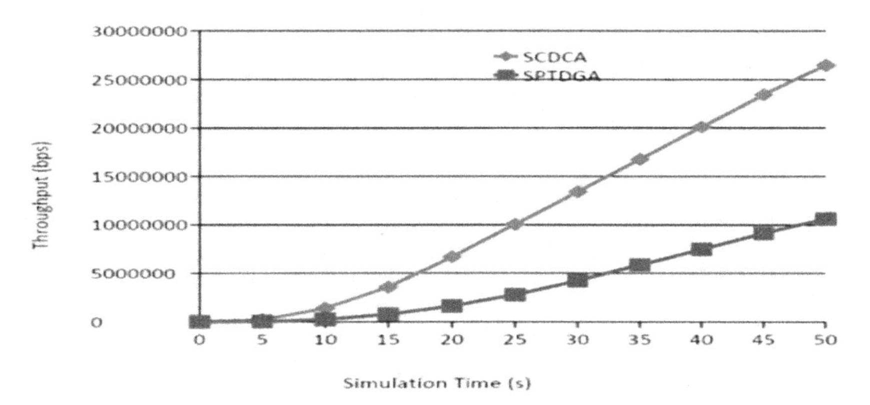

Figure 8. Residual energy

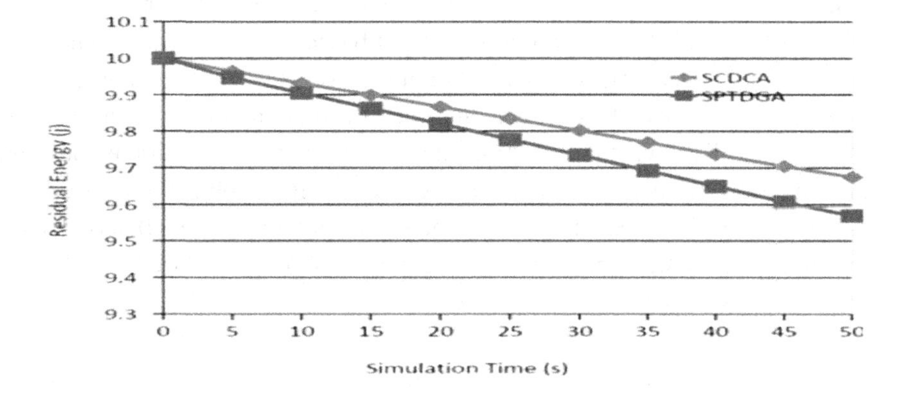

Residual Energy

The quantity of residual power in a hub at the present occurrence of instance is referred as residual or remaining power. A computation of the remaining power produces how much amount of power is utilized through the network tasks.

The remaining power of the communication network is good for the proposed SCBDCA while analyzed through the existing method SPTDGA as in figure 8.

Network Lifetime

It is described as the range of gathering iteration as far as the initial sensor hub dies because of power drain.

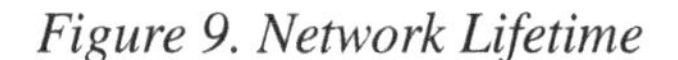

Figure 9. Network Lifetime

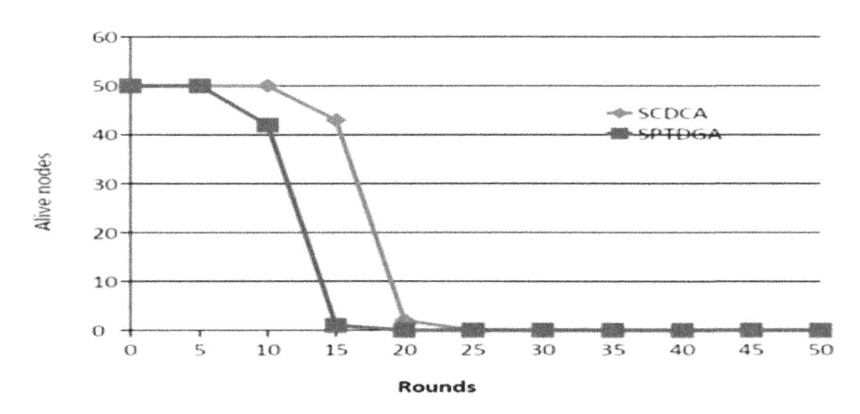

Figure 9 presents the SCBDCA and the SPTDGA line goes down abruptly once the initial hub dies. This gives that the complete network dies only after maximum consumption of power.

CONCLUSION

Writings on information congregation procedures are investigated and classified by the detected information adding route in this paper. An efficient Speed Control based Data Collection Algorithm (SCBDCA) using mobile sink is proposed for WSNs. Initially a weight based clustering method is proposed where the load of every sensor hub together takes an account of the average remaining power, the space to the BS and neighborhood hub thickness. The bounce range and the quantity of information of every hub are considered while choosing SRPs, which in addition balances the power utilization in the communication network. The innovative speed control based data collection algorithm controls the speed of the portable sink in the arrangement using the threshold value. Simulation outcome express that SCBDCA can make longer the network life span considerably compared against existing procedure SPT-DGA.

REFERENCES

Bhushan, B., & Sahoo, G. (2018). Routing Protocols in Wireless Sensor Networks. *Computational Intelligence in Sensor Networks Studies in Computational Intelligence*, 215-248. DOI: doi:10.1007/978-3-662-57277-1_10

Bhushan, B., & Sahoo, G. (2019). Secure Location-Based Aggregator Node Selection Scheme in Wireless Sensor Networks. *Proceedings of ICETIT 2019 Lecture Notes in Electrical Engineering*, 21–35. DOI: 10.1007/978-3-030-30577-2_2

Bhushan, B., & Sahoo, G. (2019). E^2SR^2: An acknowledgement-based mobile sink routing protocol with rechargeable sensors for wireless sensor networks. *Wireless Networks*, *25*(5), 2697–2721. doi:10.100711276-019-01988-7

Chang, C.-Y., Lin, C.-Y., & Kuo, C.-H. (2012). EBDC: An energy-balanced data collection mechanism using a mobile data collector in WSNs. *Sensors (Basel)*, *12*(5), 5850–5871. doi:10.3390120505850 PMID:22778617

Kinalis, A., Nikoletseas, S., Patroumpa, D., & Rolim, J. (2014). Biased sink mobility with adaptive stop times for low latency data collection in sensor networks. *Information Fusion*, *15*(1), 56–63. doi:10.1016/j.inffus.2012.04.003

Krishnan, R., & Perumal, G. (2018). Family-Based Algorithm for Recovering from Node Failure in WSN. In D. Reddy Edla, P. Lingras, & K. Venkatanareshbabu (Eds.), *Advances in Machine Learning and Data Science. Advances in Intelligent Systems and Computing* (Vol. 705). Springer. doi:10.1007/978-981-10-8569-7_31

Krishnan, R., & Perumal, G. (2019). H2B2H protocol for addressing link failure in WSN. *Cluster Computing*, *22*(S4, Suppl 4), 9687–9696. doi:10.100710586-017-1355-9

Lee, E., Park, S., Yu, F., Choi, Y., Jin, M.-S., & Kim, S.-H.(2008). A predictable mobility-based data dissemination protocol for wireless sensor networks. *Proc. 22nd Int. Conf. Adv. Inf. Netw. Appl.*, 741–747. 10.1109/AINA.2008.139

Liu, X., Zhao, H., Yang, H., & Li, X. (2013). SinkTrail: A proactive data reporting protocol for wireless sensor networks. *IEEE Transactions on Computers*, *62*(1), 151–162. doi:10.1109/TC.2011.207

Ma, M., Yang, Y., & Zhao, M. (2013). Tour planning for mobile data-gathering mechanisms in wireless sensor networks. *IEEE Transactions on Vehicular Technology*, *62*(4), 1472–1483. doi:10.1109/TVT.2012.2229309

Sugihara, R., & Gupta, R. K. (2010). Optimal speed control of mobile node for data collection in sensor networks. *IEEE Transactions on Mobile Computing*, *9*(1), 127–139. doi:10.1109/TMC.2009.113

Vupputuri, S., Rachuri, K. K., & Murthy, C. S. R. (2010). Using mobile data collectors to improve network lifetime of wireless sensor networks with reliability constraints. *Journal of Parallel and Distributed Computing*, *70*(7), 767–778. doi:10.1016/j.jpdc.2010.03.010

Wang, N., Qi, X.-G., Duan, L., Jiang, H., & Liu, X. (2014). Clustering-based routing algorithm in wireless sensor networks with mobile sink. *J. Netw.*, *9*(9), 2376–2383. doi:10.4304/jnw.9.9.2376-2383

Zhang, X., Bao, H., Ye, J., Yan, K., & Zhang, H. (2013). A data gathering scheme for WSN/WSAN based on partitioning algorithm and mobile sinks. *Proc. 10th IEEE Int. Conf. High Perform. Comput. Commun., IEEE Int. Conf. Embedded Ubiquitous Comput.*, 1968–1973.

Zhao, M., & Yang, Y. (2012). Bounded relay hop mobile data gathering in wireless sensor networks. *IEEE Transactions on Computers*, *61*(2), 265–277. doi:10.1109/TC.2010.219

Chapter 11
Analysis of Black-Hole Attack With Its Mitigation Techniques in Ad-hoc Network

Ayasha Malik
AIACTR, Delhi, India

Siddharth Gautam
GGSIPU, Delhi, India

Naghma Khatoon
Usha Martin University, India

Nikhil Sharma
https://orcid.org/0000-0003-4751-2970
HMR Institute of Technology and Management, Guru Gobind Singh Indraprastha University, India

Ila Kaushik
Krishna Institute of Engineering and Technology, India

Santosh Kumar
https://orcid.org/0000-0003-2239-164X
ITER, Siksha 'O' Anusandhan (Deemed), India

ABSTRACT

In wired and wireless communication, providing security is extremely important. It is a very challenging issue. But the flying evolution in communication technology has triggered to sturdy research interest in wireless networks. The characteristics of wireless networks make this issue even more challenging. In Ad-hoc networks, there is a huddle of autonomous nodes, which dynamically form a temporary multi-hopped, peer-to-peer radio network, without any use of predefined infrastructure. These nodes are generally mobile in nature, and to connect these nodes, the connectionless links are used. These nodes have the potential to self-organize, self-configure, and self-arrangement. Ad-hoc networks do not have fixed structure due to their dynamic nature. Ad-hoc networks are inherently prone to a number of security threats. Lack of fixed infrastructure, use of wireless link for communication, and mobility of nodes make Ad-hoc networks extremely receptive to hostile attacks, blackhole attack being one among them, which can be implemented effortlessly.

DOI: 10.4018/978-1-7998-5068-7.ch011

INTRODUCTION

A wireless network is a network that uses wireless data connection between computers to create nodes. Wireless communication is a method by which the cost of wireless communication is reduced Wireless medium plays a vital role in providing real-world communication by allowing users to take information and service electronically or digitally, despite their topographical location (Fazeldehkordi, 2016). There are two types of wireless connection/communication: infrastructure-based (contains access point) and infrastructure-less (without an access point). MANET is a network that lacks infrastructure (Jamshidi et al., 2018). The wireless connections used for interconnection may be terrestrial microwave, satellite communication, radio, and spectrum spread technologies, optical free-space communication, or many radio communication technologies used in cellular and pc systems. **Figure 1** shows the classification of wireless network.

Figure 1. Classification of wireless network

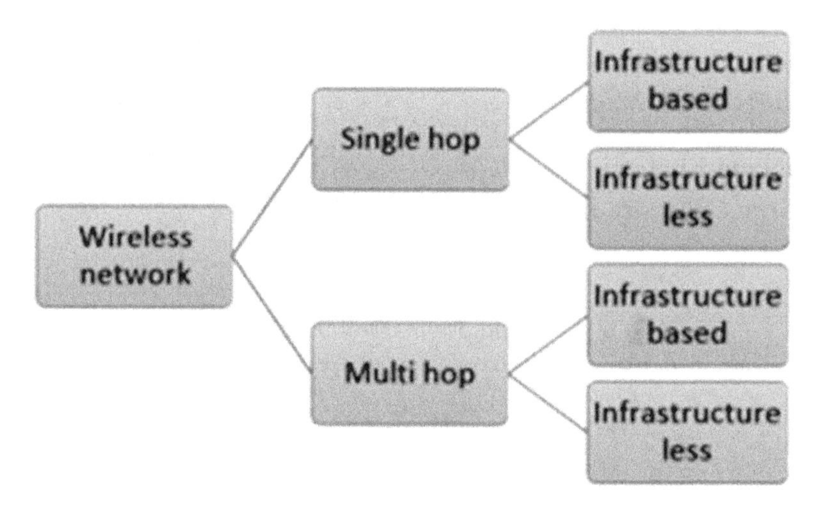

Wireless networks are classified into two categories and then classified into two categories as shown in Figure1. Wireless single-hop network- when there is a one-to-one mapping between nodes. It is further divided into the wireless (with the access point) network based on infrastructure and wireless networks (without an access point) (Liu et al., 2019). Wireless multi-hop network- when the mapping between nodes is one-to-many. A multi-hop wireless network is also divided into wireless infrastructure-based networks and wireless infrastructure-less networks as a single-hop wireless network (Bhushan & Sahoo, 2020) . **Table 1** summarizes diverse types of wireless networks.

An Ad-hoc network is a set of autonomous nodes capable of self-organizing, self-configuring, and self-arranging. These networks do not require any pre-existing or fixed infrastructure; therefore, they can be set up dynamically on demand. "Anytime, Anywhere, Any condition and anyone; communication is ready" is the goal accomplished by the Ad-hoc network because of its mobility, flexibility, and dynamic topology changes (Ren et al., 2016). An Adhoc network does not require any licensed frequency band to act, and is free of any infrastructure investment as it can dynamically form structure. These properties play a vital role in making them attractive for selected commercial applications, and military purposes

Table 1. Example of wireless networks

Type of Network	Example
Infrastructure-based Single hop wireless network	Wi-Fi, Cellular network
Infrastructure-less Single hop wireless network	Wi-Fi hotspot, Bluetooth
Infrastructure-based Multi-hop wireless network	Wireless mesh network, Sensor network
Infrastructure-less Multi-hop wireless network	Ad-hoc network, WANET

most important (Ali et al., 2018). In the Ad-hoc network, communication between nodes is done via radio waves. A node may also operate as a sender, receiver, and router in an Ad-hoc network. Communication/transmission in the Ad-hoc network is done using memo/messages, and the node can send the data through the messages to its nearest node. And these nodes don't have any other node information, i.e. whether the node is malicious or genuine (Bhushan & Sahoo, 2019).

The rest of the paper organized as follows: Section 2 illustrated Black-hole Attack along with process and nature of black hole attack using flow chart, FV-AOMDV Protocol Methodology and simulation with experimental configuration discussed in Section 3, PDR-based comparison of AODV vs FV-AOMDV, Delay explained in Section 4 and chapter conclusion presented in Section 5.

Black-Hole Attack

One of the well-known serious security threats in Ad-hoc networks is a black hole attack. It's sort of an attack on safety. In this attack a malicious (fake) node, called a black-hole node, capable of capturing all data by conveying a false routing direction, holding a falsely up-to-date or latest path to the target node, and then letting them fall without delivering them to the target node. Black-hole attack degrades the performance of the entire network, such as packet delivery rate, payload, control packets, and throughput, due to there at the same time packet drop behavior and payload due to frequent route reorganizations (Kumar et al., 2019). It is a type of active attack that attracts the node to become a victim, i.e. black-hole node forces the genuine node to establish a communication with them (Wang et al., 2017). These Black-hole nodes can perform various harmful actions on the network, such as I Behaves as a Source node by falsifying the RREQ packet, (ii) Behaves as a Destination node by falsifying the RREP / RRES packet, (iii) Reduce the number of hop counts when transmitting the RREQ packet (Amin et al., 2018). **Figure 2** show the malicious nade.

Figure 2. Malicious Node

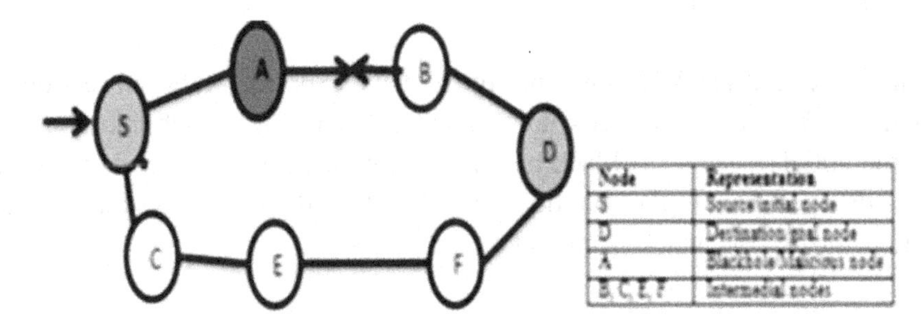

In Figure2, Suppose A is a malicious/black-hole node, the route S-A-B-D is selected as the shortest (based on hop count) for communication between S and D. After the establishment of the connection, S starts sending the data packets to A, when A receives the data packets then the node drops all the data packets. **Figure 3** presents a flowchart of a black-hole attack.

Figure 3. Black-hole attack Process

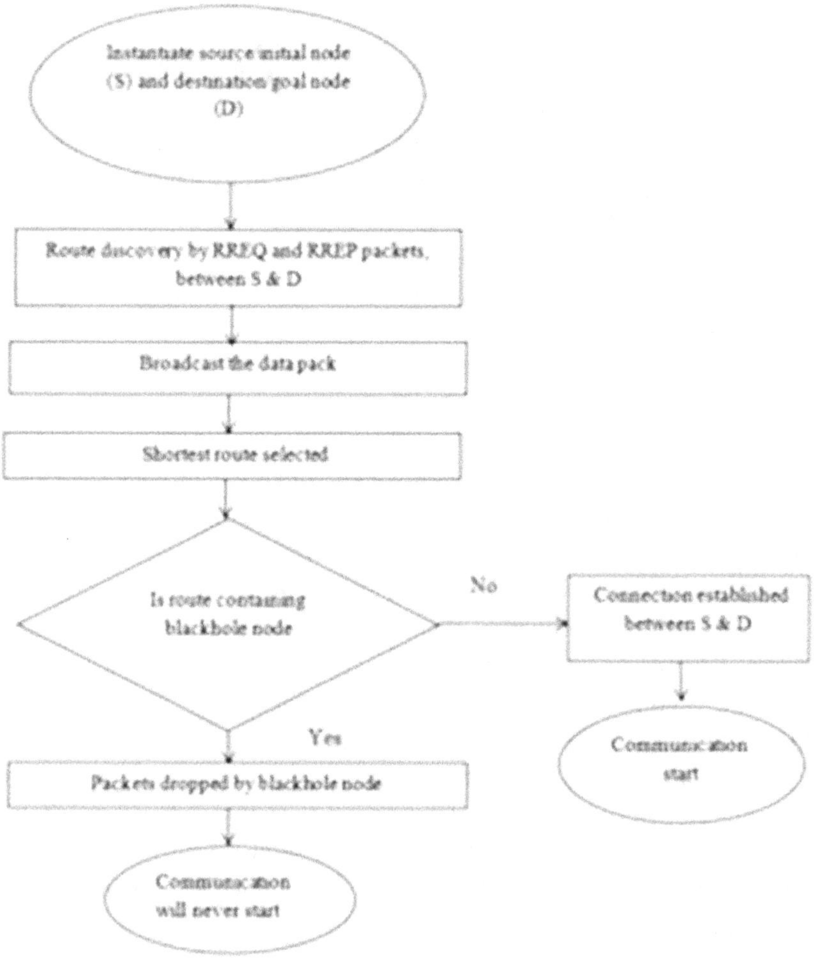

Nature of Black-Hole Attack

The nature of the Black-hole attack is shown in **Figure 4** and explored in the subsections below.

Single (Sole) black-hole attack: It is a kind of black hole attack in which fake routing information is forwarded to the initial node by the black hole node only to gain data packets from it, after all the data packets have dropped. Due to the quantity of black-hole node, i.e. always one / single, this attack is known as a single black-hole attack (Aghili et al., 2018).

Figure 4. Nature of Black-hole attack

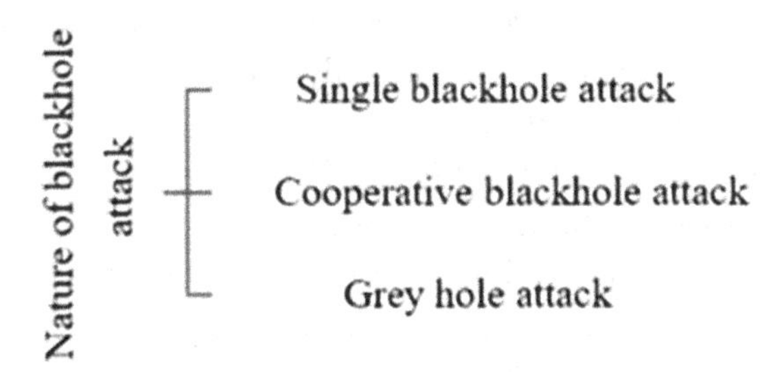

Cooperative (Multiple) black-hole attack: It works the same way as the single black hole attack, but here the combination of more than one black hole node sends the fake request for access to the data packets and then drops them. It is also referred to as a packet drop attack (Puri & Bhushan, 2019).

Grey hole attack: It is a black-hole attack of an extra-special nature. It is similar to attacking a black-hole. Except, a black-hole attack drops all the packets while the packets may or may not drop a grey hole attack. There's no fixed behavior in it. It often switches from black-hole node to the normal node, and vice-versa (Lu et al., 2016).

Prevention and Detection Techniques for Black-Hole Attack

In the present day of communication/connections, security is the pre-eminent study. It is very important to prevent the black-hole attack from the network and to provide secure communication in Ad-hoc networks. These are the different methods that were proposed to prevent black-hole attack.

PBHA (Prevention of a Black-hole Attack): It is based on the fidelity table provided by default to any genuine user when RREQ is broadcast, the source node has to wait until it receives RREP / RREP from adjacent nodes, then only that node with the highest fidelity value is selected, hence the threshold value also increases, for data transmission (Sinha et al., 2019). After receiving the ACK from the destination node the fidelity value is regularly updated by the source node. When the destination node received data, it responded to the source node via an ACK packet, and the source node must add 1 to the adjacent node's holding fidelity value. If the source node does not receive an ACK then 1 is subtracted from the holding fidelity (Abdalzaher et al., 2019).

Black-hole attack detection using threshold value: It is a method of detection based upon Destination Sequence Number (DSN). Every node in a network has a Threshold Value (TV) attached to it when the node receives an RREP packet, and then the DSN-TV comparison is made (Anand et al., 2016). If the DSN value is greater than TV then it is said that the node from which the RREP packet comes is a black-hole node. But the disadvantage of this method is that there is no regular TV update (Bhushan et al, 2017). A regular update on TV achieves the lower fake detection rate and a higher true detection rate.

Black-hole attack detection using dynamic threshold: Using the threshold value method, this method is employed to overcome the limitation of black-hole attack detection. In this approach, TV is a lively intermediate node update that is based on all the network's live nodes, and the time that has passed after knowing the last DSN (Athmani et al., 2019).

Recheck the blacklist: It is a detection method that is based on the membership time of the black-hole node. Each black-hole node in the blacklist has a membership time, meaning it only works for that particular time as a black-hole node, after which it becomes a genuine node (Banerjee et al., 2019). So, recheck the blacklist method is used to check this type of node from the blacklist and free it from the black-hole node to make communication faster, but only if the membership time expires and the node does not act as a black-hole node (Chatterjee, 2019).

IDAD (Intrusion Detection using Anomaly-based Detection): The IDAD uses host-based IDS to prevent the network from black-hole attacks. It is grounded in normal activities. IDAD assumes that it can scan any single end-user and system activity as well; identify an attacker's abnormal activity from a pool of normal activity. Hence the black-hole attack is detected by uncovering an abnormal activity of an attacker / black-hole node. A pre-collected box of abnormal activities, named as audit data, is required for that. It is hand-over to the IDAD system after assembly of audit data, and then a comparison is made between each node activity with audit data on a fly (Bhushan & Sahoo, 2017). If both activities are matched, the IDAD system first checks that node properly and if it is found to be genuine then allows further relationships to be established (Chavan et al., 2016) Working this system is based on the "trusts no peer" principle, i.e. nodes don't depend on another node to prevent intrusion from the network. It can detect novel attacks (black-hole attack variations), as well. But it takes time and needs more computational resources (Doss et al., 2018).

IDM/SD (Intrusion Detection using Misuse/Signature-based Detection): Using Signature-based Detection (El-Semary et al., 2019), intrusion detection using Misuse-based Detection method is also called Intrusion Detection. It is based on the pattern of attacks, i.e. the pattern of known attack is used for the detection of black-hole node in the Ad-hoc network pattern. It is a very precise method, i.e. it has the power in near-real-time to expose the assault immensely reliable. This method has a limitation that it can not uncover the novel attack whose signature is not known as compared to another IDS method but the other side is that this method is better in computation and delay than other IDS methods. More time and computational resources are needed to identify the black-hole node (Bhushan & Sahoo, 2017).

IDSD (Intrusion Detection using Specification-based Detection): This method combines the strength of the IDAD and the IDM / SD method by using a set of rules to uncover the black-hole node in the network. It is both swift and accurate. This approach is legal, generally successful statements are used to distinguish the nature of the legal program (Athmani et al., 2019). It only generates an alarm for the nature of the illegal program encountered in the network, not for the nature of the unusual (but legitimate) program. So that, compared to IDM / SD, the false-positive rate of IDSD reduces. As well, this system can detect unknown attacks (Faghihniya et al., 2016).

D-MBH (Detection of Multiple Black-hole) & D-CBH (Detection of cooperative Black-hole): These algorithms can detect both the (single & collaborative) nature of the black-hole attack. In the first algorithm, the source node sends all remaining nodes of the ad-hoc network a false RREQ with DSN of a non-existent node. The attacker node sends high DSN to the RREP / RRES because a higher DSN represents a fresh path. Then, after calculating the average DSN of all fake RRES / RREP, the second algorithm creates the list of multiple black-hole nodes, and finds that RRES / RREP originates from black-hole node has higher DSN than genuine RREP (Kaushik et al., 2019). This method reduces the routing overhead and computational overhead.

BBN (Back-Bone Nodes) and RIP (Restricted IP): In terms of battery power and range BBN nodes are honest and strong. Initially when the source node wants to make a data transmission, it asks for a RIP from the nearest BBN. When BBN receives the RIP request from the source node then it responds with a fresh IP address (Gai et al., 2017), randomly selected from the IP addresses that are unemployed. The initial node then broadcasts RREQ to goal and RIP simultaneously. Now if the initial node was

Figure 5. BBN (Back-Bone Nodes) and RIP (Restricted IP)

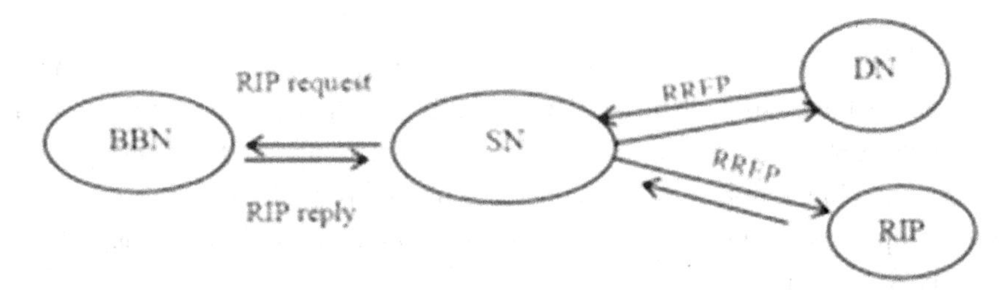

only receiving RREP / RRES for goal node and not from that RIP, then there is no malicious node in the network (Bhushan & Sahoo, 2018). And if the source node received RRES / RREP for RIP then it is clear that there is a multiple malicious node on the network as shown in **Figure 5**.

DRI (Data Routing Information) Table and crosschecking using FREQ or FREP: This method is introduced to identify the attack from the black hole. Every node in the network has a table with DRI. The DRI table keeps each node record transferred by the node and receives the data with its neighboring nodes. If the initial node does not have a path to the target node, the initial node will forward an RREQ message to its neighboring nodes to find a safe and fresh path to the target node (He et al., 2017). If any node receives this RREQ message then the node can reply or broadcast it to the entire network. If the destination node responds to the request for the route, then each intermediate node reform or insert the routing details for that objective node. When the initial node receives RREP from the target node then it will start forwarding packets with the route received and updating its DRI table based on the destination node and intermediate node (Karthik et al., 2017). The solution proposed raises the graph in terms of performance and throughput rate including less amount of packet loss but node mobility should be considered as low.

RRT (Router Request Table) and sequence numbers: In this method, SN of the source node or intermediate node is examined and a comparison is made with the SN of the source node between existing DSN, if there is a huge difference between SNs, that means it is the attacker node (Bhushan & Sahoo, 2017). This solution removes that attacker node from the RRT immediately.

DSDV (Destination-Sequenced Distance-Vector Routing): It is a protocol, widely used in ad-hoc networks. It uses DSN and dynamic TV to uncover a current route to the node of the destination. It both detects the black-hole attack in the network and prevents it. In its routing work tables it required a regular update which also consumes battery power and bandwidth; while the network remains inactive (Khanna & Sachdeva, 2019). Whenever the network changes its infrastructure, before network re-convergence, a fresh succession number is required; therefore, it is not suitable for vast, lively networks.

AODV (Ad Hoc On-Demand Distance Vector): The routing protocol was widely used in Ad-hoc networks. In AODV initial node initiates the activity of path disclosure by posting an RREQ to the whole network. Intermediate node participates in this activity by scattering this RREQ further towards the node of destination. It thus takes some time for the process to send RREP from the destination node. A black-hole or malicious node doesn't know and instantly post a false (wrong) RREP or RRES packet back to the original node after making-believe that, it has the best path to the goal (Li et al., 2018). The initial node can, therefore, know about the black-hole node, and also ignore it.

EAODV (Enhancement Ad Hoc On-Demand Distance Vector): It is an improvement on AODV. EA-ODV's main strategy is that at any point in time during data transmission the target node will convey the RREQ or RREP to the initial node. That's why; if the current or new incoming RREP or RRES received from the target node to the network, then-current RRES or RREP will override the entire existing path listed encircling black-hole or malicious nodes. At the same time, the testing of all accepted RRES or RREP is carried out using a heuristic approach from which a sort of detection and isolation is performed over RRES or RREP received. EAODV protocol is an update to AODV routing. It uses two procedures to remove the black-hole attack; change and update route logic expression and, second, add process detection and process isolation methods but the entire process takes some extra time and extra energy use (Kaushik & Sharma, 2020).

IDS-AODV/Switch the route: IDS-AODV is a prevention method. The main strategy is that originally the data is traveled along with the first organized path but if the second RREP / RRES appears then the data transmission is shifted to the second path. This approach believes that the first RREP/RRES that received by the source node is from a black-hole/malicious node, therefore pay no attention to that RREP/ RRES. This approach upgrades the packet delivery ratio but there is another side too. For example, if the second RRES/RREP received by the initial node arrives through the black-hole/malicious node, it is not avoided and the path selected which includes black-hole/malicious node (.Bhushan & Sahoo, 2019). It is also called switch the route. But there is a problem when the destination node is nearer as compare to the attacker/black-hole node then the first RREP/RRES received by the initial node is from the destination node not from the attacker/black-hole node. In this case, the data send by source through the path which is established by malicious/black-hole node.

IIDS-AODV (Improved IDS-AODV): An IDS-AODV modification. This method has been introduced to overcome the IDS-AODV limit. It is based upon the second RREP message being checked. Initial work is identical to IDS-AODV's. The difference between IDSAODV and IIDS-AODV is, when the second RREP / RRES is received by the source node in IIDS-AODV; first, a test is carried out using the BSN and RDSN. The difference between the BSN and RDSN is considered in the second RREP / RRES test and is then related to the 50 percent of the highest successive number (HSN) and the result must be less or equivalent to (HSN/2). Only if the current/second RREP / RRES satisfies the test condition will the source node be moving to this route. If this test fails, the source node will continuously send its packets along with the route that is selected by the first RREP / RRES.

Broadcast Sequence Number (BSN) is one of the RREQ message fields and was previously found by the source for any path that is on the way to the destination and Received Destination Sequence Numbers (RDSN) is a sequence number received from the RREP message.

WAODV (Watchdog AODV): Each node has a watchdog set with this method to check the node of a neighbor as the neighbors have to dispatch packets to their neighbors. If the packets have not been dispatched by the neighbor node, the watchdog will detect it and the neighbor node declares it to be a node / malicious node misbehaving. This approach can not reveal the malicious nodes in the existence of equivocal and destination-side collisions, finite transfer power, directional antennas, false or mischief behavior, and imperfect dropout.

REP (Recommendation Exchange Protocol): Each node can convey and take recommendations from neighboring nodes in this protocol. The REP protocol is based on the trust that is developed through experiences and node suggestions.

OCEAN (Observation-based Coordination Execution in Ad hoc Networks): This protocol was implemented to overcome DSR routing protocol limitations. It is a protocol extension over DSR. To detect malicious/black-hole nodes, it uses two systems; the first is the monitoring system, and the second is the reputation system (Liu et al., 2018). But OCEAN does not predict accurate results, i.e. it cannot uncover all the malicious nodes.

ODMRP (On-Demand Multiple Routing Protocol): It depends on authentication i.e. for secure communication, first the routing messages authenticated by the method of the localized certificate chain then transmitted to the node of the neighbor.

BHS-ODMRP (Black-hole Secure-On Demand Multiple Routing Protocol): It is an additive to OD-MRP. It is applied over the top of ODMRP 's path invention procedure where the safety favor is divided into more than one node and each node certifies each other in a self-ordered and well-ordered manner (Liu et al., 2019).

BDSR (Baited-Black-hole DSR): It is a DSR extension, and it can detect and prevent black-hole attack. Bait forwarded in this technique to attract a black-hole node / malicious node to respond by the wrong direction of the route. It can therefore attain proactive detection and trace the route back to the exact position of the existing black-hole. That can minimize the chance of suffering a black-hole attack after the route has been established.

Banned RREP / RRES: It is a method of prevention that works on the principle of deactivating the entire response messages returning from the intermediate nodes. In banned RREP / RRES, the intermediate nodes are blocked to forward RREP / RRES, and thus only those reply messages which originate from the actual destination node are trusted. In this technique the malicious/black-hole node is assumed to be from the intermedial node. That is why the initiation of any RREP / RRES data packet is prohibited for intermedial nodes. Wait for route increases in large networks in this approach, and may give black-hole/ malicious node power that it can forward RREP / RRES at the destination node location.

Acknowledgment of RREQ: In this technique based on cryptography techniques, when a node then receives RREQ / RRES from an unbelievable third node, ARREQ / ACK RREQ receives the initial node encoded by its shared key. So now the initial node must first decrypt it, then perform an encryption over it again, but now with the public key of destination and dispatch to the initial node (Bhushan, 2019; Luo, 2018). Node is considered as a genuine node, when the goal node gets the same message as sent. Otherwise they simply ignore RREQ / RRES.

ACO (Ant Colony Optimization): As the name suggests, this depends on the behavior of ants in the real world and the method they use to find food (Nayak & Devulapalli, 2016). This technique is derived from incidental colony conveyance by normal negotiator, known as artificial ants, when an ant moves along a path (Sharma et al., 2017); it deposits a chemical called pheromone upon it. As more and more ants travel along the same route (Ochola et al., 2017), the concentration of the route's pheromone increases. Then the route with the maximum concentration of pheromone is chosen as the optimum route (Panos et al., 2017). In the network the same method is used to find an optimal and safe route that is free from the black-hole node (Bhushan & Sahoo, 2019).

CONFIDANT (Cooperative of Nodes, Fairness in Dynamic Ad-hoc Network): It is watchdog's next genre. It uses a method similar to Pretty Good Privacy (Patil & Chaudhari, 2016). to indicate the countless amounts of faith, secret authorization, and documentation authorized. It allows the node to execute negative ratings to identify false allegations (Pham et al., 2016). But it cannot capture the dropping of incomplete packets.

Certificate Chaining: A certificate means an authorization, as the name indicates, and chaining means the number of nodes (Pruthi et al., 2019). Authentication is performed through a node chain in this method, which is self-assembled and has not used any trusted third party (Wang et al., 2019). A set of digital certificates used in certificate chaining to represent an authentication that forms a chain. The roles and responsibilities are equally divided into each network node, and a supreme node contribution level is achieved (Jaitly et al, 2017; Sharma et al., 2019). Each network node capable of issuing stored certificates and distributing them to the remaining node within their range of radio communication (Xiong et al., 2017). An authorized document/certificate binds to a node which includes its common/shared key and security framework. In a network, each single certificate is saved twice, one for the issuer node and the other for whom it is issued. A node's neighbors are responsible for updating certificates, and also adding new certificates. If there is a collision in certificates, i.e. one common/shared key for many nodes or one node with multiple common/shared keys, then there is a chance that a fake certificate will be found in the network.

DCM (Distributed and Cooperative Mechanism): The multiple black-hole attacks can be avoided, detected and mitigated by this solution (Sinha et al., 2017). With DCM the four sub-modules are included. These modules are the Local Data Assembly, Local Identification, Co-operative Identification, and Universal Respond.

Methodology and Simulation

In this chapter, we explore Ad-hoc network creation, tool configuration used, performance evaluation metrics, and analytical simulation results. First, we create an Ad-hoc network that contains 48 nodes using AODV protocol in which 47 nodes are genuine nodes and 1 node is black-hole i.e. malicious node then we analyze the network after calculating some performance metrics (Delay, Packet delivery ratio and Received packets). After that, we create another network that contains 100 nodes that are genuine nodes by some of them and some are black-hole. Then we securely transfer the data from source to destination based on ant colony optimization prevention technique based on fitness value and calculate performance metrics. The entire prevention process id depicted in **Figure 6**.

Figure 6. Prevention process

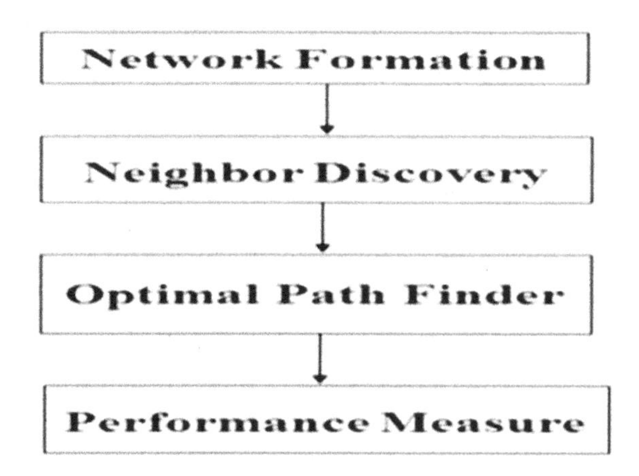

FV-AOMDV Protocol

When RREQ is generated/broadcast by the initial node, in the earliest concept of the initial AOMDV, not only one route but more than one path to the goal (destination) node is established and the data packets are sent (travel) through these paths without any information related to the quality of the routes. By implementing the rule set out above, the selection of routes is completely different. Once an RREQ is broadcast and received, the provisioning node will have three types of information to choose the optimal and shortest path: I information about the energy level of each node in the network, (ii) the distance of each route, (iii) the energy consumed during the route discovery process.

To minimize the energy consumption, the forwarder node can forward information packets via the path of the highest energy state. Also, the FV-AOMDV protocol initiates a fresh method of discovery of the path once all paths to the goal fail. The supply node can then select another path from its routing table within the event once the chosen route fails, which represents the shortest route with higher energy levels and minimum energy consumption. **Figure 7** depicts the flowchart of the prevention process.

Figure 7. Flowchart of Prevention process

Experimental Configuration

We used the simulator tool NS-2.35 for the network. For discrete events invented at UC Berkley. NS2 tool performs over wired and wireless networks; TCP simulation, network analysis, routing and multi-cast protocols, etc. Those are the languages used to write code in NS2; C, C++, OOPS TCL extension, OTCL. The parameters considered for the simulation are shown in **Table 2**. The NS2 is composed of two NS and NAM simulation tools.

Table 2. Simulation setup

Parameter	Value
Channel type	Wireless channel
Radio propagation model	TwoRayGround
Network interface type	Wirelessphy
MAC type	MAC 802.11
Interface queue type	Queue/DropTail/PriQueue
Link-layer type	LL
Antenna model	Omni Antenna
Max packet in ifq	50
Number of mobile nodes	48/100
Routing protocol	AODV/FV-AOMDV
Simulation time	100.00s
X & Y dimensions	700*700 sq.m
Simulation end	150.01s
Traffic type	TCP/UDP/CBR
Size	1024/1000/1500
Rate	1.0Mb/2.0Mb/4.0Mb
Initial Energy	100

ANALYSIS AND RESULT

Performance Matrices

The following performance metrics are taken to examine the AODV/FV-AOMDV routing protocol vs. Simulation time/Nodes.

End-to-end delay: Total time taken by data packets to hit receiver/destination from sender/source was including delay (buffering time, transmission time, re-transmission time, waiting time, propagation time).

Packet delivery ratio (PDR): Total amount of data packets received by goal/destination wealthy. It is the ratio between the received packets to the generated packets.

Received packets: It is the total amount of received data packets during the whole simulation.

Let's take an eye on reading as well as the related graphs.

Table 3 & **Figure 8** show the comparison received data packets in AODV vs. FV-AOMDV with the simulation time increases while nodes remain the same i.e. 100. As we can see the graph line of FV-AOMDV increases i.e. more data packets received by proposed protocol.

Table 3. RP_N_100

Simulation time(s)	RP (AODV)	RP (FV-AOMDV)
5	39	241
15	145	741
50	551	2524
75	850	3710
100	1161	4992
125	1471	6303

Figure 8. comparison received data packets in AODV vs. FV-AOMDV with the simulation time increases while nodes remains same i.e. 100

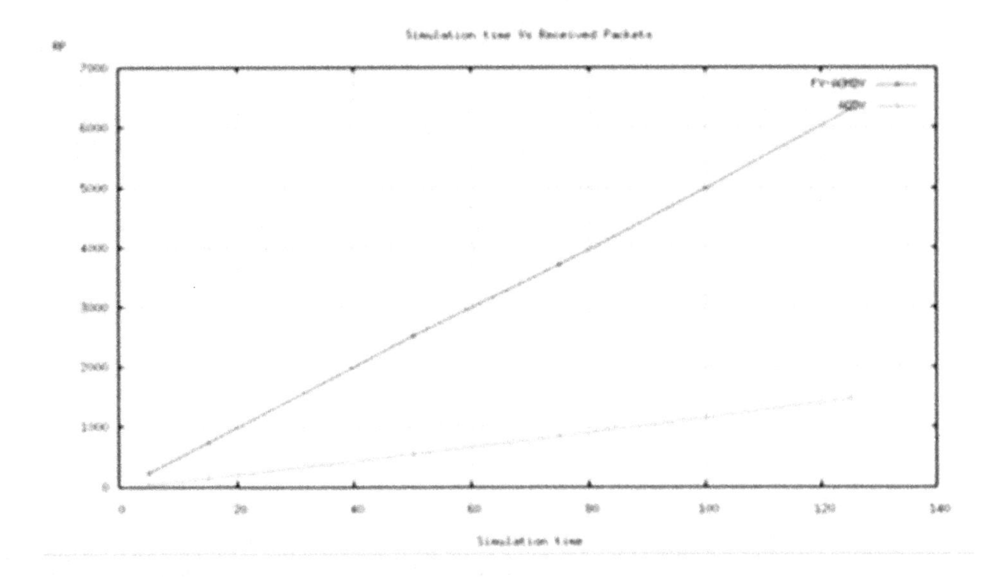

Table 4 & **Figure 9** show the comparison PDR in AODV vs. FV-AOMDV with the simulation time increases while nodes remain the same i.e. 100. As we can see the graph line of FV-AOMDV is highly increases hence we won in packet delivery ratio by our proposed protocol.

Table 5 & **Figure 10** show the comparison of delay in AODV vs. FV-AOMDV with the simulation time increases while nodes remain the same i.e. 100. In the graph, the delay is also increased when data packets follow the proposed protocol because in FV-AOMDV protocol first fitness value estimated the data packets forward to the next node, and this will be done for each selected node between source to destination hence delay increases. But it should be decreased to achieve a better result.

Table 4. PDR_N_100

Simulation time(s)	PDR (AODV)	PDR (FV-AOMDV)
5	78	95.6
15	81.9	97.88
50	88.1614	100
75	91.6955	97.99
100	93.629	98.89
125	94.7802	99.90

Figure 9. comparison PDR in AODV vs. FV-AOMDV with the simulation time increases while nodes remains same i.e. 100

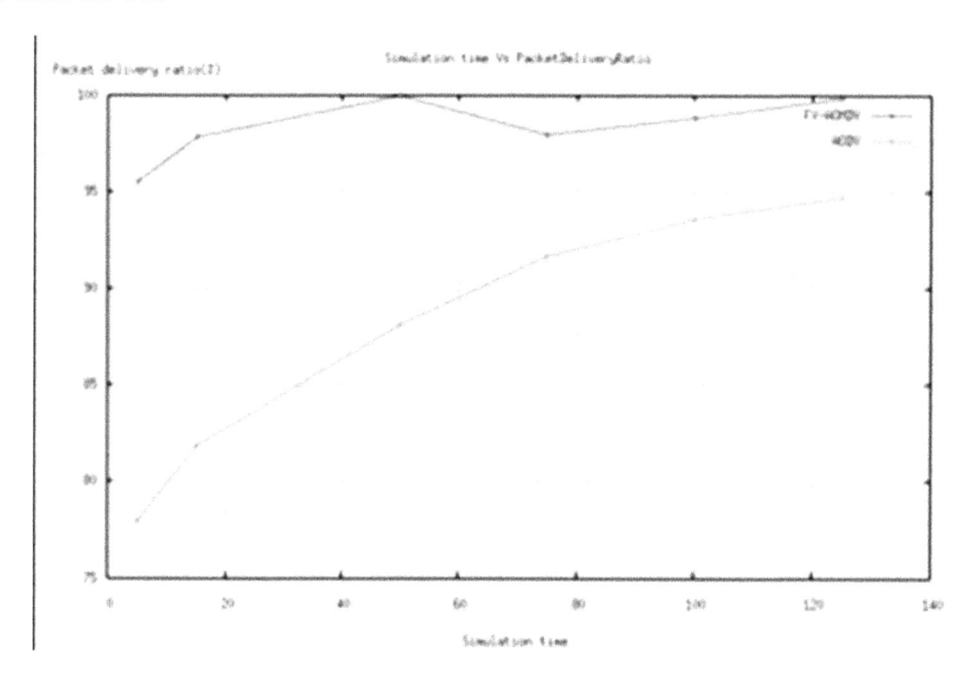

Table 5. Delay_N_100

Simulation time(s)	Delay (AODV)	Delay (FV-AOMDV)
5	35734	35914
15	107202	107832
50	357340	397410
75	536010	710110
100	714680	934312
125	893350	998242

Figure 10. comparison of delay in AODV vs. FV-AOMDV with the simulation time increases while nodes remains same i.e. 100

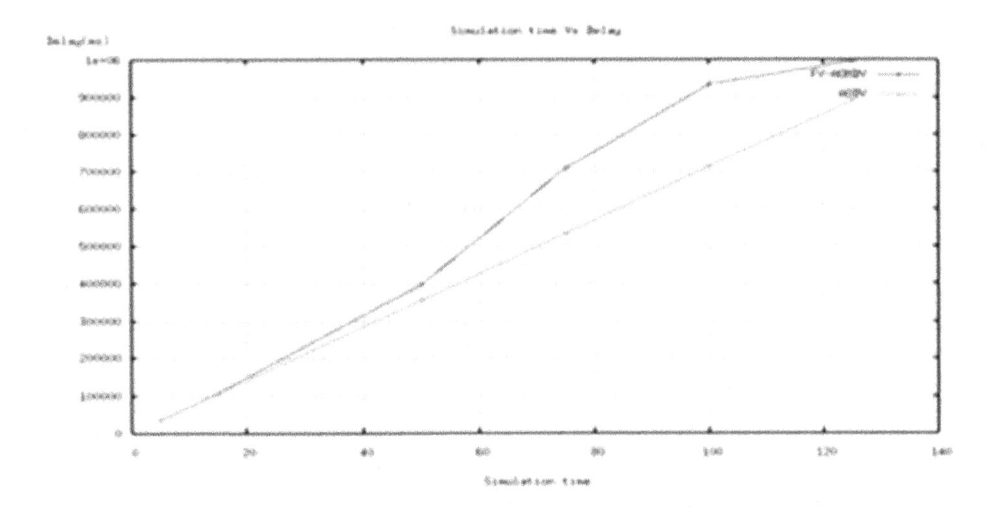

Table 6 & **Figure 11** show the comparison of received data packets in AODV vs. FV-AOMDV with the number of nodes increases while simulation time remains the same i.e. 25 seconds. As we can see graph line of FV-AOMDV is highly increases hence we won in packet delivery ratio by our proposed protocol.

Table 7 & **Figure 12** shows the comparison of PDR in AODV vs. FV-AOMDV with the number of nodes increases while simulation time remains the same i.e. 25 seconds. As we can see graph line of FV-AOMDV is highly increases hence we won in packet delivery ratio by our proposed protocol.

Table 6. RP_T_25

Number of nodes	RP(AODV)	RP (FV-AOMDV)
100	582	1262
130	738	2639
160	901	1335
190	1034	1800
230	1239	1142

Table 7. PDR_T_25

Number of nodes	PDR(AODV)	PDR (FV-AOMDV)
100	92.82	99.89
130	90.55	99.1732
160	89.83	98.9622
190	86.81	99.3377
230	85.92	99.651

Figure 11. comparison of received data packets in AODV vs. FV-AOMDV with the number of nodes increases while simulation time remains same i.e. 25 seconds

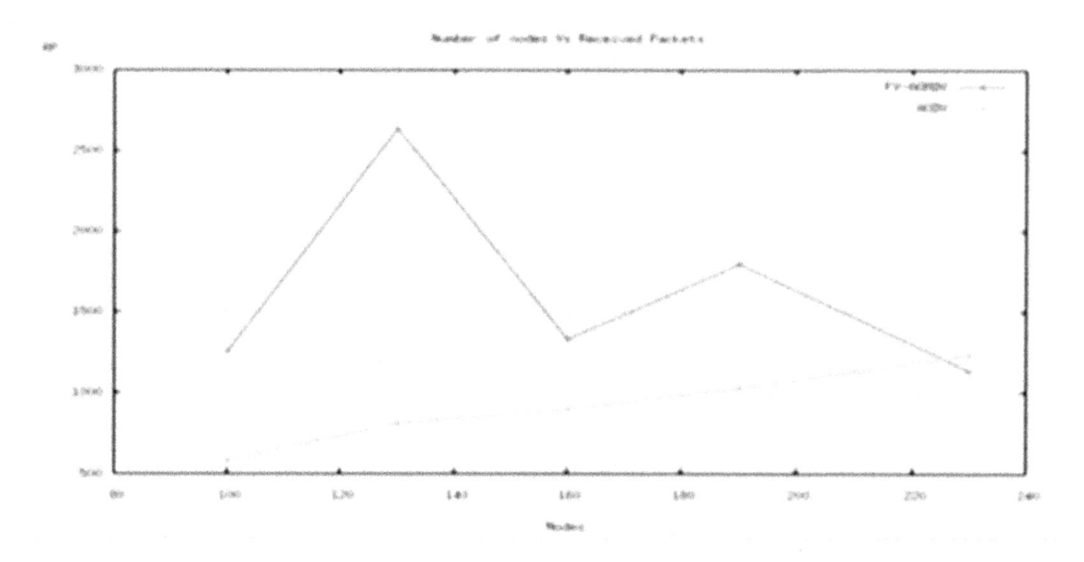

Figure 12. comparison of PDR in AODV vs. FV-AOMDV with the number of nodes increases while simulation time remains same i.e. 25 seconds

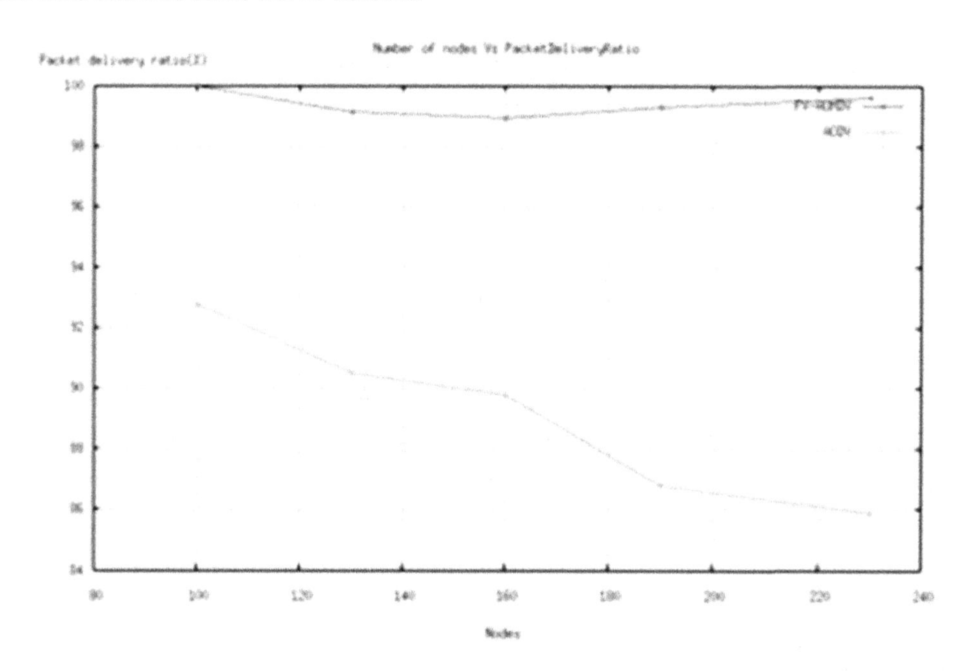

Table 8 and **Figure 13** shows the comparison of delay in AODV vs. FV-AOMDV with the number of nodes increases while simulation time remains the same i.e. 25 seconds. In the graph, a delay is also increased when data packets follow the proposed protocol because in FV-AOMDV protocol first fitness value estimated the data packets forward to the next node, and this will be done for each selected node between source to destination hence delay increases. But it should be decreased to achieve a better result.

Table 8. Delay_T_25

Number of nodes	Delay (AODV)	Delay (FV-AOMDV)
100	178670	179241
130	207867	221084
160	161814	201964
190	129204	145698
230	112800	101024

Figure 13. comparison of delay in AODV vs. FV-AOMDV with the number of nodes increases while simulation time remains same i.e. 25 seconds

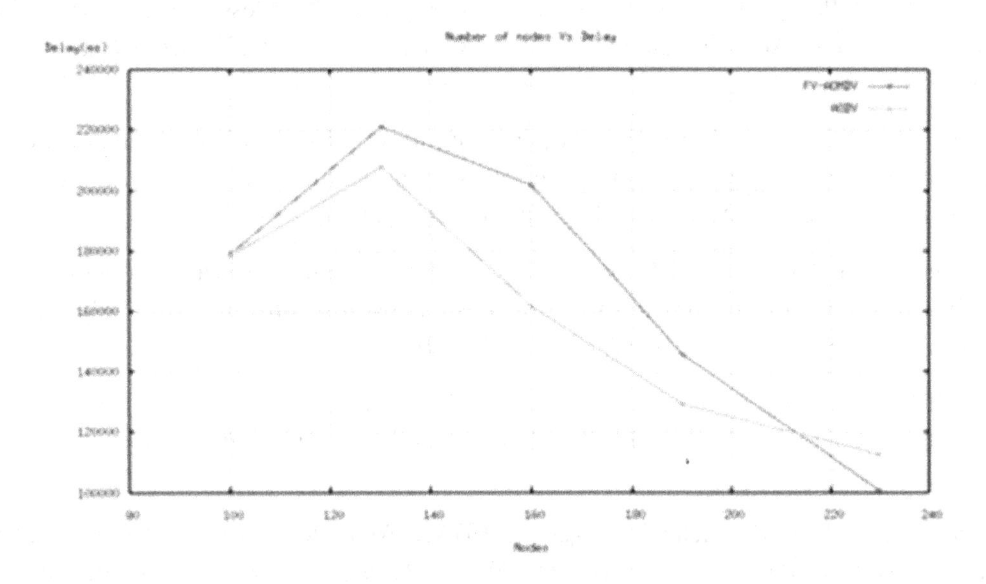

CONCLUSION

In this chapter, we have tried to discuss and present the security issues in Adhoc Network that perturb the network's normal functions. In Ad-hoc networks we discussed the Black-hole attack. The chapter presents a comprehensive survey on black-hole attack detection and prevention techniques in Ad-hoc networks, as suggested by various researchers. Then we create an Ad-hoc network that contains black hole nodes using AODV protocol and also analyzes the performance for simulation time and number of nodes. Then we secure the network through a prevention technique i.e. Optimization of the colony

Ant. And then, the network performance analyzed for simulation time and the number of nodes based on performance matrices. At last, a comparison is made between the two networks (network containing nodes of black hole and safe network). They tested and compared these networks on three performance metrics, i.e. Ratio of packet delivery, delay, and packet received. First we increase both networks simulation time and analysis, second, we increase the number of nodes to analyze the networks again. Results from the simulation showed that the prevention technique (ACO) implemented using FV-AOMDV yields a much better result than a network containing black hole nodes.

A lot of scenarios that could be put into practice with this chapter as a future scope to intensify the energy utilization and the existence of the network. That is acceptable for a reprehensive case. We can evaluate the selected routes/paths from source to destination to access an additional network resource that is the bandwidth as another fitness value and based on energy, bandwidth, and distance.

REFERENCES

Abdalzaher, M. S., Samy, L., & Muta, O. (2019). Non-zero-sum game-based trust model to enhance wireless sensor networks security for IoT applications. *IET Wireless Sensor Systems*, *9*(4), 218–226. doi:10.1049/iet-wss.2018.5114

Aghili, S., Mala, H., & Peris-Lopez, P. (2018). Securing Heterogeneous Wireless Sensor Networks: Breaking and Fixing a Three-Factor Authentication Protocol. *Sensors (Basel)*, *18*(11), 3663. doi:10.339018113663 PMID:30380595

Ali, R., Pal, A. K., Kumari, S., Karuppiah, M., & Conti, M. (2018). A secure user authentication and key-agreement scheme using wireless sensor networks for agriculture monitoring. *Future Generation Computer Systems*, *84*, 200–215. doi:10.1016/j.future.2017.06.018

Amin, R., Islam, S. H., Kumar, N., & Choo, K.-K. R. (2018). An untraceable and anonymous password authentication protocol for heterogeneous wireless sensor networks. *Journal of Network and Computer Applications*, *104*, 133–144. doi:10.1016/j.jnca.2017.12.012

Anand, C., & Gnanamurthy, R. K. (2016). Localized DoS Attack Detection Architecture for Reliable Data Transmission Over Wireless Sensor Network. *Wireless Personal Communications*, *90*(2), 847–859. doi:10.100711277-016-3231-y

Athmani, S., Bilami, A., & Boubiche, D. E. (2019). EDAK: An Efficient Dynamic Authentication and Key Management Mechanism for heterogeneous WSNs. *Future Generation Computer Systems*, *92*, 789–799. doi:10.1016/j.future.2017.10.026

Banerjee, S., Chunka, C., Sen, S., & Goswami, R. S. (2019). An Enhanced and Secure Biometric Based User Authentication Scheme in Wireless Sensor Networks Using Smart Cards. *Wireless Personal Communications*, *107*(1), 243–270. doi:10.100711277-019-06252-x

Bhushan, B., & Sahoo, G. (2017). Detection and defense mechanisms against wormhole attacks in wireless sensor networks. *2017 3rd International Conference on Advances in Computing, Communication & Automation (ICACCA)*. DOI: 10.1109/icaccaf.2017.8344730

Bhushan, B., & Sahoo, G. (2017). Recent Advances in Attacks, Technical Challenges, Vulnerabilities and Their Countermeasures in Wireless Sensor Networks. *Wireless Personal Communications*, *98*(2), 2037–2077. doi:10.100711277-017-4962-0

Bhushan, B., & Sahoo, G. (2017). A comprehensive survey of secure and energy efficient routing protocols and data collection approaches in wireless sensor networks. *2017 International Conference on Signal Processing and Communication (ICSPC)*. 10.1109/CSPC.2017.8305856

Bhushan, B., & Sahoo, G. (2018). Routing Protocols in Wireless Sensor Networks. *Computational Intelligence in Sensor Networks Studies in Computational Intelligence*, 215-248. Doi:10.1007/978-3-662-57277-1_10

Bhushan, B., & Sahoo, G. (2019). A Hybrid Secure and Energy Efficient Cluster Based Intrusion Detection System for Wireless Sensing Environment. *2019 2nd International Conference on Signal Processing and Communication (ICSPC)*. doi: 10.1109/icspc46172.2019.8976509

Bhushan, B., & Sahoo, G. (2019). ISFC-BLS (Intelligent and Secured Fuzzy Clustering Algorithm Using Balanced Load Sub-Cluster Formation) in WSN Environment. *Wireless Personal Communications*. Advance online publication. doi:10.100711277-019-06948-0

Bhushan, B., & Sahoo, G. (2019). $$E^{2} SR^{2}$$ E 2 S R 2: An acknowledgement-based mobile sink routing protocol with rechargeable sensors for wireless sensor networks. *Wireless Networks*, *25*(5), 2697–2721. doi:10.100711276-019-01988-7

Bhushan, B., & Sahoo, G. (2019). Secure Location-Based Aggregator Node Selection Scheme in Wireless Sensor Networks. *Proceedings of ICETIT 2019 Lecture Notes in Electrical Engineering*, 21–35. DOI: 10.1007/978-3-030-30577-2_2

Bhushan, B., & Sahoo, G. (2020). Requirements, Protocols, and Security Challenges in Wireless Sensor Networks: An Industrial Perspective. Handbook of Computer Networks and Cyber Security, 683–713. Doi:10.1007/978-3-030-22277-2_27

Bhushan, B., Sahoo, G., & Rai, A. K. (2017). Man-in-the-middle attack in wireless and computer networking — A review. *2017 3rd International Conference on Advances in Computing, Communication & Automation (ICACCA)*. DOI: 10.1109/icaccaf.2017.8344724

Chatterjee, K. (2019). An Improved Authentication Protocol for Wireless Body Sensor Networks Applied in Healthcare Applications. *Wireless Personal Communications*. Advance online publication. doi:10.100711277-019-07005-6

Chavan, A., Kurule, D., & Dere, P. (2016). Performance Analysis of AODV and DSDV Routing Protocol in MANET and Modifications in AODV against Black Hole Attack. *Procedia Computer Science*, *79*, 835–844. doi:10.1016/j.procs.2016.03.108

Doss, S., Nayyar, A., Suseendran, G., Tanwar, S., Khanna, A., Son, L. H., & Thong, P. H. (2018). APD-JFAD: Accurate Prevention and Detection of Jelly Fish Attack in MANET. *IEEE Access: Practical Innovations, Open Solutions*, *6*, 56954–56965. doi:10.1109/ACCESS.2018.2868544

El-Semary, A. M., & Diab, H. (2019). BP-AODV: Black-hole Protected AODV Routing Protocol for MANETs Based on Chaotic Map. *IEEE Access: Practical Innovations, Open Solutions*, *7*, 95197–95211. doi:10.1109/ACCESS.2019.2928804

Faghihniya, M. J., Hosseini, S. M., & Tahmasebi, M. (2016). Security upgrade against RREQ flooding attack by using balance index on vehicular ad hoc network. *Wireless Networks*, *23*(6), 1863–1874. doi:10.100711276-016-1259-2

Fazeldehkordi, E., Amiri, I. S., & Akanbi, O. A. (2016). *A study of black hole attack solutions on Aodv routing protocol in Manet*. Elsevier.

Fazeldehkordi, E., Amiri, I. S., & Akanbi, O. A. (2016). *A study of black hole attack solutions on Aodv routing protocol in Manet*. Elsevier.

Gai, K., Qiu, M., Ming, Z., Zhao, H., & Qiu, L. (2017). Spoofing-Jamming Attack Strategy Using Optimal Power Distributions in Wireless Smart Grid Networks. *IEEE Transactions on Smart Grid*, *8*(5), 2431–2439. doi:10.1109/TSG.2017.2664043

He, D., Chan, S., & Guizani, M. (2017). Cyber Security Analysis and Protection of Wireless Sensor Networks for Smart Grid Monitoring. *IEEE Wireless Communications*, *24*(6), 98–103. doi:10.1109/MWC.2017.1600283WC

Jaitly, S., Malhotra, H., & Bhushan, B. (2017). Security vulnerabilities and countermeasures against jamming attacks in Wireless Sensor Networks: A survey. *2017 International Conference on Computer, Communications and Electronics (Comptelix)*. DOI: 10.1109/COMPTELIX.2017.8004033

Jamshidi, M., Shaltooki, A. A., Zadeh, Z. D., & Darwesh, A. M. (2018). A Dynamic ID Assignment Mechanism to Defend Against Node Replication Attack in Static Wireless Sensor Networks. JOIV. *International Journal on Informatics Visualization*, *3*(1). Advance online publication. doi:10.30630/joiv.3.1.161

Karthik, N., & Ananthanarayana, V. S. (2017). A Hybrid Trust Management Scheme for Wireless Sensor Networks. *Wireless Personal Communications*, *97*(4), 5137–5170. doi:10.100711277-017-4772-4

Kaushik, I., & Sharma, N. (2020). Black Hole Attack and Its Security Measure in Wireless Sensors Networks. Advances in Intelligent Systems and Computing Handbook of Wireless Sensor Networks: Issues and Challenges in Current Scenarios, 401–416. doi:10.1007/978-3-030-40305-8_20

Kaushik, I., Sharma, N., & Singh, N. (2019). Intrusion Detection and Security System for Black-hole Attack. *2019 2nd International Conference on Signal Processing and Communication (ICSPC)*. doi: 10.1109/icspc46172.2019.8976797

Khanna, N., & Sachdeva, M. (2019). A comprehensive taxonomy of schemes to detect and mitigate black-hole attack and its variants in MANETs. *Computer Science Review*, *32*, 24–44. doi:10.1016/j.cosrev.2019.03.001

Kumar, D., Chand, S., & Kumar, B. (2018). Cryptanalysis and improvement of an authentication protocol for wireless sensor networks applications like safety monitoring in coal mines. *Journal of Ambient Intelligence and Humanized Computing*, *10*(2), 641–660. doi:10.100712652-018-0712-8

Li, J., Zhang, W., Kumari, S., Choo, K.-K. R., & Hogrefe, D. (2018). Security analysis and improvement of a mutual authentication and key agreement solution for wireless sensor networks using chaotic maps. *Transactions on Emerging Telecommunications Technologies, 29*(6), e3295. Advance online publication. doi:10.1002/ett.3295

Liu, Q., Yin, J., Leung, V. C. M., & Cai, Z. (2013). FADE: Forwarding Assessment Based Detection of Collaborative Grey Hole Attacks in WMNs. *IEEE Transactions on Wireless Communications, 12*(10), 5124–5137. doi:10.1109/TWC.2013.121906

Liu, X., Xiong, N., Zhang, N., Liu, A., Shen, H., & Huang, C. (2018). A Trust With Abstract Information Verified Routing Scheme for Cyber-Physical Network. *IEEE Access: Practical Innovations, Open Solutions, 6*, 3882–3898. doi:10.1109/ACCESS.2018.2799681

Liu, Z., Liu, W., Ma, Q., Liu, G., Zhang, L., Fang, L., & Sheng, V. S. (2019). Security cooperation model based on topology control and time synchronization for wireless sensor networks. *Journal of Communications and Networks (Seoul), 21*(5), 469–480. doi:10.1109/JCN.2019.000041

Liu, Z., Liu, W., Ma, Q., Liu, G., Zhang, L., Fang, L., & Sheng, V. S. (2019). Security cooperation model based on topology control and time synchronization for wireless sensor networks. *Journal of Communications and Networks (Seoul), 21*(5), 469–480. doi:10.1109/JCN.2019.000041

Lu, Y., Li, L., Peng, H., & Yang, Y. (2016). An Energy Efficient Mutual Authentication and Key Agreement Scheme Preserving Anonymity for Wireless Sensor Networks. *Sensors (Basel), 16*(6), 837. doi:10.339016060837 PMID:27338382

Luo, H., Zhang, G., Liu, Y., & Das, S. K. (2018). Adaptive Routing in Wireless Sensor Networks. *Adaptation and Cross Layer Design in Wireless Networks*, 263–299. doi:10.1201/9781315219813-9

Nayak, P., & Devulapalli, A. (2016). A Fuzzy Logic-Based Clustering Algorithm for WSN to Extend the Network Lifetime. *IEEE Sensors Journal, 16*(1), 137–144. doi:10.1109/JSEN.2015.2472970

Ochola, E., Mejaele, L., Eloff, M., & Poll, J. V. D. (2017). Manet Reactive Routing Protocols Node Mobility Variation Effect in Analysing the Impact of Black Hole Attack. *SAIEE Africa Research Journal, 108*(2), 80–92. doi:10.23919/SAIEE.2017.8531629

Panos, C., Ntantogian, C., Malliaros, S., & Xenakis, C. (2017). Analyzing, quantifying, and detecting the black-hole attack in infrastructure-less networks. *Computer Networks, 113*, 94–110. doi:10.1016/j.comnet.2016.12.006

Patil, S., & Chaudhari, S. (2016). DoS Attack Prevention Technique in Wireless Sensor Networks. *Procedia Computer Science, 79*, 715–721. doi:10.1016/j.procs.2016.03.094

Pham, T. N. D., & Yeo, C. K. (2016). Detecting Colluding Black-hole and Greyhole Attacks in Delay Tolerant Networks. *IEEE Transactions on Mobile Computing, 15*(5), 1116–1129. doi:10.1109/TMC.2015.2456895

Pruthi, V., Mittal, K., Sharma, N., & Kaushik, I. (2019). Network Layers Threats & its Countermeasures in WSNs. *2019 International Conference on Computing, Communication, and Intelligent Systems (ICCCIS)*. 10.1109/ICCCIS48478.2019.8974523

Puri, D., & Bhushan, B. (2019). Enhancement of security and energy efficiency in WSNs: Machine Learning to the rescue. *2019 International Conference on Computing, Communication, and Intelligent Systems (ICCCIS)*. 10.1109/ICCCIS48478.2019.8974465

Ren, J., Zhang, Y., Zhang, K., & Shen, X. (2016). Adaptive and Channel-Aware Detection of Selective Forwarding Attacks in Wireless Sensor Networks. *IEEE Transactions on Wireless Communications*, *15*(5), 3718–3731. doi:10.1109/TWC.2016.2526601

Sharma, M., Tandon, A., Narayan, S., & Bhushan, B. (2017). Classification and analysis of security attacks in WSNs and IEEE 802.15.4 standards: A survey. *2017 3rd International Conference on Advances in Computing,Communication& Automation (ICACCA)*. DOI: 10.1109/icaccaf.2017.8344727

Sharma, N., Kaushik, I., Singh, N., & Kumar, R. (2019). Performance Measurement Using Different Shortest Path Techniques in Wireless Sensor Network. *2019 2nd International Conference on Signal Processing and Communication (ICSPC)*. doi: 10.1109/icspc46172.2019.8976618

Sinha, P., Jha, V. K., Rai, A. K., & Bhushan, B. (2017). Security vulnerabilities, attacks and countermeasures in wireless sensor networks at various layers of OSI reference model: A survey. *2017 International Conference on Signal Processing and Communication (ICSPC)*. 10.1109/CSPC.2017.8305855

Sinha, P., Rai, A. K., & Bhushan, B. (2019). Information Security threats and attacks with conceivable counteraction. *2019 2nd International Conference on Intelligent Computing, Instrumentation and Control Technologies (ICICICT)*. doi: 10.1109/icicict46008.2019.8993384

Wang, C., Xu, G., & Sun, J. (2017). An Enhanced Three-Factor User Authentication Scheme Using Elliptic Curve Cryptosystem for Wireless Sensor Networks. *Sensors (Basel)*, *17*(12), 2946. doi:10.339017122946 PMID:29257066

Wang, F., Xu, G., & Xu, G. (2019). A Provably Secure Anonymous Biometrics-Based Authentication Scheme for Wireless Sensor Networks Using Chaotic Map. *IEEE Access: Practical Innovations, Open Solutions*, *7*, 101596–101608. doi:10.1109/ACCESS.2019.2930542

Xiong, N., Zhang, L., Zhang, W., Vasilakos, A., & Imran, M. (2017). Design and Analysis of an Efficient Energy Algorithm in Wireless Social Sensor Networks. *Sensors (Basel)*, *17*(10), 2166. doi:10.339017102166 PMID:28934171

Chapter 12
Fundamentals of Wireless Sensor Networks Using Machine Learning Approaches:
Advancement in Big Data Analysis Using Hadoop for Oil Pipeline System With Scheduling Algorithm

E. B. Priyanka
https://orcid.org/0000-0002-5318-6481
Kongu Engineering College, India

S. Thangavel
https://orcid.org/0000-0002-5318-6481
Kongu Engineering College, India

D. Venkatesa Prabu
Kongu Engineering College, India

ABSTRACT

Big data and analytics may be new to some industries, but the oil and gas industry has long dealt with large quantities of data to make technical decisions. Oil producers can capture more detailed data in real-time at lower costs and from previously inaccessible areas, to improve oilfield and plant performance. Stream computing is a new way of analyzing high-frequency data for real-time complex-event-processing and scoring data against a physics-based or empirical model for predictive analytics, without having to store the data. Hadoop Map/Reduce and other NoSQL approaches are a new way of analyzing massive volumes of data used to support the reservoir, production, and facilities engineering. Hence, this chapter enumerates the routing organization of IoT with smart applications aggregating real-time oil pipeline sensor data as big data subjected to machine learning algorithms using the Hadoop platform.

DOI: 10.4018/978-1-7998-5068-7.ch012

INTRODUCTION

IoT refers to the use of multiple connected devices to capture and use data generated by embedded sensors, actuators and other physical objects via a common network. IoT has grown quickly and will continue to do so in the coming years. The evolving technological environment will unleash new horizons and service aspects which will lead to an improved quality of life for customers and will also be beneficial for companies in their productivity. IoT applications are capable of sensing and transmitting the data at the same time (Alnasir & Shanahan, 2020). Moving into the future there will be a huge demand for these IoT devices and we will be using those devices in our homes, workplaces while traveling and in reality, all the places we can imagine. The growing number of IoT devices will also bring new opportunities and challenges. Cloud computing is a popular choice for the processing and analysis of large data volumes. Organizations can easily manage and deploy powerful clusters that run and allow distributed processing in different software environments. Scheduling is an important part of distributed computing that allows users to utilize the resources available for a faster processing time (Zhang et al., 2020).

Offshore oil and gas pipelines are vulnerable to the environment as any leak and burst in pipelines cause oil/gas spill resulting in huge negative impacts on marine lives. Breakdown maintenance of these pipelines is also cost-intensive and time-consuming resulting in huge tangible and intangible loss to the pipeline operators (Alves et al., 2018). Pipelines health monitoring and integrity analysis have been researched a lot for successful pipeline operations and risk-based maintenance model is one of the outcomes of those researches. A submarine pipeline is the major transportation way of subsea oil and gas after exploration and exploitation. However, because of scouring caused by current and wave, third-party damage and seaquake, or design defect, submarine pipelines have a relatively high probability of leakage failure given by (Baranowski et al., 2019). Once a leakage occurs, it may cause severe fire and explosion due to accident escalation, and pose a threat to human safety, environment, asset, and reputation has given brief layout by (J.Chen et al., 2013). The leakage failure risk of the submarine pipeline is unable to be eliminated, but preventive and mitigative measures can be taken to reduce the occurrence probability and consequence severity of leakage accident. Risk analysis is an efficient tool for identifying risk factors and developing strategies to prevent an accident. It includes three steps, i.e. hazard identification, frequency analysis and consequence analysis conducted. Also, an integrated risk-based assessment method developed by (Priyanka et al., 2020b) is used to predict failure probability and consequence of submarine oil pipelines, and it mainly focuses on corrosion failure of pipelines. By developing a probabilistic and numerical model for the dropped object risk assessment of the submarine pipeline, in which the collision probability of dropped object is estimated by scenario sampling while the accident consequence is simulated through finite element approach. However, the above studies mainly adopt traditional risk analysis methods or focus on the risk of single cause, and the risk analysis involving comprehensive causes and consequences of submarine pipeline leakage is not mentioned (Ye, 2020). In light of the above, it is necessary to use an integrated approach to conduct a risk analysis of leakage failure for the submarine oil and gas pipeline. The conventional risk analysis methods are known as static, it is unable to capture the variation of risk as the change occurs in the operation and environment. Also, revealed that conventional technologies use generic failure data, which makes them non-case-specific and introduces uncertainty into the results. The present work focusses on the application Hadoop platform for big data analytics to analyze real-world oil pipeline pressure data to analyze the performance of transportation (Chen at al., 2010). Besides, the scheduling algorithm is incorporated with cloud computing techniques with the data allocation cluster group to get the final decision in a more fast and robust manner.

The rest of the chapter is organized as follows: A brief description of cloud computing with data risk challenging task is presented in section 2. A proposed methodology framework of risk analysis using energy-efficient Hadoop is as shown in section 3. The accident evolution process modeling of pipelines data analysis using a scheduling algorithm is as presented in section 4. Section 5 gives the application of Hadoop in risk analysis on leakage failure of oil pipelines while the conclusion is presented in section 6.

Cloud Computing and its Challenges

In cloud computing, there are five major factors based on their participation, as shown in Figure 1. Cloud customer or cloud service user (CSC) is the one who obtains the service from a cloud provider and charges for the service as per the use. The cloud provider or cloud service provider (CSP) is the one supplying the CSC with cloud services. The cloud auditor is the one who performs an independent assessment of cloud services, information system operations, cloud implementation performance, and security. The Cloud broker is the one that communicates with CSP and CSC to put the company to pass (T.Chen et al., 2013). The cloud provider is the one delivering CSP-to-CSC access and cloud services.

Figure 1. Cloud computing major attributes

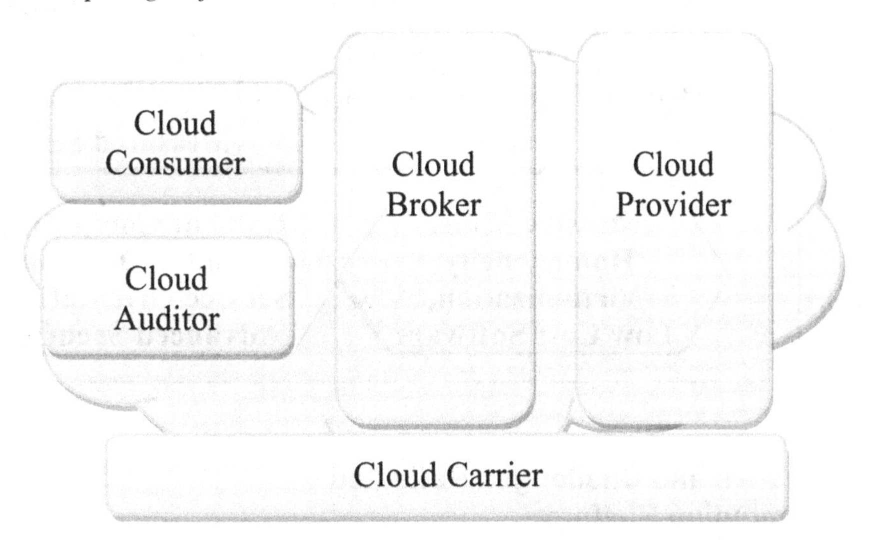

Cloud computing provides users with a network-based world vision, paving the way for exchanging calculations and resources regardless of location. Cloud computing is defined by the National Institute of Standards and Technology (NIST) as "a template for providing the appropriate and necessary internet access, a collective pool of programmable grids, storage, servers, software, and amenities that can be quickly emancipated, with little communication and provider supervision" (Devi et al., 2020). Its main contribution in data analytics focuses on Self-service on request, high-performance network access, fast elasticity, resource pooling, and calculated service. Four deployment models are also portrayed namely hybrid, collective, private and public clouds. This is then combined with three software models, namely PAAS (Platform as a Software), IAAS (Infrastructure as a Service), and SAAS (Software as a Service). NIST's cloud computing concept includes, among others, the required structure and common features depicted, such as virtualization, homogeneity, geographic distribution and service orientation (El-sayed et al., 2019). Figure 2 describes the services model with its security protocols in characteristics framework concerning common computing models.

Figure 2. Cloud computing characteristics concerning service model

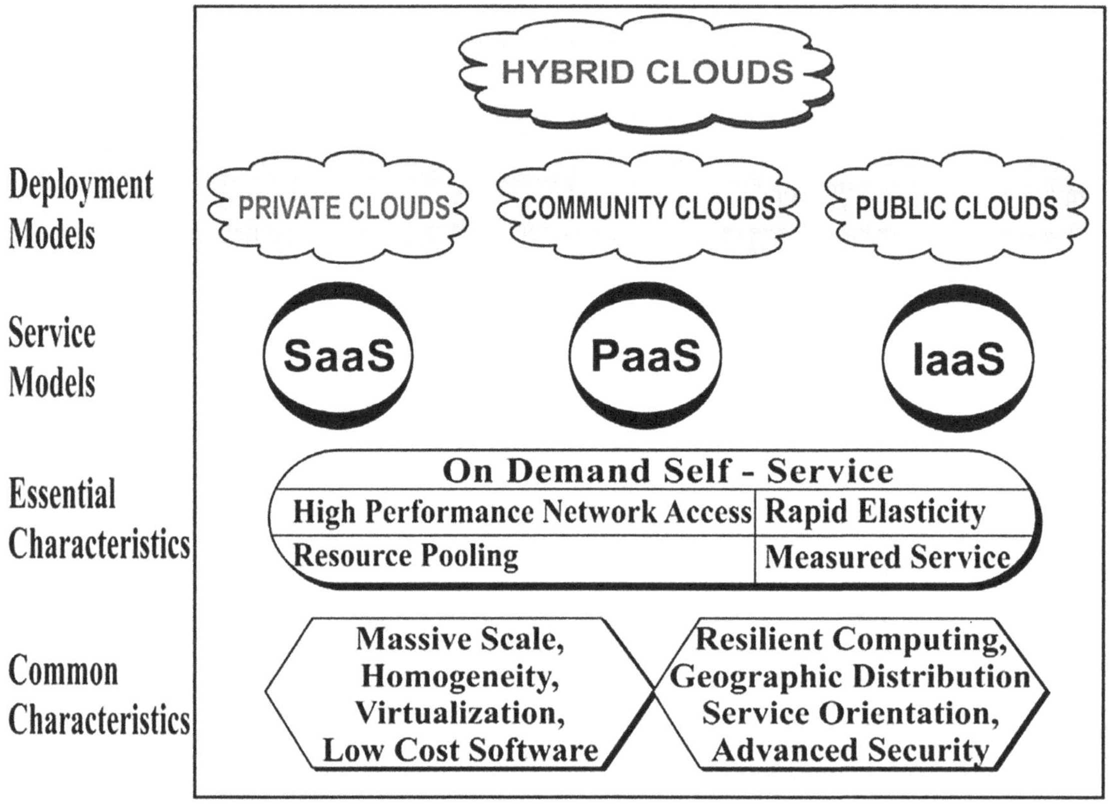

HYBRID CLOUDS

Deployment Models: PRIVATE CLOUDS, COMMUNITY CLOUDS, PUBLIC CLOUDS

Service Models: SaaS, PaaS, IaaS

Essential Characteristics: On Demand Self - Service; High Performance Network Access; Rapid Elasticity; Resource Pooling; Measured Service

Common Characteristics: Massive Scale, Homogeneity, Virtualization, Low Cost Software; Resilient Computing, Geographic Distribution, Service Orientation, Advanced Security

Data Security Issues and Challenges Reflected in the Cloud Computing Platform

In cloud computing, the data is stored outside the place of the customer (on the side of the CSP). Therefore, apart from the traditional security checks, cloud computing must employ additional security measures to ensure that data is safe and no data breaches due to security vulnerabilities. In the data-in-transmit stage, encryption is generally one of the methods to protect the data (Priyanka et al, 2020a). Data-after-delete is one of the overlooked problems and this is also called data remanence. Data remanence is the physical residual representation of the deleted data. After a storage media has been deleted, some physical characteristics may exist which allow the data to be reconstructed. Tracing the data path (data lineage) is important for cloud computing audits, particularly in the public cloud apart from the host. The three important properties of the data are confidentiality, integrity, and availability and it is popularly called a CIA triad. Another three important properties associated with people who access the data are authentication, authorization, and non-repudiation (Fasoulakis et al, 2019). Confidentiality applies to data privacy where the data belongs to CSC on no occasion is exposed to unauthorized parties. Data Integrity relates to the trust that unauthorized parties don't fiddle the data stored in the cloud. The same happens when the data is in transit. Data availability refers to the undertaking that the data should be made available to the CSC without delay whenever the data is required. Figure 3 shows the most encountered threat issues in the cloud computing platform.

Figure 3. Threat challenges in the cloud computing platform during CIA triad

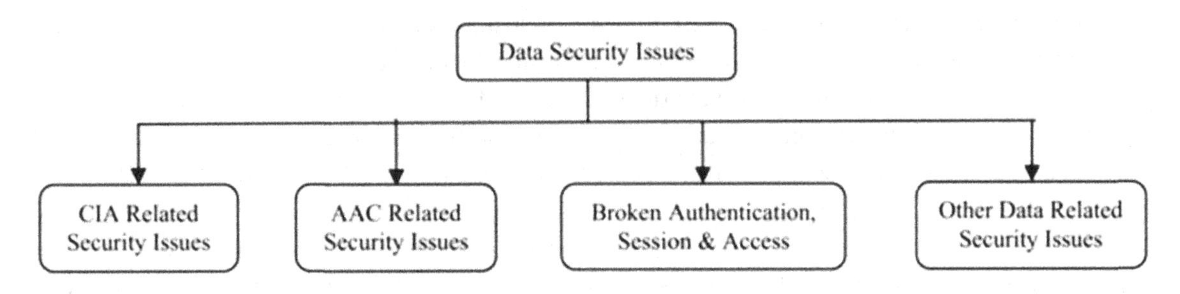

Data Level Challenges and Security Concerning Service Level Agreements (SLA)

Security overall at data level can be classified as data-in-transit and data-in-rest. Data-in-transit does not result in additional security risks compared to data-in-rest, as data is transmitted by default via TLS (Transport Layer Security), which provides a secure way of transferring data. From the hacker, the data-in-rest is more appealing. Figure 4 sets out a description of data-level risks.

Figure 4. Data level challenges in the cloud storage

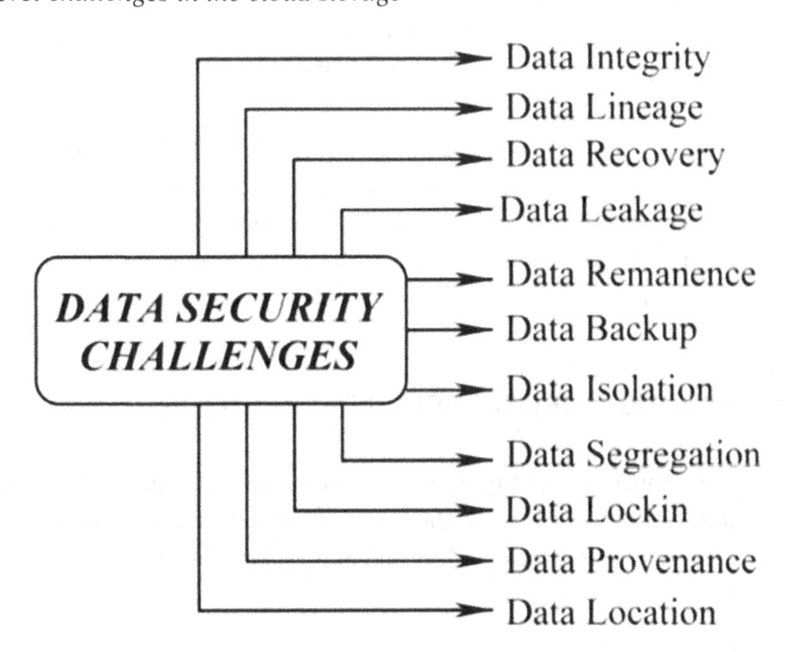

Data in-transit occurs within the crypto-cloud between entities (Gandomi et al., 2019). Communication between entities with a secure channel of communication such as Transport Layer Security can lead to the following issues:

- **Data Lineage**: The lineage of data relates to the origin of the data and where it moves over a while. Tracing the data path as a data lineage generally helps in the audit. Because of the non-linear nature of the cloud environment, it is one of the challenging and tedious issues involved in tracing (Xia et al., 2011).
- **Information Leakage**: The problem of information leakage occurs if multi-tenant access is made to the data. Due to inherent security vulnerabilities in Google Docs, it is pointed to the presence of a serious leakage of a user's private data. The risk of data leakage is substantial and requires careful handling. Several data leakage problems include Instance Messaging, WebMail, Blogs / Wikis, File Transfer Protocol (FTP) and Universal Serial Bus (USB) (Pratheep et al., 2020).

Secure data storage in the cloud and performance data integrity check when the data is accessed. Saving Encrypted file, hash file and meta-data in cloud enhances the security of cloud data stored characterized as data-in-rest.

- **Data Remanence**: At the end of its life cycle, data must be effaced safely. The overwriting is one of the traditional data sanitization solutions. Physical characteristics make it possible to restore/recover deleted data which leads to the disclosure of sensitive data. The data from failed devices can be recovered with proper skills and equipment. Upon deletion remanence of the data requires attention.
- **Data segregation**: Data segregation refers to the complete separation within a virtualized environment between cloud users. That data segregation is a multi-tenancy problem that is posed. To achieve data segregation, cloud providers will use highly secured protocols and encryption algorithms. Vulnerabilities in data segregation exist due to vulnerabilities in data authentication, unsecured storage, and SQL injection. Meeting the uses specified in a multi-tenant environment helps to mitigate the challenge of data segregation (Kaliyannan et al., 2020).

Crypto-cloud core entities have the responsibility to maintain the SLAs. Provider capacity, user performance, and service availability depend on the SLA type. To reduce risk factors, one should be mindful of the following components. Bandwidth and loss of operation, business continuity, location of data, data appropriation and integrity as well as data reliability among others. The pay-as-use model can only thrive with proper SLA's.

DATA ANALYTICS USING ENERGY-EFFICIENT HADOOP

While big data provides scientists the ability to produce, store, access, and analyze massive amounts of experimental data rapidly and cost-effectively, the use of big data technologies is experiencing unprecedented growth in a variety of research fields. More and more scientific applications are being developed using the MapReduce model of programming with the merits of Hadoop's efficient computing and storage capabilities. As more and more data-intensive applications are migrating from private infrastructure to the cloud, Hadoop has also achieved wide-ranging acceptance for big data analytics in cloud computing environments. Large-scale cloud systems offer tremendous processing power and massive storage space, but also make these applications more difficult to analyze and debug. A lightweight approach based on execution logs from both the test cloud and the real cloud to help deploy big data analytics applications in large-scale cloud environments is needed to address this issue (Glushkova et al.,2019).

Big data applications running on a cloud-based Hadoop system will simplify client-side operations and, more significantly, reduce the cost of execution. Therefore, deploying Hadoop-based Big Data Analytics applications has already been a phenomenon. The advance in data acquisition and exploration makes demands for Big Data analysis grow rapidly in almost every research field. This leads to an increasing number of big data applications running on Hadoop that provide storage and analytics to support the Internet of Things (IoT), business intelligence, social network data mining, etc. However, Hadoop's ever-widening adoption makes its energy inefficiency a prominent concern (Huang et al., 2018). Because Hadoop was originally designed for generic big data analysis on commodity clusters, little consideration is taken of its energy-use efficiency. It directly results in a major energy loss as hundreds of thousands of data analytics applications run on Hadoop.

MapReduce

MapReduce is a parallel programming framework designed for data processing in distributed environments and proposed by Google. A MapReduce job consists basically of two phases: Map and Reduce. Initially, the job's input dataset is split into several blocks while each block corresponds to a single Map task. Typically, each Map task handles a data block and generates a collection of intermediate key/value pairs. Shuffle follows the end of the map step, gathering intermediate outputs from each map task and sending them to the corresponding Reduce tasks after sorting and partitioning. At the end of the Shuffle, the temporal effects are combined on the Reducer-side. Container encapsulates resources for CPU and memory and enables more versatile configurations (Pratheep et al., 2019). Also, AM works in a container and continues to need RM capital for its corresponding work. RM then answers as the shared resource share with a list of containers. Hadoop 3.0 is now the latest version of Hadoop. However, only a few features are introduced compared to Hadoop 2.0, such as shuffling using Java Native Process, multiple standby modes and support for erasure encoding in HDFS.

Hadoop Distributed File System is the underlying distributed file system, which supports Hadoop's persistent run. You can deploy HDFS to dedicated servers or commodity machines. It can store terabytes and petabytes of data. Through its abstract interfaces, users may test HDFS as if they were running in a local file system. To simplify data integration and improve performance, HDFS adopts WORM (Write Once, Read Several times) model. Hadoop can store redundant copies of the data to increase the availability of the program (Priyanka et al., 2016). The HDFS replication mechanism allows multiple replicas to be spread on different machines to one data block. Figure 5 shows the data characterization with a segregation mechanism to enhance power in Hadoop for big data.

Figure 5. Data characterization for big data mining to enhance the power

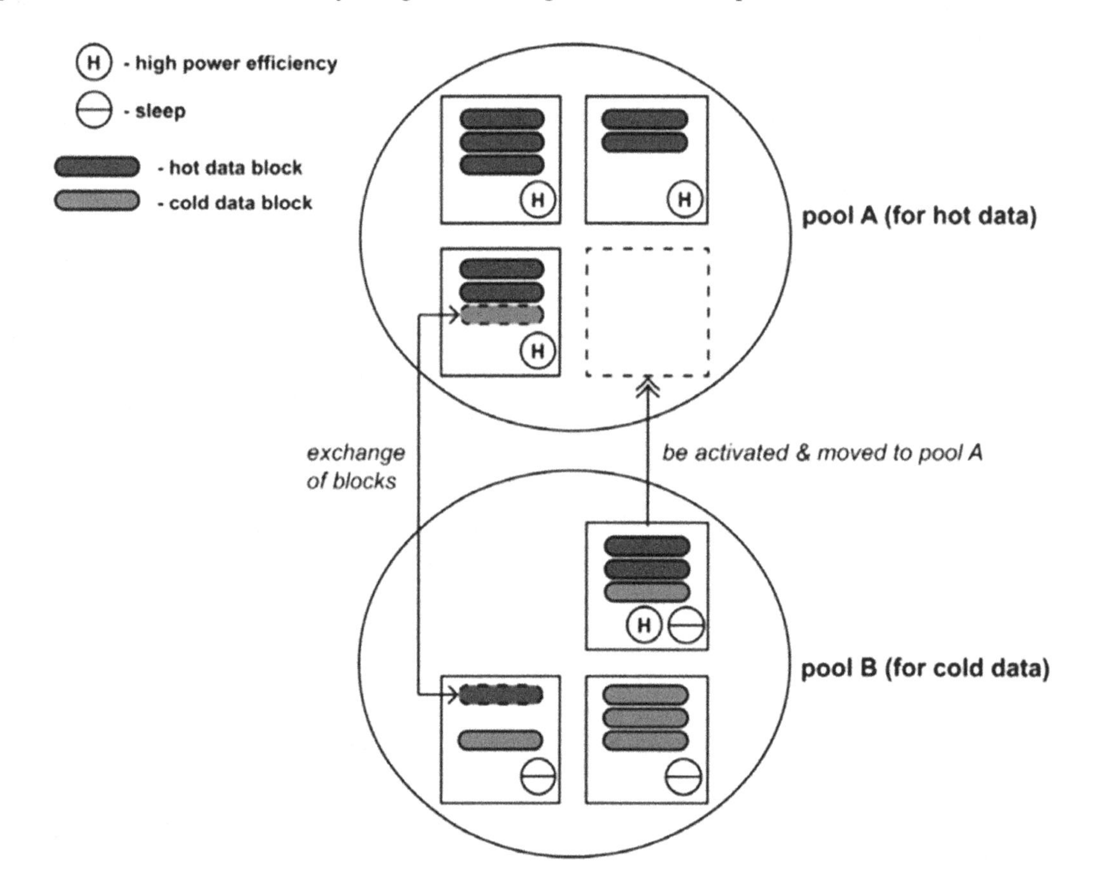

Optimization and Resource Management in Hadoop

Dynamic node management is a standard method in both homogeneous and heterogeneous environments to monitor the cluster power consumption. A set of policies function at the hardware level to dynamically control worker nodes, including shutting down servers, turning servers into low-power states and dynamically changing CPU output. Frequently accessed blocks in HDFS are located in the hot zone, which consists mainly of high-performance machines with high frequency running CPUs. The cold zone, on the contrary, consists of servers connected to large-capacity storage devices and holds cold (i.e., seldom accessed) data (Maheswari et al., 2018). Workers usually sleep in the cold zone or stand by when they are put. It is proposed that the hot zone will maintain at least 70 percent of the server total. Data blocks will be switched between zones as data hotness changes over time. Partitioning of the cluster by introducing a small cooperative zone containing high-performance worker nodes. The interactive zone is dedicated to managing jobs that have strict time limits for the response. On the other hand, batch and preemptive jobs will be allocated to another group of servers called the batch region. There are a large number of nodes in the batch region, but the workload is clustered on some of them with the rest operating in low-power states. This scheme, called BEEMR, helps cut the average response time for interactive jobs and save energy by sacrificing partial batch job performance. A Digital Shuffle technique aimed at saving energy from task execution by reducing the shuffle process of disk access (Priyanka et al., 2018e).

Unlike Hadoop's default method, the strategy creates and retains a segment table first, instead of immediately exchanging data. The table records the positions of all Map output segments, and data transfer occurs only when a Reduce function allows a segment to be retrieved. To prevent memory exhaustion caused by virtual segment mergers and the multilevel segment table, the authors use balanced subtrees to save memory space. The scheme effectively eliminates the unnecessary transfer of data over the network and thus increases the performance of MapReduce (Kapoor et al., 2019). A MapReduce job will be decomposed into Map tasks, and tasks reduced by the sum of input data and device parameters such as block size of HDFS. The task scheduler assigns all of the Map tasks and Reduce tasks to the worker nodes (physical machines and virtual machines). In Hadoop 1.0, the job scheduling function is integrated into Job Tracker and Task Scheduler is the corresponding abstract class.

Machine learning can be applied to MapReduce job profiling to avoid or reduce pre-execution overheads of time and energy. We need to develop a task profiling model that can accurately predict job execution time, resource demand and workload characteristics. The input of the model is job features such as the number of necessary virtual cores, included libraries, and the size of input data. The model can be trained using machine learning methods with job-related training data from the historical workers and their profiles. The variety of workers submitted to the Hadoop program, however, makes the accuracy of the job profiling model a major problem. Figure 6 depicts the job profiling in the MapReduce using a Machine learning algorithm (Kos et al., 2019).

Figure 6. Machine learning-based job profiling in MapReduce

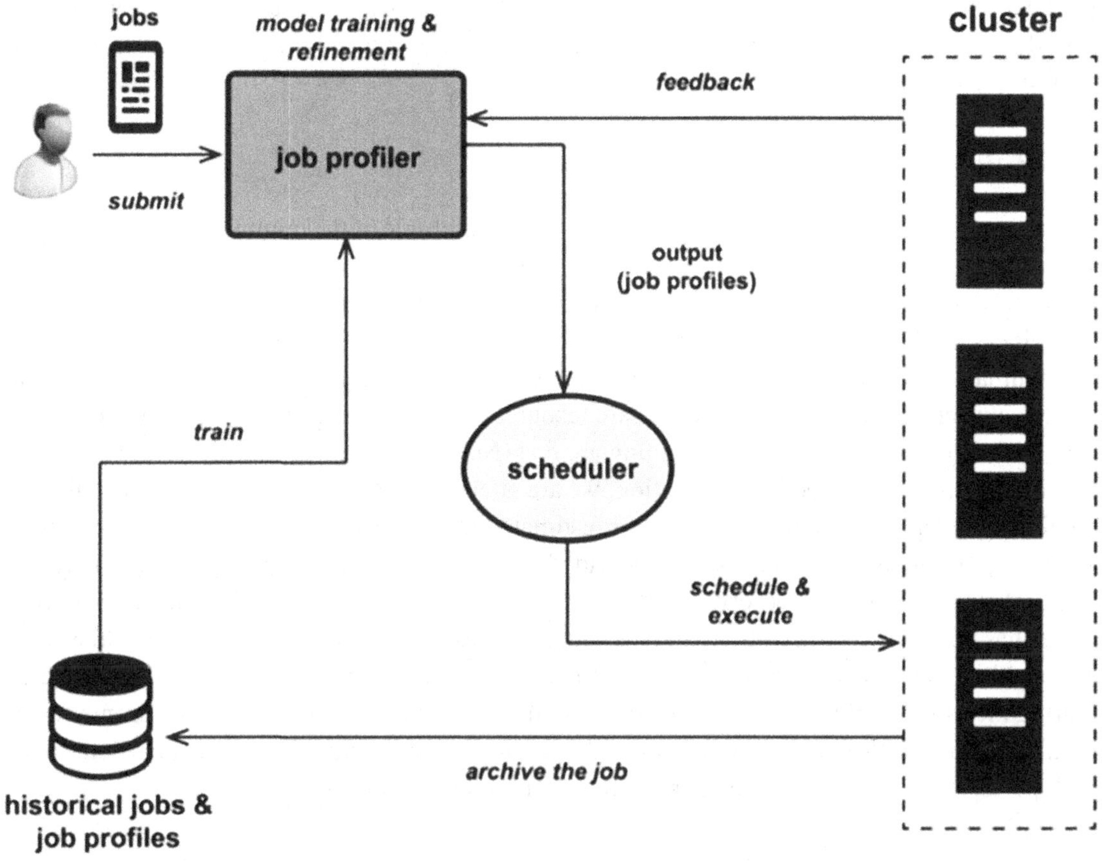

COMPUTING BY RUNNING DEADLINE SCHEDULER FOR DATA ALLOCATION

Cloud computing is a popular choice for the collection and analysis of large data volumes. Organizations can easily manage and deploy powerful clusters that run and allow distributed processing in different software environments. Scheduling is an important part of distributed computing that allows users to utilize the resources available for faster processing time. Therefore, by using a generic scheduling algorithm which takes into account deadline constraints. By creating a cost model that forecasts the remaining load of work that helps the scheduler to assign jobs according to their upcoming deadlines (Malik et al., 2019). To tackle the issue of deadlines in a dynamic environment where jobs are submitted on the fly and decisions on scheduling need to be made immediately without a predefined collection of resources. In the new Apache Hadoop environment, tackling the scheduling problem which is an open-source software framework used for distributed storage and distributed processing, inspired by the Google File System and the Google MapReduce model. The main aim of distributed computing is to provide means for fast and reliable processing of vast amounts of data. The MapReduce model has applications in various fields such as financial services, healthcare, human sciences, logistics or imaging, which could have specific completion time requirements for the jobs. The scheduling algorithm implemented assists in setting a deadline when submitting a job. Having such a runtime restriction allows users to better control the execution of the jobs submitted and to have the outcome when necessary (Priyanka et al., 2018f).

Deadline Scheduling Model

The collection of resources, W, can either be empty throughout the execution – indicating that the requested resources have been authorized and the application is currently executing its task or it may contain resource demands w that have yet to be fulfilled:

$$w= (T_c, M, S) € W$$

T_c is the number of virtual cores required, and must be a multiple of the minimum allocation, which is by default 1; M is the amount of memory allocated in MB. The sum must be a multiple of the minimum allocation that is 1024 MB by default; S is the number of containers that have T_c and M memory at least.

Containers are allocated in the Hadoop environment based on a configurable minimum set of resources, called minimal allocations, and such a system can have as many running containers as the local resources allow. In a cluster, given a set of finite hardware resources, we need to optimize the resource allocation process so that the time constraints for running are met (Nakagami et al., 2019). Given the current constraints on obtaining task-specific information, we are attempting to develop a generic scheduler based on previously completed and pending requests, estimate the remaining running time and prioritize the application accordingly. Asset requests are provided in the form of containers where a task may be run by the application master. As describing the desired resource allocation fulfilled and completed (the container has completed its execution) and the minimum resource allocation required as follows:

To calculate the remaining workload, we estimate the completion time of the pending requests by measuring the time required to complete the workload with containers running on the minimum allocation. Based on the fact that the taskmaster correctly asks for resources directly proportional to the amount of work performed, we can estimate the run time per minimum allocation t_{min}.

$$t_{min=}\frac{1}{z}\sum_{n=1}^{z}t_n\times\left(a\times\frac{T_{Cmin}}{T_{Ci}}\right)\times\left(b\times\frac{M_{min}}{M_i}\right)$$

Where a and b are application-specific factors that are dynamically determined and controlled by each of the two resources-virtual core and memory affect the runtime of the task. For example, if a container running 2 T_{cmin} cores and 2 W_{min} memory ends its execution in t (it runs a task in parallel), we can safely assume that running with only one core (V_{cmin}) would have taken 2t but running with only half of the memory may not have such a major impact on the performance (Nguyen et al., 2012).

Datacenter Communication Architecture

In version 2 of Hadoop, the scheduler resides within the daemon Resource Manager. Communicating with the daemons of Server Master and Node Manager is the responsibility of the Resource Manager. It registers itself with an event dispatcher as soon as the Resource Manager daemon begins and is ready to handle messages from the two counterparts. The architecture for data communication is shown in Figure 7.

Figure 7. Deadline scheduler data communication architecture

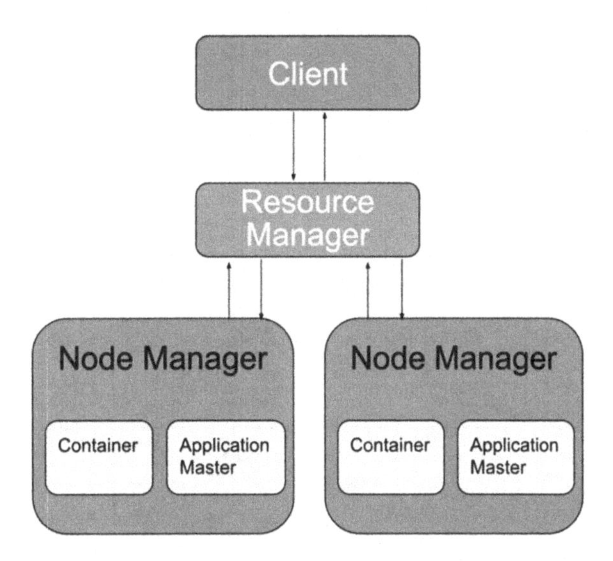

It can be implemented either as a stand-alone scheduler responsible for handling all of the cluster's resources or as a policy for queue scheduling in other schedulers that require it, such as the common Power and Equal Scheduler (Nita et al.,2015). The latter method has the benefit of reserving to the Deadline Scheduler only a portion of the cluster's capacity–as in the case of the Capacity Scheduler, where a collection of resources can be allocated to a particular queue. The algorithm can also be implemented in the Equal Scheduler as a scheduling method so that the queue it manages gets a fair amount of resources shared with other queues as well. It is important to mention that the one function that could make our algorithm incompatible with the current implementation is the transfer of the deadline from the customer to the Resource Manager in the scenario of implementing our algorithm in a production environment. The application is sent through RPC calls to the Resource Manager and Hadoop does not allow the user or application-specific information to be sent via it by default. Therefore, some of the underlying components must also be modified (Pakize et al., 2014).

The proposed scheme with the scheduling algorithm is experimented by considering containers of data in the resource allocation and validated results are given in Figure 8. A strategy for calculating the remaining work by looking at the containers completed and the memory allocated. For this experiment, we used the value 1 for the memory factor a and 0.8 for the virtual core factor b. The application shown in the graph is a MapReduce application, so the steady horizontal line from the start shows the map phase, there were about 20 map tasks in the application followed by many reduced tasks in which the average container run time has nearly doubled. The spikes in the center either account for an increased cluster load or a short sort phase, but the new reduce t. The last spike in the graph comes from the Application Manager, which occupies the first container assigned to the application and finishes its execution only when all the maps and tasks are reduced (Rasooli and Down, 2011).

Figure 8. Last 5 container estimated runtime with very less allocation

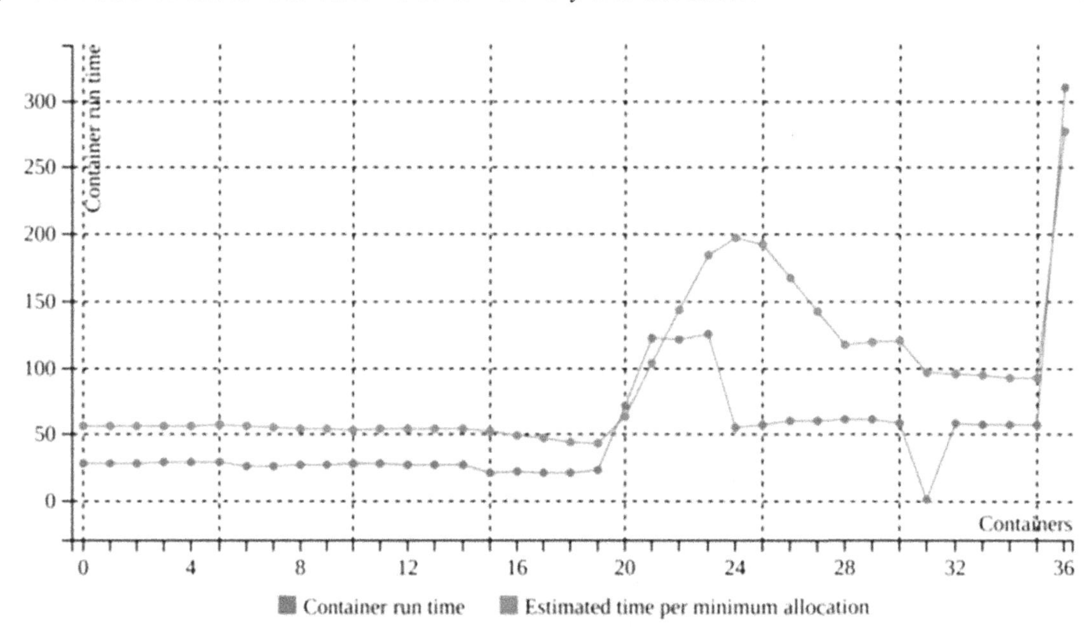

In Figure 8, a total of 5 containers have been used to calculate the minimum allocation run time, and it is confirmed that the curve is smoother while the calculation still accurately represents the step of the application. On the other hand, in the case of long-term applications, this approximation error may be changed if the application has several reduction tasks and the correct amount of work left can be recalculated and the program prioritized accordingly.

REAL-TIME VALIDATION OF BIG DATA ANALYTICS BY SPARK HOLDING HADOOP MAPREDUCE FEATURES

Each electronic device in today's world produces data whether it's the cell phones, built-in sensors, or fitness wrist bands. The increasing quantity of data contributed to the idea of Big Data. With the emergence of Big Data, IoT and Smart Cities, it became possible to gain useful insights into the data in real-time. Real-time or near-real-time analytics is data analysis as soon as it arrives or is being produced. There are many applications where it's not important to evaluate the data in real-time so batch processing can be done. But there are various applications where this data analysis is very important in real-time or near real-time (Shankar et al., 2015).

With the advent of Big Data, real-time data processing has become a requirement in different applications where action needs to be taken as soon as the data arrives or is detected. For more than ten years, the concept of real-time data analytics and business analytics has been around but businesses have been slow to adopt this dream. Part of the reason for this is the lack of real-time data analytics technologies and tools available. Traditional data-warehousing methods were primarily geared to batch processing that had high latencies and was extremely costly (Subramaniam et al., 2019). The cost of storing data has decreased considerably over the past few years. All of these technological advances and the availability of inexpensive data processing software and real-time data analytics have allowed large amounts of data to be analyzed. The number of powerful open-source projects has appeared in recent years that provide the ability to do data analytics in real-time or near real-time (Sun et al., 2012). Apache Spark is among the most popular and widely used. Throughout this real-time experimental validation, we will use the power of Apache Spark to conduct real-time data analytics on the data generated by IoT. Apache Spark is a lightning-fast computing cluster technology, optimized for fast computing. This is based on Hadoop MapReduce and extends the MapReduce model to use this effectively for more computation types, including interactive queries and stream processing. The implemented framework has three main components of the integrated technology: IoT Sensors, Big Data Management and Analytics.

Figure 9. Implemented framework for data analytics in an oil pipeline system

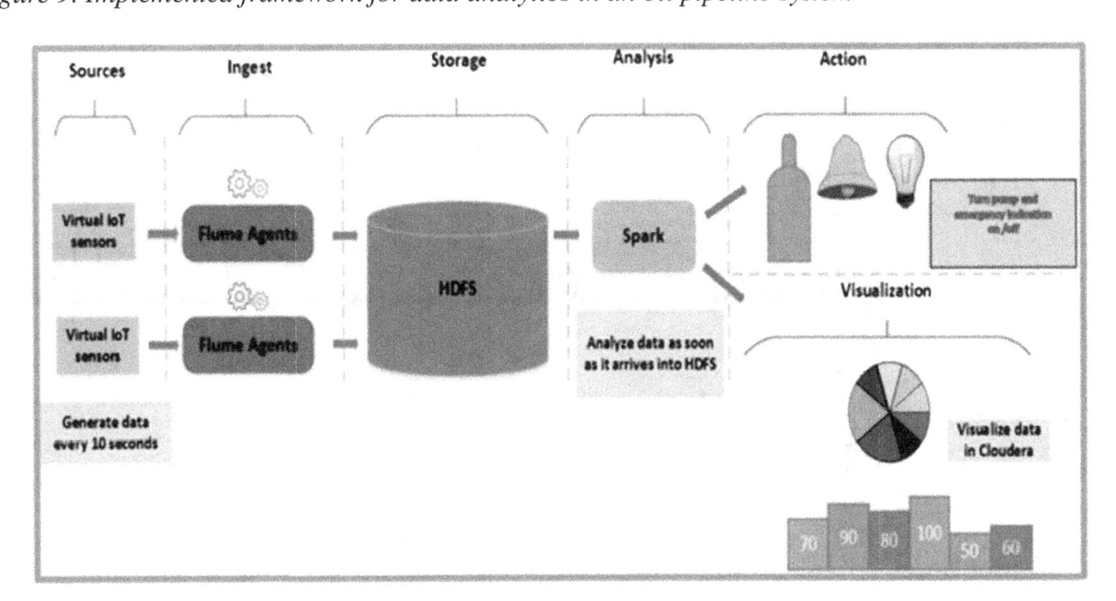

In Figure 9, the implemented conceptual layout of data processing and control in the oil pipeline system is given in block diagram format. Furthermore, these sensors produce a large amount of data (Big Data), which is then transmitted to a TCP (Transmission Control Protocol) port where an Apache Flume agent already runs and listens to this port (Tariq et al., 2019). The Apache Flume agent is set up with the real-time data streams created by IoT as the source and the HDFS as the sink so that after listening from the TCP port, the data is stored on HDFS. In real-time validation, the Virtual Machine for the Apache Hadoop framework (Cloudera Big Data platform) is utilized to store the data. Thirdly, PySpark scripts (Analytics) were followed to acquire and to analyze the data stored on HDFS in real-time.

Since the developed architecture used the Cloudera Hadoop distribution system, IoT sensor data were processed using a TCP connection. It is designed to send data by one sensor at a time with an interval of every 10 seconds. The data communication in the hub terminal is designed by the token-passing scheme so that successive sensors send data one by one without any congestion traffic (Xia et al., 2020).

Pressure Data Analysis in an Oil Pipeline System

The management system is essential to ensure the long-distance pipeline is run smoothly. Here, a concept is proposed for a smart pipeline system and its implementation on a multi-product pipeline network where a pipeline monitoring and accidental leak handling system is implemented for integrity management based on real-time Big Data and Distributed Computing (Priyanka et al., 2018a, 2018b). A novel data-chain and module-chain architecture are provided in the management of pipeline integrity. New

Figure 10. Configured IoT pressure sensors interconnections in an oil pipeline system

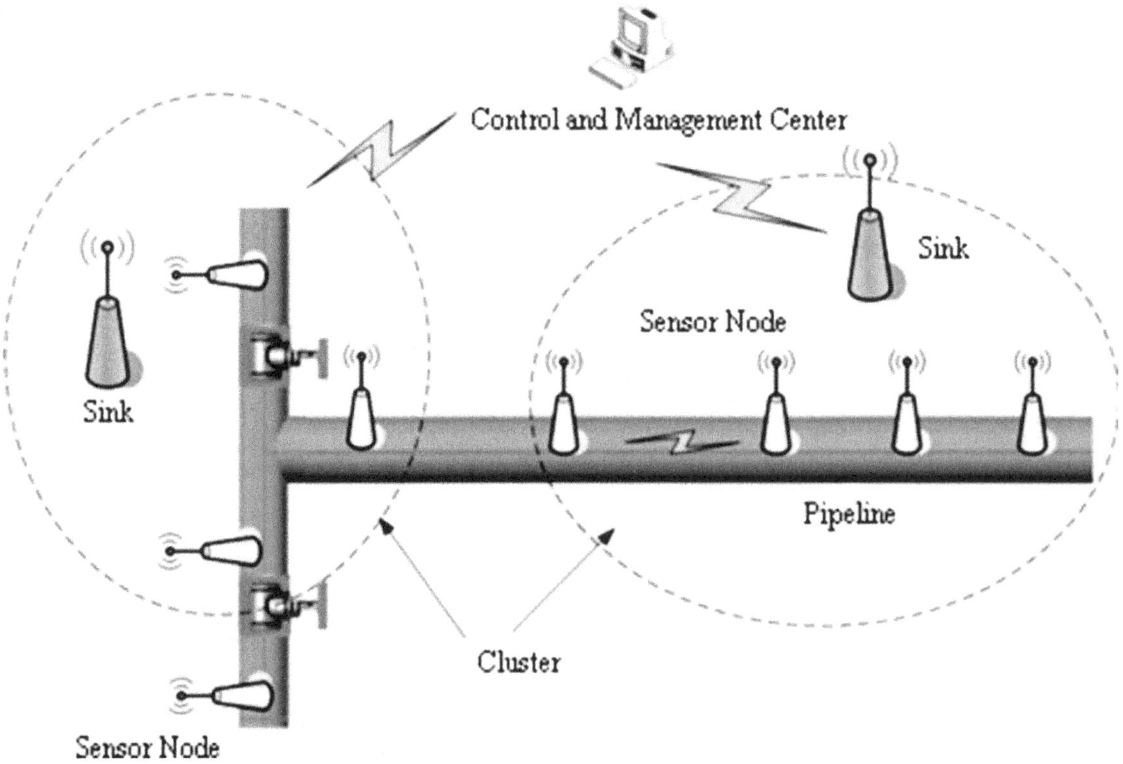

Table 1. Summary statistics of the pressure data by a different sensor on a particular day at different time

Pressure sensor data in psi recorded on 01st hour on June 20						Pressure sensor data in psi recorded on 22nd hour on June 20					
Sensor 1	Sensor 2	Sensor 3	Sensor 1	Sensor 2	Sensor 3	Sensor 1	Sensor 2	Sensor 3	Sensor 1	Sensor 2	Sensor 3
21.02	22.34	19.51	37.72	39.91	41.02	16.34	24.56	25.61	37.95	42.01	44.82
21.23	22.47	19.77	37.91	40.21	41.13	16.78	24.72	25.67	38.01	42.31	44.84
21.44	22.6	20.03	38.1	40.51	41.24	17.22	24.88	25.73	38.33	42.61	44.86
21.65	22.73	20.29	38.29	40.81	41.35	17.66	25.04	25.79	39.01	42.69	44.88
21.86	22.86	20.55	38.48	41.11	41.46	18.1	25.2	25.85	38.71	42.77	44.9
22.07	22.99	20.81	38.67	41.41	41.57	18.54	25.36	25.91	40.01	42.85	44.92
22.28	23.12	21.07	38.86	41.71	41.68	18.98	25.52	25.97	39.09	42.93	44.94
22.49	23.25	21.33	39.05	42.01	41.79	19.42	25.68	26.03	41.01	43.01	44.96
22.7	23.38	21.59	39.24	42.31	41.9	19.86	25.84	26.09	39.47	43.09	44.98
22.91	23.51	21.85	39.43	42.61	42.01	20.3	26	26.15	42.01	43.17	45
23.12	23.64	22.11	39.62	42.91	42.12	20.74	26.16	26.21	39.85	43.25	45.02
23.33	23.77	22.37	39.81	43.21	42.23	21.18	24.56	26.27	43.01	43.33	45.04
23.54	23.9	22.63	40	43.51	42.34	21.62	24.72	26.33	40.23	43.41	45.06
23.75	24.03	22.89	40.19	43.81	42.45	22.06	24.88	26.39	44.01	43.49	45.08
23.96	24.16	23.15	40.38	44.11	42.56	22.5	25.04	26.45	40.61	43.57	45.1
24.17	24.29	23.41	40.57	44.41	42.67	22.94	25.2	26.51	45.01	43.65	45.12
24.38	24.42	23.67	40.76	44.71	42.78	23.38	25.36	26.57	40.99	43.73	45.14
24.59	24.55	23.93	40.95	45.01	42.89	23.82	25.52	26.63	46.01	43.81	45.16
24.8	24.68	24.19	41.14	45.31	43	24.26	25.68	26.69	41.37	43.89	45.18
21.65	23.25	21.59	41.33	45.61	43.11	24.7	25.84	26.75	47.01	43.97	45.2
21.86	23.38	21.85	41.52	45.91	43.22	25.14	26	26.81	41.75	44.05	45.22
22.07	23.51	22.11	43.42	46.21	43.33	25.58	26.16	26.87	48.01	44.13	45.24
22.28	23.64	22.37	43.61	46.51	43.44	26.02	24.56	26.93	43.84	44.21	45.26
22.49	23.77	22.63	43.8	46.81	43.55	26.46	24.72	26.99	49.01	44.29	45.28
22.7	23.9	22.89	43.99	47.11	43.66	26.9	24.88	27.05	44.22	44.37	45.3
22.91	24.03	23.15	44.18	47.41	43.77	24.34	25.04	27.11	50.01	44.45	45.32
23.12	24.16	23.41	44.37	47.71	43.88	23.78	25.2	27.17	44.6	44.53	45.34
23.33	24.29	23.67	44.56	45.01	43.99	26.22	25.36	27.23	51.01	44.61	45.36
23.54	24.42	26.17	44.75	45.31	44.1	25.66	25.52	27.29	44.98	44.69	45.38
23.75	24.55	23.15	44.94	45.61	44.21	28.1	25.68	27.35	52.01	44.77	45.4
23.96	24.42	23.41	45.13	45.91	44.32	27.54	25.84	27.41	45.36	44.85	45.42
24.17	24.55	23.67	45.32	46.21	46.81	27.98	26	27.47	53.01	44.93	45.44
24.38	24.68	23.93	45.51	46.51	47.11	20.42	26.16	27.53	45.74	45.01	45.46
24.59	24.68	24.19	45.7	46.02	47.41	20.86	27.22	27.59	54.01	45.09	45.48

computational modules are proposed for emergency treatment to locate a leak, measure the amount of the leakage and execute the shutdown procedure. To obtain a pipeline shutdown scheme a new optimal model is developed that aims at minimal ALV (accumulative leakage volume). Several tests based on multi-product pipelines are introduced to check the Framework's efficacy and assess the modules ' applicability. The results show that estimates of the starting time, position, coefficient and volume of leakage could be achieved within the span of a negative pressure wave passing through the selected pipeline. In the meantime, both the pre-forming of the pipeline shutdown process and the obtaining of the ALV could be achieved. The findings could provide guidelines for developing a smart pipeline network, handling the pipeline shutdown, predicting the powerful range of leakage and subsequent incident investigations (Priyanka et al., 2019a, 2019b). Hence in the proposed work, pressure data are acquired and processed by placing three pressure transmitters in each substation of the oil pipeline system shown in Figure 10.

Once the data is stored in HDFS, the default visualization tools available in the Cloudera environment can be used to display it. In the form of tables, the virtual sensor data stored in HDFS can be imported into Hue and analyzed to gain useful insights into the data.

Figure 11 displays a visual representation of the phase of displaying the acquired pressure sensor data received from multiple oil pipeline substations. Table 1 gives the corresponding numerical average data recorded on the 01 and 22 hours of the June 20 received from the pressure sensors mounted at different substations to analyze the occurrences of the failure rate in the oil pipeline during long run transportation.

Figure 11. The graph shows cloud-stored pressure data taken for analyzing the pipeline failure

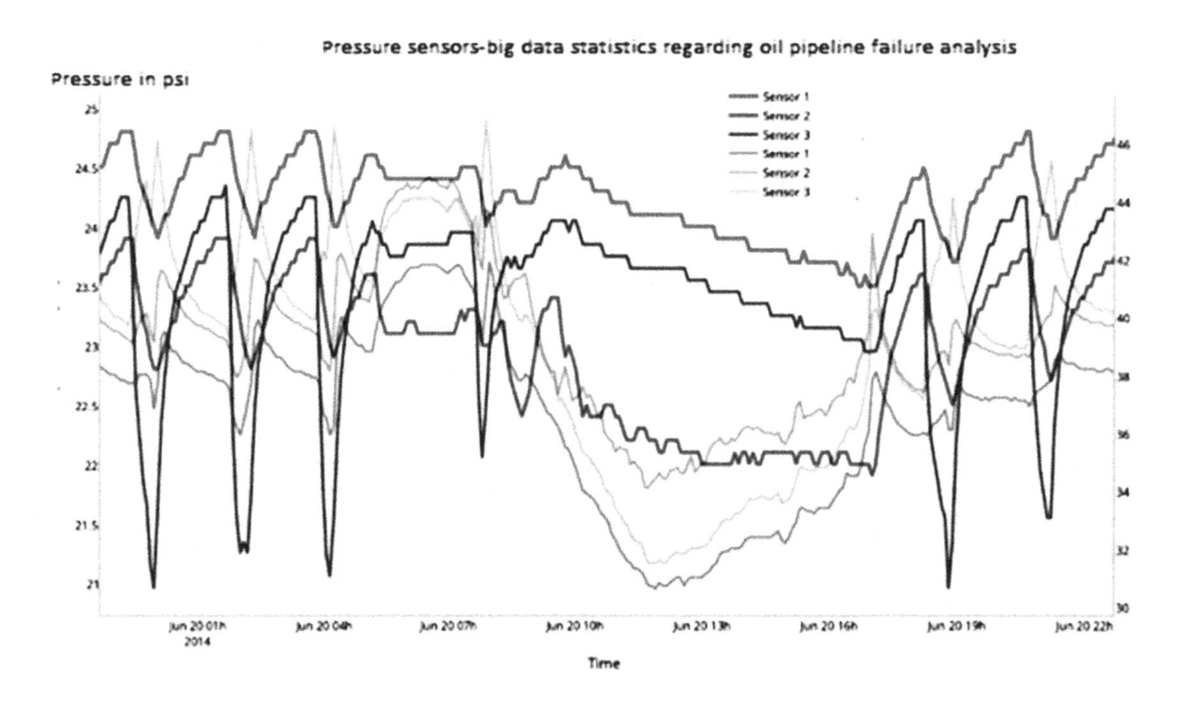

Machine Learning Techniques for Leak Detection in the Oil Pipelines

The so-called Negative Pressure Wave (NPW) based approach has been widely adopted in real applications among the different approaches for leak detection. In the pressure curve, NPW may be an indicator of pipeline leakage. The basic idea of the NPW-based methods is that, as leakage happens, there is a sudden drop in the pressure (i.e. the Negative Pressure Wave) flowing from the leak point to the upstream and downstream ends and so that we can detect leaks by detecting pressure changes at both ends. The main purpose is to allow the method to detect leakage errors from a pressure curve that cannot be detected effectively by conventional negative wave-based methods but can be easily detected by human visual perception (Priyanka et al., 2018c, 2018d). Inspired by other object detection applications of support vector machine (SVM), we are investigating the use of the SVM learning system for NPW detection in this paper and show that it provides better performance compared to the Wavelet-based methods we have tested.

Figure 12. Flowchart of the SVM for leak identification in oil pipelines

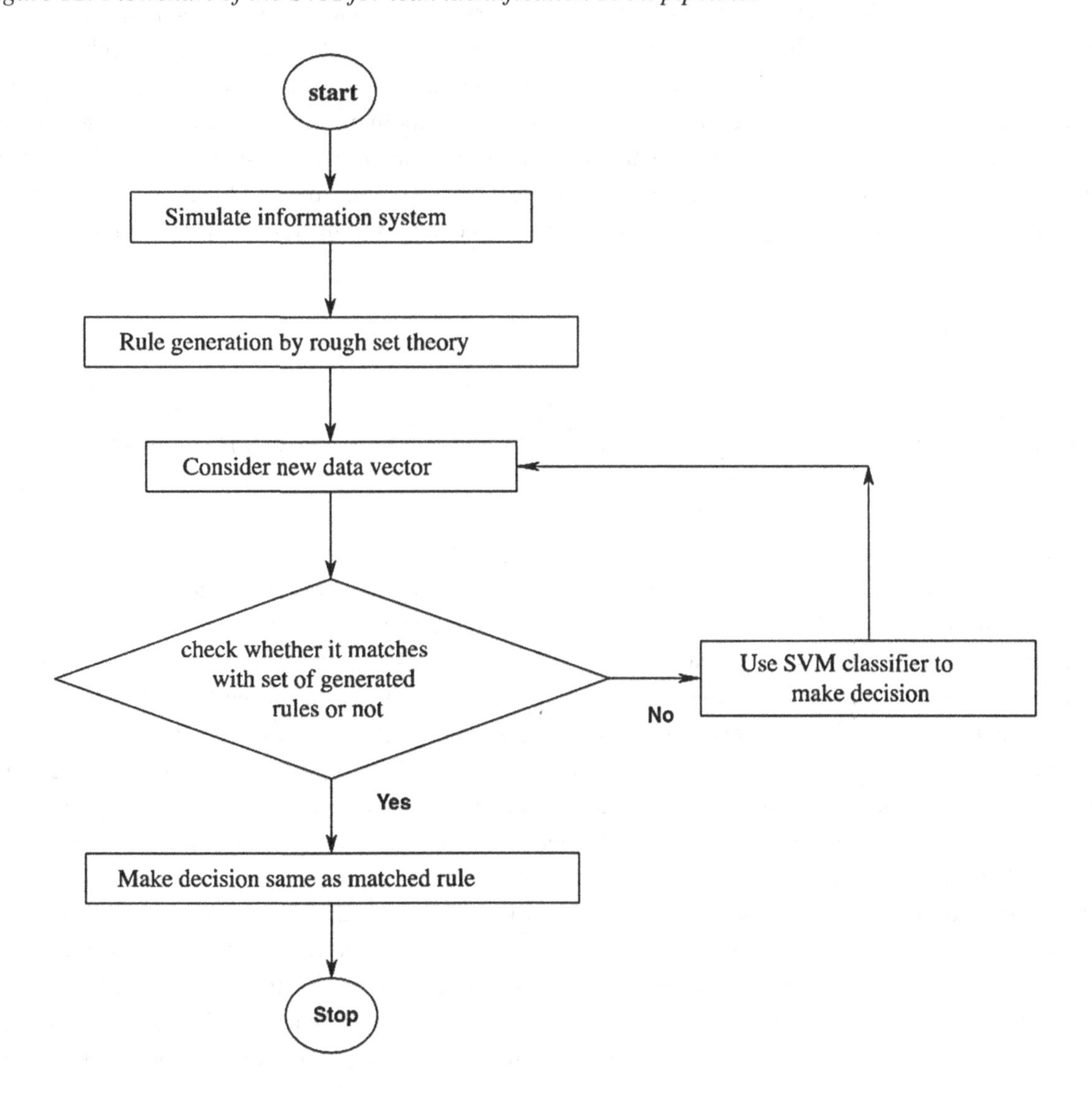

SVM is a learning method based on the modem mathematical learning theory, SVM learning is based on the systemic risk minimization principle. Rather than reducing learning error directly, it seeks to minimize the generalization error bound. Consequently, when applied to data outside the training set, an SVM can perform well. Figure 12 shows the flowchart of the proposed technique for leak identification in the oil pipeline system using the Support Vector Machine algorithm. In the proposed techniques holds the NPW detection is considered to be a function of classification of two groups performed at each specific pressure curve (Maheswari et al.,2017). The two groups are current NPW and absent NPW. A nonlinear classifier is equipped with an SVM formulation, using supervised learning to automatically detect the existence of NPW in the pressure curve. By this approach, its quite flexible and can easily detect tiny or slow leaks out of noise.

The leakages along the crude oil and liquid fuel pipeline are causing financial loss and damage to the environment. This research brings with it a novel leak detection system that uses data trained on rough set theory and supports vector machine. The rough set theory was combined to create a set of rules that examine the existence of any anomalies for the first time. Instead, the classifier proposed by SVM examines the remaining unresolved instances. In summary, the advantages of the technique presented are as follows: it gives very high detection precision, it can be used in all cases where the techniques of mass balance and pressure point analysis are used in particular, the thorough simulation of the pipeline is not necessary and it is very useful to detect leaks or fold pipelines where the precise structure and the internal surface state are not present. This work can be expanded in the future by adding a few more measures to prevent losses caused by leaks such as leak size, response time and location of leaks. Also, the effect on detection accuracy of other process variables (such as temperature, friction factor, etc.)

CONCLUSION

This chapter summarizes a modern Big Data approach that handles data of any volume and variety to predict the risk occurring or failure analysis using SVM incorporated with Hadoop. The proposed architecture of Hadoop with a scheduling algorithm for data allocation assigns the proper Big Data technologies for traditional BI and reporting real-time analytical processing, and discovery analytics for structured, semi-structured, and unstructured data. It supports both physics-based modeling and empirical approaches such as ANN. The underlying Big Data approach removes the limitations of relational databases in managing different data standards. In off-premises and hybrid Cloud, and also with the Internet of Things platform solves a much wider range of oil pipeline industry problems than is possible with traditional information technologies. The experimental results are performed on a big dataset that is generated from a sophisticated real-time oil pipeline sensor data results show that the method used in the proposed framework outperforms the traditional ones in terms of classification accuracy and other statistical performance evaluation metrics. Also, this chapter addresses the problems and challenges of using geometric representation-learning techniques and current Big Data networking technologies to manage Big Data classification. In particular, this paper addresses issues related to the combination of supervised learning techniques, data allocation, data risk challenges in the cloud computing and Big Data technologies (e.g., Hadoop and Cloud) concerning data congestion in the classification of network traffic.

In the future work, the present architecture will be extended to Create Engineering Data Warehouse (relational or Hadoop) from various source systems, as to complement to field data historian, to bring data together and support a combined theoretical modeling and empirical analytics environment. Further, the other parameters like corrosion, temperature and flow rate will be undertaken to analyze risk occurrences rate in the pipeline using deep neural network techniques and the results are compared with supervised machine learning algorithms to compute the better prediction suitability of the experimentation.

ACKNOWLEDGMENT

The author thanks the Scientific and Industrial Research Council (CSIR), Pusa, New Delhi, India for providing financial support to the experimental research work under Senior Research Fellowship-Direct (08/678(0001)2k18 EMR.

REFERENCES

Alnasir, J. J., & Shanahan, H. P. (2020). The application of hadoop in structural bioinformatics. *Briefings in Bioinformatics*, *21*(1), 96–105. PMID:30462158

Alves, T., Das, R., Werth, A., & Morris, T. (2018). Virtualization of SCADA testbeds for cybersecurity research: A modular approach. *Computers & Security*, *77*, 531–546. doi:10.1016/j.cose.2018.05.002

Baranowski, Z., Kleszcz, E., Kothuri, P., Canali, L., Castellotti, R., Marquez, M. M., . . . Duran, J. C. L. (2019). Evolution of the Hadoop Platform and Ecosystem for High Energy Physics. In *EPJ Web of Conferences* (*Vol. 214*, p. 04058). EDP Sciences. 10.1051/epjconf/201921404058

Chen, J., Wang, D., & Zhao, W. (2013). A task scheduling algorithm for Hadoop platform. *Journal of Computers*, *8*(4), 929–936.

Chen, Q., Zhang, D., Guo, M., Deng, Q., & Guo, S. (2010, June). Samr: A self-adaptive mapreduce scheduling algorithm in heterogeneous environment. In *2010 10th IEEE International Conference on Computer and Information Technology* (pp. 2736-2743). IEEE. 10.1109/CIT.2010.458

Chen, T. Y., Wei, H. W., Wei, M. F., Chen, Y. J., Hsu, T. S., & Shih, W. K. (2013, May). LaSA: A locality-aware scheduling algorithm for Hadoop-MapReduce resource assignment. In *2013 International Conference on Collaboration Technologies and Systems (CTS)* (pp. 342-346). IEEE. 10.1109/CTS.2013.6567252

Devi, T. K., Sakthivel, P., Pratheep, V., & Priyanka, E. (2020). *120 MHz, 2.2 mW Low Power Phase-Locked Loop using Dual Mode Logic and its Application as Frequency Divider*. Academic Press.

El-sayed, T., Badawy, M., & El-Sayed, A. (2019). Impact of Small Files on Hadoop Performance: Literature Survey and Open Points. *Menoufia Journal of Electronic Engineering Research*, *28*(1), 109–120. doi:10.21608/mjeer.2019.62728

Fasoulakis, M., Tsiropoulou, E. E., & Papavassiliou, S. (2019). Satisfy instead of maximize: Improving operation efficiency in wireless communication networks. *Computer Networks*, *159*, 135–146. doi:10.1016/j.comnet.2019.05.012

Gandomi, A., Reshadi, M., Movaghar, A., & Khademzadeh, A. (2019). HybSMRP: A hybrid scheduling algorithm in Hadoop MapReduce framework. *Journal of Big Data*, *6*(1), 106. doi:10.118640537-019-0253-9

Glushkova, D., Jovanovic, P., & Abelló, A. (2019). Mapreduce performance model for Hadoop 2. x. *Information Systems*, *79*, 32–43. doi:10.1016/j.is.2017.11.006

Huang, W., Wang, H., Zhang, Y., & Zhang, S. (2019). A novel cluster computing technique based on signal clustering and analytic hierarchy model using hadoop. *Cluster Computing*, *22*(6), 13077–13084. doi:10.100710586-017-1205-9

Jin, J., Luo, J., Song, A., Dong, F., & Xiong, R. (2011, May). Bar: An efficient data locality driven task scheduling algorithm for cloud computing. In *2011 11th IEEE/ACM International Symposium on Cluster, Cloud and Grid Computing* (pp. 295-304). IEEE.

Kaliyannan, G. V., Palanisamy, S. V., Priyanka, E. B., Thangavel, S., Sivaraj, S., & Rathanasamy, R. (2020). Investigation on sol-gel based coatings application in energy sector–A review. *Materials Today: Proceedings*. Advance online publication. doi:10.1016/j.matpr.2020.03.484

Kapoor, R., Gupta, R., Kumar, R., & Jha, S. (2019). New scheme for underwater acoustically wireless transmission using direct sequence code division multiple access in MIMO systems. *Wireless Networks*, *25*(8), 4541–4553. doi:10.100711276-018-1750-z

Kos, A., Milutinović, V., & Umek, A. (2019). Challenges in wireless communication for connected sensors and wearable devices used in sport biofeedback applications. *Future Generation Computer Systems*, *92*, 582–592. doi:10.1016/j.future.2018.03.032

Maheswari, C., Priyanka, E. B., & Meenakshipriya, B. (2017). Fractional-order PIλDμ controller tuned by coefficient diagram method and particle swarm optimization algorithms for SO2 emission control process. *Proceedings of the Institution of Mechanical Engineers. Part I, Journal of Systems and Control Engineering*, *231*(8), 587–599. doi:10.1177/0959651817711626

Maheswari, C., Priyanka, E. B., Thangavel, S., & Parameswari, P. (2018). Development of Unmanned Guided Vehicle for Material Handling Automation for Industry 4.0. *International Journal of Recent Technology and Engineering*, *7*(4S), 428–432.

Malik, M., Neshatpour, K., Rafatirad, S., Joshi, R. V., Mohsenin, T., Ghasemzadeh, H., & Homayoun, H. (2019). Big vs little core for energy-efficient hadoop computing. *Journal of Parallel and Distributed Computing*, *129*, 110–124. doi:10.1016/j.jpdc.2018.02.017

Nakagami, M., Fortes, J. A., & Yamaguchi, S. (2019, July). Job-Aware Optimization of File Placement in Hadoop. In *2019 IEEE 43rd Annual Computer Software and Applications Conference (COMPSAC)* (Vol. 2, pp. 664-669). IEEE. 10.1109/COMPSAC.2019.10284

Nguyen, P., Simon, T., Halem, M., Chapman, D., & Le, Q. (2012, November). A hybrid scheduling algorithm for data intensive workloads in a mapreduce environment. In *2012 IEEE Fifth International Conference on Utility and Cloud Computing* (pp. 161-167). IEEE. 10.1109/UCC.2012.32

Nita, M. C., Pop, F., Voicu, C., Dobre, C., & Xhafa, F. (2015). MOMTH: Multi-objective scheduling algorithm of many tasks in Hadoop. *Cluster Computing*, *18*(3), 1011–1024. doi:10.100710586-015-0454-8

Pakize, S. R. (2014). A comprehensive view of Hadoop MapReduce scheduling algorithms. *International Journal of Computer Networks & Communications Security*, *2*(9), 308–317.

Pratheep, V. G., Priyanka, E. B., & Prasad, P. H. (2019, October). Characterization and Analysis of Natural Fibre-Rice Husk with Wood Plastic Composites. *IOP Conference Series. Materials Science and Engineering*, *561*(1), 012066. doi:10.1088/1757-899X/561/1/012066

Pratheep, V. G., Priyanka, E. B., Thangavel, S., Gousanal, J. J., Antony, P. B., & Kavin, E. D. (2020). Investigation and analysis of corn cob, coir pith with wood plastic composites. *Materials Today: Proceedings*. Advance online publication. doi:10.1016/j.matpr.2020.02.288

Priyanka, E. B., Krishnamurthy, K., & Maheswari, C. (2016, November). Remote monitoring and control of pressure and flow in oil pipelines transport system using PLC based controller. In *2016 Online International Conference on Green Engineering and Technologies (IC-GET)* (pp. 1-6). IEEE. 10.1109/GET.2016.7916754

Priyanka, E. B., Maheswari, C., Ponnibala, M., & Thangavel, S. (2019b). SCADA Based Remote Monitoring and Control of Pressure & Flow in Fluid Transport System Using IMC-PID Controller. *Advances in Systems Science and Applications*, *19*(3), 140–162.

Priyanka, E. B., Maheswari, C., & Thangavel, S. (2018d). IoT based field parameters monitoring and control in press shop assembly. *Internet of Things*, *3*, 1–11. doi:10.1016/j.iot.2018.09.004

Priyanka, E. B., Maheswari, C., & Thangavel, S. (2018e). Online monitoring and control of flow rate in oil pipelines transportation system by using PLC based Fuzzy-PID Controller. *Flow Measurement and Instrumentation*, *62*, 144–151. doi:10.1016/j.flowmeasinst.2018.02.010

Priyanka, E. B., Maheswari, C., & Thangavel, S. (2018f, December). Proactive Decision Making Based IoT Framework for an Oil Pipeline Transportation System. In *International conference on Computer Networks, Big data and IoT* (pp. 108-119). Springer.

Priyanka, E. B., Maheswari, C., & Thangavel, S. (2019a). Remote monitoring and control of LQR-PI controller parameters for an oil pipeline transport system. *Proceedings of the Institution of Mechanical Engineers. Part I, Journal of Systems and Control Engineering*, *233*(6), 597–608. doi:10.1177/0959651818803183

Priyanka, E. B., Maheswari, C., & Thangavel, S. (2020b). A smart-integrated IoT module for intelligent transportation in oil industry. *International Journal of Numerical Modelling: Electronic Networks, Devices and Fields*, e2731.

Priyanka, E. B., Maheswari, C., Thangavel, S., & Bala, M. P. (2020a). Integrating IoT with LQR-PID controller for online surveillance and control of flow and pressure in fluid transportation system. *Journal of Industrial Information Integration*, *17*, 100127. doi:10.1016/j.jii.2020.100127

Priyanka, E. B., & Subramaniam, T. (2018a). Fuzzy Logic Forge Filter Weave Pattern Recognition Analysis on Fabric Texture. *International Journal of Electrical and Electronic Science*, *5*(3), 63–70.

Priyanka, E. B., & Thangavel, S. (2018b). Advanced Enhancement Model of Bionics Fish Cilia MEMS Vector Hydrophone-Systematic Analysis Review. *International Journal of Applied Physics, 3*.

Priyanka, E. B., & Thangavel, S. (2018c). Optimization of Large-Scale Solar Hot Water System Using Non-traditional Optimization Technique. *J Environ Sci Allied Res, 2018*, 5-14.

Rasooli, A., & Down, D. G. (2011, November). An adaptive scheduling algorithm for dynamic heterogeneous Hadoop systems. In *Proceedings of the 2011 Conference of the Center for Advanced Studies on Collaborative Research* (pp. 30-44). Academic Press.

Shankar, R. K., Priyanka, E. B., & Saravanan, B. (2015). Performance analysis of gasoline direct injection in two-stroke spark-ignition engines. *Int J Adv Res Electr Electron Instrum Eng*, *4*, 4940–4947.

Subramaniam, T., & Bhaskaran, P. (2019). Local intelligence for remote surveillance and control of flow in fluid transportation system. *Journal*, *74*(1), 15–21. http://iieta. org/journals/ama_c

Sun, X., He, C., & Lu, Y. (2012, December). ESAMR: An enhanced self-adaptive mapreduce scheduling algorithm. In *2012 IEEE 18th International Conference on Parallel and Distributed Systems* (pp. 148-155). IEEE.

Tariq, H., Al-Sahaf, H., & Welch, I. (2019, December). Modelling and Prediction of Resource Utilization of Hadoop Clusters: A Machine Learning Approach. In *Proceedings of the 12th IEEE/ACM International Conference on Utility and Cloud Computing* (pp. 93-100). 10.1145/3344341.3368821

Xia, J., Li, C., Lai, X., Lai, S., Zhu, F., Deng, D., & Fan, L. (2020). Cache-aided mobile edge computing for B5G wireless communication networks. *EURASIP Journal on Wireless Communications and Networking*, *2020*(1), 15. doi:10.118613638-019-1612-0

Xia, Y., Wang, L., Zhao, Q., & Zhang, G. (2011). Research on job scheduling algorithm in hadoop. *Journal of Computer Information Systems*, *7*(16), 5769–5775.

Ye, L. (2020). Study on embedded system in monitoring of intelligent city pipeline network. *Computer Communications*, *153*, 451–458. doi:10.1016/j.comcom.2020.02.004

Zhang, K., Chuai, G., Gao, W., Liu, X., Maimaiti, S., & Si, Z. (2019). A new method for traffic forecasting in urban wireless communication network. *EURASIP Journal on Wireless Communications and Networking*, *2019*(1), 66. doi:10.118613638-019-1392-6

Chapter 13
Security in Rail IoT Systems:
An IoT Solution for New Rail Services

Francisco Parrilla Ayuso
Indra Sistemas S.A., Spain

David Batista
Indra Sistemas S.A., Spain

Daniel Maldonado
Indra Sistemas S.A., Spain

Jon Colado
Indra Sistemas S.A., Spain

Sergio Jiménez Gómez
Indra Sistemas S.A., Spain

Jorge Portilla
https://orcid.org/0000-0003-4896-6229
Universidad Politécnica de Madrid, Spain

Gabriel Mujica
Universidad Politécnica de Madrid, Spain

Jaime Señor
Universidad Politécnica de Madrid, Spain

ABSTRACT

Indra Sistemas S.A. have designed and developed a safety and secure solution system for the rail transportation environment based on a distributed architecture under the domain of the Industrial IoT that enables V2V, V2I, and I2I communications, allowing peer-to-peer data sharing. UPM has designed and implemented a HW-based security infrastructure for extreme edge devices in IoT. The implementation takes advantage of HW accelerator to enhance security in low resources devices with a very low overhead in cost and memory footprint. Current security solutions are problematic due to centralized control entity. The complexity of this kind of system resides in the management, in a decentralized way, of the security at each point of the distributed architecture. This chapter describes how the system secures all the infrastructure based on a distributed architecture without affecting the throughput and the high availability of the data in order to get a top-performance, in compliance with the strengthen safety and security constrains of the rail environment's regulations.

DOI: 10.4018/978-1-7998-5068-7.ch013

INTRODUCTION

The Internet of Things (IoT) has appeared strongly in the market since the last decade, scaling through different environments and trying to make easier the life of EU citizens. The rail environment is suffering a process of adaptation to the new technologies that cannot ignore the power of IoT networks to build innovative solutions in that field.

The railway market is a regulated environment which must guarantee the safety and security of the rail systems deployed, by applying a set of rules clearly defined by the railway regulation organizations (i.e. CENELEC (European Committee for Electrotechnical Standardization, 2010)). In this context, the rail domain is an attractive market to develop wireless technologies which may replace classic wired systems with innovative solutions.

The Internet of Things (IoT) as one of the major enablers of the digital transformation trend Europe will enable both obtaining information from the IoT systems and providing data to them. These concepts are possible by using wireless sensors and actuators in a solid manner. Moreover, these systems are having major benefits for usage, such as increased flexibility, mobile applications, weight reduction, adaptability for changes and the recently added trustworthiness by reason of the next projects.

The DEWI (Dependable Embedded Wireless Infrastructure) project was the starting point to demonstrate the feasibility of the deployment wireless solutions for the rail domain concerning safety applications.

The SCOTT (Secure Connected Trustable Things) project continues the DEWI works by implementing security and trustability in the communications in several domains, including the rail sector. This project adds new end-to-end secured, trustworthy and interoperable wireless capabilities between trains (V2V communications) and between train and infrastructure (V2I/I2V communications) (X2Rail-1, 2018) to solve the hazardous situations relating to safety that can occurs in typical rail lines. Moreover, the introduction of cloud-based platforms in the context of SCOTT project, improves the efficiency and reduce the cost of the rail services.

The systems needs to be continuously evolved and adapted to their new environment ensuring and increasing their trustworthiness, quality and user experience. The DevOps movement advocates a set of software engineering best practices and tools, to ensure Quality of Service whilst continuously evolving complex systems and foster agility, rapid innovation cycles, and ease of use. Current DevOps solutions also lack mechanisms to ensure end-to-end security and privacy, mechanisms able to take into consideration open context and actuation conflicts and perform continuous deployment and evolution of IoT systems across, IoT, edge, and cloud spaces.

To solve this, ENACT (Development, Operation, and Quality Assurance of Trustworthy Smart IoT Systems) project intends to introduce the DevOps movement by generating enablers for the monitoring of operations in a smart IoT Systems context, solving the former mentioned issues for IoT systems. Moreover, ENACT assesses the feasibility of IoT services in the domain of train control for the rolling stock and the on-track equipment logistics and maintenance, combining IoT architecture and platforms with cloud resources.

Shift2Rail European initiative has launched several innovations programs to accelerate the integration of new and advanced technologies into innovative rail product systems. In this context, one of the projects which makes use of wireless technologies is X2Rail-1 (Start-up activities for Advanced Signalling and Automation Systems). This project proposes, among other innovative solutions, the use of wireless communications in signalling systems, especially for the control of wayside objects in a smart way (X2Rail-1, 2016).

Joining the efforts of X2Rail-1, DEWI, SCOTT, and ENACT projects, a secure platform for the integration of new rail services making use of IoT technologies has been design and implemented by Indra. This platform covers all edge data collection, wireless communication systems, and Cloud services. The present document intends to explain the features of this platform and how it accomplishes the needed requirements concerning security for safety-related systems.

The document is organized in chapters to provide a complete view of the security aspects that a Rail IoT system requires:

1. **Introduction:** Introduction to the rail environment issues that generate the need to use IoT in a secure manner.
2. **Background:** Introduction to the current state of the art related with security into the IoT environment in particular for rail purposes. It is also describe the current works and projects that are solving security issues for IoT.
3. **Main Focus of the Chapter:** The core of the document where all the rail secure IoT system components are explained including the IoT system layers and services implementation.
4. **Solution and Recommendations:** The infrastructure and functionalities to provide a secured rail solution using IoT.
5. **Future Research Directions:** The trends that are identified for a rail IoT system aligned with the general tendencies identified for all the IoT systems.
6. **Conclusions:** Summary of the facts extracted once the system is defined, implemented, and tested in a real rail environment.

1.1. Major Contribution

The proposed solution is supported on an ontology and data model which defines the properties of the elements belonging to the distributed architecture and the relations among them. By guaranteeing that all these elements accomplish the ontology, it is possible to define a predictable traffic pattern to characterize the network and making use of AI techniques (i.e. Machine Learning) monitor and detect attacks from outside.

The focus of the chapter is the description of this secure platform, attending to the different parts that composes it. The solutions and recommendations subchapter present and analyses the strong security features implemented to guarantee a Secure Connected and Trustable Internet of Things Platform for the deployment of innovative rail services, focusing on providing a distributed architecture to the rail signalling systems.

2. BACKGROUND

The security in IoT Networks is a field which have been studied recently, applied to cybersecurity issues. In this context, the European Union Agency for Network and Information Security (ENISA) (European Union Agency for Network and Information Security, 2017), in 2017 provides Baseline Security Recommendations for IoT in the context of Critical Information Infrastructures. This document is intended to provide guidelines to promote harmonization of IoT security initiatives and regulations and define secure software/hardware development lifecycle for IoT, among other objectives. The gap analysis performed in this document listed several important gaps:

- There is not a consensus in the EU and non-ethical restrictions
- The clients have usually a lack of knowledge
- The security is not usually considered in an in-design stage and it is not considered during the project to scale the security
- Lack of standardization among partners
- Trend to believe that the security in IoT does not generate revenues

In order to analyse IoT Networks and protocols from cyber security perspective, Luzuriaga et al. (Luzuriaga et al. 2015) shows some brief extra advices to secure an IoT system, focusing in the analysis of AMQP and MQTT. This study is used as a reference to start analysing the implications if the use of MQTT and AMQP reinforced with the analysis of security applied to IoT Networks in Rail environment. Furthermore, concerning railway industry, SCOTT project analyses the state of the railway technology from a safety and security point of view (SCOTT Project, 2018) and ENACT project introduces the DevOps philosophy to improve the trustworthiness in the IoT systems (ENACT, 2018).

In this context, wireless and IoT technology will be disruptive into the railway market, and these technologies must adapt to the regulations for classic systems. There are several regulations to be applied to the railway systems, but among them, CENELEC EN-50159 provides a list of defences concerning data transmission apply to wireless networks. This standard provides the basic requirements to ensure safety and security to safety-related communication in railway transmission systems. These defences can be covered by performing a brainy analysis of the security applied to IoT and find the way to match these new open possibilities with the security requirements of the railway market.

3. MAIN FOCUS OF THE CHAPTER

This chapter is intended to study how the secure platform has been implemented and how have been integrate the IoT architecture in the Rail Transportation enabling V2V, V2I and I2I communications of safety data through a distributed network.

Focusing on the different functionalities that it performs and the type of technology challenges that are involved, the main goal is to achieve a secure and safe solution to develop innovative rail services. These services goes from a train wagons composition tracking (Train Integrity (Barkovskis et al. 2016)), support the On-Track signalling or trains travelling together in a synchronize manner (Virtual Coupling) to the train signalling systems.

First of all, it is necessary to describe the architecture over which the Industrial IoT is based and the explanation of how is integrated with the Secure Platform developed by Indra Sistemas S.A. The proposed architecture is separated in three layers:

- **Edge**: It is formed by the devices in charge of collect data and shape the operational core of each rail scenario. This devices are deployed all over the rail scenario in order to provide interconnectivity data and analytics to the Cloud's Industrial IoT system, enabling the multi-purpose feature

of this platform. It forms a mesh network based on a modular and interconnected architecture, enabling a fast and efficient data processing and collection. Detailed explanation into the section 3.2.
- **Fog**: It is formed by a set of devices that receives the pre-processed data from the edge in order to storage, compute, and send to Cloud services that data. These devices are located near to each edge modules oversee. The goal of the fog is to secure, manage, orchestrate and distribute dynamically the information flow of each service and application all over network in order to get a top performance and ensure the data integrity at the safety system. Detailed explanation into the section 3.3.
- **Cloud**: The hybrid Cloud, composed by an IaaS (Infrastructure as a Service) and On-premise environments, stores and process the data collected by the edge. The Cloud distribution allows to centralize services for Data and Security Monitoring, devices Orchestration and Simulation and Testing. This architecture also provides scalability, interoperability and an easy deployment of the implementation. Detailed explanation into the section 3.4.

This multi-purpose implementation defines the security protocols at each layer of the architecture in order comply with the regulations of a rail secure and safety environment.

This chapter is structured in fourth important subchapters:

1. Secure Platform for new Rail Services (Section 3.1).
2. Edge architecture, computing and securing (Section 3.2).
3. Fog architecture, computing and securing (Section 3.3).
4. Cloud architecture, computing and securing (Section 3.4).

3.1. Secure Platform for New Rail Services

To find a solution to this issue, a Rail secure scenario is configured by creating a full communication system in order to transmit messages between all network elements in a railway environment. The use of wireless technologies in railway safety environment have been studied to cover new services making use of IoT technologies. These new services are covered by two main functionalities at the Industrial IoT edge: the data collecting, access control and the data computing.

IoT data collecting is commonly implemented in railway systems through sensors and actuators. These sensors and actuators can be allocated in a train composition (On-Board) or on the rail track. They are connected composing nodes which create a physical communication layer across the local infrastructure (On-Board or On-Track), gathering, processing and transmitting data in a distributed environment.

The security must be guaranteed in vertical and horizontal distribution along the network until it is correctly delivered to the end users.

The Secure Rail Scenario proposed is supported by the works performed in SCOTT Project and ENACT Project, following the AIOTI recommendations (Alliance for Internet Things Innovation, 2019) for IoT. The scenario is shown in the Figure 1.

Concerning communication and as mentioned in the first part of the chapter, it was selected the use of Advanced Message Queuing Protocol (AMQP) (ISO/IEC 19464:2014, 2019) and Message Queuing Telemetry Transport (MQTT) (ISO/IEC 20922:2016) as transmission protocols and following an ontology fully compliant with the ISO/IEC 29182 – Sensor Network Reference Architecture (ISO/IEC

Figure 1. IoT Solution Proposed Scenario

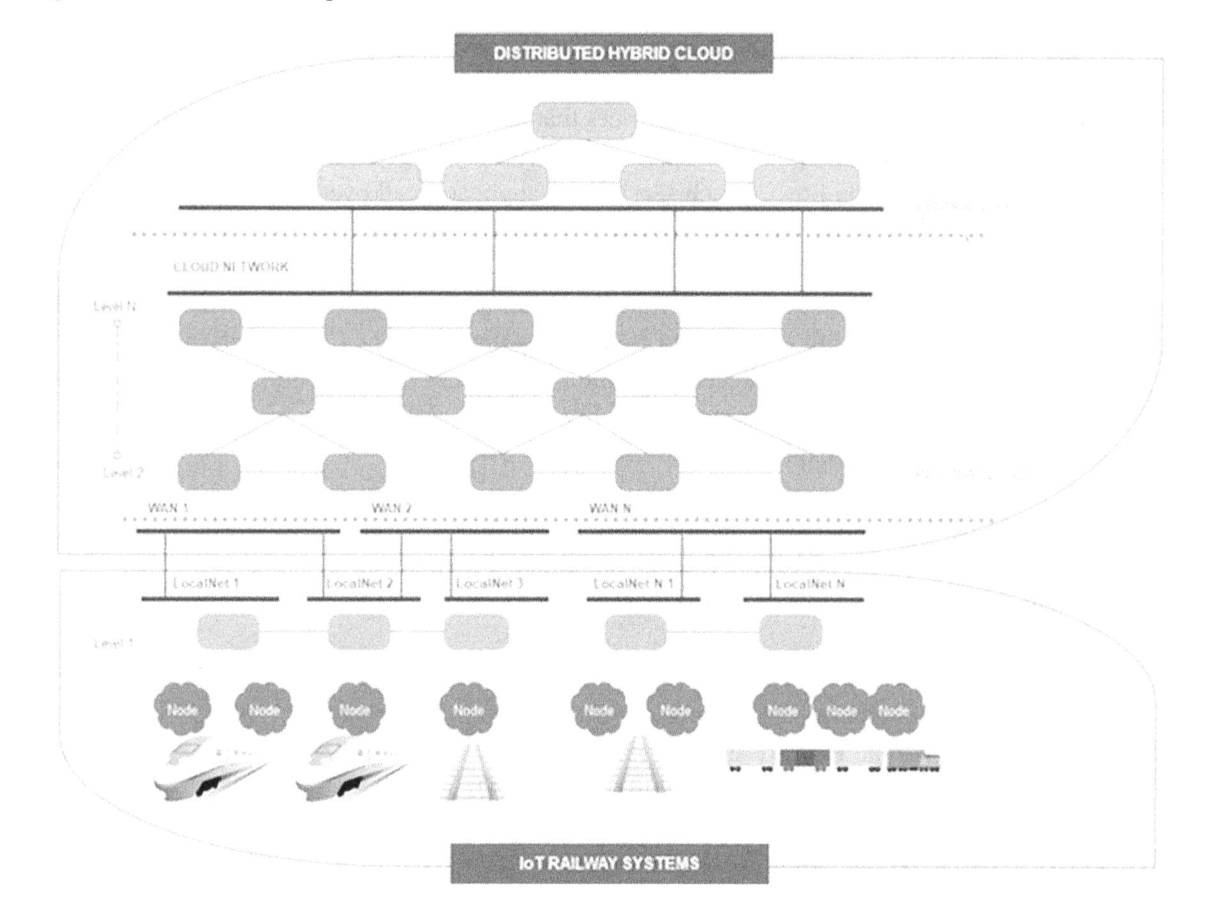

29182-1:2013). Concerning communication, it was selected the use of AMQP/MQTT for data model, due the reliability and interoperability of both. Some of the security capabilities of AMQP protocol are:

- Flow-controlled communication enables delivery options, such as at-most-once, at-least-once, or exactly once.
- Delivery authentication is provided by SASL (Simple Authentication and Security Layer) (Melnikov, A., Zeilenga, K, 2018); Encryption is provided by TLS (Transport Layer Security) (Dierks and Rescorla, 2020), which is the successor to SSL (Secure Sockets Layer) (Turner and Polk, 2018), obsoleted in 2011. These protocols have been complemented with extra analysis of security in different layers to cover the security limitations of AMQP/MQTT.

The communications have been divided in two levels:

- **Level 1.** For short range communication, among the sensors and the WSN Coordinator, the protocol used is MQTT in combination with the specific developments carried out in order to increase the security in the railway systems.

- **Level 2.** For medium and long-range communication, AMQP and the featured developed already mentioned are used.

At this point, the regulation that applies to the communications described below is the CENELEC EN-50159 (European Committee for Electrotechnical Standardization, 2010), which indicates a set of defences to be implemented for assuring the message transmissions is not modified in the network. MQTT/AMQP cover part of these defences, and the rest of them are guaranteed by the security provided by the implemented platform. These defences are:

- **Sequence Number:** AMQP/MQTT does not include Sequence Number for messages. The developments added must cover this point by implementing in the message a sequence number to keep it into the messages transmitted through the network and reach the destination.
- **Timestamp:** AMQP/MQTT does not include Timestamp for messages. With the secure analysis to be performed, all messages must include a timestamp (Unix Epoch time) in order to assure the detection of old messages. This broadly used timestamp indicates when the message was created, so the destination can reject messages older than a specific configured time.
- **Timeout:** AMQP/MQTT does not include specific timeout for messages. To guarantee the security timeout is configured in the system for determining when messages should be received and rejecting messages not received on time.
- **Source and Destination identifiers:** AMQP/MQTT includes a Source and Destination identifier based on subscription, but does not identify specific source/destination into subscriber. This identification can be covered by including CRC calculation.
- **Feedback messages:** AMQP implements a delivery guarantees process for assuring messages has been delivered to destination through the network.
- **Identification procedure:** Due to AMQP implements TLS security, the identification procedure is assured through certificates that assure the identification between senders and add trustily to the network avoiding interferences to the messages.
- **Safety Code:** AMQP does not implement a mechanism for assuring data integrity. By including the use of CRC in the header and the payload, this defence can be covered.
- **Cryptographic techniques:** AMQP protocol includes a set of security features (TLS, Authentication/Security mechanism as SASL) that assure messages exchanges by using this protocol are delivered to destination system and cannot be altered during the transmission. To ensure the security of the developments, TLS has to be considered and explained as an encryption solution.

3.2. Edge

The edge defined in the secure scenario for IoT Networks consists in two parts:

- The first part are edge devices, composed by the Wireless Sensor Network (WSN) based on WSN provider and WSN Coordinator.
- The second part is edge layer, composed by the Communication Middleware (CMW).

In the Table 1 protocols used to communicate WSN provider and WSN Coordinator also to communicate WSN Coordinator with the network devices are presented. This protocol distribution is based on the OSI layer model.

Table 1. OSI Layer Model Protocol Distribution

	WSN provider to WSN Coordinator connection	WSN Coordinator to the Network Device connection
Physical and Data Link	802.15.4 (6LoWPAN)	802.11.b/g/n (WIFI)
Network	RPL	IP (with DCHP + DNS)
Transport	TCP/UPD	TCP + TLS
Session	--	LDAP
Presentation	JSON	JSON
Application	MQTT	MQTT/TLS

Edge Architecture

The WSN provider is composed by smart devices forming a mesh network of sensors and actuators and it is in charge of capture and generate data under the orchestration of the WSN Coordinator.

These devices are based on the platform developed at CEI-UPM, the Cookie node, which intend to be a modular and flexible solution to the implementation of a WSN. The version used for the proposed network has been designed with special focus on the security requirements needed at the edge level. This means that it can run multiple tasks of a common cryptographic scheme, without showing a meaningful drawback to the general performance of the node in processing time and power consumption.

In order to implement those security features directly inside the nodes of the edge, a Trusted Platform Module (TPM) has been added to the set of components attached to the I2C bus of the node. The TPM is a hardware accelerator that includes several cryptographic algorithms that can be mixed to build a complete security scheme, such as authentication procedures or encryption of data blocks. In general, these tasks represent a heavy load in terms of power consumption, so moving them from software to hardware improves the autonomy of the node. Also, the speed achieved by these accelerators increases the overall efficiency.

Another advantage of the TPM is that it creates an isolated environment, where the information stored inside is said to be trustable. The chip is protected against side-channel attacks, e.g. power analysis from external measurements. This means that, if properly configured, secret keys can be stored securely inside the TPM, solving one of the main challenges of practical cryptography. Indeed, this solution is highly preferable in comparison to using the cryptographic accelerators that can be found inside the main microcontroller of the node, which does not guarantee this isolated environment.

The CMW, developed by Indra Sistemas S.A., is the gateway of the fog in the edge and is in charge of validate, process and distribute the data send by WSN Coordinators.

Each WSN Coordinator can only communicate with its related CMW using MQTT protocol. In turns, each CMW can communicate with its related WSN Coordinators and with all the CMW across the system using AMQP protocol. This communication is based on OSI layer protocol showed at Table 1.

All the associations among WSN provider, WSN coordinator and CMW forms each module of the edge. These modules collect, process and distribute data and can be deployed On-Track and On-Board based on the scenario proposed at section 1 of this chapter.

Edge Computing

The architecture explained above enables each CMW individually process the data collected by the WSN Coordinators.

The edge takes on less computational charge, due to its mission is not distributing the information, but collect it. Once the data is transmitted from the edge devices to the CMW, this entity manages the validation, processing and distribution of the data.

The WSN Coordinator is also in charge of orchestration the data collection from the WSN providers and the aggregation and packaging of its data.

The low computational charge of the edge allows low energy consumption and concentrates the physical resources specifically in the data collection of the WSN devices.

Edge Securing

The security in the edge is based on the authentication of users and devices of the entire network and the cyphering of communication between WSN Coordinator - WSN providers and WSN Coordinator - CMW.

The proposed solution collects all the security protocols and functionalities needed to build a full secure chain that have to be ensured from the data collecting to the last level of the distribution layer in a railway environment. The security mechanisms implemented involved both user/device access and message encryption and validation.

The access of the WSN Coordinators to CMW is secured with Radius implementation to authenticate the user passwords making use of PEAP protocol. This mechanism receives the necessary information for user authentication from a LDAP directory server.

Moreover, at Network level, the access of the users can be securely guaranteed by providing ID and IP connection to the different edge nodes and elements, through DHCP and DNS protocols. DNS is in charge to assign a name to each IP and in turns, DHCP assign the IP to a MAC, taking into account the information provided by LDAP.

At Transport and Application level, the use of TLS and encryption algorithms can assure that the information point to point is encrypted. In addition to the encryption, the information contained in each message is protected using CRC, both in the header and in the payload.

The authentication and message encryption at the lowest level of the network is done via the TPM included in the Cookie nodes. A combination of two well differentiated schemes is used: asymmetric and symmetric. Asymmetric schemes are efficient when authentication procedures are needed, but symmetric alternatives perform message exchange better. Thus, the TPM is selected to be able to complete an asymmetric authentication and share a new pair of symmetric keys, once the nodes are checked to be trustable.

The asymmetric authentication uses Elliptic-Curve Cryptography (ECC), since the same security level provided by other algorithms, e.g. RSA, can be achieved with smaller key lengths (see Table 2). Therefore, it is one of the best alternatives for WSNs with limited memory resources. Also, larger times needed to break the ciphering are obtained with reduced key sizes in ECC compared to RSA (see Fig-

Table 2. RSA and ECC Security Level

Security Level	RSA	ECC
80	1024	160
112	2048	224
128	3072	256
192	7680	384
256	15360	512

Figure 2. RSA and ECC algorithm comparison

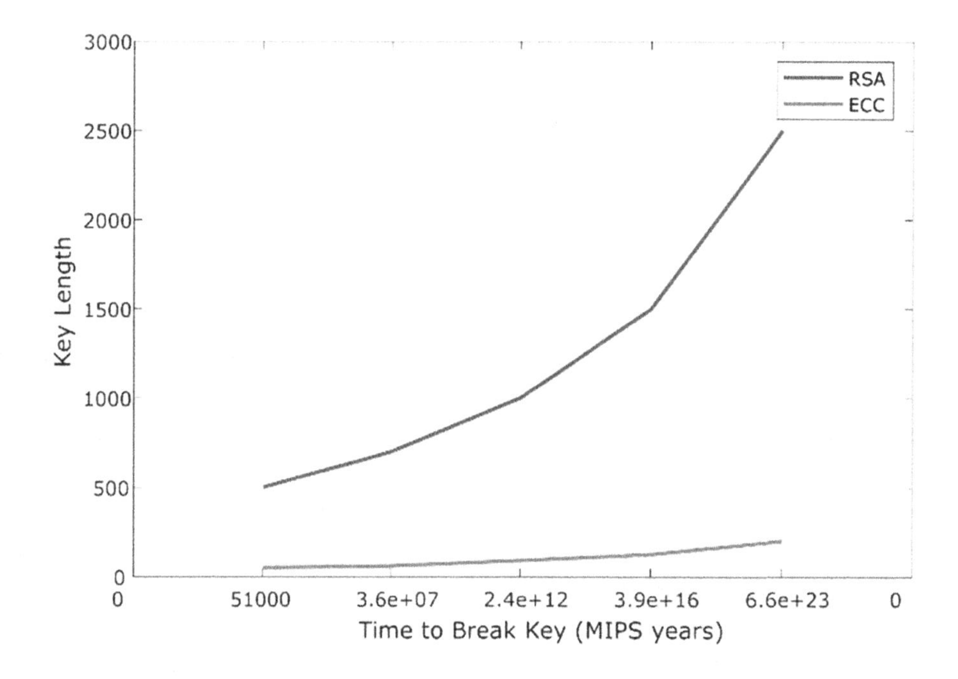

ure 2). While configuration of the nodes is done, an ECC private key is generated for each one inside its TPM, and it is stored in a memory slot which does not allow external access for reading operations

The full authentication is performed in two phases. In the first one (Figure 1), the public key of the nodes is checked using a chain of trust of certificates. Two different certificates are used for each node in this step, one that identifies de node itself, and another one that identifies the root signer. Both are properly stored in each TPM in a compressed format, reducing memory usage of the microcontroller. In the second phase, the private key of the new node is checked by requesting to sign a random number with the private key associated with the previously confirmed public key. ECDSA algorithm is used in this case, and a successful result of this operation means that the evaluated node is trustable. Both phases of the authentication request are fully accelerated through the hardware of the TPM.

After the authentication is successfully finished, the security scheme must be changed to a symmetric cipher to get the previously mentioned optimization. The TPM provides acceleration for the ECDH key exchange algorithm Lang, 2015), which ensures that both parties get the same result for the symmetric

Figure 3. ECDSA Authentication

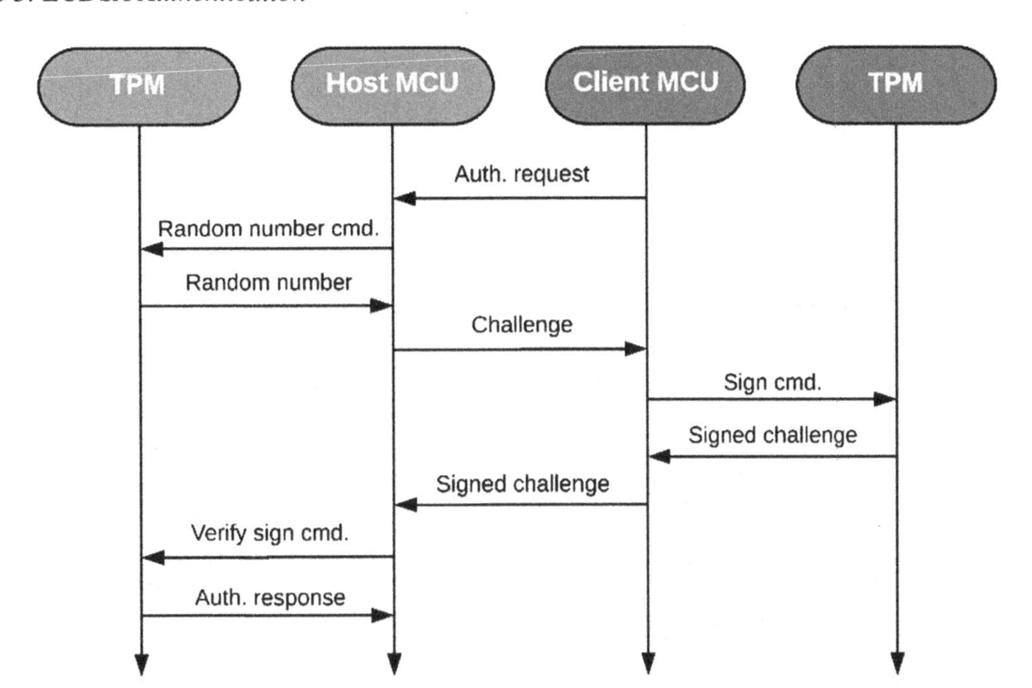

key, even if no extra information is shared. As an extra step for high security, the generated key is hashed with a Key Derivation Function (KDF) supported by the TPM (see Figure 4). The final shared secret can now be read from the memory of the TPM or, as a better alternative, let the chip to secure the key within its isolated environment.

Symmetric cyphering is performed with the usual AES algorithm, with a key size of 128 bits. The TPM only supports acceleration for cyphering plaintext blocks of 16 bytes, which means that external software is needed to implement a useful cypher for larger amounts of data. AES-CBC and AES-CTR modes are implemented in the microcontroller by ensuring coordination with the internal AES engine of the TPM. Also, AES-GCM mode is available, through the Galois field multiplier provided by the TPM. An authentication code may be appended to the resulting cypher texts by hashing, in order to check for errors at the arrival of the message. Those authentication codes are included within the cypher text in the case of AES-GCM, improving the overall performance of the symmetric scheme.

3.3 Fog

The fog defined in the secure scenario for IoT Networks consists in the association of networks of CMW.

In the following table (Table 3), the protocols used to communicate each CMW are presented. This protocol distribution is based on the OSI layer model

Fog Architecture

Fog's architecture is composed by enhanced CMW forming a decentralized network of CMW. The fog layer works between Edge layer and Cloud layer bringing all cloud services near to the edge layer.

Figure 4. Authentication validation

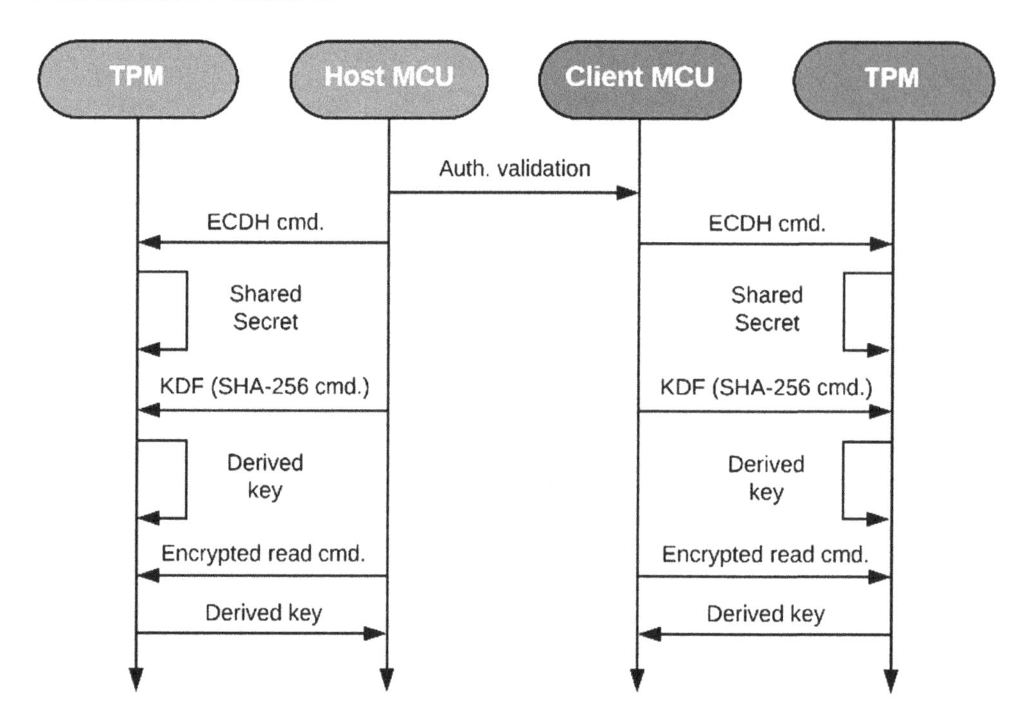

Table 3. OSI Layer Model Protocol Distribution

	WSN Coordinator to the Network Device connection
Physical and Data Link	802.11.b/g/n (WIFI)
Network	IP (with DCHP + DNS)
Transport	TCP + TLS
Session	LDAP
Presentation	JSON
Application	AMQP/TLS

The version used for the proposed architecture has been designed according to the secure requirements in the railway environment.

Fog architecture designed by Indra Sistemas S.A. provides solution to a complex multipurpose environment as is the Railway Industrial IoT scenario. This architecture is a modular architecture which meet with the requirements of a fog architecture as fast data distribution, secure, orchestration and management of resources and services all over the CMW network.

Also, this architecture enables a peer-to-peer data sharing with no single point of failure and scales the edge to cloud achieving an interconnected Railway system.

Fog Computing and Data Distribution

The Fog is the key element of all this architecture due to it joins data collection in the edge and cloud services. Accordingly, CMW must ensure data high availability and top performance.

The fog layer acts with the goal of orchestrate, distribute and validate the data supply by the edge and optimize data latency and share the workload with Cloud layer in order to maximize the efficiency of data processing and distribution.

All the compositions of WSN coordinators – CMW forms different modules at low layer in the global architecture making possible a distributed fog computing by all the CMW. This means that each CMW individually process the data collected by its related WSN Coordinators distributing the workload over all the CMW network achieving scalability and efficient data distribution to the system, making it more reliable and faster, reducing the latency, the lag and bandwidth.

Thanks to this system-level distribution by CMW network it is possible to distribute the cloud security services to each zone of the edge. CMW is in charge of process the security metadata of each edge module.

In this way, always exist an active communication between the edge and the cloud which the CMW network, distributed all over the system, takes the responsibility of each module security, being able to take decisions over its related zones.

By doing so, the fog layer is able to release centralized Cloud Security Services' workload increasing the dynamic feature of the system.

The goal of data processing by CMW is to validate the data integrity making use of the function of CRC calculation and timestamp, discarding messages that does not follow the specific payload structure and messages out of time. Moreover, the use of CRC validation in the header avoids extra message processing by detecting corrupted message without analysing the message content, optimizing processing data, reducing processing time and prevent attacks committed by replicating existing messages.

In results, the decentralized architecture and data processing enables a real time V2V, V2I and I2I communication for new rail services based on the scenario proposed.

Fog Securing

As was explained in the previous subchapter, CMW is the key element of the system. This means that CMW ensures the security at point to point communications between the devices which share safety data.

Security in the fog is managed by the CMW and has three parts:

Authentication: Each connection request from any edge device with each CMW needs to be authenticated with X.509 Public Key Infrastructure (PKI) with a preconfigured shared Certificate Authority (CA) and Elliptic Curve DSA (ECDSA) for authentication providing a way to verify that the peer can be trusted to mitigate against Man in the middle attacks. All the certificates created are based on the X.509 PKI standard and identified by a common name. Moreover, user existence and related password are checked against a LDAP server located at each CMW.

Access Control: The Access Control is configured by domain using a shared certificate signed by CA. When a client establishes a connection with the CMW, it specifies some access parameters. These access parameters are checked against the LDAP server.

The first level of access control is complete when the security handshake between the WSN devices and CMW is done. Once this happens the user and password and other specific user parameters are checked against LDAP.

Data Integrity: Thanks to the TLS implementation, the P2P communications between most elements in the network is encrypted using AES128-GCM and AES256-GCM for encryption with protected key distribution. Also, the message reception is protected by CRCs. In the case of bad CRC the message is discarded preventing either malicious or erroneous data further inserting itself into the system.

Due to decentralized architecture of the fog, the security services are performed/served by each CMW distributing the security over all the fog network bringing close to the edge the security services.

This decentralized security control of the edge facilitates the detection of possible attacks.

3.4 Cloud

The Cloud infrastructure consist of two components that form a hybrid solution. A public Cloud component that enables the public Cloud storage, security, and third party entities integration capabilities, besides the privacy and the confidentiality that the rail services require. This architecture permits centralizing the security, safety, network, and DevOps services alongside ensuring the public and the private part. Moreover, the hybrid Cloud structure permits the platform scalability, interoperability and an easy deployment of the implementation.

Cloud Architecture

The present subchapter explains the distributed and decentralized Cloud environment implementation for the Rail IoT systems.

The architecture consists of a hybrid cloud distributed in the form of N-matrix. In this matrix N, different layers are differentiated from each other. These layers decompose from what is called the central point of the cloud to the physical layers of On-track and On-board, as indicated in Figure 1.

The cloud solution presented is classified as hybrid because it is composed of different types of cloud. In this case, it is composed of a private cloud and a public cloud. This distribution provides the following advantages: control and management of resources, scalability, flexibility or interoperability.

Thanks to N-matrix structure implementation, a decentralized structure is achieved, where the different lower layers or zones are completely independent of each other and, at the same time, are included in a bigger one. By having layers or zones independent of each other, it is possible to put the focus in the desired place and be able to detect and correct incidents that may occur in a more efficient way. In this way, areas that are alien to that incidence are not affected.

Cloud Computing

This Cloud solution provides common services to the different elements that composed the system. This entity is based on a Data Distribution Service (DDS), which balances the computing load between the same layer elements allowing that the system has a lower latency and manages the messaging in a more efficient way.

Thanks to the processing of the edge data by the fog, this layer has more processing capacity to provide other services. In this way, the cloud layer is able to distributed the information (messages, users and configuration files) overall the system, maximizing performance and ensuring the high availability of information.

The high availability of information allows the use of the enablers developed in the ENACT project to provide better quality to the system.

Cloud Securing

The security in the cloud part of the system presented in this solution maintains and replicate the same mechanisms as those mentioned in the Edge and Fog Securing sections. In addition to the security already implemented in the edge and the fog previously explained, different capabilities are added to provide extra security to the system.

Thanks to the Security and Privacy Monitoring and Control tool (ENACT, 2018), it is possible to detect anomalies in the system. By having traffic monitored and a prediction of how this should behave, the non-expected parameters can be detected, identify possible errors and even external attacks over the system. The prediction of how the communications system has to respond is derived from the history that data persistence provides, previously mentioned.

In this context, the centralized user authentication system (LDAP) is responsible for revoking the user and isolating the area in which that anomaly or deviation was found in the expected traffic, if necessary.

Adding to LDAP, NTP and DNS services are also centralized in the cloud.

The NTP is currently used as a reliable method of synchronizing timing measurements and timestamps in the messaging system. It is a hierarchical system divided in "strata" or levels of dependency, where the first strata in the system currently depends on the Global level until the Local layer. A message without the time synchronized with the rest of the system will be ignored and reported.

DNS Service implements and control the domain name resolution service of the entire system to allow working with interchangeable names (FQDN) instead of static networks IPs. This means that a device has to be registered in the domain to communicate with the other devices.

4. SOLUTION AND RECOMMENDATIONS

All the security mechanisms implemented in lower levels are vertically and horizontally maintained when the messages are inserted and transmitted along the distributed network, guaranteed that the messages can be adequately sent to the end users or the monitoring systems.

As mentioned in the main focus of the chapter, interoperability is the main key in a Distribution Data Service in order to get a top performance and a robust, scalable architecture. With the inclusion of strong security features it is possible to guarantee a Secure Connected and Trustable Internet of Things solution concerning a complete Secure Platform.

The Central Cloud defined in the main chapter has to support the edge services relating authentication, timestamps measurements (using NTP) or data integrity among others. One of the main disruptive solutions is the integration core component between both the Central Cloud and edge, the Communication Middleware. This Communication Middleware introduces the security mechanisms that enables spreading the system authority along the system in a secure manner. Moreover, DevOps can be used to provide useful tools to manage the behaviour of different components of the network through enablers, which improve several functionalities involved in a railway environment, making a special emphasis on the security aspects.

These enablers, among others that can be deployed, are developed in the central cloud and provide robustness to the rail solution by implementing centralized services which collects metadata relating the rail infrastructure On-Track and On-Board (latency, bandwidth). By collecting this metadata, it is possible to build patterns which can be used to monitor the performance and actuation of the different elements of the system. This patterns are used to analyse anomalous variations in the use of communication channels in real-time, identifying the source of this variations and prevent possible attacks by user revocation. This provides more robustness and security to the solution.

In addition to these developments, the inclusion of Artificial Intelligence in the monitoring process in the cloud make possible that the patterns are continually adapting to the data gathered, processed and received.

5. FUTURE RESEARCH DIRECTIONS

This section is intended to clarify the different future actions and tendencies that are planned into the Security in Rail IoT system's roadmap that it is aligned with the shifts that all the IoT systems will follow:

- **5G technology:** With the arrival of 5G Technologies, the IoT will develop a high potential in several market segments. 5G will provide low latency, high speed and data transmission in real-time capacity, which empowers the IoT Networks to provide more efficient services and will a high interconnectivity. Rail market has to consider the irruption of 5G as a competitive advantage for the enhancing of current state of innovation projects and researches.
- **Artificial Intelligence (AI):** Several new projects objective is the IoT and AI integration into a multi-domain environment making use of a distributed intelligent processing. This project is intended to provide the railway market of safe and secure communications to develop smart rail services based on Metropolitan Area Networks (MANET), being able to broadcast relevant information to different types of vehicles and infrastructures existing in the environment. This use case hides a huge improvement in the explained cloud and fog services as it permits spreading the control and supervision to the entire infrastructure including the edge.

The former actuation lines combined, into the IoT environment, permit dealing with the possibility to integrate rail segment into a smart city context, developed into the InSecTT Project (InSecTT Project, 2020-2023) framework, which connects all to all using a distributed architecture in a multi-domain environment thanks to the 5G X2X interfaces. These interfaces establish communications both among layers and into each layer.

Moving the control to edge using the 5G technology permits, in a safety and secure manner, take rail decision into the train itself. This decision would be supervised by the centralized cloud infrastructure and the other trains. Therefore, the decisions would be coordinated and coherent along the rail infrastructure supporting the new emerging autonomous and intelligent rail systems. A complete example about this improvements is the possibility to give to the trains the signalling management in a local level, enabling the interaction between the trains and the wayside objects (WO).

6. CONCLUSION

The modernization of the railway market demands the use of new technologies to provide new rail services to be developed over a distributed architecture to decentralize the communications. The secure platform presented in the chapter is been demonstrated as an optimal solution to integrate wireless IoT technologies in the rail environment, accomplishing with the security requirements established by the regulations concerning safety-related systems.

As it is shown along the chapter, the decentralized architecture of the Edge and Fog layers allows to isolate each zone from other decreasing the computational charge and the latency and improving the processing and distribution of the data. Then, this data are collected, centralized and managed by the Cloud.

Moreover, the platform is also able to provide security in every layer of the platform, ensuring in every point of the architecture the integrity of the data and even preventing against attacks making use of different capabilities.

In conclusion, the use of new technologies in the railway market which allows the deployment of innovative rail services is possible in a distributed environment, guaranteeing the safety and security and making easier the life of EU citizens.

REFERENCES

Alliance for Internet Things Innovation. (2019). *AIOTI Vision and Recommendations: European IoT challenges and opportunities.* Retrieved December 11, 2019, from https://aioti.eu/aioti-vision-and-recommendations-european-iot-challenges-and-opportunities-2019-2024/

Barkovskis, N., Salmins, A., Parrilla, F., Ozols, K., & Moreno, M. A. (2016). WSN based on accelerometer, GPS and RSSI measurements for train integrity monitoring. *4th International Conference on Control, Decision and Information Technologies (CoDIT).*

DEWI Project. (2014-2017). *Dependable Embedded Wireless Infrastructure.* http://www.dewiproject.eu/

Dierks, T., & Rescorla, E. (2020). *The Transport Layer Security (TLS) Protocol Version 1.3. RFC8446.* Network Working Group.

ENACT. (2018). *D1.1 Use case definition and requirements and validation and evaluation plan* (Grant Agreement N° 780351). ENACT consortium.

ENACT Project. (2018-2021). *Development, Operation, and Quality Assurance of Trustworthy Smart IoT Systems.* https://www.enact-project.eu/

European Committee for Electrotechnical Standardization. (2010). *CENELEC: EN 50159. Railway applications -communication, signalling and processing systems.* European Union: CENELEC.

European Union Agency for Network and Information Security. (2017). *Baseline Security Recommendations for IoT in the context of Critical Information Infrastructures.* ENISA.

InSecTT Project (2020-2023). Intelligent Secure Trustable Things.

ISO/IEC 19464:2014 (2019). Information technology – Advanced Message Queuing Protocol (MQP) v1.0. ISO/IEC Information Technology.

ISO/IEC 20922:2016 (2016). Information technology - Message Queuing Telemetry Transport (MQTT) v3.1.1. ISO/IEC JTC 1 Information Technology.

ISO/IEC 29182-1:2013 (2013). Information technology - Sensor networks: Sensor Network Reference Architecture (SNRA). International Standard ISO/IEC.

Kerry Maletsky. (2015). *RSA vs ECC Comparison for Embedded Systems. Technical Report*. Atmel Corporation.

Lang, J. (2015). *The Elliptic Curve Diffie-Hellman*. ECDH.

Luzuriaga, J. E., Perez, P., Boronat, P., Cano, J. C., Calafate, C., & Manzoni, P. (2015). A comparative evaluation of AMQP and MQTT protocols over unstable and mobile networks. Valencia: Department of Computer Engineering (Universitat Politecnica de Valencia). doi:10.1109/CCNC.2015.7158101

Melnikov, A., & Zeilenga, K. (2018). *Simple Authentication and Security Layer (SASL). RFC4422*. Network Working Group.

SCOTT Project. (2018). *D23.2 Dependable Communication and Safety Building Blocks*. Call H2020-ECSEL-2016-2-IA-two-stage - Innovation Actions (IA). Secure COnnected Trustable Things.

SCOTT Project. (2017-2020). *Secure COnnected Trustable Things*. https://scottproject.eu/

Turner, S., & Polk, T. (2018). *Prohibiting Secure Sockets Layer (SSL) Version 2.0. RFC6176*. Network Working Group.

X2Rail-1. (2016-2020). *Start-up activities for Advanced Signalling and Automation Systems*. https://projects.shift2rail.org/s2r_ip2_n.aspx?p=X2RAIL-1

X2Rail-1. (2016). *D7.1 Analysis of existing lines and economic models*. European Union: Shift2Rail.

X2Rail-1. (2018). D3.1 User & System Requirements (Telecommunications).

X2Rail-1. (2018). D7.2 Railway requirements and Standards application conditions.

KEY TERMS AND DEFINITIONS

AI (Artificial Intelligence): It is a branch of the computed science focus on the designs and implementation of smart machines oriented to solve issues that are typically required human intelligence.

AMQP (Advanced Message Queuing Protocol): Open standard application layer protocol with message orientation, queuing, routing and security as main features.

Cloud Computing: Availability to provide several services, especially data storage, without direct user management.

Communication Middleware: Part of the system in charge of processing the services and works as the core for the integration of all the different systems.

Edge Device: Intelligent objects in an IoT system which control the data flow between two networks.

Fog Computing: Network architecture that uses edge devices to extend different services from the outer edges where data is created to the cloud where the data is located and accessible for different users.

IoT (Internet of Things): A system of interconnected devices with the capacity of establish communication along a network without needing human interaction.

MQTT (Message Queue Telemetry Transport): Open protocol server/client based on publish-subscribe network protocol which transport messages between smart sensors and devices.

Chapter 14
The Internet of Things in the Russian Federation:
Integrated Security

Anna Zharova
Higher School of Economics University, Russia & Institute of State and Law RAS, Russia

Vladimir M. Elin
National University of Oil and Gas "Gubkin University", Russia

ABSTRACT

The chapter presents a study on ensuring the information security of the Russian Federation in the field of the internet of things (IoT), an analysis of Russian state policy in this field and methods for its implementation in terms of technical and legal regulation, and the directions of state development in field creating a system of regulation of relations in the field of IoT. To present the general picture of the state's opposition to the risks and threats arising from the use of the IoT, a comparison is made of the risks and threats used by the FSTEC of Russia and ENISA. The authors disclose Russian approaches to ensuring information security, reflected in state strategic documents, including the strategy adopted in 2015 in the field of ensuring information security by switching to their own information technologies. In conclusion, recommendations are made for government bodies and users to ensure integrated information security.

INTRODUCTION

Over the past four years, Russia has adopted strategic programs to develop fields related to information technology (IT). These include artificial intelligence (AI), strategic computer technologies, information and communication technologies (ICTs) and systems, narrow-band wireless networks of the Internet of things (IoT), and the digital economy ecosystem. AI is used to identify criminals and sources of danger, risks, and threats associated with the IoT, as well as ensure information security. Since 2015, the Russian Federation has ensured its information security through a course on the import of substitution software and IT used by state structures. In addition, commercial organizations have been working in the field of critical information infrastructure since 2019.

DOI: 10.4018/978-1-7998-5068-7.ch014

When implementing these tasks, difficulties arise regarding the development of IT, Russian standards, and the regulatory framework for information security. Difficulties also impact user identification and authentication of IoT devices.

As a result of the development and use of ICTs, the number of connected devices on the Internet (and those entities that use such devices) are increasing. This creates a problem related to ensuring information security.

There is an exponential growth of information available on the Internet that is of interest to third parties. According to the Kaspersky Security Network (KSN):

Kaspersky Lab solutions repelled 843,096,461 attacks that were carried out from Internet resources hosted in 203 countries of the world. 113,640,221 unique URLs to which the web antivirus worked were recorded. Attempts to launch malware to steal money through online access to bank accounts are reflected on the computers of 243,604 users. Encryptors attacks are reflected on the computers of 284,489 unique users. File antivirus detected 247,907,593 unique malicious and potentially unwanted objects, etc. (Kaspersky, 2019)

A larger number of connected devices makes it more difficult to ensure integrity, accessibility, and confidentiality. First, devices record and process restricted information, including personal data. Second, the Russian Federation does not have requirements for mandatory certification of Internet devices. This creates several systemic problems at various levels of the network architecture. For example, most gadgets collect and store personal information in their clouds or on servers (Zharova et al., 2017). Therefore, there is concern related to information security of storage on remote devices and information systems. Thus, questions of placement, storage rules, IT interaction, and information security need careful legal and technical regulations.

In the Russian Federation, the levels of legal and technical regulations are independent. In fact, in most situations, they do not interact with each other. In this work, the authors aim to prove that ICT security is possible through close interaction of these levels. The authors demonstrate this idea using Russian information security solutions.

The following section provides an analysis of Russian state policy in the field of IT security. It studies implementation methods in terms of technical and legal regulations. The authors offer approaches to the problems of ensuring IoT security by generalizing the problems presented in technical research and offering solutions through legal regulation. They developed practical proposals for the use of regulations of the FSTEC of the Russian Federation together with the ENISA recommendations for the providing the Integrated security of the Internet of things. The study discussed the problem of eliminating technologies with preinstalled malware by developing a system of standards that include integrated security methods. In conclusion, authors substantiate the need for the interaction of legal and technical regulation and analyze the Russian state policy in ensuring the security of IT and methods for implementation from the point of view of technical and legal regulation.

IOT IN THE RUSSIAN FEDERATION

Prerequisites for Ensuring Information Security

In 2016, business intelligence (BI) analysts predicted that the global IoT market would significantly outperform mobile electronics, portable equipment, and personal computers by 2019 (The 'Internet of Things', 2015). By 2022, this market would reach $561.04 billion (CNews conducts, 2019). In the Russian Federation, smart technologies are actively being introduced into city management, citizens will be able to use smart devices as part of the DaaS approach. Since 2019 a concept for identifying the user Internet of things is being developed [the Ministry of Communications approved the concept]. The main areas of the Russian Federation's IoT market involved the development of AI, digital twins, cloud technologies, and blockchain (IoT-2019, 2019). The Government of the Russian Federation associates an increase in the intelligence of end devices with the achievement of indicators like prevention of crime and epidemics, improved governance, and security of the state. To realize such opportunities, methodological recommendations were developed in 2017 regarding the implementation and use of industrial IoT to optimize control and supervisory activities (approved by the Protocol of the meeting of the project committee of 09.11.2017 No 73 (13) (Guidelines, 2017).

The government planned to obtain the first results at the end of 2019. However, problems in the field of ensuring information security halted the implementation of the above tasks.

INFORMATION SECURITY REGULATION

Information Security System of IoT by Means of Standardization

In the 2030 forecast of scientific and technological development of the Russian Federation, IoT (or the semantic Web) is considered a stage in the development of the Internet. In this stage, all devices or "things" have access to the Internet (The Forecast of development until 2030, 2014).

There is no independent legal act aimed at regulating relations that arise from the use of IoT in the Russian Federation. Information security of the IoT is regulated by separate legal norms related to Internet relations. In addition, Russian legislation does not systematically regulate information security. However, over the past four years, this area of legal and technical regulation has been developing.

Progress related to the development of the information society depends on the level of scientific and technological development, as well as the pace of IT development. This generates a fragmentation of the economy in the areas of developing technologies and uneven distribution. Such fragmentation includes problems related to the availability of standards for ensuring the interaction of technologies or an insufficient number. For an example, refer to Qin et al (2019).

However, the lack of standards may not be associated with the speed of technology development. Instead, it relates to the lack of strategic state acts aimed at setting development tasks for state structures. This includes the creation of their system of technical regulation standards and development of technology development forecasts.

The authors agree that research in the field of IoT focuses on application prospects and security (Khan & Salah, 2018), standardization of IoT-device protocols (Banafa, 2016), and interoperability as a single technology platform (Sachi et al., 2016).

The lack of reliable standards is relevant for both the Russian Federation and other developed countries. For example, in European states, problems with their choice were revealed in the context of diversity and absence of general agreements between states on the application of a standard (Bauer et al., 2013; Goertzel, 2016). These problems are noted by most experts in the technical field. A survey conducted by IT security firm, Trend Micro, indicated that more than half of IT decision makers stated that standardization was a key requirement for IoT projects (Significant lack, 2016). The connection of the physical and virtual world has led to the creation of a new world: cyber-physical systems. This new world consists of heterogeneous devices, the number of which is growing exponentially. The amount of collected information, such as user information, is also growing.

Understanding the complexity of the standardization problem and diversity of participants in this process, the Government of the Russian Federation approved directions on October 29, 2019, to implement the Information Security federal project within the national Digital Economy of the Russian Federation program (The Resolution of the Government, 2019). The following directions were identified:

- Development of requirements for operators of the industrial Internet and draft security standards for cyberphysical systems, including IoT devices
- Creation of a system of industry regulations regarding the use of cyberphysical systems (like IoT devices), the establishment of requirements for the identification of participants in information interaction, and the registration of equipment networks of IoT devices
- Development and adoption of a set of information security standards to minimize risks and threats to the safe functioning of public communication networks
- Analysis of existing and promising information security tools
- Development of a model of information security threats for personal devices to collect biometric data and develop a roadmap for ensuring information security when citizens of the Russian Federation use this class of technical equipment
- Creation of information security incident processing technology, using AI to increase the level of automation of decision-making processes and reduce incident response times
- Measures to develop the domestic broadcasting infrastructure and ensure the safety of its functioning
- Information and reference system that allows businesses to determine their compliance with the requirements of Russian and international legislation and industry in the development of national and international standards in the field of information security

Moreover, to solve the problem of information security of remote or cloud management, the Government of the Russian Federation approved the creation of a state-unified cloud platform. This platform defines a system of legal mechanisms that ensures the effective transition of information systems of public authorities and local governments to the ecosystem of accredited services and information and telecommunications infrastructure providers (Order of the Government, 2019).

While researching problems associated with information security, the authors concluded that it is incorrect to subscribe to the concept of separate regulation of Internet relations at legal and technical levels (Zharova, 2020). For others, the "classical branches of law" approach was acceptable. However, the authors found that breaking levels of regulation leads to inefficiency in information security.

Information security problems will not be solved through the availability of standards, their diversity, or the developed and approved methodology for their application. A clear state strategy is needed to reduce threats and security risks. On October 14, 2019, the Ministry of Digital Development, Telecommunications, and Mass Media of the Russian Federation created a roadmap for the development of end-to-end digital technology (or wireless technology) as a strategic public policy instrument to identify priorities and prospects for the development of Russian Federation technologies by 2024. Also, in 2019, the Federal Service for Supervision of Communications, Information Technologies, and Mass Communications (Roskomnadzor) identified goals for its annual activities. These included timely responses and relevant measures to respond to challenges and threats associated with active technological changes, as well as the widespread dissemination of IoT, big data processing systems, AI, and next generation communications networks (Public declaration, 2019).

To understand the processes of information security in the Russian Federation, it should be noted that it is not mandatory to have either the application of requirements of technical regulation to information security or the compliance of information technologies with information security requirements. Therefore, the Digital Economy of the Russian Federation national program identified several tasks to develop and adopt a national system of standards in this area (Law “On Technical Regulation”, 2002).

Requirements for the organizational and technical support of the stable functioning of the public communications network were approved by order No. 113of the Ministry of Information Technologies and Communications of the Russian Federation in September 27 (2007). This order defined organizational and technical support for the stable functioning of the public communications network as a set of requirements and measures aimed at maintaining integrity and stability through a communications network and technical standards. The functional and physical compatibility of communications included user (terminal) equipment, as well as the unity of measurements in a communications network. The trust

Table 1. Activities and timing of implementation

Events	Purpose	Deadline
Development and adoption of a pool of information security standards	Minimize risks and threats to the safe operation of public communication networks	December 31, 2021
Creation and functioning of mechanisms for information and analytical support and coordination of participation of Russian experts in the activities of major international organizations	Develop standards in the field of cryptography and IT security	December 31, 2021
Development, adoption, harmonization, and implementation of information security standards	Provide information security in systems that implement cloud, fog, or quantum technologies, virtual and augmented reality systems, and AI technologies Create national standards for big data processing Draft security standards for cyberphysical systems	December 31, 2021
Development of requirements for operators of the industrial Internet	Develop requirements and methods for verification of cyberphysical systems, including the IoT	December 31, 2021
Analysis and assessment of potential vulnerabilities, threats, and information security risks related to existing information security standards in systems implemented on cloud, fog, quantum, virtual technologies, AI technologies, augmented reality technologies, and distributed registry technologies	Define the list of necessary standards and resource support	December 31, 2021

requirements for operating security (OS) are formed based on the requirements components from the national standard of the Russian Federation GOST R ISO/IEC 15408-3-2013 (2013).

The Digital Economy of the Russian Federation program defined the task of developing requirements for ensuring control of processing and access to personal, big user data by the end of 2021. This included social networks and other means of social communication, as well as legal requirements to revoke or reduce the amount of consent to personal data processing (for example, requirements for industrial Internet operators and security standards for cyberphysical systems or IoT devices) (The National Program, 2019).

The list of activities and the timing of their implementation are presented in Table 1.

Thus far, the Russian approach to building a system of information security standards has focused on adapting International Organization for Standardization (ISO) standards related to national conditions. At the same time, the Federal Service for Technical and Export Control (FSTEC) of Russia performs special and controlled functions in the field of state security on the issue of information security in information and telecommunications infrastructure systems (Decree of the President, 2004). To solve these problems, the FSTEC issues normative legal acts on information security issues, as well as develops and approves methodological documents within its competence.

It is necessary to correlate the regulatory practice in Russia with international standards of technical regulation in the field of cybersecurity. The results of the analysis are presented in Table 2.

To ensure information security, the Russian Federation uses a methodology that supports the application of industry standards and creates a new control body in the field of certification of cryptographic standards (the national certification center). Thus, to ensure the stability of the interaction of devices in the Russian segment of the Internet by the end of 2021, a national certification center should be put into operation. This center should be equipped with the necessary equipment, ensuring the creation of certificates using Russian cryptographic standards.

Problems related to authentication and trust will be solved by developing industry standards and a trust chain of authentication based on the public key infrastructure and access control. This will also ensure interoperability.

A strategic state approach to ensuring information security of IT with a complex application of legal, organizational, and technical impact mechanisms has already been developed. Interaction has also been established between state authorities and management, noting a clear distribution of social responsibility and the powers of participants.

A combination of activities (for example, the development of industry standards), a trusted authentication chain based on public key infrastructure and access control, and interoperability will allow us to approach solutions related to authentication and trust problems (Decree of the Government, 2018).

The following strategic goals of Russia's 2020 development include:

- Creation of a system of industry regulation on the use of cyberphysical systems, including IoT devices
- Establishment of requirements for the identification of participants in information interaction, as well as registration of equipment of IoT device networks
- Development of evaluation criteria for the compliance of operating systems with domestic or trusted systems, including domestic processor platforms
- Development and approval of a regulatory legal act on the identification of users of communication and other services that ensure the interaction of participants in these services, as well as the identification of the user's IoT

Table 2. Compliance of the provisions of international standards in the field of cybersecurity with the norms of the Russian Federation

International Standards	Russian Approach
ISO/IEC 27000	
The ISO/IEC 27000-series comprises information security standards published jointly by the ISO and the International Electrotechnical Commission (IEC).	The series of standards is applied on a voluntary basis in areas where there is no state regulation (i.e., personal data, banking, critical information infrastructure)
Overview and Vocabulary	
ISO/IEC 27000:20016. This standard provides an overview of information security management systems (ISMS). These form the subject of the ISMS family of standards, as well as defines related terms. As a result of implementing ISO/IEC 27000:2016, all types of organization (e.g., commercial enterprises, government agencies, and nonprofit organizations) are expected to obtain the following: (1) overview of the ISMS family of standards; (2) introduction to ISMS; (3) brief description of the Plan-Do-Check-Act (PDCA) process; and (4) understanding of terms and definitions in use throughout the ISMS family of standards.	Concepts like confidentiality and cybersecurity do not correspond to the definitions of Federal Law No. 149-FZ, "On information, information technologies, and information protection," and Federal Law No. 187-FZ, "On the security of critical information infrastructure of the Russian Federation."
ISMS Requirements	
ISO/IEC 27001:2016. This specifies the requirements for establishing, implementing, operating, monitoring, reviewing, maintaining, and improving a documented ISMS within the context of the organization's overall business risks. It specifies requirements for the implementation of security controls customized to the needs of individual organizations or parts thereof.	ISO/IEC 27001:2016. This can be correlated with the approach defined in the rules for categorizing critical information infrastructure objects of the Russian Federation, as well as in the list of indicators of the criteria for the significance of critical information infrastructure objects of the Russian Federation and their values. This was approved by the Government of the Russian Federation on February 8, 2018 (No. 127).
Code of Practice for Information Security Management	
ISO/IEC 27002:2013. This establishes guidelines and general principles for initiating, implementing, maintaining, and improving information security management in an organization. The objectives outlined provide general guidance on the commonly accepted goals of information security management. ISO/IEC 27002:2013 contains good practices of control objectives and controls in the following areas of information security management: security policy; organization of information security; asset management; human resources security; and physical and environmental security.	The methodology for determining current threats to information security in key information infrastructure systems was approved by the FSTEC on May 18, 2007. It defines the practice of information security in key information infrastructure systems of the Russian Federation.
ISMS Implementation Guidance	
ISO/IEC 27003:2017. This standard focuses on critical aspects needed for the successful design and implementation of an ISMS in accordance with ISO/IEC 27001:2016. It describes the process of ISMS specification and design from inception to the production of implementation plans. It describes the process of obtaining management approval to implement an ISMS, defines a project to implement an ISMS (referred to in ISO/IEC 27003:2017 as the ISMS project), and provides guidance on how to plan the ISMS project. This provides a final ISMS project implementation plan.	The basic model of information security threats in key information infrastructure systems was approved by the FSTEC on May 18, 2007. It is used for modeling information system security threats.
Information Security Management Guidelines for a Telecommunications Organization	
ISO/IEC 27011:2016 based on ISO/IEC 27002:2013. The scope of this Recommendation I International Standard is to define guidelines supporting the implementation of information security management in telecommunications organizations. The adoption of this Recommendation I International Standard will allow telecommunications organizations to meet baseline information security management requirements of confidentiality, integrity, availability, and any other relevant security property.	GOST R 52448-2005 Protection of information. This ensures the security of telecommunication networks. Generalities GOST R 53109-2008 System of Information Security of Communication Network of General Use. Passport of the organization of communication on information security. Generalities

There are several approaches to ensuring information security of IoT. Some authors summarize their experiences of information security of IoT (Simon et al., 2015), noting that information threats include cybercrime (Hugh, 2020), cybercrime economy associated with the credentials of hacked servers, theft of money through the SWIFT interbank network, critical information infrastructure, targeted attacks, theft of personal data databases, transfer of information by smart devices, and the inefficiency of cryptographic protection against internal attacks (Venus et al., 2019).

The authors note that most of the IoT research in the technical field is devoted to building models to prevent IT vulnerabilities. They offer and justify various options (or integrated solutions) for data protection in IoT (Ming et al., 2018).

Some authors believe that IoT solutions should be built using industry-wide open standards. They suggest using the IEEE 802.1 AR standard for device identification, which "supports cryptographic binding of the secure device identifier (DevID) to the device and allows authentication of this device" (Luca et al., 2019).

While researching the problems of information security, the authors concluded that the concept of separate regulation of Internet relations at the legal and technical levels is incorrect. For "classical branches of law," a separate approach is acceptable. However, when considering problems related to the development of the ICT industry and information security, the gap between levels of regulation leads to inefficiency of information security.

The authors offer approaches to the problems of ensuring IoT security by generalizing the problems presented in technical research and offering solutions through legal regulation.

The system of information security standards in the Russian Federation is still in formation. Therefore, the authors turned to the European experience to analyze the existing practice. The European and Russian approach to cybersecurity is very similar. Russia does not have its own systematic threats to IoT. Therefore, the authors use threat sources provided by the European Union Agency for Network and Information Security (ENISA). In Good Practices for Security of IoT in the context of Cement Manufacturing, an analysis studied existing European practices for ensuring cybersecurity of the IoT (Good Practices, 2018).

Using Good Practices for Security of IoT in the Context of Smart Manufacturing, it can be concluded that the main threats to information security of IoT include: nefarious activity; abuse; unfair activity; manipulation through data and devices; eavesdropping; interception; hijacking (i.e., listening, interception, hacking); information collection and hacking; unintentional or accidental damages; errors in configuration/administration/application; outages associated with the loss of power/communications/services; disasters or destructive external effects of natural and manmade nature; physical attack (i.e., theft, vandalism, and sabotage) directly on equipment; failures or malfunctions (i.e., accidental hardware failures, failures of the provider's services, problems in software development, introduction of vulnerabilities); and legal issues or deviations from the requirements of laws and contracts.

Using these threats, the following matrices can be built: "threats" and "threat sources"; "threats" and "attacks"; "attacks" and "vulnerabilities"; etc.

In Russia, the question of the source of threats to information security is determined by the basic model of threats to the security of personal data when processed in the information systems of personal data UTV. FSTEC of the Russian Federation is classified as follows: internal intruder; external intruder; built-in hardware bookmark; autonomous hardware bookmark; software bookmark, a program like Trojan horse; software virus; malware that spreads over the network; malicious programs (i.e., password matching, remote access, etc.) ((FSTEC of Russia, 2013).

The authors present a matrix of related "threat" and "sources of threats." When constructing the matrix, they applied the concept of "is the cause," which established a binary relation. This is presented in Table 3. In this case, value "1" should be defined as "actual" reason; "0" is "irrelevant" reason.

Table 3. Relationship matrix of "sources of threats" and "threat"

Sources of threats Threat	Insider	External intruder	Built-in hardware bookmark	Standalone hardware bookmark	Program bookmark	Software virus	Malware spreading over the network	Other malicious programs (password matching, remote access, etc.)
Nefarious activity / Abuse	1	0	1	1	1	1	1	1
Eavesdropping / Interception / Hijacking	0	1	1	1	1	1	1	1
Unintentional damages (accidental)	1	0	0	1	1	1	1	1
Outages	1	1	1	1	1	1	1	1
Disaster	0	0	1	1	1	1	1	1
Physical attack	1	1	0	0	0	0	0	0
Failures / Malfunctions	0	0	0	1	1	1	1	1
Legal	1	1	0	0	0	0	0	0

Based on the results of identifying current relationships, the authors assessed the risk of negative impact for the following risk categories: the risk of causing financial damage (loss of values or increased costs); the risk of causing industrial damage (inability to perform their duties); the risk of causing im-

Table 4. Risk assessment matrix

Damage Threat	Financial damage	Production damage	Image damage
Nefarious activity / Abuse	H	H	H
Eavesdropping / Interception / Hijacking	H	L	M
Unintentional damages (accidental)	M	M	L
Outages	H	H	L
Disaster	H	H	H
Physical attack	M	L	H
Failures / Malfunctions	M	H	L
Legal	M	L	H

age damage (loss of public trust); and the risk of causing image damage (loss of public trust). The risk assessment determines the high (H), medium (M), or low (L) level of negative impact for each threat to information security objects when using IoT technology. The negative impact risk assessment matrix relates damage to a specific threat (see Table 4).

The authors defined a set of organizational and technical measures aimed at countering threats to the IoT and ensuring information security. A set of measures by the FSTEC was proposed to address information security (Order of the FSTEC No 17, 2013):

- Identification and authentication of access subjects and access objects
- Access control of access subjects to access objects
- Limitation of the software environment
- Protection of machine media
- Check the security event
- Antivirus protection
- Intrusion detection (prevention)
- Control (analysis) of data security
- Ensure the integrity of the information system and data
- Ensure the availability of data

Table 5. Matrix of compliance of high-level risks with measures for prevention

High-level threats / Measures for Prevention	Nefarious activity / Abuse	Eavesdropping / Interception / Hijacking	Outages	Disaster	Physical attack	Failures / Malfunctions	Image damage from legal threat
Identification and authentication of access subjects and access objects	1	1	0	0	1	0	0
Access control of access subjects to access objects	1	1	0	0	0	1	1
Limitation of the software environment	1	1	1	1	1	0	0
Protection of machine media	1	1	1	0	0	0	0
Check the security event	1	1	1	1	1	1	1
Antivirus protection	0	1	1	1	1	0	0
Intrusion detection (prevention)	1	1	1	1	1	1	1
Control (analysis) of data security	1	1	1	1	1	1	1
Ensure the integrity of the information system and data	1	1	1	0	0	0	0
Ensure the availability of data	1	0	0	0	0	1	1
Protect the virtualization environment	1	1	1	1	1	1	1
Protect technical means	1	1	1	1	1	1	0
Protect the information system, its means, communication, and data transmission systems	1	1	1	1	1	1	1
Identify incidents	1	1	1	1	1	1	1
Configuration management of the information system and data protection system	1	1	1	1	1	1	0

- Protect the virtualization environment
- Protect technical means
- Protect the information system, its means, communication, and data transmission systems
- Identify incidents
- Configuration management of the information system and data protection system

To identify measures to ensure information security of IoT technology, it is necessary to create a matrix of correspondence between high-level threats and measures for prevention (see Table 5).

Similarly, various correspondence matrices are constructed for medium- and low-level threats, as well as for relations of "threat" and "attack," "attacks" and "vulnerabilities," etc. Based on the matrices, it is possible to determine a list of measures to ensure the information security of IoT.

Information Security of IOT by Means of Transition to IT

A significant threat to the security of information systems is the use of technologies with preinstalled harmful software implemented at the level of software development or harmful bookmarks implemented at the level of hardware development. This problem is relevant for all states. However, the Russian Federation has its own specifics because most of their incoming IT belongs to foreign developers. This means that Russian users depend on foreign developers when developing standards and methods to ensure information security and architecture resistance to attacks impacting access control, stability, data integrity, authentication, and confidentiality.

IoT systems produce and process a large amount of data and information. This includes information related to restricted access, which can be used to establish a person's identity. Therefore, controlling the development process of IT in IoT should be of paramount importance for the state.

The functionalities of IoT are implemented in the process of developing services and applications. This is confirmed by the methodological document developed by the FSTEC of Russia for information protection measures. In this document, information security threats are associated with establishing the possibility of violating the confidentiality, integrity, and availability of information in the information system. In addition, it violates at least one of the security properties, acts as an onset of unacceptable negative consequences or damage to the owner of the information or operator, and breaches the confidentiality of personal data. Definitions of sources of information security threats are determining factors in identifying information security threats in information systems.

Such a source of threats results in an inability to control the development of IT in Russia. In fact, information security is a component of all state security (Information protection, 2014). It cannot be ensured without creating its own IT. Switching to domestic software and protecting information infrastructure were identified in two federal projects of the Digital Economy of the Russian Federation national program (Digital Economy, 2019).

The state policy of transitioning to Russian IT has been implemented since 2015. Thus, by Order No. 96 of the Ministry of Communications of Russia on April 1, 2015 (2015), a plan was approved to import substitution of software. To ensure this task, an autonomous nonprofit organization must be created to organize collective software development for segments with a high level of dependence on foreign software (GOST R 54593,2011). However, in 2015, the Ministry of Economic Development of Russia issued a negative opinion on the project of the Ministry of Communications of Russia in this field.

Despite this, Order No. 96 has not been canceled. Based on the Order, the organization of collective software development for segments with a high level of dependence on foreign software is ensured by a nonprofit organization tasked with giving grants to developers in the creation of IT.

In addition, a Decree of the Government of the Russian Federation on November 16, 2015 No. 1236 (2015) banned the use of software originating from foreign states for the procurement of state and municipal needs. The purchase of foreign software is allowed only if there is justification for the impossibility of observing the prohibition. This is confirmed by judicial practice (Determination of the Supreme Court, 2018). Criteria for software classes were defined to determine whether the software is Russian or foreign (Ministry of Communications, 2015).

Legal entities guilty of violating the laws of the Russian Federation and other regulatory legal acts on the contract system in the procurement are brought to administrative responsibility on Art. 107 on the contract system.

Cases regarding the cancellation of the purchase of the operating system, which is part of the set of goods purchased by the customer, falls under the ban on the admission of foreign software. This was established by Government Decision No. 1236 (2015).

Restrictions on the use of foreign IT are also provided for relations with the participation of state organizations and legal entities of the critical information infrastructure sectors. Thus, on October 1, 2019, the Government of the Russian Federation submitted proposals to ensure technological independence and security of critical information infrastructure through predominantly domestic software. Art. 12.1 of the Federal Law, "On Information, Information Technologies, and the Protection of Information," defines the specifics of state regulation in the field of Russian software and databases. It also defines the procedure for creating a unified register of Russian software and databases.

The Government of the Russian Federation determined that standards for processing large data arrays, information security standards in systems implementing cloud, fog, quantum technologies, virtual and augmented reality systems, and AI should be developed, adopted, harmonized, and implemented by the end of 2019. In addition, new interstate standards and amendments to existing standards in the field of information security for the Eurasian Economic Union (EAEU, 2019) countries should be approved.

A successful example of the implementation of these tasks is the Russian software titled "Unified Geodata Information Space." Created in 2017, it provides a unified organization of data storage, information processing principles, and high-level information security. In 2018, Rostelecom developed the RusGIS software, a comprehensive infrastructure solution and geoinformation support tool for making managerial decisions regarding a thematic focus, scale, and level from municipal and regional departments to federal executive bodies (Concept of the development, 2019).

SOLUTIONS AND RECOMMENDATIONS

The growing volume of vulnerabilities and threats is associated with an increase in IT use, a lack of regulatory standards, and the emergence of attack tactics and criminal actions. These problems were investigated, regulatory uncertainties were identified, and directions of state development to create a system of regulation of relations in the field of IoT in the Russian Federation were presented. The authors substantiate the need for the interaction of legal and technical regulation and analyze the Russian state policy in ensuring the security of IT and methods for implementation from the point of view of technical and legal regulation. At the same time, the compliance of international standards of technical regulation in the field of information security with Russian standards of the same sphere was determined. Finally, it presents an analysis of the strategic approach to building a system of normative regulation of information security.

Successful implementation of the activities of the Digital Economy of the Russian Federation program will allow for the formation of a stable information and communication infrastructure of state and municipal bodies. This includes the uninterrupted operation of government services and organizations, remote identification services, and legal storage of significant information in a digital platform for the provision of state and municipal services.

The problem is the lack of reliable standards, which does not allow for the reliable and stable interaction of the virtual and physical environments in the application of cyberphysical systems. A solution to this problem is possible only when the system of standards takes into account risks and threats. This problem is relevant for most developed countries.

A threat to the security of information systems in Russia is the widespread use of foreign-developed technologies. The risk of the existence of pre-installed malware is clear. Less obvious, but no less significant, is the dependence on foreign developers for standards and methods to ensure information security and resistance to attacks while maintaining the "access control", "stability", "data integrity", "authentication" and confidentiality of Russian users. 2015 can conditionally be defined as the beginning of the solution to this problem, but only since 2019 has the Russian Federation been actively implementing the transition to its own ICT.

Recommendations for resolving problems can be divided into 2 categories: strategic and situational.

At the strategic, state level it is necessary:

- to determine the requirements for industry regulation of IoT and other cyberphysical systems;
- to develop and adopt standards for processing large amounts of data, and to develop information security standards in systems that implement cloud, fog, quantum technologies, virtual and augmented reality systems, artificial intelligence;
- to identify and authenticate the subjects of IoT-relations;
- to develop requirements for industrial Internet operators by developing industry standards and trust authentication based on public keys and access control, and ensuring interoperability;
- to develop requirements and security standards for cyber-physical systems, including IoT, and requirements for the registration of equipment of IoT networks; to create certificates using Russian cryptographic standards;
- to develop protocols for processing information security using artificial intelligence.

At a lower (situational) level, a number of measures should be developed for subjects of IoT relations:

- to monitor threats and vulnerabilities of information systems of limited access information;
- to develop threat models for IoT and personal devices;
- to create a set of organizational and technical measures to counter IoT threats and ensuring information security, including the identification and authentication of subjects and access control for them, limiting the software environment, protecting hardware and virtualization environments;

For operational interaction:

- an information and reference system should be developed for the subjects of IoT relations to determine their compliance with the requirements of Russian and international legal and technical regulations;
- it is necessary to ensure control over the processing and access to confidential information;
- it is necessary to establish interaction between government and management bodies with a clear distribution of powers and social responsibility in the development of industry standards and a trusted authentication chain based on public key infrastructure and access control.
- The successful implementation of the activities of "The Digital Economy of the Russian Federation" will allow the formation of a stable ICT infrastructure of state, including the uninterrupted operation of government services and organizations, the remote identification services and the legal provision of the storage of information in a digital platform for the provision of state and municipal services.

ACKNOWLEDGMENT

This section was supported by RFBR project No. 20-011-00077 "Legal regulation of a person's digital profile on the Internet".

REFERENCES

Banafa, A. (2016). *IoT Standardization and Implementation Challenges*. IEEE. https://iot.ieee.org/newsletter/july-2016/iot-standardization-and-implementation-challenges

Bauer, M., Boussard, M., Bui, N., De Loof, J., Magerkurth, C., Meissner, S., Nettsträter, A., Stefa, J., Thoma, M., & Walewski, J. W. (2013). *IoT Reference Architecture*. Retrieved from https://link.springer.com/chapter/10.1007/978-3-642-40403-0_8

CNews conducts the Internet of things 2019 conference in Moscow. (2019). Retrieved from https://events.cnews.ru/events/internet_veschei_2018.shtml

Decree of the Government of the Russian Federation of February 8, 2018 No. 127 "On approval of the Rules for categorizing objects of critical information infrastructure of the Russian Federation, as well as the List of indicators of criteria for the significance of objects of critical information infrastructure of the Russian Federation and their values". (2018). *Reference legal system "Consultant Plus"*.

Decree of the Government of the Russian Federation of November 16, 2015 No 1236. (2015). https://legalacts.ru/doc/postanovlenie-pravitelstva-rf-ot-16112015-n-1236/

Decree of the President of the Russian Federation of August 16, 2004 No 1085 "Issues of the Federal Service for Technical and Export Control". (2004). Retrieved from https://fstec.ru/

Federal Law of December 27, 2002 No 184-FZ "On Technical Regulation", *Collection of Legislation of the Russian Federation,* December 30, 2002, No. 52 (part 1), Art. 5140. (2002). https://ivprom.ru/lib/245/

Goertzel, K. (2016). Legal Liability for Bad Software. *Crosstalk*, *2016*(September/October). http://static1.1.sqspcdn.com/static/f/702523/27213494/1472233517737/201609 – Goertzel.pdf

Good Practices for Security of IoT in the context of Smart Manufacturing. (2018). https://www.enisa.europa.eu/publications/good-practices-for-security-of-iot?fbclid=IwAR1q-chv88kZRsIESHtGTEw-bA0Mbx8mb9hV1Euqy-Y--IHVYvLuFhGuvi6o

GOST R 54593-2011. National standard of the Russian Federation. Information Technology. Free software. General Provisions (approved and enforced by the Order of Rosstandart of December 06, 2011 No 718-St). (2011). *Reference legal system "Consultant Plus"*.

GOST R ISO / IEC 15408-3-2013 "Information technology. Methods and means of security. Criteria for assessing the security of information technology. Requirements for trusting OS security. (2013). http://docs.cntd.ru/document/1200105711

Guidelines for the implementation and use of the industrial Internet of things to optimize control (supervisory) activities "(approved by the minutes of the meeting of the Project Committee of 09.11.2017 No 73 (13)). (2017). Retrieved from https://legalacts.ru/doc/metodicheskie-rekomendatsii-po-vnedreniiu-i-ispolzovaniiu-promyshlennogo-interneta-veshchei/

Hugh, T. (2020). *What Are Cyber Threats: How They Affect You and What to Do About Them*. Retrieved from https://preyproject.com/blog/en/what-are-cyber-threats-how-they-affect-you-what-to-do-about-them/

Information protection measures in state information systems. (2014). Retrieved from https://fstec.ru/component/attachments/download/675

IoT-2019: main trends in the Russian market. (2019). Retrieved from https://iot.ru/promyshlennost/iot-2019-glavnye-trendy-na-rossiyskom-rynke

Kaspersky Security Network. (2019). https://securelist.ru/it-threat-evolution-q1-2019-statistics/94021/

Khan, M. A., & Salah, K. (2018). IoT security: Review blockchain solutions and open challenges. *Future Generation Computer Systems*, *82*, 395–411. doi:10.1016/j.future.2017.11.022

Luca, C., Antonio, M., & Dario, M. (2019). IoT Manager: An open-source IoT framework for smart cities. *Journal of Systems Architecture*, *98*(September), 413–423. doi:10.1016/j.sysarc.2019.04.003

Ming, T., Zu, J., Zhusong, L., Aniello, C., & Francesco, P. (2018, January). Multi-layer cloud architectural model and ontology-based security service framework for IoT -based smart homes. *Future Generation Computer Systems*, *78*(Part 3), 1040–1051. doi:10.1016/j.future.2016.11.011

Order of the FSTEC of Russia dated February 11, 2013 No 17 "On approval of the requirements for the protection of information not constituting state secrets contained in state information systems". (2013). Retrieved from https://fstec.ru/normotvorcheskaya/akty/53-prikazy/702

Order of the Government of the Russian Federation of August 28, 2019 N 1911-r "On approval of the Concept of creating a unified cloud platform". (2019). *Reference legal system "Consultant Plus"*.

Order of the Ministry of Communications of Russia of December 31, 2015 No 622 "On approval of the rules for using the classifier of programs for electronic computers and databases". (2015). Retrieved from https://legalacts.ru/doc/prikaz-minkomsvjazi-rossii-ot-31122015-n-622/

Passport of the national project. "National Program" Digital Economy of the Russian Federation "(approved by the Presidium of the Presidential Council for Strategic Development and National Projects, Minutes dated June 06, 2019 No 7). (2019). https://legalacts.ru/doc/pasport-natsionalnogo-proekta-natsionalnaja-programma-tsifrovaja-ekonomika-rossiiskoi-federatsii/

Public declaration of the goals and objectives of the federal service for supervision in the field of communication, information technologies and mass communications (Roskomnadzor). (2019). https: Rkn.gov.rn

Qin, Zhu, Ni, Gu, & Zhu. (2019). Blockchain for the IoT and industrial IoT: A review. *IoT*. .2019.100081 doi:10.1016/j.IoT

Sachi, N., Mohanty, C. R., Sheeba, R., Deepak, G. K., Shankar, S. K., & Lakshmanaprabu, A. K. (2019). An efficient Lightweight integrated Blockchain (ELIB) model for IoT security and privacy. *Future Generation Computer Systems*. Advance online publication. doi:10.1016/j.future.2019.09.09.050

Significant lack in awareness of IoT security among IT leaders, study finds. (2018). Retrieved from https://www.IoT technews.com/news/2018/nov/21/significant-lack-awareness-IoT -security-among-it-and-security-leaders-study-finds/

Simon, E., Henry, D., & Morris, J. S. (2015). *IoT – Enabled Analytic Applications Revolutionize Supply Chain Planning and Execution*. November 2015, IDC #259697.

The Determination of the Supreme Court of the Russian Federation of February 20, 2018 No 306-KG17-22931 in the case of No A65-25932 / 2016. (2018). https://legalacts.ru/sud/opredelenie-verkhovnogo-suda-rf-ot-20022018-n-306-kg17-22931-po-delu-n-a65-259322016/

The Forecast of scientific and technological development of the Russian Federation for the period until 2030. (2014). Retrieved from https://www.garant.ru/products/ipo/prime/doc/70484380/

The 'Internet of Things' will be the world's most massive device market and save companies billions of dollars. (2015). Retrieved from https://www.businessinsider.com/the-internet-of-things-market-growth-and-trends-2015-2

The Ministry of Communications approved the concept of the development of the Internet of things in Russia. (2019). Retrieved from https://iot.ru/promyshlennost/minkomsvyaz-odobrila-kontseptsiyu-razvitiya-interneta-veshchey-v-rossii

The National Program. "Digital Economy of the Russian Federation" (approved by the Presidium of the Presidential Council for Strategic Development and National Projects, Minutes No. 7 dated June 4, 2019). (2019). https://digital.gov.ru/ru/activity/directions/858/

The Order of the Ministry of Information Technologies and Communications of the Russian Federation of September 27, 2007. (2007). On the approval of the requirements for the organizational-technical support of sustainable operation of the communication network. *Reference legal system "Consultant Plus"*.

The Resolution of the Government of the Russian Federation of October 29, 2019 No 1382 "The rules for granting subsidies from the federal budget to achieve the results of the federal project "Information Security" of the national program "Digital Economy of the Russian Federation". (2019). *Reference legal system "Consultant Plus"*.

Venus, M., Amir, M. R., Aso, M. D., & Amir, S. (2019). Trust based recommendation systems in IoT: a systematic literature review. *Human-Centric Computing and Information Sciences, 9*, 21. 01901838 doi:10.118613673

Zharova, A. (2020, June). The protect mobile user data in Russia. *Iranian Journal of Electrical and Computer Engineering*, *10*(3), 3184–3192. doi:10.11591/ijece.v10i3.pp3184-3192

Zharova, A., Elin, V., & Panfilov, P. (2017). Personal Data in Cloud. Russia Experience. *Proceedings of the 28th DAAAM International Symposium*, 1136-1142. DOI: 10.2507/28th.daaam.proceedings

Chapter 15
Artificial Intelligence and SHGs:
Enabling Financial Inclusion in India

Vinay Kandpal
https://orcid.org/0000-0003-1823-4684
UPES, Dehradun, India

Osamah Ibrahim Khalaf
Al-Nahrain University, Baghdad, Iraq

ABSTRACT

For inclusive growth and sustainable development of SHG and women empowerment, there is a need to provide an environment to access quality services from financial and non-financial agencies. While banks cannot reach all people through a 'brick and mortar' model, new and advanced banking technology has enabled financial inclusion through branchless banking. By using artificial intelligence in banking, banks have a cost-effective and efficient solution to provide access to services to the financially excluded. Digital technology improves the accessibility and affordability of financial services for the previously unbanked or underbanked individuals and MSMEs. A big data-driven model can also be helpful for psychometric evaluations. Several psychometric tools help evaluate the applicant's answers which aid to capture information that can help to predict loan repayment behavior, comprising applicants' beliefs, performance, attitudes, and integrity.

INTRODUCTION

Artificial Intelligence (AI) is creating a buzz across the world. In the sphere of finance and technology, artificial intelligence has demonstrated its merits by creating an option of faster, cheaper and better transactions. Artificial Intelligence in the financial and technology sector has already made inroads in India in the past decade. It is said that this will alter and bring in huge transformation for a variety of applications.

According to studies by KPMG, "the Indian Fintech market is expected to grow at a CAGR of 22 percent over the next five years, with the transaction value crossing $73 billion by 2020".

DOI: 10.4018/978-1-7998-5068-7.ch015

The Internet is more or less accessible by the majority of the people in India, which depicts an amazing opportunity of artificial intelligence to step in. Reports suggest that, "Financial services are influencing more than 300 million users by enabling them access to credit, while the same users are completely ignored by traditional players due to insufficient credit history" It was recently found in a study that, 52% of the customers utilise fintech services in our country. In the year 2018, India emerged as a great market for the fintech options. India is home to 1.3 billion people with diversity in language, culture, belief and lifestyle. This at times may pose as a challenge. Hence, it is important that fintech services are available in easy and lucid manner in several languages so that people from every nook and corner of the country can avail it with ease. Fintech believes in providing a satisfactory user experience and it is a mandatory criteria.

All the companies should strive for using artificial intelligence which will transform the way services are dispensed. The seamless service through artificial intelligence will bring in accuracy, automated and immense speed. These factors will mean that it is preferred by the user as the visits to the bank will reduce and transactions will be just a click away. In order to access credit from any financial institution it is understood that fintech will aid in completing all the formalities digitally. There is no reason to visit the bank and fulfill the paperwork. "The borrowing process happens in a snap, and the overall experience is now seamless" (Anantharaman, 2019).

LITERATURE REVIEW

Financial Inclusion and Fintech

The ease in the access of banking due to internet and mobile banking has made the job easier for the customers. But the scenario is different in the rural areas where people love to visit banks and are not proficient with the internet banking mode. Studies have reported that checking of the last transactions are the most popular activities in internet banking. Despite the digital boom in the country, internet banking has not penetrated the rural parts of the country. Mobile banking is detested by not only rural but certain sections of the urban population due to security issues. As the authors have mentioned, "The penetration of mobile banking was only 0.2 per cent in India until 2009, and this was expected to increase to 2 per cent by 2012. The expected benchmark was achieved with 22.51 m active m-banking users in 2012-2013, which was only 2.5 min in 2009" Sharma, (2016).

Mbogo, (2013) in his study on entrepreneurs who were dealing with microbusiness conducted a study on how and why they were using M-PESA services. The results revealed that the ease in the cash transfer as well as the convenience and strong security options were the reasons for availing the app. As the study mentioned, "This way, M-PESA reduces costs and risk and thus has a positive impact on wealth generation".

Plyler et al., (2010) in his survey on the Kenyan population for the socio-economic effects of using M-PESA found that ease in the transfer of money to people living far away has had significant changes in their social bonding. The need to visit banks or visit relatives to deliver money is all in the past now. The study revealed that the family structure has weakened. According to the authors, "in some cases this has even led to family breakups. Private and foreign players in metropolitan cities seem to be involved in the major chunk of online banking-related business activities".

As per McKinsey (2013) "there was 130 percent increase in the use of Internet banking in India during 2007-2011 and a 15 per cent decline in branch banking. Nevertheless, in India, only 7 per cent account holders use Internet banking for their transactions".

Sharma (2016) and Sassi & Goaied, (2013) in their study found that the penetration of ICT and the development of finance has led to significant improvement and prosperity in Mena.

Joseph & Varghese (2014) conducted a study on the Indian economy and its development. The researchers took up five nationalised banks and five private banks. The parameters to understand the development were the increase in branches, ATM Services and the usage of credit and debit cards. The results showed that "the usage of debit cards has increased tremendously throughout the study period and banks focused more on semi-urban areas and rural areas. The study also found that the number of people with access to the products and services offered by the banking system is very limited despite inclusive banking initiatives in the country".

Kamboj (2014) in his study found out that there is optimism in the development of the financial sector when there is a significant growth in the number of branches and the ATMS.

Sriram & Sundaram (2016) and Joshi (2014) studied the awareness among the urban poor of Nagpur regarding the financial services provided by the banks. The results showed that "financial products like current account, demand loan, direct debit facility, credit card and mobile banking is low. Aggregate awareness of banking services offered was found to below 41%. Lack of cooperation, improper guidance, and lack of transparency are the main reasons for this level of awareness". Shah & Dubhashi (2015) found out that "the technological reforms pertaining to banking sector such as e-commerce, mobile phones, email, ATMs and plastic money were available only in towns and cities, which leads to limited access to financial products and services in rural and semi urban regions".

Onaolapo (2015) mentioned that the financial services apart from banks can be utilised for the smooth access of ICT based financial options like post office and so on. The author mentioned that "The delivery mechanism under this system needs to depend on mobile banking services and other ICT based services like Point-of-sale device networks to communicate between retail agent, financial service provider and the customer in a branchless banking system".

Rojas-Suarez (2016) studied the expansion of the financial facilities in countries like Kenya and Mexico. The study revealed that the regulatory framework as well as the models to operate had a huge contribution for growth in the usage of mobile financial services. The researcher stated that, "in Kenya, more than 50% of the population uses mobile money payments compared to only 2% of the Mexican population. The author identifies in these countries two models that contribute to the growth of mobile money: Mobile Network Operator- and bank-led models. While Mexico has adopted a bank-led model of mobile payments, which operates under strict regulations and is mainly serving those with bank accounts, Kenya has adopted a model with less regulation led by Mobile Network Operators". He further expressed that the restrictive policy of Mexico refrained the people from availing mobile banking whereas the deregulated framework of Kenya saw an exponential growth in the use of mobile financial services.

Ranade (2017) mentioned that the financial services along with the Jan Dhan Aadhar mobile phone can speed up the usage of financial services over the phone. The researcher suggested, "caution in terms of privacy and ownership of data that would be generated in the process". (Kandpal, 2020),

The report by Digital Financial Inclusion and Consumer Capabilities in India (2017) A handbook for financial service providers conducted a study to understand the digital literacy among the consumers and the service providers through survey and focus group discussions. The results showed that the respondents "drew attention to the shift in challenge of digital finance from willingness and ability to its access and regular usage. There was a need to systematically develop the local ecosystem to bring more users to accept digital finance".

The most critical cloud-computing task is scheduling to provide customers with the best service possible. The scattering of assets in cloud topology faces classification problems that cannot be resolved with existing asset allocation methods.

Moreover, the financial data which is very diverse can be used to understand the pattern of savings, payments for designing financial services which will be helpful for a much diverse section of the society. But as Singh & Naik (2018) expressed, "however, these potentials remained untapped as the digital finance ecosystem was yet to be scaled up beyond urban financially integrated individuals". The results showed that the bee algorithm gives the optimum coverage percentage compared to the genetic algorithm and takes less time to use the system resources and implementation of the algorithm and also requires a smaller number of sensors. Khalaf of this new emerging technology and its superiority compared to other designed technologies in the field of live streaming and telecommunications are discussed.

SHGs, Women Empowerment and Financial Inclusion

Roy (2006) in the study expressed the pay disparity between men and women based on the gender in the micro enterprises. He stated that, "women have been excluded from the formal financial services on the ground of their property less status, castes and discriminatory practices".

Lahiri Chavan et al., (2009) in her study analysed the need to provide financial facilities like bank credit to women so that they will be uplifted from their poor economic condition. Her study revealed that the relationship between the transaction, bank credits and women is comparatively low and in an indirect manner. There are intermediaries who work to establish relationships between them like MFI, NBFC and several Non Governmental Organisations. As the researcher quoted, "Women are not aware about the rules, regulations and procedures of a bank so they are more hesitant to approach banks directly. They turn to someone they know and trust from their community who could help them to get the loan". The researcher has also noted that when the competition is high for providing financial services banks tend to satisfy every customer irrespective of any gendered notion. This helps in the increase of the profit of the bank.

Kumar, Vipin; Kumar, (2016) and Prasad, (2009) expressed a similar opinion too, "it is believed that as the financial markets and banking sectors evolve in developing countries, the higher level of competition among banking institutions will curb gender disparity in credit decisions."

Badajena, S, N; Gundimeda, (2010) in their study understood the impact that the self help group has over establishing financial inclusion in the sixteen states. The studies revealed interesting results that a larger section of the society was unable to be covered in spite of the wide range networking of the banks. The study which was conducted in the year 2008 found that women begin their businesses with the help of the capital earned from local resources. The author has expressed that, "To encourage the entrepreneurial habit among women, banks and financial institutions should provide easy access of credit to poor women. Most poor women don't have the right to spend their own earnings and don't have ownership of assets which makes them economically dependent on others".

Lokhande (2014) in the study mentioned that banks are not easily available to provide bank facilities like credit to women, especially the poor women. The researcher said that if credit facilities are provided then the overall economy will flourish, as it will enhance the ownership and the growth in assets. She suggested the establishment of a Bhartiya Mahila Bank by the government. She mentioned, "This increased participation in economic activities by women can lead to their economic empowerment in the truest sense. The establishment of an all women bank will play a great role in triggering the access of banking services and credit facilities by women and extend a vast opportunity for women employment in public sector banks as the bank is predominantly for women customers, by the women employees".

Tiwari Madhu, (2014) mentioned that it is not unknown that the motive of a bank is to earn profits and therefore a balance should be struck between profit and responsibility towards uplifting women. She expressed that the majority of the women in India are not socially and economically privileged. Therefore, banks must show their innovative approach towards boosting women's participation in economic activities.

Rajeshwari, (2015) showcased the challenges and the needs for economic empowerment of women. She expressed the need for decision making among the women which is ably assisted by the Self Help Groups (SHG). Not just entrepreneurial ventures but employment will usher in a self confidence amongst women. To achieve this policies should be put in place and aid in the equal employment options. As the author mentioned, "This will ensure equal participation of women in economic activities which would eventually lead to sustainable development of the economy".

Shakya, (2016) carried out a study on the "impact of micro credit on its borrowers in Nepal". The study revealed that there was "an increase in income and savings of the female borrowers. The economic and social conditions of such women also became better".

Porkodi & Aravazhi, (2013) analysed the role microfinance has in empowering people for realising the financial options. There are a lot of challenges in the efficiency of the MFIs therefore they want the government to take an effort in mobilising the savings. There has been an increase in the need for rural finance therefore MFIs can help in providing micro-credit with better facilities in the sphere of fintech. The author mentioned, "This is of utmost importance in order to upgrade MFIs from thrift and credit institutions to capacity building and livelihood- sustaining associations of people. NGOs have played a commendable role in promoting Self Help Groups linking them with banks. There is, therefore, a need to evolve an incentive package, which should motivate these NGOs to diversify into other backward areas".

Singh, Pratibha; Tewari, Poonam; Verma, (2017) analyzed that SHGs benefit its members by providing regular saving, employment opportunities, access to credit, participation in local government, and change in family decision-making. Present study shows that of the 400 SHGs surveyed in six districts of Uttarakhand only 24% remained active after eight years of their formation. These groups were engaged in individual income generating activities only and a large number of them discontinued or were dormant i.e. involved only in collection of money and micro-finance. An important reason remains the lack of follow up by agencies responsible for their formation. The overall picture emerged that women SHGs in Uttarakhand were not very effective in improving their economic status and in generating any new enterprise.

Chunera, (2014) revealed that the major constraints perceived by the SHGs were lack of training, poor financial condition of the members, delay in sanctioning of loan, unprofessional attitude of the members, less education and time management.

Singh, A, (2014) mentions that for the SHGs, education is a crucial parameter. She said that "it is not a matter of aged and young women to be a leader but women are expected to reflect the participative / democratic style of leadership". Her study revealed that familial bonds and relations play an important part in becoming a vocal leader in the Self Help Group. The motive of the SHG is to initiate financial inclusion, employment options from the grass root level.

As ILO (2014) recommended, "This is a direct derivation of SEWA's core philosophy of "empowering through organizing", that finds its ultimate expression at ground-level in the promotion of the SHGs.

Fostering these groupings is viewed as particularly important in rural areas, where women are often more isolated and less cognizant of their economic, social, and political potential, as well as their fundamental."

Kaur, (2014) found out that the rural population is deprived of various facilities like employment, transportation, education and financial services as well. These issues can have a remedy only when a group initiative is undertaken.

Yadav et al., (2015) in their study suggested the promotion of SHG to initiate participation of the community and create awareness about various other facilities. They recommended that "in a formal banking system, it is essential to present a collateral to get financial assistance".

Esmaeil Zaei et al., (2018) said that, "Generally, rural women have little or no property or assets to present such collateral. Hence, micro credit facilitated a solution to this problem by providing collateral free loans at their doorstep".

A pilot project was conducted by SEWA with assistance from Micro Finance (CMC) in the 2006-2013 period. This study wanted to evaluate the impact rainfall insurance had in Gujarat. It was found that around 65% of the farmers were dependent on rainfall for their irrigation. As the report mentioned, "index-based weather insurance is a valid product for farmers – in theory - to hedge against these variables".

Munns, G., Sawyer, W., & Cole, (2013) has expressed by supporting the SEWA study that due to less awareness on financial facilities and issues of pricing, people took up the agricultural insurance policies by the members of SEWA.

PROGRESS OF MFIS AND DIGITAL INNOVATIONS IN INDIA

Initiatives by MFIs and different Institutions Towards Financial Inclusion and Rural Development

Microfinance was the first major disruption in this space. The numbers tell all-as of March 2019, there are over 93 million microfinance accounts in India. The industry boasts a gross loan portfolio (GLP) of $26.38 billion. Another aspect of microfinance in India is the Self-Help Group (SHG) movement. This movement helped SHG become "world's largest and most successful network of women-owned community-based microfinance institutions". The Self Help Group Bank Linkage Programme (SHG-BLP) is an innovative and a path-breaking model by the National Bank for Agriculture and Rural Development (NABARD) in 1992. The primary objective is for delivering affordable banking services and facilitating financial inclusion. Today, SHG-BLP caters to 120 million households through more than 10 million SHGs. Deposits span over $3.24 billion and annual loan off-take value is $8.16 billion. (NABARD, 2018) [41]

Initiatives by Government Towards Financial Inclusion and Fintech

The JAM trinity (Jan Dhan, Aadhaar and mobile phones) is another major initiative. Launched in 2014, Jan Dhan has certainly proved its mettle, providing over 370 million bank accounts. Consequently, over 80 per cent of India's adults are now financially included, compared to a mere 35 per cent in 2011. Notably, 53 per cent of Jan Dhan account holders are women and 58 per cent reside in rural and semi-urban areas. To provide a global perspective, Jan Dhan upholds several of the United Nations' Sustainable

Development Goals (SDGs). It comprises of the SDG 8, expanding financial inclusion, SDG 5 – giving women equal access to financial services and SDG 10 – promoting economic inclusion for all.

Universal financial access attained through Jan Dhan certainly changed the game. For one, the government obtained the opportunity to reduce corruption and losses incurred whilst transferring subsidies and benefits to citizens. The Direct Benefit Transfer (DBT) and Aadhaar tools proved very effective to achieve this. The DBT enabled the government to directly transfer subsidies and benefits of various social welfare schemes to the beneficiaries' bank accounts. Meanwhile, the Aadhaar identification tool, with over 1.2 billion holders, were linked to the beneficiaries' bank accounts, which, of course, had its own set of benefits. Currently, DBT supports 437 schemes of 56 ministries and has processed over $119.68 billion since inception. DBT and other governance reforms have brought about estimated savings of $19.78 billion. Not just that, it has helped India make significant progress in SDG 16 (Peace, Justice and Strong Institutions) by reducing corruption.

It does not end there, of course. Merely storing funds in bank accounts is not enough. Customers would, naturally, require touch-points to cash-out the funds, which is more vital since India is still primarily a cash-based economy. The Micro ATMs introduced by the National Payments Corporation of India (NPCI) in 2015 offered a viable solution. The aim is simple-expand the banking system's last-mile reach, particularly in rural areas. Since its inception, over 577 million financial transactions valuing $21.29 billion have been processed by micro ATMs. Moreover, mobile handsets have proliferated significantly, naturally implying that these devices have emerged as important transaction channels as well.

These initiatives have, of course, had a cascading impact overall. Here is how-collectively, the Jan Dhan, Aadhaar, the proliferation of mobile handsets and DBT have simplified the implementation of government-driven schemes. These schemes, needless to say, drive the SDGs as well. For instance, the Ujjwala Scheme is aimed at providing clean cooking fuel to those rural women who have traditionally relied on firewood, coal and dung cakes. An equally vital aim is to prevent premature deaths, due to polluted air in the household-estimated at 480,000 deaths per year! Under the scheme, 80 million households below the poverty line have been provided a free gas connection. This was followed by the "Give it up" campaign, where over 11 million affluent customers surrendered their LPG subsidy. These initiatives removed 38 million "ghost" or "fake" beneficiaries from the system. The $6.3 billion saved was then distributed to customers in rural areas without access to clean cooking gas. The Ujjwala Scheme directly or indirectly impacts multiple SDGs including SDG 3 (Good Health and Well Being), SDG 5 (Gender Equality), SDG 7 (Affordable and Clean Energy) and SDG 16 (Peace, Justice and Strong Institutions).

Another example is the Micro Units Development & Refinance Agency (MUDRA). This institution is aimed at providing loans to micro and small enterprises, with credit needs under $14,000. Under the MUDRA scheme, 155.6 million loans amounting to $101.26 billion have been disbursed. About 75 per cent of the recipients are women. The MUDRA scheme impacts the SDG 8 (Decent Work and Economic Growth) and the SDG 9 (Industry, Innovation and Infrastructure). The National Payments Corporation of India (NPCI) threw its hat in the ring once again, with the introduction of two vital schemes. I allude, of course, to the Immediate Payment Service (IMPS) and the Unified Payments Interface (UPI). The former (for which I was, as a happy coincidence, involved in establishing the standards) provides 24×7 real-time interbank electronic fund transfer. The Unified Payments Interface (UPI) helps the user to use multiple bank accounts in just one mobile application (of any participating bank). Of course, several banking options like "seamless fund routing and merchant payments" are under one umbrella. UPI allows users to create a unique virtual address and use it for transferring money and making payments from devices such as mobile handsets. The IMPS and UPI have 498 and 141 participating banks respectively. Over the last one year, (from September 2018 to August 2019) IMPS and UPI have collectively processed 10.26 billion transactions, valuing $465.66 billion. Milton Friedman once said that "the poor stay poor, not because they're lazy, but because they have no access to capital." Khanvilkar, Amol, (2016) [42]has expressed in his study that despite so many efforts a huge section of the population are not under the formal banking system. The only option available to them is the microfinance institution which offers the rural masses financial services.

SHGs Work Towards Financial Inclusion and Women Empowerment

As per the policy of E-Shakti, an initiative by NABARD, "In tune with the changing scenario of the banking sector's engagement with technology and to address the financial exclusion of the segment of poor SHGs who have not been able to access technology, NABARD conceived a project for the digitization of SHGs called EShakti. Digitization of Self Help Groups was conceptualized to build credible credit histories of SHGs and their members and bring them into the fold of financial inclusion and mainstream banking systems". India was one of the early adopters of digital financial services in the developing world. Internet banking made its foray in the country in the second half of the nineties, while mobile banking made its debut a decade later. Let's not forget, however, India offers an interesting study in contrasts. While a small part of the country was digitally evolving, a significant portion remained untouched by the digital revolution. This stark reality compelled governments, banks and financial institutions to take drastic measures. These entities therefore collaborated and, subsequently, created disruptions to accelerate financial inclusion in the country.

This project was launched in the year 2015 in the districts of Ramgarh and Dhule in Maharashtra. This E-Shakti scheme is an innovative approach as it offers financial inclusion through digital mode. This project has been successfully in progress in 22 states and 1 Union Territory and is working well in 100 districts. This project has aided the members of SHG to take their own financial decisions and use financial options through digital mode. This has reduced the burden of several things like "quality of bookkeeping, multiple membership of SHG members, patchy credit histories". The financial facilities are available easily and the convenience of the users.

The E-Shakti website mentions that, "As on 31 March 2019, EShakti pilot project has on boarded 4.34 lakh SHGs involving 47.91 lakh rural households in 100 districts across 22 States and 1 UT covering the entire length and breadth of the country. The credit linkage under the pilot districts has increased from 1.65 lakh (38%) to 2.22 lakh (51%) thus proving the success of the project in catering to the credit needs of the rural poor". In the year January 2019, an initiative was undertaken to link SHG members for the financial inclusion and several Social Security schemes available in the districts where the pilot projects are being undertaken. There were awareness programmes held for the SHG members about PMJDY, PMSBY, PMJJBY and APY and were counselled to avail these by enrolling in the policies. The report mentions, "As on 31 March 2019, 0.64 lakh Jan Dhan accounts were opened and 2.53 lakh PMSBY insurance policies, 0.48 lakh PMJJBY insurance policies and 0.08 lakh APY pension policies were enrolled across the nation."

Artificial Intelligence and its Application in Financial Inclusion

Superfluid Labs CEO and data evangelist Timothy Kotin explains "the significant impact data and artificial intelligence can have in Africa, where data has traditionally been much harder to come by.

From fintech to policy to microfinance, the ability to process large volumes of consumer data can have far-reaching impact across the continent".

One of Artificial Intelligence (AI)'s most important applications is in financial technology, where over the past year global investment has risen 38 percent. Machine learning could lighten a way out of poverty for the world's two billion unbanked adults by helping mainstream lenders accept loans using hundreds of non-traditional data points. AI is capable of adding value across Africa at the individual, small business, and large corporate level alike. In developing markets, where businesses fail to attract potential buyers in the absence of granular information regarding them, data is just as important [as it is in the West] and more difficult to come by. Comprehensive, current, and open public opinion data across Africa may have wider policy effect as well as more conventional, commercial value. Startups like mSurvey in Kenya have stepped in, using mobile phones to build up-to-date profiles of local consumers. Companies on the continent must build their own AI-driven data marketplaces to fully unlock Africa's consumer data, enabling consumers to opt-in and share their consumption habits.

ITU's Digital Financial Services Focus Group released 85 policy guidelines for digital financial services and 28 thematic findings positive. With more than 60 organizations representing more than 30 nations, the Focus Group was the first effort to put together all the players operating in the interests of financial inclusion, including ICT regulators and central bankers. The 'Financial Inclusion Global Initiative' (FIGI), led by ITU, the World Bank Group and the Committee on Payments and Market Infrastructures, and with financial support from the Bill & Melinda Gates Foundation, is a three-year plan of concerted action to advance digital finance research and promote digital financial inclusion in developing countries. FIGI would enable national authorities in developed and emerging markets to use ICTs to expand the financial system's scope, while managing related risks. The above 85 ITU policy recommendations for digital financial services are enacted by FIGI.

Banking, Fintech and Issues Regarding Cyber-Attacks

The use of Artificial Intelligence (AI) cognitive technology brings the benefit of digitization to banks and lets them compete with the Fintech companies. Indeed, according to a joint study conducted by the National Market Research Institute and Narrative Science, "about 32 per cent of financial service providers are now using AI innovations such as Predictive Analytics, Voice Recognition, among others". Artificial Intelligence is banking's future as it brings the power of sophisticated data processing to combat fraudulent transactions and improve enforcement. In a few seconds, AI algorithm performs anti-money laundering activities that otherwise take hours and days.

Digital transformation is a redefinition of industries and a revolution in the way businesses work. In the technology-driven world, each industry assesses options and adopts ways of generating value. The banking sector is undergoing groundbreaking changes: the client-centricity increase above all. Exposed to new technology in their daily lives, tech-savvy consumers expect banks to offer seamless experiences. In order to meet these expectations, banks have expanded their sector landscape to include retail, IT and telecom services such as mobile banking, e banking and real-time money transfers. While these innovations have allowed consumers to access most of the banking services at their disposal anytime, anywhere, they have also added costs to the banking sector. Along with banking and sectors such as IT, telecom and retail, the transfer of sensitive information through virtual networks that are vulnerable to cyber-attacks and fraud has increased. Such events not only influence bank profitability but also hamper the confidence and partnership of banks with customers. The emergence of threats to cyber security in banking transactions has tightened government regulations. Though these regulations are useful for tracking electronic financial transactions, the capacity of banks to keep up with digital transformation has been curbed. Banks are unable to invest in technology, as they must maintain a capital adequacy ratio as set out in the criteria for international regulatory structure.

FUTURE RESEARCH TRENDS

In the financial service industry, artificial intelligence is still in the early days. In finance, AI will become omnipresent, resulting in more challenges including legal, ethical, fiscal, and social hurdles. AI will also continue to give the global financial environment new complexities. AI systems will become more complex as more and more data becomes available and computational capacity increases. Much still needs to be achieved and more work need to be done about education about AI and its role towards financial inclusion.

CONCLUSION

The need for finance for any purpose for the matter should never be restricted to the urban people who have credit cards and keep a track of their credit score. Fintech has transformed the lives of every individual with banking facilities at just a click away. Artificial Intelligence has made inroads to our society to help people to use banking facilities in a convenient, easy and transparent manner through digital aids. This reduces the burden on the traditional banking set up and saves time for the consumers. Fintech has helped people in starting up their own entrepreneurial ventures. As it is mentioned in Yourstory.com, "Fintech, powered by AI, helps in assessing credit scores of users based on various factors like digital footprint and other alternative data points, thus lending to NTC customers and therefore significantly increasing the number of 'credit worthy' people in the country by introducing them to the market, safely and affordably".

Artificial Intelligence has become an important component for all the financial companies and is a very good opportunity for the users to avail and be aware of the facilities offered by the banking sector. This has reduced the gap between the urban and the rural to a considerable extent with the government taking several initiatives. Thus, financial inclusion can be a reality in the days to come.

REFERENCES

Anantharaman, A. (2019). *How Artificial Intelligence is enabling financial inclusion in India*. https://yourstory.com/2019/05/artificial-intelligence-enabling-financial-inclusion

Badajena, S. N., & Gundimeda, H. (2010). *Self help group bank linkage model and financial inclusion in India*. http://skoch.in

Chunera, A. (2014). Capability assessment of women self help groups: A study in Nainital district of Uttarakhand. G.B. Pant University of Agriculture and Technology, Pantnagar - 263145 (Uttarakhand).

Digital Financial Inclusion and Consumer Capabilities in India: A handbook for financial service providers. (2017). http://www.ifmrlead.org/wp-content/uploads/2017/09/DFS/IFMR JPM DFS Handbook_PDF.pdf

Esmaeil Zaei, M., Kapil, P., Pelekh, O., & Teimoury Nasab, A. (2018). Does Micro-Credit Empower Women through Self-Help Groups? Evidence from Punjab, Northern India. *Societies (Basel, Switzerland)*, *8*(3), 48. doi:10.3390oc8030048

ILO. (2014). *The role of economic and social councils and similar institutions in promoting social protection floors for all through social dialogue. Report of the ILO-AICESIS Conference.*

Joseph, J., & Varghese, T. (2014). *Role of Financial Inclusion in the Development of Indian Economy*. www.iiste.org

Joshi, V. K. (2014). Financial Inclusion : Urban-Poor in India. *SCMS Journal of Indian Management*, *11*(4), 29–37.

Kamboj, S. (2014). Financial inclusion and growth of Indian economy: An empirical analysis. *International Journal of Business and Management*, *2*(9), 175–179.

Kandpal, V. (2020). Role of regulators in intensifying financial access to the untouched segment of society in developing country. *Corporate Governance and Organizational Behavior Review*, *4*(1), 8–14.

Kaur, H. (2014). *A study on role of financial institutions in financial inclusion with special reference to micro financing in North India*. Punjabi University.

Khalaf, O. I., Abdulsahib, G. M., & Sadik, M. (2018). A Modified Algorithm for Improving Lifetime WSN. *Journal of Engineering and Applied Sciences (Asian Research Publishing Network)*, *13*, 9277–9282.

Khanvilkar, A. S. (2016). *Challenges faced by Microfinance Institutions*. https://www.nelito.com/blog/challenges-faced-microfinance-institutions.html

Kumar, V.; Kumar, S. (2016). Mahila Bank in India- A Catalyst for Economic Empowerment of Women (A study of Women Customers in Delhi & Semi Urban Area). *Conference Proceedings of Confernce on Stand-up India - International Conference on "Entrepreneurship and Women Empowerment", THE NATIONAL INSTITUTE OF ENTREPRENEURSHIP AND SMALL BUSINESS DEVELOPMENT (NIESBUD)*, 9–38.

Lahiri Chavan, A., Arora, S., Kumar, A., & Koppula, P. (2009). How Mobile Money Can Drive Financial Inclusion for Women at the Bottom of the Pyramid (BOP) in Indian Urban Centers. In N. Aykin (Ed.), *Internationalization, Design and Global Development* (pp. 475–484). Springer Berlin Heidelberg. doi:10.1007/978-3-642-02767-3_53

Lokhande, M. A. (2014). Financial inclusion initiatives in India-A review. *Intercontinental Journal Of Finance Research Review*, *2*(12), 22–33.

Madhu. (2014). BhartiyaMahila Bank (BMB) An empowerment to Indian Wome. *Radix International Journal of Banking, Finance and Accounting*, *3*(11).

Mbogo, M. (2013). The Impact of Mobile Payments on the Success and Growth of Micro-Business: The Case of M-PESA in Kenya. *Journal of Language. Technology & Entrepreneurship in Africa*, *53*(9), 1689–1699. doi:10.1017/CBO9781107415324.004

McKinsey. (2013). *Retail banking in Asia: Actionable insights for new opportunities*. https://goo.gl/4QjSLw

Munns, G., Sawyer, W., & Cole, B. (2013). *Exemplary teachers of students in poverty*. Routledge. doi:10.4324/9780203076408

NABARD. (2018). *Status Of MicroFinance in India 2018-19*. Author.

Onaolapo, A. R. (2015). Effects of Financial Inclusion on the Economic Growth of Nigeria. *International Journal of Business and Management Review*, *3*(8), 11–28.

Plyler, M., Haas, S., & Nagarajan, G. (2010). *Community Level Economic Effects of M-PESA in Kenya: Initial Findings*. Financial Services Assessment - Iris Center, University of Maryland College Park. http://www.fsassessment.umd.edu/publications/pdfs/Kenya-MPESA-Community.pdf

Porkodi, S., & Aravazhi, D. D. (2013). Role of Micro Finance and Self Help Groups in Financial Inclusion. *International Journal of Marketing*, *2*(3), 137–149. http://indianresearchjournals.com/pdf/IJMFSMR/2013/March/13.pdf

Prasad, E. (2009, Dec.). Growth in Asia. *Finance & Development*, 19–22.

Rajeshwari, M. S. (2015). A Study on Issues and Challenges of Women Empowerment in India. *Journal of Business and Management*, *17*(4).

Ranade, A. (2017). *Role of "Fintech" in Financial inclusion and new business models*. Academic Press.

Rojas-Suarez, L. (2016, Oct.). Financial Inclusion in Latin America: Facts and Obstacles. SSRN Electronic Journal. doi:10.2139srn.2860875

Roy, M. A., & Wheeler, D. (2006). A Survey of Micro-Enterprise in Urban West Africa: Drivers Shaping the Sector. *Development in Practice*, *16*(5), 452–464. doi:10.1080/09614520600792432

Sassi, S., & Goaied, M. (2013). Financial development, ICT diffusion and economic growth: Lessons from MENA region. *Telecommunications Policy*, *37*(4), 252–261. doi:10.1016/j.telpol.2012.12.004

Shah, P., & Dubhashi, M. (2015). Review Paper on Financial Inclusion-The Means of Inclusive Growth. *Chanakya International Journal of Business Research*, *1*(1), 37. doi:10.15410/cijbr/2015/v1i1/61403

Shakya, K. (2016). *Microfinance and Woman Empowerment*. Academic Press.

Sharma, D. (2016). Nexus between financial inclusion and economic growth: Evidence from the emerging Indian economy. *Journal of Financial Economic Policy*, *8*(1), 13–36. doi:10.1108/JFEP-01-2015-0004

Singh, A. (2014). Leadership in female SHGs: Traits/abilities, situational or forced? *The International Journal of Sociology and Social Policy*, *34*(3/4), 247–262. doi:10.1108/IJSSP-10-2013-0110

Singh, C., & Naik, G. (2018). Financial Inclusion After PMJDY: A Case Study of Gubbi Taluk, Tumkur. *SSRN Electronic Journal*. doi:10.2139srn.3151257

Singh, P., , & Tewari, P., & Verma, D. (2017). Status of Self Help Groups in Uttarakhand. *International Journal of Basic and Agricultural Research, 15*(3), 242–247.

Sriram, M., & Sundaram, N. (2016). Financial inclusion in India: A review. *International Journal of Applied Engineering Research, 11*(3), 1575–1578.

Yadav, M., Das, A., Sharma, S., & Tiwari, A. (2015). Understanding teachers' concerns about inclusive education. *Asia Pacific Education Review, 16*(4), 653–662. Advance online publication. doi:10.100712564-015-9405-6

Chapter 16
Applicability of WSN and Biometric Models in the Field of Healthcare

Nikhil Sharma
https://orcid.org/0000-0003-4751-2970
HMR Institute of Engineering and Technology, Delhi, India

Bharat Bhushan
https://orcid.org/0000-0002-9345-4786
HMR Institute of Technology and Management, Delhi, India

Ila Kaushik
Krishna Institute of Engineering and Technology, India

Siddharth Gautam
HMR Institute of Technology and Management, Delhi, India

Aditya Khamparia
https://orcid.org/0000-0001-9019-8230
Lovely Professional University, India

ABSTRACT

Health is considered as the most important ingredient in human life. Health is wealth is the most frequent used proverb. A healthy person can perform its entire task with full enthusiasm and great energy and can solve all problems as mind is a powerful weapon, which controls all our functioning. But now due to change in our lifestyles, we are becoming prone to all kinds of health hazards. Due to unhealthy mind, we are not able to perform any tasks. Humans are becoming victims of many diseases and one of the most common reason for our degradation in health is stress. In this chapter, the authors present role of WSN and biometric models such as two factor remote authentication, verifying fingerprint operations for enhancing security, privacy preserving in healthcare, healthcare data by cloud technology with biometric application, and validation built hybrid trust computing perspective for confirmation of contributor profiles in online healthcare data. A comparison table is formulated listing all the advantages and disadvantages of various biometric-based models used in healthcare.

DOI: 10.4018/978-1-7998-5068-7.ch016

INTRODUCTION

Health is one of the most essential state in human life. No task is carried out without being healthy, whether it is personal or professional life. In order to have proper functioning in all aspects one must be mentally and physically fit. Healthy living has the opportunity, motivation and capability to perform and act in all aspects. Some common factors such as specific diet requirements must be considered important for being healthy (Barbara et al.,2016). Proper diet plan, exercise, sound sleep are main ingredients of healthy life. But now a days, due to excessive work load and eating habits humans are becoming more vulnerable to diseases. Due to advancement in field of technology no physical work is carried out by humans due to which they are becoming unhealthy and prone to diseases. As increase in the number of humans visiting hospitals, taking appointments from doctors, lot of time is being wasted. Now as people have less time, they are not interested to have long span of time. For that purposes biometric have done tremendous work in health sector (Alpaslan, 2016). Everything is now automated which saves time. If person needs to visit doctor, patients' card is created which acts as a smart card having all the details of the patient such as his id, name, age, phone number, gender, address. Using this card, authenticity of the patient is being maintained and every time he need not have to get all his details entered. Just during his next visit card is being scanned and all the updates are being carried out. For treatment of iris and retina, biometric techniques are used which in very less amount of time are used for doing any eye surgery (Khosrowjerdi,2016). Laser based techniques are also used for this purpose. Security plays a very crucial role in any type of secure system. Main features of security such as secrecy which means the intended message must be kept secret during transmission, integrity which means there must be no modification in between while sending and receiving messages, availability which means the message must be available in future for further references (Amjad et al.,2017). In order to have security in the system, many biometric systems are being installed which uses face, palm recognition techniques to have authentication within the system and the outside intruders cannot steal any confidential information from the system (Bakke,2017). In the later section of the chapter, different biometric models such as two factor remote authentication, for security purposes effective fingerprint technique using biometric is adopted, privacy preserving in healthcare, Healthcare information using cloud technology with biometric application & Biometric Validation (Chattopadhyay,2016) based combined technique for trust computing perspective for authentication of providers profile in online healthcare data are studied. A comparison table is designed listing all the Merits and Demerits of the biometric system used in health care.

Internet of Things (IoT) plays a vital role in today's era. Almost each and every product which we use in our day to day life is being mapped with latest use of technologies which is concerned with IoT. With the use of this technology we are becoming advanced in every field (Athanasiou & Lymberopoulos, 2016). IoT applications are used in every field of different sectors which reduced human intervention, saves time and the work is carried out in a very healthy environment. The bridge that binds the gap between user's behaviour and software is known as Natural User Interfaces (NUIs) (Chen et al., 2019). Many recent features are used under this emerging technology. Using smart windows and refrigerators concept online information, can be accessed. Various biometric applications such as gestures on touch screen and touch pads, voice recognition used in homes and cars by using home automation and operations on smartphones. With recent trends and innovations in technologies where mouse and keyboard working are considered time consuming, natural user interfaces came into picture (Din et al., 2019). As the use of natural interfaces increases, risk to sensitive information also increases. The key features of security such as secrecy which means the important information must be made confidential to unauthorised users, integrity which means the information received must be consistent to both the ends i.e. sending and receiving, availability which means information must be made available to the users (Roy et al., 2017). One of the common examples for secure transaction is to enable e payment transactions. Innovative technology of user interface uses ability to capture high dimensional data. Earlier days, only few mouse clicks, mouse movements and key strokes with keyboard were captured and analysed (Premarathne et al., 2015). Whereas with recent technology a multiple key strokes, mouse clicks, touch area, touch locations, touch pressure etc. Users information can be extracted using natural interface. The captured information using biometric captures all information like palm detection, face detection, iris detection, fingerprint etc. (Sae-Bae et al., 2019) which can be used over the internet along with on a local device. Various studies have been done to access the uses of authentication mechanisms. For that reason's various authentication-based algorithms came into picture (Challa et al., 2017). Many techniques were proposed for the patterns which changes over time and various threats and vulnerabilities existing during that period were not addressed. In order to overcome all these above stated drawbacks, various security-based methods were introduced such as cryptography-based techniques, public key-based cryptosystem, authentication methods, dynamic key exchange etc. (Chatterjee et al., 2018). Security is considered as one of the most important concerns in any security-based model. The key elements of security such as availability, integrity and confidentiality must be preserved for carrying out any successful security model (Beldad et al., 2010). Patient's data is considered as one of the most important factors in any health-related system. Health information system is becoming more powerful these days where information is to be shared between patient and providers in order to provide diagnostics, self-care (Singhal et al., 2016). Computer based models were introduced for feeding patient's necessary details such as name, phone number, address, age, gender etc. Now in direction to preserve the privacy and integrity of the information certain security-based models were introduced such as cryptography-based models, setting up of firewalls, access management, backup policies etc. (Mohan et al., 2016). In order to have successful health information system one of the adopted methods is systematic mapping technique which maps certain set of steps such as analysing of health-related problem, types of research conducted for the disease, solution to the problem, recovery time. Some of the security properties comprises of no password verifier table which explains that no table is maintained for storing user password, friendliness of password gives the ease to user to have a user friendly password according to their comfort, no derivation of password by server administration, no loss of smart card attack means that unauthorized user have no access to the patient's card (Jaswal et al., 2019), resemblance to known key attack implies all the necessary combinations of password to have illegal access to the password, pairing of smartcards without having change in the identities of the user, user key agreement defines secret channel of data transfer using a key value between sender and receiver, no synchronization of clock explains that for verification of any patient's data clock must not be synced (Sodhro et al., 2018), means at any point of time patient's record can be maintained, when login is to be accessed, if user enters wrong password it can be easily notified, mutual authentication between user and server, in order to have proper tractability user activities can be properly traced (Gope et al., 2016). Natural user identification technique uses concept of authentication using actuators and sensors.

Three main components in this method are: sensors, actuators, and credentials (Krasteva et al., 2017).

There are three major components such as Actuators, Sensors and Credentials which are explained as follows:

SENSORS

The different types of NUI sensors are explained as follows:

- *Touch Surface:* It is used for capturing information using touch screens which captures fingerprints, hand motion in a two-dimensional plane. The output is shown in form of time line graphs, timestamp values. The touch surfaces are now days used in smartphones, tablets, and smart watches large interactive displays etc.
- *Camera:* It act as sensor node for capturing an image or for displaying videos from different input sets gathered from different locations. Brain-computer interface, microphone and Motion sensor. A 3D sensing camera captures highly sensitive image with more precision and reflects each small point with high sensitivity (Pirbhulal et al., 2018). For using control device, an eye tracking interface, user requires to change gaze direction which is hands on operation. These are commonly used in user based electronic devices.
- *Motion Sensor:* These sensors are used for capturing person's gestural information. These types of sensor are frequently embedded with various electronic gadgets like Instrumented gloves, smartphones, smartwatches, tablets, and increased-reality headsets to wireless game controllers etc. In any three-dimensional coordinated system, these are used to locate position, orientation and space of the object.
- *Microphone:* This device allows hand free operation which the device and captures voice. The interface has been embedded in many electronic devices and is used as an interface between user and devices (Dubey et al., 2016).
- *Brain Computer Interface:* It act as a bridge between user and machine which does not require any interaction between the physical movements. To measure electrical of brain's activity, one of its application is used. The activity can be seen in form of waveform with varying timelines. The waveform pattern can be sent in smart phone using recent technologies. **Figure 1** shows the different types of NUI Sensors.

Figure 1. The different types of NUI Sensors

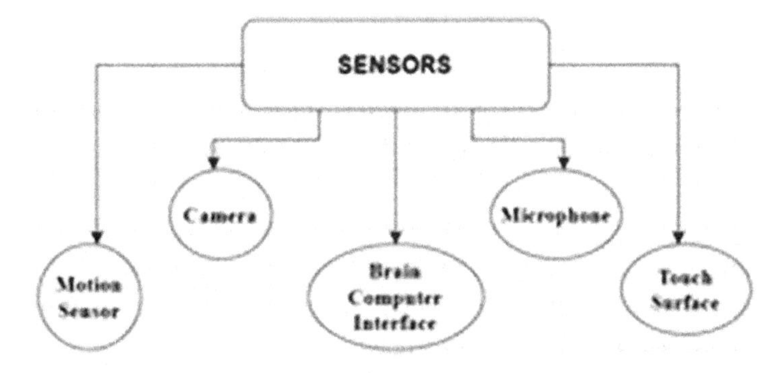

Actuators

In any communication model, different human parts can be used as positioned for interacting between natural user interface. **Figure 2** shows the different types of positioners used in the process of communication are body, eye, head and brain (thought), finger, hand, vocal fold (voice) (Son et al., 2009).

Figure 2. Types of Positioners used in the communication process

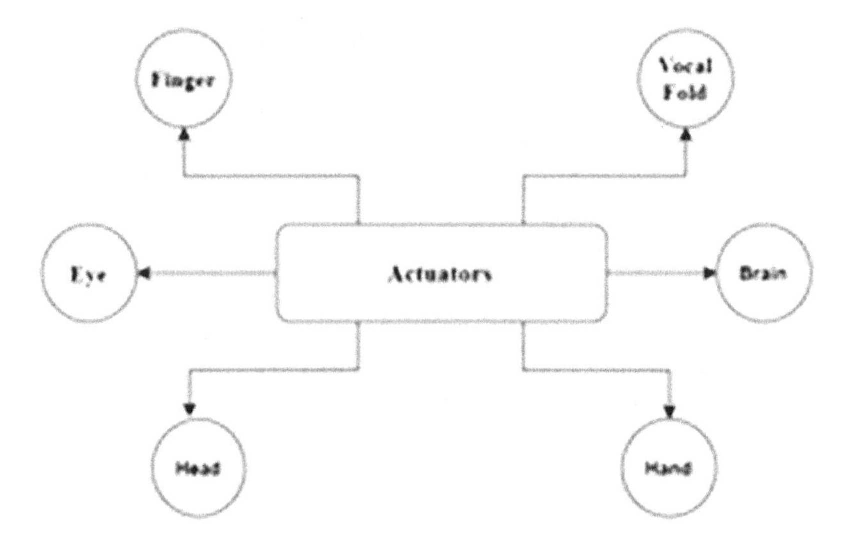

- *Finger:* Any type of non-verbal communication carried out by finger in the communication process. The interaction process can be deceived using touch surface, camera, or motion sensor. Using different locations of fingers such as in fingertips different sensors are recorded over timeline series with different instant of time.
- *Hand:* This is also a type of non-verbal communication, where the message is being exchanged using hand gesture. Mainly hand gestures are of two types: moving motion and stationary motion. The signs can be captured using red green blue colour spectrum or can also be used with touch surface or motion sensors (Devi et al., 2017).
- *Head:* This is another type of non-verbal communication involving movement of head which can be captured by a camera, motion sensor, or touch surface. When a motion sensor is connected via head it captures head orientation and velocity. The intensity of velocity can be measured via using video streaming algorithms.
- *Eye:* Applications of eye used in biometric such as iris pattern, retina pattern, eyeball tracking is captured using special cameras. This provides fast, easier and convenient method for user to interact with the system. All the important information can be tracked on two-dimensional coordinate axis using timeline series with varying parameters.
- *Vocal Fold:* On daily basis communication, we use natural mechanism for communicating, by use of vocal cords. Voice recognition system is the technology which uses software programs which have the ability to decode human voice is used to operate device, perform tasks or write something without involvement of mouse, keyboard, or press any button (Alaiad et al., 2017).

- *Brain:* Any unintentional thought or intentional thought with any specific characteristics can be used to authenticate users. Recording of any neural activity can be recorded using electroencephalography.

Credentials

Credentials are some qualities, experiences that add meaning to something or it provides authenticity to the data. In order to have exchange of any confidential information, authenticity is required which states that only authorize users have the right to access information. Validation identifications that do not need a user to carry anything can be classified into one of the following classes: Physiological Biometrics, Secrets, Behavioural Biometrics, and combination (Abdulnabi et al., 2017). **Figure 3** shows different types of Credentials.

Figure 3. Different types of Credentials

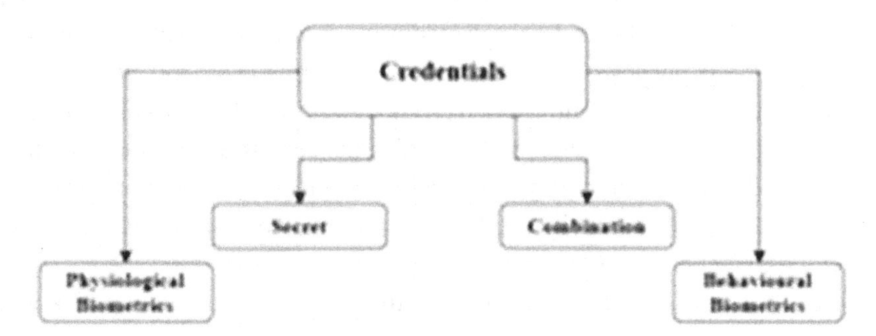

- *Secret:* it is defined as any user specific information which is used as an important credential for authenticity. One of its recent examples is entering of text password using physical keyboard. This specification with natural user interface can be used as an input with other mechanisms.
- *Behavioural Biometrics:* New types of users are enabled using this technique. Characteristics such as signatures, gestures, speech, swipes are distinct among individuals and act as behaviour biometric.
- *Physiological Biometrics:* Some of the biometric applications do not take into consideration use of behaviour, rather they rely on physiological biometric information such as iris, face, fingerprint.
- *Combination:* Different combinations of a behavioural biometric, physiological biometric and secret are used in any authentication system.

The rest of the chapter is organised as follows: Section 2 shows the literature review, Role of Wireless Sensor Networks in Healthcare Applications describes in Section 3, Section 4 discussed various Security Measures Using Biometric Fingerprint Schemes, Section 5 explain the role and utilization of Cloud Technology with Biometric Applications for healthcare records, Two Factor Remote Authentication In Healthcare illustrated in Section 6, Section 7 elucidated role of Biometric Authentication Based System For Validating Online Healthcare Information followed by the conclusion in Section 8.

LITERATURE REVIEW

In order to have secure communication and confidentiality between data transfer, various security-based mechanisms were introduced. One of the used security mechanisms is password-based authentication technique. In this technique an additional database is maintained for storing password of various authenticated users. From the table where the password is being saved is matched, if it is matched then the user is considered authenticated otherwise no information is being shared. In order to decrease the overhead of maintaining an additional database one of the alternative solutions used is using of fingerprint (Patil et al., 2017). Applications of biometric are widely used in all security-based models. One of the loop hole in any security-based model is different types of attack such as key impersonation, man in middle attack, replay etc which decreases the overall efficiency of the system (Venkatesan et al., 2018). Some of the cryptographic based techniques such as hashing, public key and private key are used to provide authentication in the system. In wireless based network many outside adversaries affect the sensor node by capturing its necessary information (Sodhro et al., 2018). An elliptic based cryptography for light weighted authentication protocol is being used. One of the main concerns in wireless network is energy constraint. As wireless nodes have limited energy resources, it is very important to preserve the energy (Sodhro et al., 2018). For this purpose, many energy-based routing protocols were introduced. For any IoT based application, mutual authentication-based protocol was introduced (Kalaivani et al., 2016). This technique provides authenticity of message on both sides i.e. sender and receiver (Kalid et al., 2018). IoT based medical care decreases the replay and disclosure attack frequency occurring in the system. In order to increase the reliability, scalability, efficiency and readability, IoT based technologies can be merged with cloud-based technology (Car et al., 2017).

While designing a Wireless sensor network for healthcare and multimedia applications, power consumption will be the essential factor. Multiple techniques are used to minimize the power consumption in WSNs. A wake-up radio in multimodal sensing that automatize sensor nodes proposed by (Jelicic et al., 2014). In WSNs, it helps to reduce the power consumption as well as response latency to manage. This methodology gets applied for high power consumption sensors in two tier Wireless sensor network for video/camera monitoring applications using low power consuming infrared sensor nodes. The big health application system presented by (Ma et al., 2017), based on health internet of things and big data. The author proposed the cloud to terminates fusion big health application system. The system comprised of multiple layers such as transport, perception and big health cloud layer. The major challenge between these layers is the power consumption for the shorter distance wireless communication.

Wireless multimedia sensor networks (WMSNs) is an infrastructure less implanted gadgets that permits retrieving audio, video streams, scalar sensor information and still images from the physical environments. (Mekonnen et al., 2017) suggested a combination of different methods multilevel WMSNs prototype which comprises of low power hardware systems. WMSN uses a network structure which includes different modes i.e. a shutdown mode, a sleep mode, and wake up mode. (Wong et al., 2006) presented a hash based dynamic authentication strategies to counter different attacks such as Key impersonation, forgery, Denial of Service (DoS), password disclosure, replay and man in the middle etc.

The secret password-based authentication system suggested by (Lamport, 1981). By applying elliptic curve cryptosystem, (Chang et al., 2016) devised a lightweight authentication system. For attaining the possession of forward confidentiality, (Chang et al, 2016) devised an ECC-based authentication. (Yeh et al., 2011) build a two-factor system based on ECC authentication. Based on DH and RHA, (Watro et al., 2004) presented a secure authentication protocol for WSNs. (Yoon et al., 2013) suggested a biometric based user authentication strategy to mitigate security vulnerabilities like DoS, poor repairability and sensor impersonation attack. For handling the data integrity in IoT based WSNs using the knowledge of base stations, (Hameed et al., 2018) introduced a security system based on integrity techniques. For supporting Cloud computing, mobile edge and IoT services like feature adaptability, scalability and reliability, (Al-Turjman et al., 2018) build a cloud desegregated structure. For context aware IoT, a hash-based RFID authentication devised by (Deebak et al., 2019). For IIoT domain, an ECC based au-

Table 1. Various Security Techniques and Their Limitation. (Deebak et al., 2019)

Ref.	Security Techniques Employed	Formal Analysis Model	Simulation Used	Limitations
(Car et al.,2017)	Seamless Key Agreement Framework [For Mobile-Sink and IoT-Based Cloud-Centric Network]	Partial	No	Difficulty in recovering from different prospective attacks such as key impersonation, password disclosure, DoS (Denial of Service), and official password guessing.
(Wong et al., 2006)	Secure Authentication for Medicine Anti-Counterfeiting System	Yes	No	Difficulty in recovering from different prospective attacks such as message eavesdropping, Denial of Service (DoS), and Smartcard forgery.
(Roy et al., 2018)	Anonymous User Authentication Using Chaotic Map [With Biometrics and Fuzzy Extractor]	Yes	No	Several loopholes were address during working of this mechanism such as storage cost, large computation and partial perfect confidentiality.
(Li et al., 2017)	Improved Secure Authentication [With Data Encryption and User Anonymity]	No	No	Difficulty in recovering from different prospective attacks such as key impersonation, Denial of Service (DoS), and privileged insider.
(Al Turjman et al., 2018)	Seamless Identity Provisioning Framework [With Mutual Authentication Approach]	No	No	Difficulty in recovering from different prospective attacks such as Message Eavesdropping, Denial of Service (DoS), smartcard forgery, and man in the middle.
(Gope et al., 2018)	Lightweight Privacy Preservation Protocol [Using Physically Uncloneable Functions (PUFs)]	Yes	No	Several loopholes were address during working of this mechanism such as storage cost, perfect confidentiality, and large computation.
(Li et al., 2018)	Lightweight RFID Mutual Authentication [Reader with Cache]	No	No	Difficulty in recovering from different prospective attacks such as message eavesdropping, reader impersonation, and tag forgery.
(Li et al., 2018)	Secure 3PAKE Protocol Using Chebyshev Chaotic Maps [With Random Oracle Model]	No	No	Difficulty in recovering from different prospective attacks such as key impersonation, password disclosure and offline password guessing.
(Wazid et al., 2017)	Secure Lightweight Three-Factor Remote User Authentication [Using Smartcard, Password and Personal Biometrics]	Yes	No	Difficulty in recovering from different prospective attacks such as DoS, smartcard forgery and message eavesdropping.
(Deebak et al., 2019)	Hash-Based RFID Authentication [For Context-Aware IoT]	Yes	Yes	Difficulty in recovering from different prospective attacks such as data forgery, DoS, and privileged insider.

thentication system devised by (Li et al., 2018) which uses biometrics characteristics to authenticate the service approach. For IoT networks, secure lightweight authentication scheme which employs password, biometric and smart card as three components to satisfy with key agreements possession introduced by (Wazid et al., 2018). To prove the authenticity of dosage forms, a newly authentication techniques gets devised by (Wazid et al., 2017) for medical spurious systems. For IIoT, a seamless mutual authentication strategy suggested by (Al-Turjman et al., 2018). **Table 1** shows security techniques suggested by different authors and their limitations.

ROLE OF WIRELESS SENSOR NETWORKS IN HEALTHCARE APPLICATIONS

Wireless sensors network also play key role in healthcare applications. Sensors are directly placed to the patient body for monitoring them. In emergency cases, like heart attacks, sudden falls, temperature (by transferring secured signals), low oxygen levels, video or images to the appointed unimpeachable unit, sensors are employed to identify these kinds of serious situation because even few seconds might save any human's life. Thus, for transmission, information gathering and reception real time monitoring requires energy sources (Jain et al., 2016). In WSNs all gadgets are operating using batteries, thus power is the challenging condition almost in every application of WSNs. The total energy utilization of the entire network gets minimized when the energy consumptions of sensors gets reduced which extends the durations of entire networks. Compressive sensing is an approach in which before transmission the information's such as video, signals or images gets compressed and transfer to the reception where this data decompressed.

The clustering is another effective approach where WSN gets split up into a group of clusters. Each group is having a cluster Head (CH). Cluster Heads liaises with each other to share the collected information from forbearing's sensors to Base Station (BS) that liaises with server. All these communication approaches introduced above are based on Binary Transmissions schemes. Instead of using binary transmissions schemes, the quaternary schemes can be used effectively so that energy can be more efficiently utilized. Using this scheme, instead of using a binary logic, multi valued logic gets employed through which energy efficient structure of WSNs can be build. In (Saleh et al., 2018), the design of sensor networks gets modified so that instead of using binary (two symbols), the quaternary (four symbols) can be manipulated by the sensor networks (SNs). **Figure 4** shows WSNs architecture based on quaternary.

Figure 4. WSNs Architecture based on Quaternary

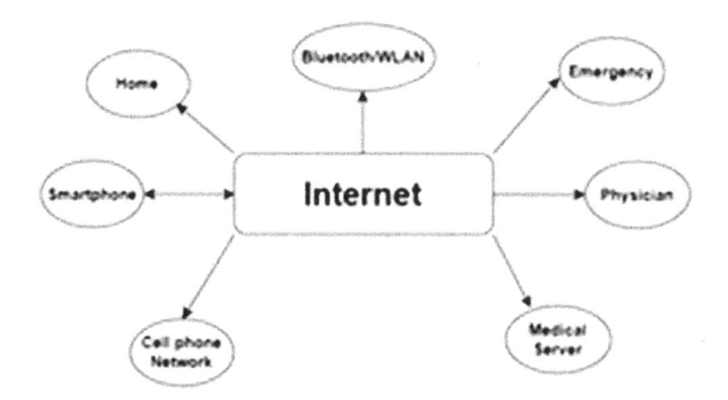

(Saleh et al., 2015) suggested a quaternary architecture through which the transmission of information in WSNs can be modified from binary symbols to the quaternary symbols for healthcare and multimedia applications. The two main modules are used for converting a binary link into a quaternary interconnecting link i.e. a binary to quaternary modulator and quaternary to binary demodulator as shown in **Figure 5**.

Figure 5. WSNs Architecture with the Quaternary Transceiver

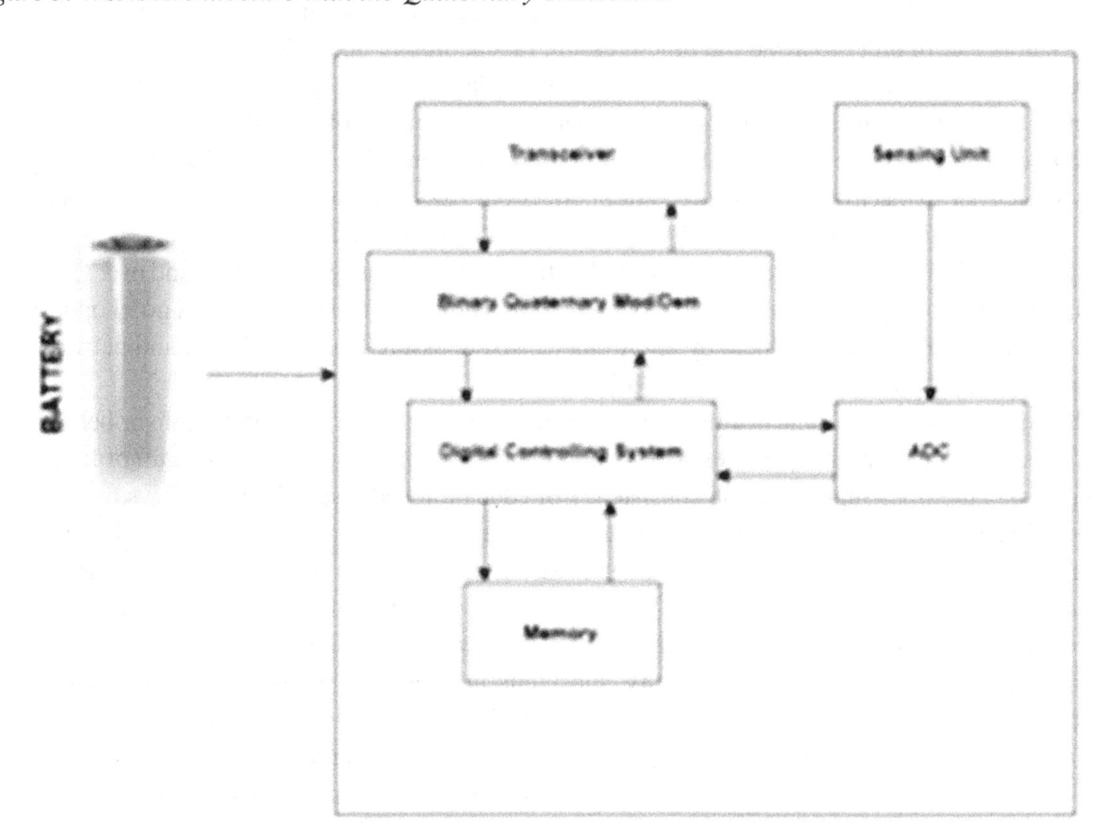

In WSNs, LEACH protocol is used to reduce the energy consumption. It contains only meaningful information while transferring data. In LEACH Protocol, the clustering approach used which divides the whole Wireless sensor network into a group of clusters. Each cluster is having a cluster head (Hussain et al., 2006). The author also designed the Neural Network Static RAM (NN-SRAM) architecture for gathering and storing data. The NNSRAM used to store binary information and remains valid till the power is supplied in the network. NN-SRAM helps in improving the lifetime of the system in WSNs as it reduces the energy consumption in the network (Lee et al., 2016). Sensor networks are stored data temporarily that leads in power consumption in wireless sensor network system. Therefore, instead of storing data temporarily in Sensor networks, NN-SRAM (Neural network storage units) (Haider et al., 2015) used which helps in decreasing the power consumption in a network and improves its life span. In WSNs, the energy utilization in NN-SRAM-CBEES (Neural network static random-access memory with clustering-based energy efficient systems) will be minimized by limiting the storage cost that get increased because of using neural network and also by limiting the transferring/receiving cost between sensor nodes and their cluster heads. The non-cluster head of sensor networks must be turned off as long as possible.

SECURITY MEASURES USING BIOMETRIC FINGERPRINT TECHNIQUE

In today's world as health-related issues are increasing rapidly due to change in the lifestyle, unhealthy eating habits etc. In order to avoid the time pressure of standing in queue for routine check-ups everywhere in health sector, biometric applications are used like recording the personal information of the patient in smart cards which reduces the time burden, and authenticity of the user is maintained. In order to avail authenticity of the user, certain security-based parameters are introduced Nowadays, secrecy, safety& security in electronic healthcare system effect our health Rapidly-Health Records procedure a good communication between patients & any person of the Medical, pharmacy or nursing profession. The secrecy and safety issues recently arising e-Health area are primarily focused around user validation, user information privacy, and patient secrecy protection. Biometric technology has a good opportunity to deal successfully privacy problem by providing trusted & assured user validation and permission to obtain the health information (Jain et al., 1997).

The emergency of electronic health has transformed the health care industry to deliver better health care and service at affordable price, eHR is a digitally archived health care with the aim of looking after a person's lifetime. Supporting the consistency of education and research, and always make certain privacy. Issues of secrecy and security in the eHealth area primarily focus about user validation, data integrity, information privacy and patient's privacy security. Because of conscious nature of health data, dignity, data privacy and patient security. With the conscious nature of health information of patients, the greatest task is toward the enhancement of the effective safety model that can give assurance of information confidentially and dependability, substantiating that only undertaking employees can access their respective health information (Lin et al., 2016). Old validation and access control measure may not match the eHealth record system where roles are not properly addressed to all stakeholder (administrations, patients, health professionals). Presenting healthcare professional needed access to all information as possible that helps in supporting decision making. Forbearing also wanted to gain the access over their personal data related to eHealth records with data privacy. SURF (Speeded Up Robust Features) aims at providing loop variant, technique for descripting and detecting points. This technique is also considered as advanced description of SIFT (Scale Invariant Feature Transformation). SURF aims its popularity in speediness, robustness and different orientation transformation against different locations. For finding space of different scale Gaussian space is calculated using difference of Gaussian. Combination of difference of Gauss and box filter goes beyond expectations in SURF. Using integral images convolution with box filter can be easily calculated, which is one of the major advantages of this approximation technique. For both the metrices such as scale and location, a fixed value calculation i.e. determinant is calculated of Hessian matrix. (Jahan et al., 2017) in his work concluded that biometric has its applications in every field due to its advancing feature in security. One of its application used in healthcare can be used at low cost and is highly secure. Traditional applications lack important security matrices such as authentication, integrity and confidentiality which is provided in biometric systems. Biometrics technology can be integrated into an extensive variety of healthcare applications to assure security and privacy. **Figure 6** shows Schemes used for Verification of users using fingerprint biometric for access control.

Figure 6. Schemes used for Verification of users using fingerprint biometric for access control

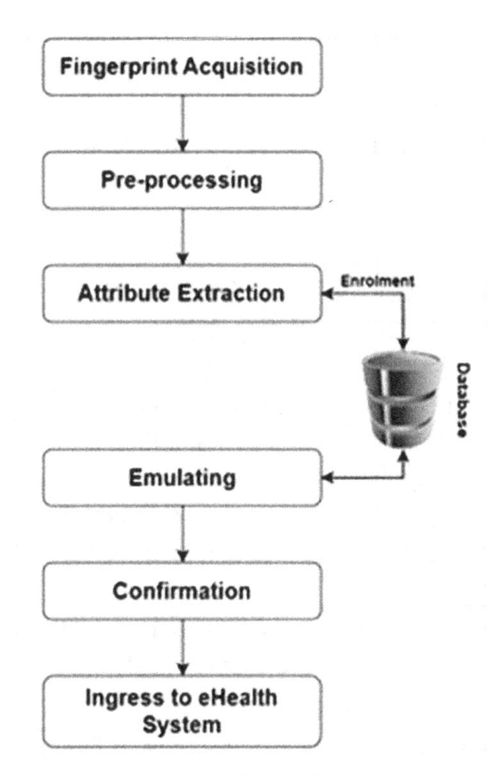

To defend the solitude and freedom of the patient's materials a secure verification system must be used to ingress health reports. Any person or institutions wishing to ingests a second hand health record will require a definite approval of that victim. Including freedom perils and reclaiming health history in the EHR arrangement has expanded the demand for a trusted buyer certification in the E-Health realm plan. Certification is considered an important aspect of security in the Healthcare realm which intension to clarify the user's integrity. The user wants to desire maintenances from the cloud. Universal verification methods like usernames, passwords and avenue cards are not useful in the E-Health situation. The incident is being adrift, lifted, erased and mistaken. Secondly, biometric based authentication and approach control have proven capability of providing the crucial safety and confidentiality E-Health applications. This paper presents an extensive audit of biometrics techniques addressing e-security. In the paper, we nominate a potent and active biometric plan to use user authentication and authority fingerprint biometrics to hike confidentiality and safety eHealth system.

Conventional authentication system was based On ID card. It May include badge or encryption that can be off-track, or shattered, on the other hand, attributes can't be misplaced or forgotten. They are hard to hand up dispense. In Addition, the database is needed to authenticate the persons presence, it is crucial to Produce and it is unlikely for the user to cancel it. In order to need HIPAA guidelines, patients and health Executive both required to have approach to medical records. Keeping in mind the demands of health professionals both and patients, biometric validation is capable of meeting security and privacy necessity.in confirmation method, it demonstrates the legitimacy of an individual by their Finger print in the identity method, it provides the recognition of the individual among the enrolled in a database. Without the judgement of an individual, the fingerprint recognition system bid to match it with fingerprint in the entire fingerprint record. Among all the others biometrics characters, fingerprints biometrics attend a vital and very important part in the field health due to its exclusivity and accessibility. The first step of the fingerprint validation process for catching the traces of fingers pattern using a sensor of digital copy. The catcher picture is called a live scan. The live scan is an electronically managed factor to generate a biometric prototype.

USE OF CLOUD TECHNOLOGY WITH BIOMETRIC APPLICATION FOR HEALTHCARE RECORDS

Healthcare records are prerequisite to each medical professional and patient too, a technique called biometric safety system in health-related sector for availability via the use of programs connected via internet rather than system will be reviewed. By the use of cloud technology as a protected storage means to hoard essential healthcare data have an important part. Currently, this technology is majorly used for storage of data so it may well impregnate wherever and anytime. Security based on biometric is a significant aspect in this technology so that data supremely vital data cannot be retrieved by anyone except you have been acknowledged contact by the system by registration. Biometrics offer improved safety in are zones to which it is required. Healthcare nevertheless, is also in requirement of safety when it deals with storage of data and further essential data. Biometrics offers security and agreeableness anywhere this technology is used (Haghighat et al., 2015). In the field of healthcare, the records and data about patient's health are one of the greatest valued personal information, doctor needs to access them quickly,

Figure 7. Use of Cloud technology with biometric application for Healthcare records

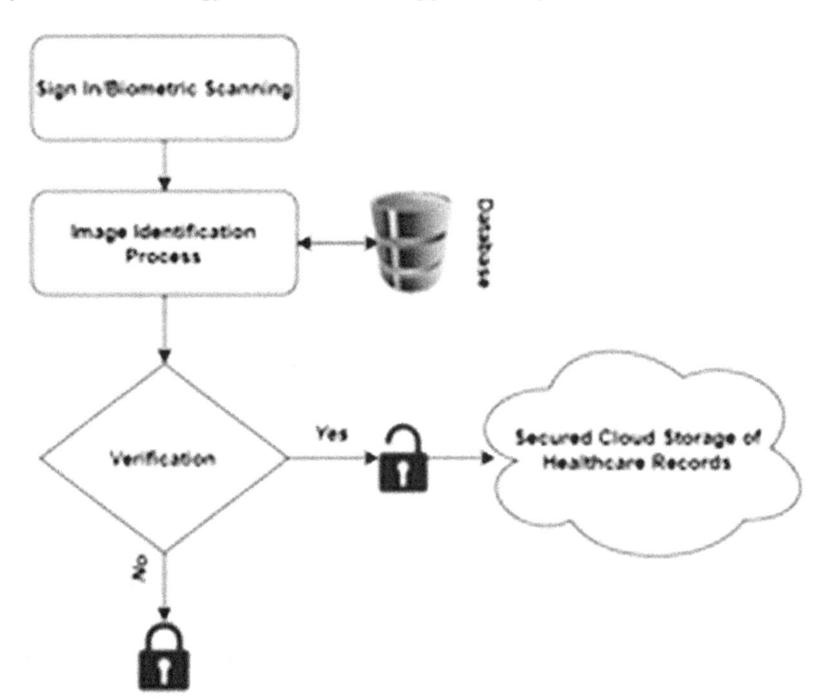

and they have to be precise and faultless. A failure in security and appropriate counting of information can mean a great variation amongst timely and precise analysis or permit health falsification. Biometrics though can be the response to all those breaches. **Figure 7** shows use of Cloud technology with biometric application for Healthcare records (Choi et al., 2015).

Applications based on use of biometric- Identification and Validation are the two most significant areas of security. Breaching of any of these two factors are not considered secure for any security-based system. The different types of authentication process in any secure system are: something you have (e.g. token device, badge), something you are and something you know (e.g. password). Different applications of biometric includes: Fingerprint biometric system, face recognition system, palm recognition, smart card for containing all the details of the patient.

- *Cloud Technology:* Cloud based computing technology which is based on internet connections, also called demand technology, where user can access and share data and resources to computers connected on wide range. For a shared pool of computing resources this model act as enabling technology. It orders to have effective utilization of resources; this technology is used to scale its expenditure.
- *Cloud Security and Privacy:* Three important aspects of security: Confidentiality, integrity and availability must be ensured in any security-based model. Confidentiality implies that information must be kept confidential while transferring from one source to another. Integrity implies that the content of information must be preserved while reaching at receiver side. No alteration can be done in between the content of the data. Availability implies that data or information must be available for future use.
- *Cloud storage:* For storage of any digital data over the network, cloud storage is used. The physical environment is operated by company, physical storage spans over multiple servers located at different locations where data is being stored and retrieved from the server. The responsibility of storage provider is to keep data available, authenticated and protected from unauthorized users. Individuals and management purchase or rent storage quantity from the providers to keep data of users, association, or application. In order to have security in the system, many biometric systems are being installed which uses face, palm recognition techniques to have authentication within the system and the outside intruders cannot steal any confidential information from the system Using the technology which uses programs from net rather than system for storing data and additional safety, and the biometric system for user testimony, it will enhance the safety on retrieving important health related data online. Medical experts and health staffs from several healthcare services can access this data online.

For use of security in health care-based sector, biometric applications are being widely used. For accessing patient's data, it is very much difficult to keep track records of all the information. For these reasons, biometric applications are used. They reduce the time boundations and involves less human intervention. Patient now uses smart cards provided by the hospitals which comprises of their unique id, name, gender, sex, address, phone number etc., which provides more authenticity to the patient (Huang et al., 2016).

TWO FACTOR REMOTE AUTHENTICATION IN HEALTHCARE

On procedure to command approach of health accompany info stocked in general database, effective confirmation structure want to be deliberate. The early influence need easy & efficient observable biometrics key-stroke scrutiny exemplary although the next influence need undisclosed PIN system. In influence first confirmation, key-stroke determination is expected where basic info is calm early; then treated info are stocked and trustworthy record are producing certainly (Das, 2011). Conduct determination of the projected system exhibition its ability to validate end-users. Assistance worker may not be efficient to get back delicate health info reserved in server in its infant mode to control secrecy of info. Biometrics certification is one of the possible explanations in server rule being it handle alone trait of end-customer for testimony & approach authority. Biometrics feature may be corporal or observable. Physiological info involves few countenances that endure hardly for several particular such as vein, DNA, fingerprint, nails, iris, teeth pattern etc. (Apama et al., 2017). Biometrics confirmation over key-stroke examine is an accurate procedure for integrity facts. Key-stroke motion boost its demand between the observable biometrics' confirmation approach in assigned atmosphere, because of its clarity & expense influence. Use of key-stroke disclose the classify method of an entity by using alone regular keyboard & easy program. It measures that the exclusive custom are developing in period, ensuing the ability with console, mental status or physical & more container. Reputation of the method could also be ordered as average that means that crowd are not efficiently curious by the way of biometrics info association. Several protection alive systems should tool two-factor confirmation, which underrate defy of info crack caused by illegal approach to delicate info at distant health scheme (Pirbhulal et al., 2015). Couple cause of confirmation is planned attending to improve safety matched with basic attempt & charge in the server. both patients & health specialist want to contribute confirmation evidence before pervade isolated health info of patients. The extensive offering of the paper is to suggest a two-factor confirmation system using key-stroke inquiry & PIN that can be used to remote core. **Figure 8** shows Flowchart of Pin Generation (Bhattasali & Saeed, 2014).

The equitable of this effort is to resolve some of the dominant objection in great free content-based key-stroke inquiry & to suggest two element faraway confirmation core as a possible explanation. It contains examine on extant remote healthcare ideal, access control system & key-stroke gesture procedure for verification. Server established sensor system permit early conclusion of subject. Healing handler have to genuine themselves to notice health status of patients (Azeta et al., 2017). In server based isolated health core, server utility provider manages biometric info where info is stocked in encode plan. Key-stroke study is located on the usual analytical determination or arrangement recognition method. Arrangement has been erected from entity typing style. All of these analysis target on the key-stroke motion input from a measure console (Zhao et al., 2015). In pressure- located user confirmation scheme, the distinct time outstanding is convert into the frequency territory. Typing method is also evaluate during use of hand-held electronic gadget just like wireless cell-phone. After examine differing current entirety on key-stroke gesture, it is projected to be used in server located faraway health core along with privacy PIN approach to diminish expectation of false disclosure. main dispute of key-stroke confirmation is wrong positives & wrong negative that either chunk connection to legal consumer or allocation access to illegal consumer. Key-stroke biometrics can also margin deviation in the typing manner due to occasional classify, using sweaty hand, single hand for typing after a long period. False agreement gives ascent to zero-effort charge, where hypocrite can calmly enter into the rule. False refusal opposes accurate user to come into the organisation which may disturb order ability obliquely. The fair of planned effort is to

Figure 8. Flowchart of Pin Generation

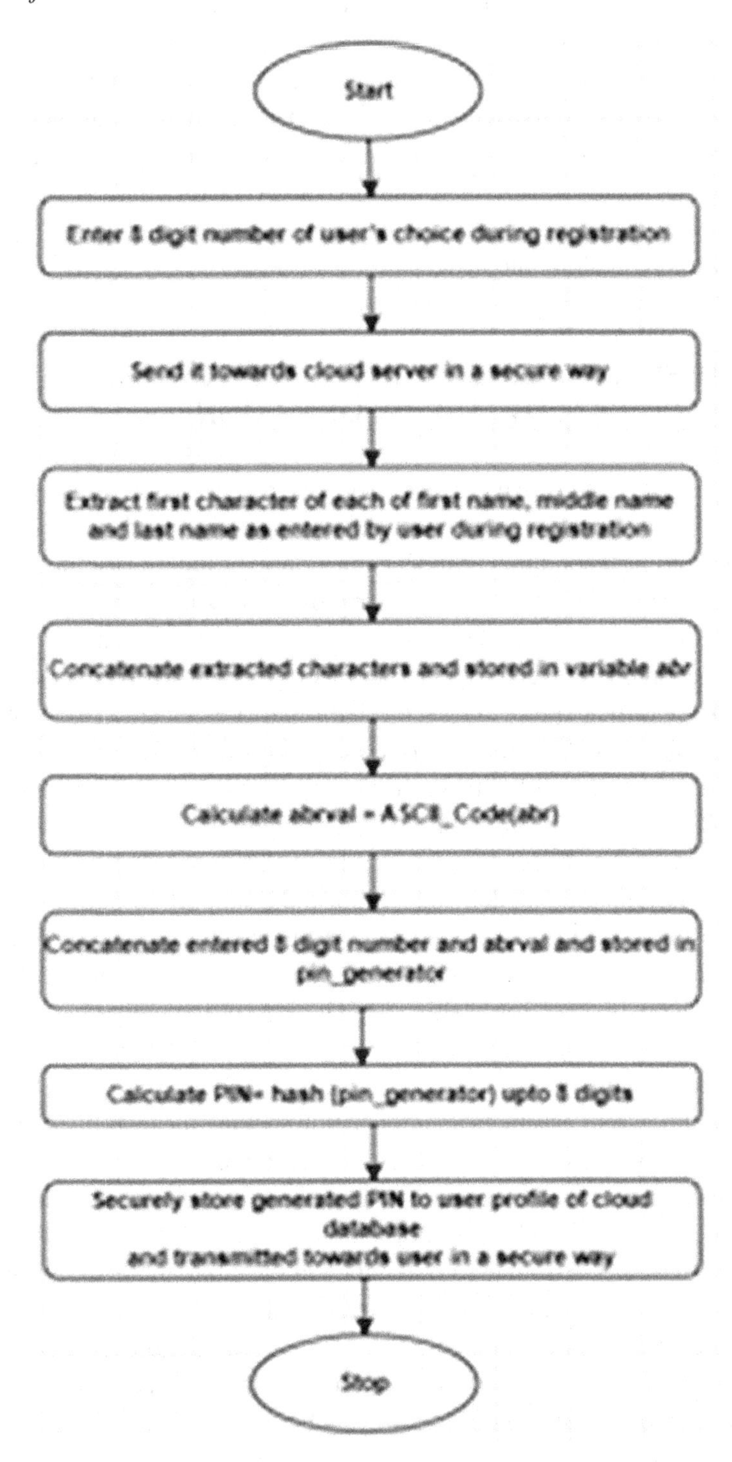

command access of EHR/EMR of patients in a simple technique (Zhang et al., 2014). Faraway health core arranges dissolve biometric confirmation to cut the mistreat of ability within the system. Usual procedure of verification is much more performing exhaustive and may be mistreat by any hypocrite in server condition. It is pretended that assistance support by server provider is assure give to service level compromise.

The journalist has recommended here the custom of key-stroke motion for scheme approach, which does not charge & needs no extra tools except the console. The two major causes confirmation is advised here, where hidden proof is linked with key-stroke inquiry to improve precision level of confirmation. The conclusion show that correctness level is improve in two factor confirmation distinguished to using only one factor key-stroke. The projected key-stroke sense is being changed by considering various operation particular limit & healthy info bent.

BIOMETRIC AUTHENTICATION BASED SYSTEM FOR VALIDATING ONLINE HEALTHCARE INFORMATION

Onset of the Internet and the rise in the usage of cyberspace by clients for online healthcare Information (OHI), various analyst in discrete protocols are engaged on protest and risk of cyberchondria, which is relevant to cyber psychological condition of indecision, concern, superiority, and reliability. OHI-based faith analysis has not once used biometrics to confirm multi-dimensional trust constructions, together with visualization, influence, awareness, and social status. Moreover, this study path is not handled by proof of OHI prospect at provider level through proof institutional Profile and relationship. Accordingly, in order to boost accuracy through proof at trust level, this paper expectation a conception and nominate an offbeat trust-computing model, motorized by visual acknowledgment-based biometric verification of medical profiles. Introductory proof of theory prototype for the proposed approach, an observation is executed that establish the viable execution of the trust-computing pattern. The investigational reaction implementation gather over this model are joint as part of this paper. This model will offer more new researches with the trust-computing model, and will form the support for outlook conviction related research at OHI to address cyberchondria. The Internet has an intellectual shock on the way people pursuit for health care data, and transmission. According to a domestic audit on Internet usage, eight in seven American adults examine networked for health data. The idea of the online fact expectation is specifically crucial in the situation of fitness advice obtaining. While record numbers of American adults are unstable to self-diagnose on the Internet, look for prescription alternative and select a physician, many people face many disputes. Crisis arising from information seeking incapability to discover correct data, irregular instruction or report, inability to find health data, and psychological emotion processes are all samples of some of these disputes. For example, cyberchondria maturity increased concern associated with health litter ensuing from online health information. Study in this area advise that few individuals are mainly liable to online-related trouble to find data. On the clone moment, position of health worry is easily relevant to the regularity and the event of online health information exploration. This seasonal design of agony in facts looking have damaging belonging on the singular, who is in most cases earlier agony outside of a deeply perceptive or Incessant health circumstance that incite the primary seeking to find something. One valuable constituent of trustworthy has to act with consumer refusal or option of a specific place of activity, its opinion to logic such user will imply with place of activity they vision as trustworthy and refuse those, they doubtfulness.

The connection among the handling of images and consumer trustworthy of health web-sites. The analysis in (Chattopadhyay et al., 2017) desire that images boost trustworthy by discover a sensibility of intimacy or advocacy. Consumer occurs to be affected by the trait of the spot or by the occupancy of well-known statue or trustworthy icons. OHI desire that the feature of facts ready for use on a user-based site plays act ahead with the seen profile added knowledge of the physicians. They again acumen that OHI assurance is compelled by the range to which the breadwinner emerges to be weak-known with connectable accomplished a joint public integrity (He et al., 2015). Cyberchondria is not plainly hypochondriacs adopting the Internet, but a bit an obvious shape of online data looking for and develop increase health worry (Awasthi et al., 2013). Cyberchondria has been deliberate as an obvious spiritual clutter and a complex idea with doubtfulness of medical artist as one of its key characteristics. Current analysis has form that cyberchondria is existent additional carefully line up with other mode of dubious Internet purpose than usual hypochondria. Cyberchondria includes an analytical answer to a scarcity of data. Additional Individually, behavioural proof of hypochondria and cyberchondria are where the key dissimilarity halfway the two become seeming. As Vice-versa, hypochondriacs assure themselves they are agony from an exact medical situation, clear material response to their seen problem, and employ in conduct that accomplish their matter. The basic stimulus for hypochondria is the opinion that one has a particular medical situation, the cyberchondria is generally compelled by a want to receive greater data around an explicit circumstance.

The subject of cyberchondria hand-over produce base for social-science researcher, and also particularly, for giving critic. The brunt of cyberchondria can be verified from an informative outlook over the lens of mechanics expertise, health ideas, and social analysis. Some model of tricky or driving need of the Internet has social shock. Therefore, there are entity usually looking for out health data online only to be mistaken or mislead. The basic mater is a wry or faulty idea of private prospect outstanding to online misreport and misleading (Wu et al., 2012). Present-day consumer ability has produced approach toward health concerns as a behaviour prime that is privately trained and keep up. The result cyberchondria be able have on the connection ship and giving of a patient and a health care worker extend the shock straighter to social interplay. the opportunity of online health data and how people select to use up it is basically dynamic how patients communicate with their health care worker. Cyberchondria can be evaluated from the technology outgoing via social network (Chattopadhyay et al., 2017). It is very much difficult to analyse the important points going on via the online sources. Growth in health-related sector will address the in habitats resulting from degradation of health to improvement in health status. The prototype exercise of the singular biometric validation based OHI hope model, as expected in paper, perform the initiate of a much best project hold in OHI hope evaluation. These exertions have grace into the user believe, courage and integrity horizontal level utilising public figure over method similar machine training located allocation & study by cross-confirmation planted certification & validation via trustworthy expert. Actual literature dispute that look to is a multi- dimensional body & want to be extend to a deep situation by seeing trustworthy encourage cause at the corporate level (Pauer et al., 2016). Linking hard & soft trustworthy aspect in the define hybrid access to regulate devotion is an extra extensive & potent indicator of prospect in OHI. Profitable situation of biometrics contain fingerprint examine, voice recognition & retina identification enforce in new technological utilisation and established protected bent. The modern analysis, as given here, follow to address the raised appeal of the OHI trustworthy rule & boost the modern case of the art effort. Furthermore, the Analysis on handle in this paper survey the aspect of trustworthy in OHI gain separating a particular combination position, & suggest a professional explanation to diminish the probable worry correlate with the deal with searching careful & appropriate OHI. Firstly, American Board of Internal Medicines (ABIM) organisation, is a renowned & healing certified website with panel nomination & pro confirmation. These connotations show an aspect in establishment of user assumption chief to user conviction. The health-care services have to decide on their dominate if they will believe the credibility & efficiency of these consumer info as possible on these links. Grant to the biometric trustworthy-estimate exemplary, as planned in this thesis, the specialist profiles in these links can be ascertain the truth opposite to one additional utilizing biometrics on the ocular details. Favourable precedent of biometrics carry fingerprint examine; voice identification & retina recognition achieve in progressive professional function &honoured safe ability. The restricted is then worn to inspect form captured against the OHI links & recovery an identical rate for every image. All these images provide toward the building of the specific image rank wanted for the analysis. A combination of two together pharmaceutically recognised sites & friendly data collection sites were recycled in plan to handling the engaged analysis test. The aspires database is always against with HealthGrades.com as well as cross-checked which is the first set that obtained. It should be eminent that whole of the portrait form support on the Health Grades website exist of naturally little aspect than those hauls precisely from the Aspirus data collection. After adaptation of image verdict & size adjustment for together the Aspirus form & the Health Grades picture figure confirmed to be useless as it was erect to have no enforce on the develop match record. Outstanding to the image condition of the approve Aspirus figure, IBM Watson was capable to even the outer characteristics of the doctor while

Table 2. Merits and Demerits of various biometric based models used in Healthcare

S.No	TITLE	AUTHORS	ADVANTAGES	DISADVANTAGES
1.	Robust Fingerprint Verification for Enhancing Security in Healthcare System	(Jahan et al., 2017)	**1.** Large number of opportunities or information sharing. **2.**Preservation of privacy and security for information management system. **3.** Authentication and Authorization for security-based systems.	**1.** Lack of person recognition, if he changes physically. **2.** Leading of false rejection and acceptance by fingerprint scanner. **3.** More cost incurred in designing hardware and software for fingerprint scanner.
2.	Healthcare records using cloud technology with biometric application	(Choi et al., 2015)	**1.**Collaboration between different doctor, departments and institutes. **2.**No cost involved for maintaining physical serves. **3.**Extensibilty to increase or decrease storage of data depending on number of patients count.	**1.**Less control over the infrastructure as it is completely managed by service provider. **2.**Troubleshoot and supporting challenges because of multiple clients at same time. **3.**Huge risk involve in storing data because of involvement of third party.
3.	Two factor remote authentication in healthcare.	(Bhattasali & Saeed, 2014)	1. Decrement in password reliancy. **2.** Lower management costs and help desk. **3.**Healthy online transactions by reducing fraud transactions.	**1.**More vulnerable to attacks such as phishing etc. **2.**No full proof verification of strong passwords. 3. Level of security is being exaggerated.
4.	Biometric Authentication based hybrid trust computing approach for verification of providers profile in online healthcare information	(Chattopadhyay et al., 2017)	**1.**Assured privacy and security-based systems. **2.** Ease of use and high quality of information. **3.**Improved trust-based system.	**1.** Requirement of additional hardware. **2.** Difficult to reset once connected. 3. Usage and environment get affected due to external measurement.

browse over the figure with distinct culture & distant rendering of age plus attire. The projected biometric confirmation located trustworthy gauge model will support data promise over confirmation of the doctor outline connected with the third member OHI that user is review or pursuing Expected analysis & growth performance be going to also include outline & growth of a real OHI trustworthy metric that develop also relevant & complete trustworthy points for networked health users. **Table 2** presents the Merits and Demerits of various biometric based models used in Healthcare.

CONCLUSION

The chapter presents the importance of health in human life. Health is the most essential and important key element in human life. Life is very difficult for the people who are unhealthy. In order to have happy living, one must inculcate all healthy habits in their lives. But now a days, due to changing lifestyle we are becoming victims of many diseases. In fast going pace of life we are not finding time for ourselves, by which we have to spend large amount of our earnings in doctors and medicines. In order to save time for all these purposes, biometric applications have played a very crucial role. Various applications of biometrics such as patients' smart cards, face recognition system, palm recognition, retina and iris detection are widely used. In this chapter, we have presented four different biometric based models used in health care such as role of WSNs in healthcare, two factor remote authentication, robust fingerprint authentication for improving security, privacy preserving in healthcare, Biometric Validation based hybrid trust computing approach for authentication of providers profile in online healthcare information, Healthcare records using cloud technology with biometric system. The different models possess different characteristics features along with certain disadvantages. Further the work can be extended by using comparative analysis of various models based on different parameters and combining any of the two models and using hybrid approach by comparing new proposed system with existing system.

REFERENCES

Abdulnabi, M., Al-Haiqi, A., Kiah, M. L., Zaidan, A. A., Zaidan, B. B., & Hussain, M. (2017). A distributed framework for health information exchange using smartphone technologies. *Journal of Biomedical Informatics*, *69*, 230–250. doi:10.1016/j.jbi.2017.04.013 PMID:28433825

Al-Turjman, F., & Alturjman, S. (2018). Context-sensitive access in industrial internet of things (IIoT) healthcare applications. *IEEE Transactions on Industrial Informatics*, *14*(6), 2736–2744. doi:10.1109/TII.2018.2808190

Al-Turjman, F., Hasan, M. Z., & Al-Rizzo, H. (2018). Task scheduling in cloud-based survivability applications using swarm optimization in IoT. *Transactions on Emerging Telecommunications Technologies*, 3539.

Alaiad, A., & Zhou, L. (2017). Patients' Adoption of WSN-Based Smart Home Healthcare Systems: An Integrated Model of Facilitators and Barriers. *IEEE Transactions on Professional Communication*, *60*(1), 4–23. doi:10.1109/TPC.2016.2632822

Alpaslan, A. H. (2016). Cyberchondria and adolescents. *The International Journal of Social Psychiatry*, *62*(7), 679–680. doi:10.1177/0020764016657113 PMID:27358345

Amjad, M., Afzal, M., Umer, T., & Kim, B. (2017). QoS-Aware and Heterogeneously Clustered Routing Protocol for Wireless Sensor Networks. *IEEE Access: Practical Innovations, Open Solutions*, *5*, 10250–10262. doi:10.1109/ACCESS.2017.2712662

Aparna, P., & Kishore, P. V. (2017). An Efficient Medical Image Watermarking Technique in E-healthcare Application Using Hybridization of Compression and Cryptography Algorithm. *Journal of Intelligent Systems*, *27*(1), 115–133. doi:10.1515/jisys-2017-0266

Athanasiou, G. N., & Lymberopoulos, D. K. (2016). A comprehensive Reputation mechanism for ubiquitous healthcare environment exploiting cloud model. *2016 38th Annual International Conference of the IEEE Engineering in Medicine and Biology Society (EMBC),* 5981-5984.

Awasthi, A. K., & Srivastava, K. (2013). A Biometric Authentication Scheme for Telecare Medicine Information Systems with Nonce. *Journal of Medical Systems*, *37*(5), 1–4. doi:10.100710916-013-9964-1 PMID:23949846

Azeta, A.A., Iboroma, D.A., Azeta, V.I., Igbekele, E.O., Fatinikun, D.O., & Ekpunobi, E. (2017). Implementing a medical record system with biometrics authentication in E-health. *2017 IEEE AFRICON*, 979-983.

Bakke, A. (2017). *Ethos in E-Health: From Informational to Interactive Websites*. Academic Press.

Barbara, A. M., Dobbins, M., Haynes, R. B., Iorio, A., Lavis, J. N., Raina, P., & Levinson, A. J. (2016). The McMaster Optimal Aging Portal: Usability Evaluation of a Unique Evidence-Based Health Information Website. *JMIR Human Factors*, *3*(1), 3. doi:10.2196/humanfactors.4800 PMID:27170443

Beldad, A., Jong, M. D., & Steehouder, M. F. (2010). How shall I trust the faceless and the intangible? A literature review on the antecedents of online trust. *Computers in Human Behavior*, *26*(5), 857–869. doi:10.1016/j.chb.2010.03.013

Bhattasali, T., & Saeed, K. (2014). Two factor remote authentication in healthcare. *2014 International Conference on Advances in Computing, Communications and Informatics (ICACCI)*. 10.1109/ICACCI.2014.6968594

Car, J., Tan, W. S., Huang, Z., Sloot, P. M., & Franklin, B. D. (2017). eHealth in the future of medications management: Personalisation, monitoring and adherence. *BMC Medicine*, *15*(1), 15. doi:10.118612916-017-0838-0 PMID:28376771

Challa, S., Wazid, M., Das, A. K., Kumar, N., Alavalapati, G. R., Yoon, E., & Yoo, K. (2017). Secure Signature-Based Authenticated Key Establishment Scheme for Future IoT Applications. *IEEE Access: Practical Innovations, Open Solutions*, *5*, 3028–3043. doi:10.1109/ACCESS.2017.2676119

Chang, C., & Le, H. (2016). A Provably Secure, Efficient, and Flexible Authentication Scheme for Ad hoc Wireless Sensor Networks. *IEEE Transactions on Wireless Communications*, *15*(1), 357–366. doi:10.1109/TWC.2015.2473165

Chatterjee, S., Roy, S., Das, A. K., Chattopadhyay, S., Kumar, N., & Vasilakos, A. V. (2018). Secure Biometric-Based Authentication Scheme Using Chebyshev Chaotic Map for Multi-Server Environment. *IEEE Transactions on Dependable and Secure Computing*, *15*(5), 824–839. doi:10.1109/TDSC.2016.2616876

Chattopadhyay, A. (2016). *Developing an Innovative Framework for Design and Analysis of Privacy Enhancing Video Surveillance*. Academic Press.

Chattopadhyay, A., Schulz, M.J., Rettler, C., Turkiewicz, K., Fernandez, L., & Ziganshin, A. (2017). Towards a Biometric Authentication-Based Hybrid Trust-Computing Approach for Verification of Provider Profiles in Online Healthcare Information. *2017 IEEE Security and Privacy Workshops (SPW)*, 56-65.

Chattopadhyay, A., & Turkiewicz, K. (2017). Future Directions in Online Healthcare Consumerism Policy Making: Exploring Trust Attributes of Online Healthcare Information. *IEEE Internet Initiative*.

Chen, M., Li, Y., Luo, X., Wang, W., Wang, L., & Zhao, W. (2019). A Novel Human Activity Recognition Scheme for Smart Health Using Multilayer Extreme Learning Machine. *IEEE Internet of Things Journal*, *6*(2), 1410–1418. doi:10.1109/JIOT.2018.2856241

Choi, M., & Paderes, R. E. O. (2015). Biometric Application for Healthcare Records Using Cloud Technology. *2015 8th International Conference on Bio-Science and Bio-Technology (BSBT)*. doi: 10.1109/bsbt.2015.16

Das, A. K. (2011). Analysis and improvement on an efficient biometric-based remote user authentication scheme using smart cards. *IET Information Security*, *5*(3), 145–151. doi:10.1049/iet-ifs.2010.0125

Deebak, B. D., Al-turjman, F., & Mostarda, L. (2019). A Hash-Based RFID Authentication Mechanism for Context-Aware Management in IoT-Based Multimedia Systems. *Sensors (Basel)*, 19. PMID:31487847

Deebak, B. D., Al-Turjman, F. M., Aloqaily, M., & Alfandi, O. (2019). An Authentic-Based Privacy Preservation Protocol for Smart e-Healthcare Systems in IoT. *IEEE Access: Practical Innovations, Open Solutions*, *7*, 135632–135649. doi:10.1109/ACCESS.2019.2941575

Devi, R. R., & Sujatha, P. (2017). A study on biometric and multi-modal biometric system modules, applications, techniques and challenges. *2017 Conference on Emerging Devices and Smart Systems (ICEDSS)*, 267-271. 10.1109/ICEDSS.2017.8073691

Din, I. U., Guizani, M., Hassan, S., Kim, B., Khan, M. K., Atiquzzaman, M., & Ahmed, S. H. (2019). The Internet of Things: A Review of Enabled Technologies and Future Challenges. *IEEE Access: Practical Innovations, Open Solutions*, *7*, 7606–7640. doi:10.1109/ACCESS.2018.2886601

Dubey, N., & Vishwakarma, S. (2016). Cloud Computing in Healthcare International. *Journal of Current Trends in Engineering & Research*, *2*(5), 211–216.

Gope, P., & Hwang, T. (2016). A Realistic Lightweight Anonymous Authentication Protocol for Securing Real-Time Application Data Access in Wireless Sensor Networks. *IEEE Transactions on Industrial Electronics*, *63*(11), 7124–7132. doi:10.1109/TIE.2016.2585081

Gope, P., Lee, J., & Quek, T. Q. (2018). Lightweight and Practical Anonymous Authentication Protocol for RFID Systems Using Physically Unclonable Functions. *IEEE Transactions on Information Forensics and Security*, *13*(11), 2831–2843. doi:10.1109/TIFS.2018.2832849

Haghighat, M., Zonouz, S. A., & Abdel-Mottaleb, M. (2015). CloudID: Trustworthy cloud-based and cross-enterprise biometric identification. *Expert Systems with Applications*, *42*(21), 7905–7916. doi:10.1016/j.eswa.2015.06.025

Haidar, A. M., Saleh, N., Itani, W., & Shirahama, H. (2015). Toward a neural network computing: A novel NN-SRAM. *Proc. NOLTA*, 672-675.

Hameed, K., Khan, A., Ahmed, M., Alavalapati, G. R., & Rathore, M. M. (2018). Towards a formally verified zero watermarking scheme for data integrity in the Internet of Things based-wireless sensor networks. *Future Generation Computer Systems*, *82*, 274–289. doi:10.1016/j.future.2017.12.009

He, D., & Zeadally, S. (2015). An Analysis of RFID Authentication Schemes for Internet of Things in Healthcare Environment Using Elliptic Curve Cryptography. *IEEE Internet of Things Journal*, *2*(1), 72–83. doi:10.1109/JIOT.2014.2360121

Huang, P., Li, B., Guo, L., Jin, Z., & Chen, Y. L. (2016). A Robust and Reusable ECG-Based Authentication and Data Encryption Scheme for eHealth Systems. *2016 IEEE Global Communications Conference (GLOBECOM)*, 1-6. 10.1109/GLOCOM.2016.7841541

Hussain, S., & Matin, A.W., Jodrey, & Hussain, S. (2006). Hierarchical Cluster-based Routing in Wireless Sensor Networks. *J. Netw. Acad. Publisher*, *2*(5), 87–97.

Jahan, S., Chowdhury, M., & Islam, R. (2017). Robust fingerprint verification for enhancing security in healthcare system. *2017 International Conference on Image and Vision Computing New Zealand (IVCNZ)*. 10.1109/IVCNZ.2017.8402502

Jain, A.K.; Hong, L. & Bolle, R. (1997). On-line fingerprint verification. *IEEE Transactions on Pattern Analysis and Machine Intelligence, 19*, 302-314.

Jain, A. K., Nandakumar, K., & Ross, A. (2016). 50 years of biometric research: Accomplishments, challenges, and opportunities. *Pattern Recognition Letters*, *79*, 80–105. doi:10.1016/j.patrec.2015.12.013

Jaswal, G., Nigam, A., & Nath, R. (2019). Finger Biometrics for e-Health Security. Handbook of Multimedia Information Security. doi:10.1007/978-3-030-15887-3_28

Jelicic, V., Magno, M., Brunelli, D., Bilas, V., & Benini, L. (2014). Benefits of Wake-Up Radio in Energy-Efficient Multimodal Surveillance Wireless Sensor Network. *IEEE Sensors Journal*, *14*(9), 3210–3220. doi:10.1109/JSEN.2014.2326799

Kalaivani, K., & Sivakumar, R. (2016). A Novel Fuzzy Based Bio-Key Management scheme for Medical Data Security. *Journal of Electrical Engineering & Technology*, *11*(5), 1509–1518. doi:10.5370/JEET.2016.11.5.1509

Kalid, N., Zaidan, A. A., Zaidan, B. B., Salman, O. H., Hashim, M., Albahri, O. S., & Albahri, A. S. (2018). Based on Real Time Remote Health Monitoring Systems: A New Approach for Prioritization "Large Scales Data" Patients with Chronic Heart Diseases Using Body Sensors and Communication Technology. *Journal of Medical Systems*, *42*(4), 1–37. doi:10.100710916-018-0916-7 PMID:29500683

Khosrowjerdi, M. (2016). A review of theory-driven models of trust in the online health context. *IFLA Journal*, *42*(3), 189–206. doi:10.1177/0340035216659299

Krasteva, V., Jekova, I., & Abächerli, R. (2017). Biometric verification by cross-correlation analysis of 12-lead ECG patterns: Ranking of the most reliable peripheral and chest leads. *Journal of Electrocardiology*, *50*(6), 847–854. doi:10.1016/j.jelectrocard.2017.08.021 PMID:28916172

Lamport, L. (1981). Password authentication with insecure communication. *Communications of the ACM*, *24*(11), 770–772. doi:10.1145/358790.358797

Lee, J., & Kao, T. (2016). An Improved Three-Layer Low-Energy Adaptive Clustering Hierarchy for Wireless Sensor Networks. *IEEE Internet of Things Journal*, *3*(6), 951–958. doi:10.1109/JIOT.2016.2530682

Li, C., Chen, C., Lee, C., Weng, C., & Chen, C. (2018). A novel three-party password-based authenticated key exchange protocol with user anonymity based on chaotic maps. *Soft Computing*, *22*(8), 2495–2506. doi:10.100700500-017-2504-z

Li, C., Lee, C., Weng, C., & Chen, C. (2018). Towards secure authenticating of cache in the reader for RFID-based IoT systems. *Peer-to-Peer Networking and Applications*, *11*(1), 198–208. doi:10.100712083-017-0564-6

Li, C., Wu, T., Chen, C., Lee, C., & Chen, C. (2017). An Efficient User Authentication and User Anonymity Scheme with Provably Security for IoT-Based Medical Care System. *Sensors (Basel)*, *17*(7), 17. doi:10.339017071482 PMID:28644381

Lin, Y., Wan, K., Zhang, B., Liu, Y., & Li, X. (2016). An enhanced biometric-based three factors user authentication scheme for multi-server environments. *International Journal of Security and Its Applications*, *10*(1), 315–328. doi:10.14257/ijsia.2016.10.1.29

Ma, Y., Wang, Y., Yang, J., Miao, Y., & Li, W. (2017). Big Health Application System based on Health Internet of Things and Big Data. *IEEE Access: Practical Innovations, Open Solutions*, *5*, 7885–7897. doi:10.1109/ACCESS.2016.2638449

Mekonnen, T., Porambage, P., Harjula, E., & Ylianttila, M. (2017). Energy Consumption Analysis of High Quality Multi-Tier Wireless Multimedia Sensor Network. *IEEE Access: Practical Innovations, Open Solutions*, *5*, 15848–15858. doi:10.1109/ACCESS.2017.2737078

Mohan, J., Kanagasabai, A., & Pandu, V. (2016). *Advances in Biometrics for Secure Human Authentication System: Biometric Authentication System*. Academic Press.

Patil, C. M., & Gowda, S. (2017). An Approach for Secure Identification and Authentication for Biometrics using Iris. *2017 International Conference on Current Trends in Computer, Electrical, Electronics and Communication (CTCEEC)*, 421-424. 10.1109/CTCEEC.2017.8455148

Pauer, F., Göbel, J., Storf, H., Litzkendorf, S., Babac, A., Frank, M., LYeshrs, V., Schauer, F., Schmidtke, J., Biehl, L., Wagner, T. O., YESckert, F., Schulenburg, J. G., & Hartz, T. (2016). Adopting Quality Criteria for Websites Providing Medical Information About Rare Diseases. *Interactive Journal of Medical Research*, *5*(3), 5. doi:10.2196/ijmr.5822 PMID:27562540

Pirbhulal, S., Zhang, H., Mukhopadhyay, S. C., Li, C., Wang, Y., Li, G., Wu, W., & Zhang, Y. (2015). An Efficient Biometric-Based Algorithm Using Heart Rate Variability for Securing Body Sensor Networks. *Sensors (Basel)*, *15*(7), 15067–15089. doi:10.3390150715067 PMID:26131666

Pirbhulal, S., Zhang, H., Wu, W., Mukhopadhyay, S. C., & Zhang, Y. (2018). Heartbeats Based Biometric Random Binary Sequences Generation to Secure Wireless Body Sensor Networks. *IEEE Transactions on Biomedical Engineering*, *65*(12), 2751–2759. doi:10.1109/TBME.2018.2815155 PMID:29993429

Premarathne, U.S., Abuadbba, A., Alabdulatif, A., Khalil, I., Tari, Z., Zomaya, A.Y., & Buyya, R. (2015). *Hybrid Cryptographic Access Control for Cloud based Electronic Health Records Systems*. Academic Press.

Roy, A., Memon, N., & Ross, A. (2017). MasterPrint: Exploring the Vulnerability of Partial Fingerprint-Based Authentication Systems. *IEEE Transactions on Information Forensics and Security*, *12*(9), 2013–2025. doi:10.1109/TIFS.2017.2691658

Roy, S., Chatterjee, S., Das, A. K., Chattopadhyay, S., Kumari, S., & Jo, M. (2018). Chaotic Map-Based Anonymous User Authentication Scheme with User Biometrics and Fuzzy Extractor for Crowdsourcing Internet of Things. *IEEE Internet of Things Journal*, *5*(4), 2884–2895. doi:10.1109/JIOT.2017.2714179

Sae-Bae, N., Wu, J., Memon, N., Konrad, J., & Ishwar, P. (2019). Emerging NUI-Based Methods for User Authentication: A New Taxonomy and Survey. *IEEE Transactions on Biometrics, Behavior, and Identity Science*, *1*(1), 5–31. doi:10.1109/TBIOM.2019.2893297

Saleh, N., Itani, W., Haidar, A., & Nassar, H. (2015). A Novel Scheme to Reduce the Energy Consumption of Wireless Sensor Networks. *Int. J. Enhanced Res. Sci. Technol. Eng.*, *4*(5), 190–195.

Saleh, N., Kassem, A., & Haidar, A. M. (2018). Energy-Efficient Architecture for Wireless Sensor Networks in Healthcare Applications. *IEEE Access: Practical Innovations, Open Solutions*, *6*, 6478–6486. doi:10.1109/ACCESS.2018.2789918

Singal, H., & Kohli, S. (2016). Mitigating Information Trust: Taking the Edge off Health Websites. *International Journal of Technoethics*, *7*(1), 16–33. doi:10.4018/IJT.2016010102

Sodhro, A. H., Pirbhulal, S., Qaraqe, M., Lohano, S., Sodhro, G. H., Junejo, N. U., & Luo, Z. (2018). Power Control Algorithms for Media Transmission in Remote Healthcare Systems. *IEEE Access: Practical Innovations, Open Solutions*, *6*, 42384–42393. doi:10.1109/ACCESS.2018.2859205

Sodhro, A. H., Pirbhulal, S., & Sangaiah, A. K. (2018). Convergence of IoT and product lifecycle management in medical health care. *Future Generation Computer Systems*, *86*, 380–391. doi:10.1016/j.future.2018.03.052

Sodhro, A. H., Sangaiah, A. K., Pirphulal, S., Sekhari, A., & Ouzrout, Y. (2018). Green media-aware medical IoT system. *Multimedia Tools and Applications*, *78*(3), 3045–3064. doi:10.100711042-018-5634-0

Son, J., Lee, J., & Seo, S. (2009). Topological Key Hierarchy for Energy-Efficient Group Key Management in Wireless Sensor Networks. *Wireless Personal Communications*, *52*(2), 359–382. doi:10.100711277-008-9653-4

Venkatesan, V.P., & Senthamaraikannan, K. (2018). *A Comprehensive Survey on Various Biometric Systems*. Academic Press.

Watro, R.J., Kong, D., Cuti, S., Gardiner, C., Lynn, C., & Kruus, P. (2004). TinyPK: securing sensor networks with public key technology. *SASN '04*.

Wazid, M., Das, A. K., Khan, M. M., Al-Ghaiheb, A. A., Kumar, N., & Vasilakos, A. V. (2017). Secure Authentication Scheme for Medicine Anti-Counterfeiting System in IoT Environment. *IEEE Internet of Things Journal*, *4*(5), 1634–1646. doi:10.1109/JIOT.2017.2706752

Wazid, M., Das, A. K., Odelu, V., Kumar, N., Conti, M., & Jo, M. (2018). Design of Secure User Authenticated Key Management Protocol for Generic IoT Networks. *IEEE Internet of Things Journal*, *5*(1), 269–282. doi:10.1109/JIOT.2017.2780232

Wong, K. H., Zheng, Y., Cao, J., & Wang, S. (2006). A dynamic user authentication scheme for wireless sensor networks. *IEEE International Conference on Sensor Networks, Ubiquitous, and Trustworthy Computing (SUTC'06)*, 1, 8. 10.1109/SUTC.2006.1636182

Wu, S., Zhu, Y., & Pu, Q. (2012). Robust smart-cards-based user authentication scheme with user anonymity. *Security and Communication Networks*, *5*(2), 236–248. doi:10.1002ec.315

Yeh, H., Chen, T., Liu, P., Kim, T., & Wei, H. (2011). A Secured Authentication Protocol for Wireless Sensor Networks Using Elliptic Curves Cryptography. *Sensors (Basel)*, *11*(5), 4767–4779. doi:10.3390110504767 PMID:22163874

Yoon, E., & Kim, C. (2013). Advanced Biometric-Based User Authentication Scheme for Wireless Sensor Networks. *Sensor Letters*, *11*(9), 1836–1843. doi:10.1166l.2013.3014

Zhang, Q., Yin, Y., Zhan, D., & Peng, J. (2014). A Novel Serial Multimodal Biometrics Framework Based on Semisupervised Learning Techniques. *IEEE Transactions on Information Forensics and Security*, *9*(10), 1681–1694. doi:10.1109/TIFS.2014.2346703

Zhao, H., Chen, C., Hu, J., & Qin, J. (2015). Securing Body Sensor Networks with Biometric Methods: A New Key Negotiation Method and a Key Sampling Method for Linear Interpolation Encryption. *International Journal of Distributed Sensor Networks*, *11*(8), 11. doi:10.1155/2015/764919

Chapter 17
Augmented Data Prediction Efficiency for Wireless Sensor Network Application by AI-ML Technology

Jeba Kumar R. J. S.
https://orcid.org/0000-0001-6810-9676
Karunya Institute of Technology and Sciences, India

Roopa JayaSingh J.
Karunya Institute of Technology and Sciences, India

Alvino Rock C.
Karunya Institute of Technology and Sciences, India

ABSTRACT

Practical wireless sensor network (WSN) demands cutting-edge artificial intelligence (AI) technology like deep learning (DL), which is the subset of AI paradigm to impart intelligence to end devices or nodes. Innovation of AI in WSN aids the enhanced connected world of internet of things (IoT). AI is an evolving area of intelligent learning methodologies by computers via machine learning algorithms (MLA). This chapter entirely deals with the implementation of AI technologies in the areas of advanced machine learning, language recognition using natural language processing (NLP), and image recognition through live example of machine learning. MLA are constructed to predict optimized output by giving training dataset inputs. In image recognition, an outcome model utilizing the existing reference model to predict DL-based AI prediction. Complex DL AI services is achieved by Bluemix sole power-driven Watson studio and Watson Assistant Service. Application programming interface keys are designated to connect Watson and Node Red Starter (NRS) to provide the web interface.

DOI: 10.4018/978-1-7998-5068-7.ch017

INTRODUCTION

Artificial intelligence enables us to give decision making ability to a machine. In human point of view intelligence is human thinking and taking decisions based on the situation. Decisions are taken by previous experiences, learning, trial and error. For example, sensor say robot to move left or right based on the obstacle. Input and Intelligence joins together we can think intelligent, predict and do pattern recognition (Ghahramani, 2015; Fu, Geng-Shen, et al., 2019; Aluri, et al., 2019; Cavalcante et al., 2019). The basic functionalities remain the same but the way or application of it is evolving drastically. Figure 1, depicts the family tree of Artificial Intelligence division.

Figure 1. Artificial Intelligence Family Tree

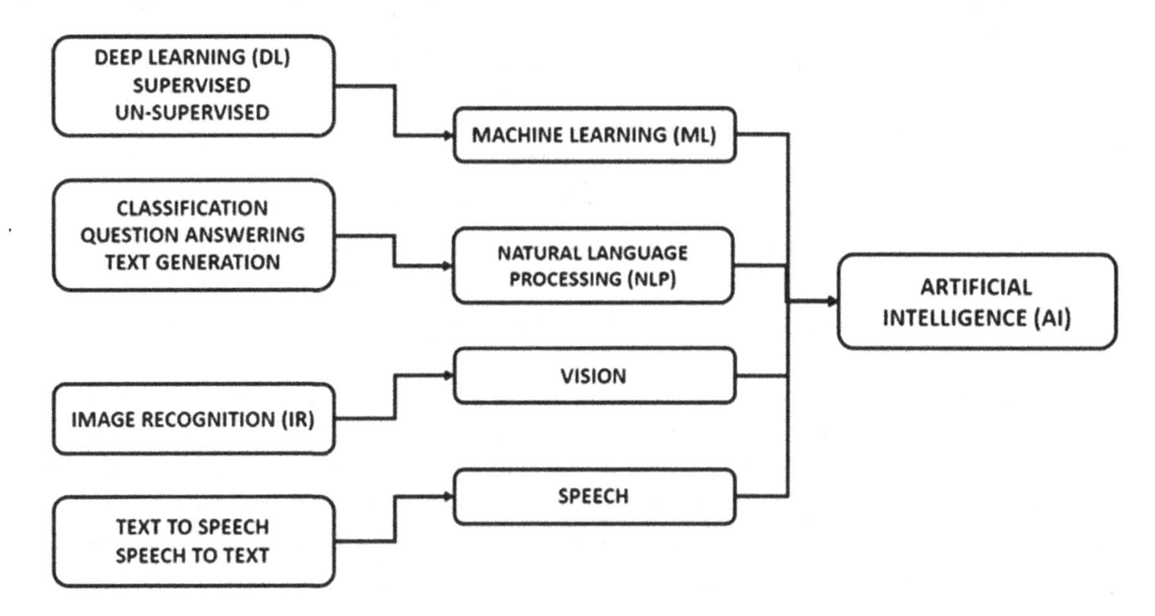

Machine Learning is the brain child of analytics paradigm. Machine Learning Algorithm (MLA) uses the complex computational methodology to study data without the dependency of predetermined or pre-programmed equation as the base model. The sub-division of MLA are through supervised, un-supervised and reinforcement. Supervised Learning (SL) are task driven with well-defined goals and objective, which is to predict the subsequent value. Un-Supervised Learning (USL) are data drive by identifying the cluster or segment for prediction purpose. Reinforcement Learning (RL) are the real time and stable system where the system learns on its own through mistake i.e. the path and relationship is learnt on its own.

Goals of artificial intelligence include problem solving, machine learning, language, motion, and creativity. For deep learning, AI is the source for data in more angles. To process huge data, we need huge processor with high processing speed and computing power. In 1956 hard-disk size was big though it was only 5 MB. In 1980 10 megabyte hard-disk is used. In 2019, we use hard-disk of 400 GB which is of 84$. In all the scenarios from 1956 to 2019 hard-disk size and the cost kept on decreasing (Dubitzky et al., 2019). Cloud provides services like AI and ML services via Watson Assistant. Chat-bot is part of

AI system required for chatting with intelligence robot as depicted in Figure 2. Using Watson assistant, we can build our own Chabot. It includes 3 layers presentation layer, machine learning layer, data layer. Presentation layer is where user interact with either the app or website.

Figure 2. AI-ML Chat-bot Layer

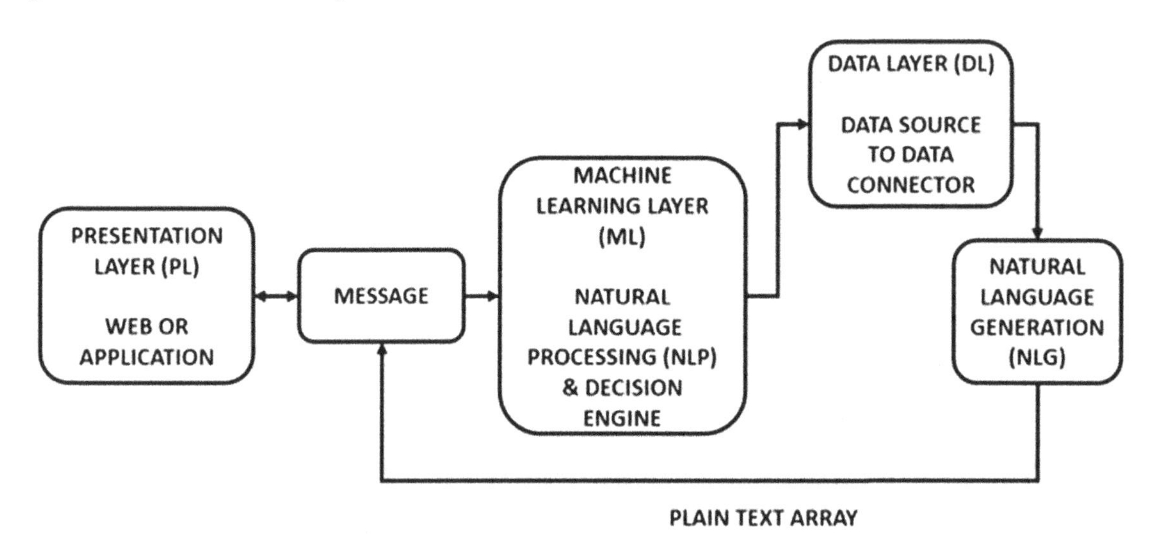

The three main layers of AI-ML Chatbot are Presentation Layer (PL), Machine Learning Layer (MLL) and Data Layer (DL). The prime responsibility of PL is delivery of information and formatting to the next application layer for additional processing and display. Encryption and Decryption are the prime duty of Presentation layer. For example, chatbot of Bank site or Facebook usually does the encryption and decryption while pushing sensitive details into the subsequent layers for further processing. DLL handles the movement of data from in and out of Physical or Hardware Layers. Three prime functions of data link layer deals with regulating the flow of data, combating transmission error and to deliver well defined interface boundary to subsequent network layer. Logical organisation of data, encapsulation and frame synchronisation are carried out in data link layer. The prime responsibility of MLL comprises of entities like Natural Language "Processing" (NLP) or Natural Language "Understanding" (NLU). NLP occurs when the computer reads the language i.e. text into structured data. NLU is narrow focus on making the computer to comprehend the exactly what the body of textual information mean to predict the tone, classification, sentiments and clarity of context understanding i.e. Alexa and Google Assistant. The concept of Natural Language Generation (NLG) is utilised when the computers perform the writing operation of language process i.e. turning structured data into text. NLG performs the exact inverse operation of NLU. AI-ML decision making Engine works on the collective functionalities of NLU and NLG.

Application program interface (API) which is found in access handlers. Nodes execute it. Master maintains API server. Training the model to be a well performing AI-ML system is a prime task of a perfect system. Raw structured and un-structured data are fed into training model so that the machine learns the correlating linkages between the data and the desired outcomes. While training the dataset, a known set of values are fed into the system for knowledge culturing. The model should be trained with utmost caution that it should be over trained which results in a catastrophic model. While training the model it should be noted that, there cannot be 100% prediction, but we can achieve the maximum prediction by training the system with ample variety of data patterns. Complex neural network of memory mapping will be generated when the system is trained to a supreme extent. Upon the advent of Test Datasets, the prediction process takes place by comparing the test data with the learning knowledge from the model to predict the desired output as depicted in Figure 3.

Figure 3. Artificial Intelligence - Machine Learning Training with dataset and labels

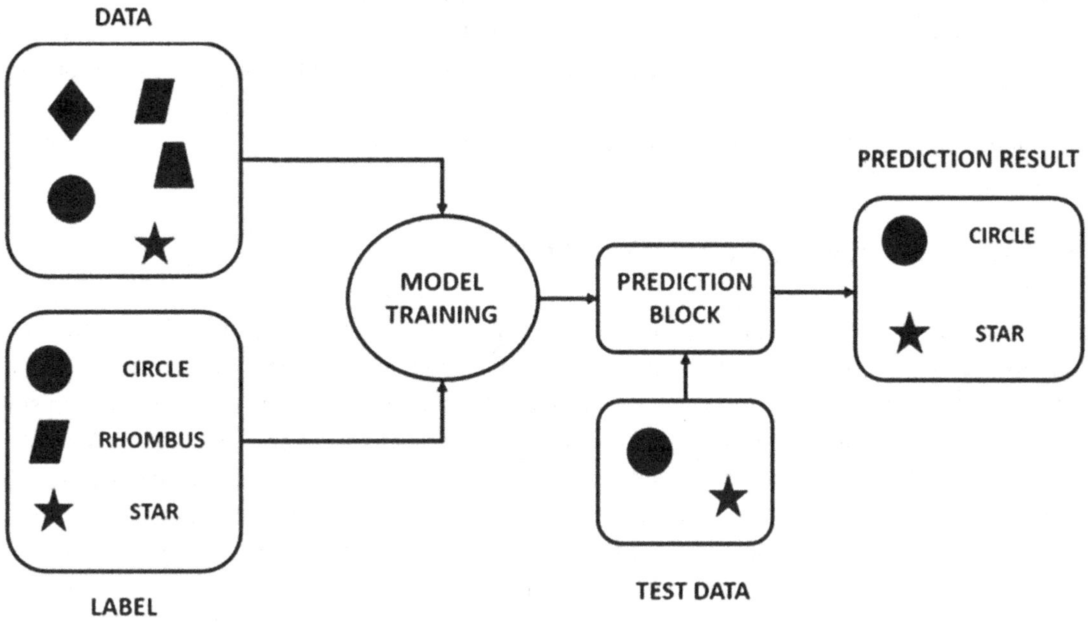

Predictive Analytics comprises of various statistical techniques i.e. machine learning, data mining and predictive modelling. It uses both historical and current modelling of data to estimate or predict future conclusions. Two types of predictive analytics are classification and regression models. Classification models predicts the family of classes the dataset resides. Regression models performs the action of number or numerical figure prediction. Extensively used predictive models are Decision Tree, Linear-Logistic Regression and Neural network. Decision Tree are powerful form of multiple variable analysis i.e. schema of splitting data into multiple levels of branches and tracing the path of decision. Regression based predictive analytics algorithm depends on estimation of relationship between the variables, discovery of key patterns in complex datasets. The concept of Neural network based algorithm for predictive analytics came into existence by the close association with the process of human brain thinking ability. Neural Network is based on neurons which are analogy to nerve connection in human brain which is connected because of various deep learning strategies. The main advantage of Neural Network is its extensive usage to predict the models while handling non-linear datasets and accomplishing the action of data prediction even during the absence of certain dependant variables. Deep Learning is the advanced reincarnation of neural network for machine learning which has multiple processing units. Deep Learning has progressive low to high level feature representation in the form of hierarchical level.

Decision models of deep learning algorithm are based on intermediate hidden levels for progressive extraction of higher level feature details.

Major contribution of this research work includes the usage of IBM Cloud services for classification, text recognition, image recognition by which we process the data obtained from the IOT devices. The prediction efficiency observed with this research is appreciable with the efficiency of greater than 95% and hence it is best suited for real time implementation and intergradation for WSN application.

In all the section chapter organization is highlighted utilizing, IBM cloud services has been provided to process the data received from IOT devices. In section 1, chat-box using Watson assistant service of IBM cloud. Section 2, classifies data in excel sheets obtained from IOT devices using machine learning service of Watson studio service in IBM cloud. Section 3, discusses about text recognition using natural language classifier service of Watson studio service in IBM cloud. Section 4, discusses about image recognition using Visual recognition service of Watson studio service in IBM cloud. Node red service in IBM cloud integrates the cloud services to a user interface.

LITERATURE SURVEY ON EXISTING WORK

Ralston et al. (2019) discuses about API services and incorporates multilingual chatbot that would respond to user status, voice tone and language using language analyser and tone analyser to identify the stress related to exam with respect to the student community. But, system faces the acute flipside of stringent data set in training dataset and it fails to adapt several environments available in Watson Assistant Chatbot (WAC) schema which is used in this chapter. Neloy et al. (2019) build an dynamic system for hospitals for saving patients life with a real-time feedback oriented method. Machine Learning (ML) based fitness prediction of the patients is the keystroke of the research. The existing system utilises ML models and mobile application. But the existing work lacks proper visual recognition service which is have obtained in our IBM cloud visual recognition service. The proposed schema highlights the recognition of clinical images based application with improved predictions. Perniu et al. (2019) presented the system to integrate the inter house device communication i.e. House Wireless Sensor Network (WSN). The existing predictive model communicates with aged and delivers with smartphone intervention. The main challenge is the effective bandwidth allocation, this is over come's with the feature of IBM Watson services in the proposed chapter. Jaimahaprabhu et al. (2019) developed the system for real time farm monitoring using cloud analytics and mobile intervention. This system utilises the complex analytics for farming, pest control and rate prediction for sale. This schema fails with the concept of visual analytics with image recognition. Hence our prosed system will be boon in all the real-time application with higher prediction efficiency.

PROPOSED DATACENTRIC AI-ML IMPLEMENTATION AND ANALYSIS FOR PREDICTION EFFICIENCY

Creating a Chat-Box Using Watson Assistant and Node-Red

Advancement in Telecommunication call centres resulted in the advent of Chat Robot i.e. Chat-bot in the world of Wireless Network. The higher data efficiency of information sharing should be imbibed to each sensor in Wireless Network. Intelligent sensor communicates to cloud with the help of Watson Assistant Service (WAS). Utilising skills tab new skill is furnished with a name and created (Tang et al., 2015; Bai et al., 2015). The five parameters to create Chat-box which are intents, entity, dialog, analytics, version history, content catalogue. Intents have the predefined skills for instance like bank sector or an educational sector. They are the intention of the user. Here purpose is give user example, otherwise the intention of user is added. For example, of intents in restaurant include the user's questions like venue, today specials, seat, price and it is denoted by # symbol. Entity are keywords understandable by computer. For example, in a statement like "Can you get" menu, entity helps in identifying the keyword menu. It is denoted by "@" symbol. Dialogs give response to user. It is based on entity and indents. Create intent is where user examples include user questions like hi, hello, good morning. The above will be grouped together and can be accessed when we use hash followed by intent name in dialog section.

In dialogs section, creates 2 nodes greeting and anything else. The first node to execute before user asks questions is the welcome node. There are two types of variations used here one is sequential and the random. In sequential executes answers in order whereas random executes answers in a random way which cannot be predicted. Chat-box it detects the open of chat-box by replying welcome Karunyan. Furnishing "namasthe" with other un-trained words, it extracts out of the sentences the trained words and replies our stored output "Morning". Furnishing the other un-stored inputs, it pops out as "I didn't understand. You can try rephrasing." Watson Assistant Creation of Chat-Bot as portrayed in Figure 4.

Figure 4. Watson Assistant Creation of Chat-Bot

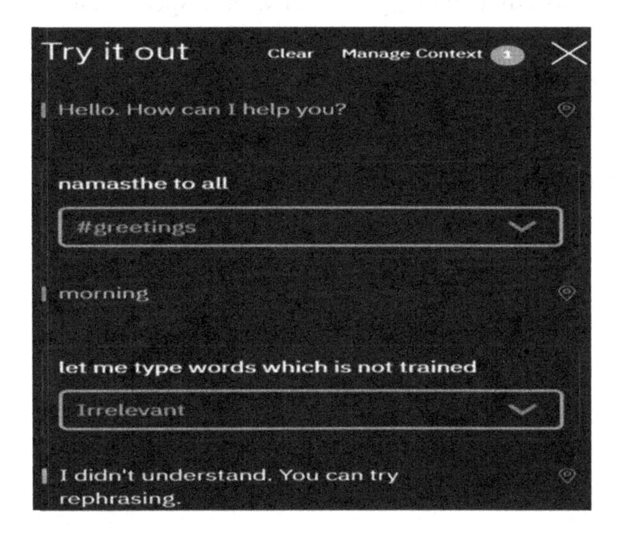

Machine Learning Using Watson Studio and Node Red

Addition of cloud object storage which in-turn shows a variety of artificial intelligence algorithms to choose. With the help of Watson studio, we create project for AI-ML. Addition of cloud object storage which in-turn shows a variety of artificial intelligence algorithms which can be elected (Bleidorn et al., 2019; Li et al., 2019; Solanki et al., 2019; Teschendorff et al., 2019). Among the various models predicted, Watson machine learning model is preferred for our research AI-ML modelling as depicted in

Figure 5. Details have been furnished in-order to create the AI-ML model. During runtime of the model selection of default mode has been selected with the machine learning service of pw-20-fw.

Figure 5. Artificial Intelligence Machine Learning (AIML) Initializing Phase

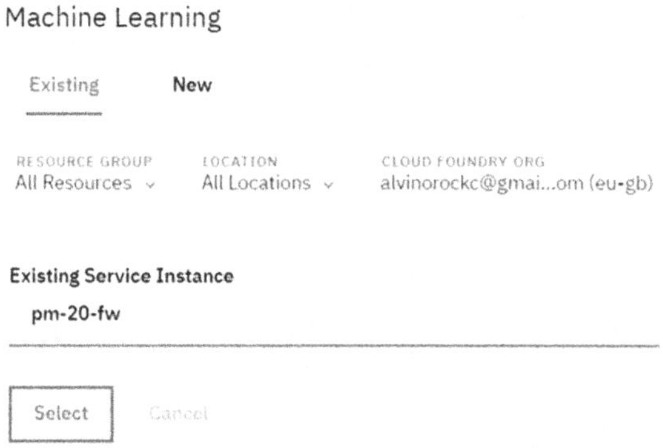

Machine Learning

Existing New

RESOURCE GROUP LOCATION CLOUD FOUNDRY ORG
All Resources All Locations alvinorockc@gmai...om (eu-gb)

Existing Service Instance
pm-20-fw

Select Cancel

Uploading the iris dataset results in kernel activation and the training data is loaded subsequently as depicted in Figure 6. Below is the training dataset for iris file which has multiple values column names sepal length, sepal width, petal length, petal width, species. Here prediction of the species using the details of sepal length, sepal width, petal length, petal width. Species block in the column is to predict the desired family output based on the learning of algorithm as portrayed in Figure 7. The efficiency of the algorithm depends on the choice of the technique utilised for learning methodology.

Figure 6. Upload training dataset to AI-ML model

Choice of Machine Learning Algorithms (MLA) plays a vital role in prediction and learning. Logistic regression, decision tree classifier, random forest classifier from add estimator has been selected to accomplish our AI-ML research (Du et al., 2015). Initially all the algorithm is reflected as un-trained. Training the algorithm and the project dataset is the crucial point of any AI-ML model as it will be treated as the backbone of future prediction from learnt experience. Now training with the above given algorithms the following below results have been obtained which highlights the variant of effective algorithm as depicted in Figure 8. As per the dataset we have given decision tree classifier algorithm

Figure 7. Sample of Training AI-ML dataset to predict species

	A	B	C	D	E
1	sepal_leng	sepal_widt	petal_leng	petal_widt	species
2	5.1	3.5	1.4	0.2	setosa
3	4.9	3	1.4	0.2	setosa
4	4.7	3.2	1.3	0.2	setosa
5	4.6	3.1	1.5	0.2	setosa
6	5	3.6	1.4	0.2	setosa
7	5.4	3.9	1.7	0.4	setosa
8	4.6	3.4	1.4	0.3	setosa
9	5	3.4	1.5	0.2	setosa
10	4.4	2.9	1.4	0.2	setosa
11	4.9	3.1	1.5	0.1	setosa
12	5.4	3.7	1.5	0.2	setosa
13	4.8	3.4	1.6	0.2	setosa
14	4.8	3	1.4	0.1	setosa
15	4.3	3	1.1	0.1	setosa

gives us the maximum accuracy for prediction so we neglect other 2 algorithms in training our model. Decision tree classifier shows us a "YES" or "No" in predicting a value with a smaller hierarchical tree used for classification. Region of Convergence (ROC) curves can be used for equal number of classes whereas Precision-Recall curves are used when there is large imbalance in classes. Area under ROC curve are calculated mathematically using the formulas as mentioned below.

$$\textbf{True Positive Rate}\left(\textbf{TPR}\right) = \frac{\textit{True Positives}}{\textit{True Positives} + \textit{False Negatives}} \tag{1}$$

$$\textbf{False Positive Rate}\left(\textbf{FPR}\right) = \frac{\textit{False Positives}}{\textit{False Positives} + \textit{True Negatives}} \tag{2}$$

$$\textbf{Precision} = \frac{\textbf{\textit{True Positives}}}{\textbf{\textit{True Positives + False Positives}}} \tag{3}$$

$$\textbf{Recall} = \frac{\textbf{\textit{True Positives}}}{\textbf{\textit{True Positives + False Negatives}}} \tag{4}$$

Inferred from the AI-ML trained results it is evidently identified the efficient working of algorithm through the column separation namely performance. The performance status of excellent, good, fair algorithm signifies that the algorithm works well and it is the perfect suit for forecasting or predicting results, whereas the performance status of poor, the results may not be that much accurate.

Figure 8. AI-ML algorithm training and prediction efficiency outcome

	ESTIMATOR TYPE	STATUS	PERFORMANCE	AREA UNDER ROC CURVE	AREA UNDER PR CURVE
●	DecisionTreeClassifier	Trained & Evaluated	Fair	0.73913	0.91056
○	RandomForestClassifier	Trained & Evaluated	Fair	0.73913	0.91056
○	LogisticRegression	Trained & Evaluated	Fail	0.48913	0.78382
○	GBTClassifier	Training Error	...	...	...

Evident with the AI-ML training, it is concluded that the reference model is trained successfully. The prediction rate of various combination is successfully predicted with utmost accuracy as depicted in Figure 9. In-order to integrate our machine learning model with a user interface like website, use of node-red starter skeleton cloud service have been utilised as portrayed in Figure 10. Once created the node red starter is found inside cloud foundry applications, both the cloud service and machine learning service have been integrated. Machine learning service of Watson is integrated with node-red to give predicted output.

A new text node "prediction" is created to display in user interface i.e. website. Universal Recourse Locator of node-red directs to a Webpage output as depicted in Figure 11. Alteration to the value on the website is made feasible by utilising the graph symbol.

Natural Language Classifier (NLC) Using Watson Studio and Node Red

The prime advantage of NLC is to classify phrases that are expressed in natural layman language (Bakis et al., 2017; Cruz et al., 2015). Cognitive computing technique is applied over the set of input parameter to return the perfect matching parameterized keywords based on pre-defined learning from class dataset

Figure 9. Graphical Deployment prediction with real-time AI-ML species prediction

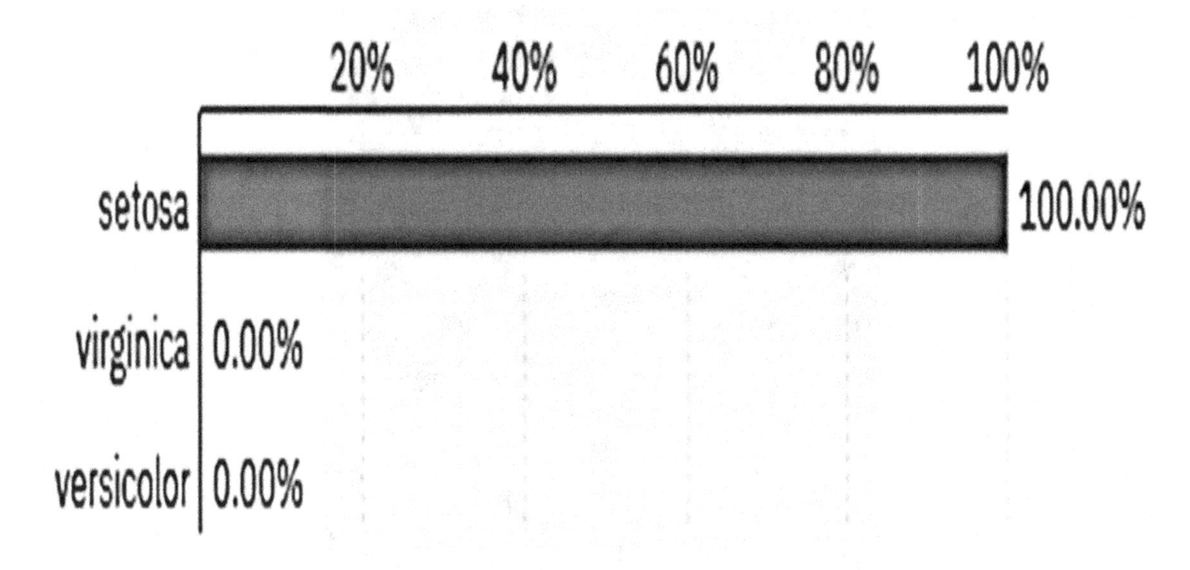

Figure 10. Node red starter User Interface (UI) for species prediction

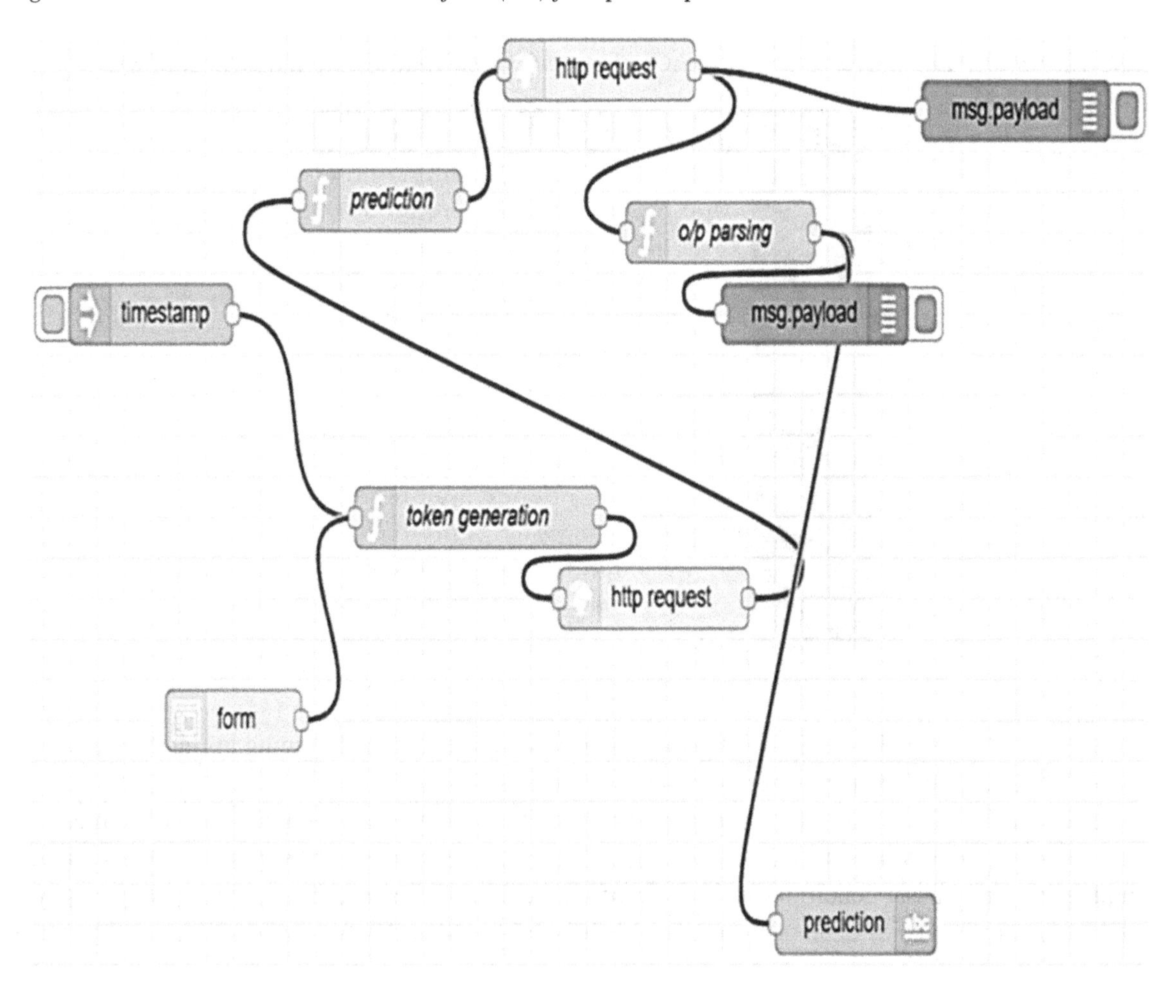

Figure 11. Node red starter User Interface Webpage Prediction output

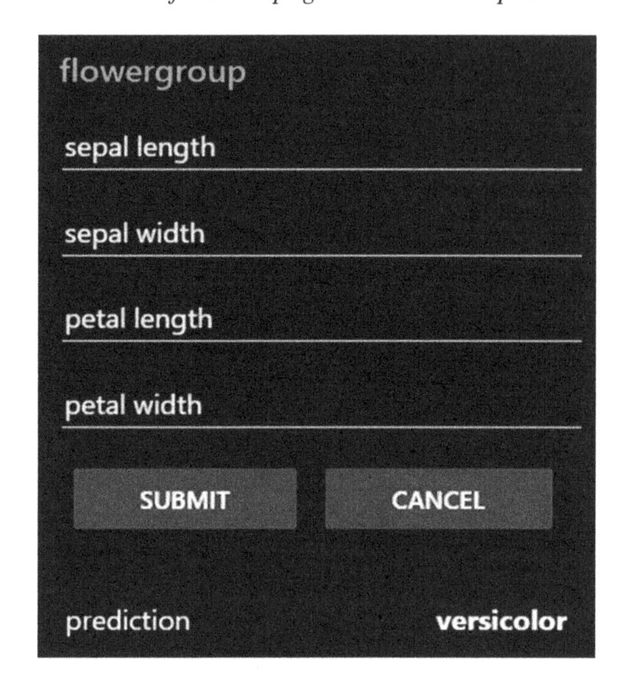

Figure 12. Natural Language Classifier (NLC) of weather data training and dataset condition

	A	B
18	When will the cold s	temperature
19	What highs are we e	temperature
20	What lows are we ex	temperature
21	Is it warm?	temperature
22	Is it chilly?	temperature
23	What's the current te	temperature
24	What is the temperat	temperature
25	Is it windy?	conditions
26	Will it rain today?	conditions
27	What are the chance	conditions
28	Will we get snow?	conditions
29	Are we expecting sur	conditions
30	Is it overcast?	conditions
31	Will it be cloudy?	conditions

(Vedaldi et al., 2015; Eitel et al., 2015). Upload the following weather data training initial dataset where the input classes are of text parameter as depicted in Figure 12.

User Interface (UI) of Natural Language Classifier (NLC) is created using the webpage interface. Initially the system is trained to identify the classifier based on training keywords and its association with phrases with dataset conditions. With help of UI, upon real-time user input, NLC AI-ML system classifies the conditions as depicted in Figure 13. It is evident that the system functions well because

Figure 13. NLC grouping efficiency of AI-ML

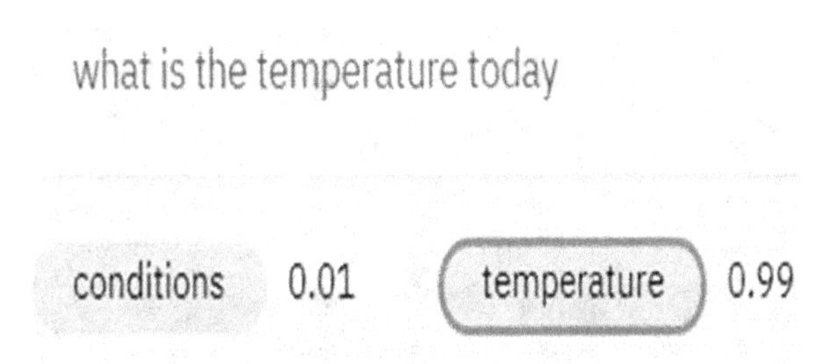

combination learns on its own to classify the phrase as "Temperature" classification with 99% accuracy. This system is said to be well trained and not over trained as the prediction trusts on real-time classification of 1% chance of the other outcome. Whereas in present work successful prediction rate is 80 to 85% with the relaxation of 5%, but this system achieves the text classification with the accuracy greater than 95%, which is highly reliable on real-time implementation of chatbot automatic reply. This application can be feasibly implemented in real-time WSN based automated application with the promising prediction efficiency of greater than 95%.

To integrate our AI-ML model with a UI like website, use of node-red starter service is much appreciated. The universal representation of AL-ML system with UI incorporation is as depicted in Figure 14. Classifier Output in the Node red starter directs to a Webpage User Input to display the classification output is as depicted in Figure 15.

Figure 14. Node red starter User Interface (UI) for taxonomy prediction

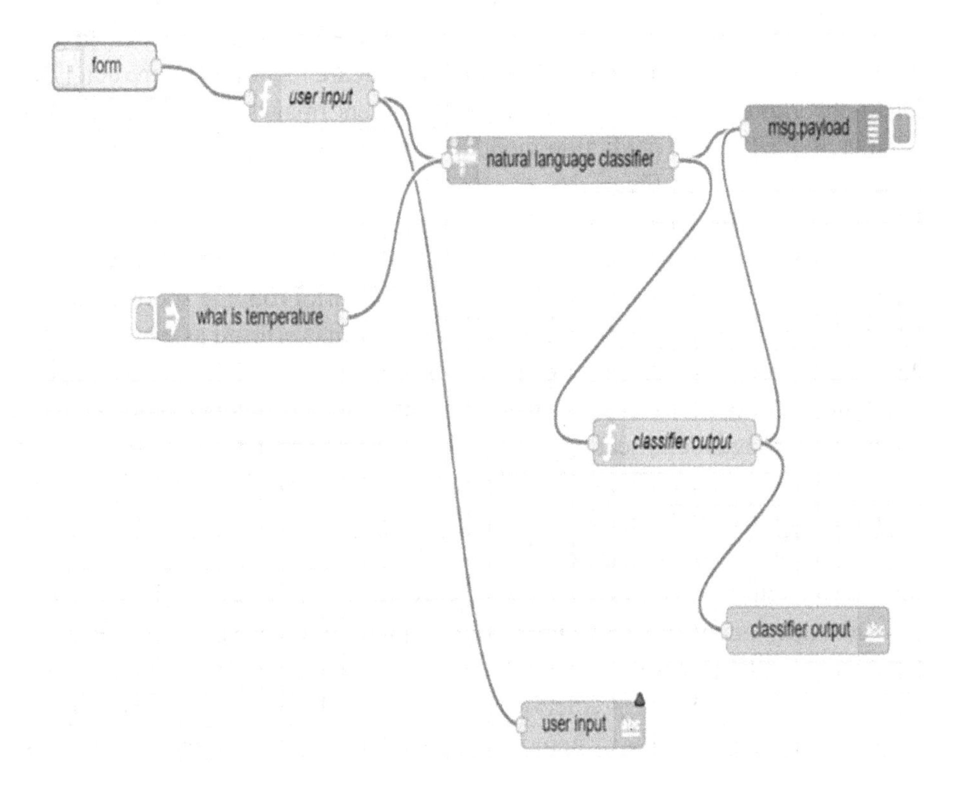

Figure 15. Node red starter Webpage UI Classifier prediction output

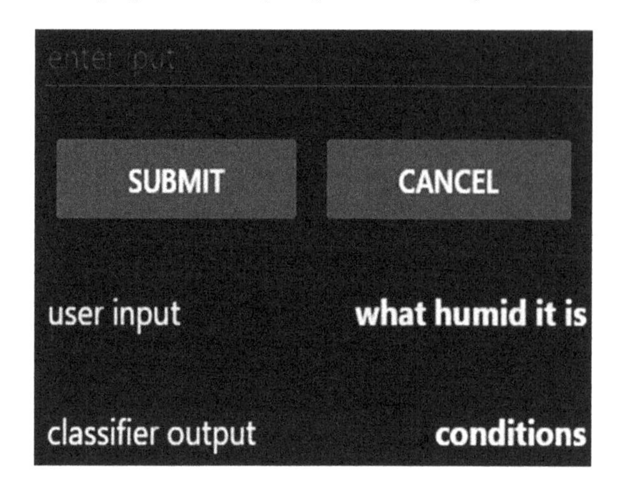

AI-ML Image Recognition Using Watson Studio and Node-Red

Image Processing and recognition sensor need Artificial Intelligence and Machine Learning technology to communicate within the sensor network effectively. With the advent of Internet of Things (IoT), each image capturing sensor can be given intelligence so that they can feature out the image components effectively and predicts the future foreign images with maximum prediction rate. Visual recognition training of images is based on pixels ratio study. Convolutional Neural Networks (CNN) is used to extract feature from image. Watson Visual Recognition Project (VRP) model has been made enabled to study the image recognition of training data set images. In visual recognition service are furnished with pre-trained models like general model, food model, face model, explicit model and text model. In general model thousands of pictures as dataset and it finds out the picture that are requested to predict. Explicit model is used by social networking portal to identify the adult content images in cybercrime of information certification. Zhu et al. (2017), live prediction model images of generic house-hold furniture are fed to the system to predict as depicted in Figure 16.

Face model predicts gender, face location in image by pixels clustering. The system is trained to predict the sexual classification of the learning image by utilising Convolutional Neural Network (CNN) and facial triangulation method for feature extraction (Chen et al.). From the set of trained images, the Frown Pixel Clustering (FPC) to identify the shrinkage in the skin is utilised as the prime factor to identify the age slab of the image. The given set of images is compared with the multiple age factor slabs and the slab which holds the majority of pixel matching values are highlighted as depicted in Figure 17. The complex prediction of age identification is achieved with the prediction rate of 77%. The existing work identifies the sexual status as Male or Female but, not the age of the personality. This schema of FPC will be a boon to Central Intelligence Bureau (CIB) to estimate the persons age factor in common place, which enhances the investigation process.

Food model identifies the type of food. The major classification is based on the training like assortment of Fruits, vegetables, sandwich, wrap, burger, rice bowl training images (Gongal et al., 2015; Dong et al., 2015; Khalid et al., 2020). Live example of veg wrap image is fed into the system for food type prediction as depicted in Figure 18. The prediction result successfully highlights with the threshold value of 1 for the condition of wrap. But, the system identifies it also as a sandwich because of stuff which

Figure 16. Extrapolation model for House Hold Image recognition in AI-ML System

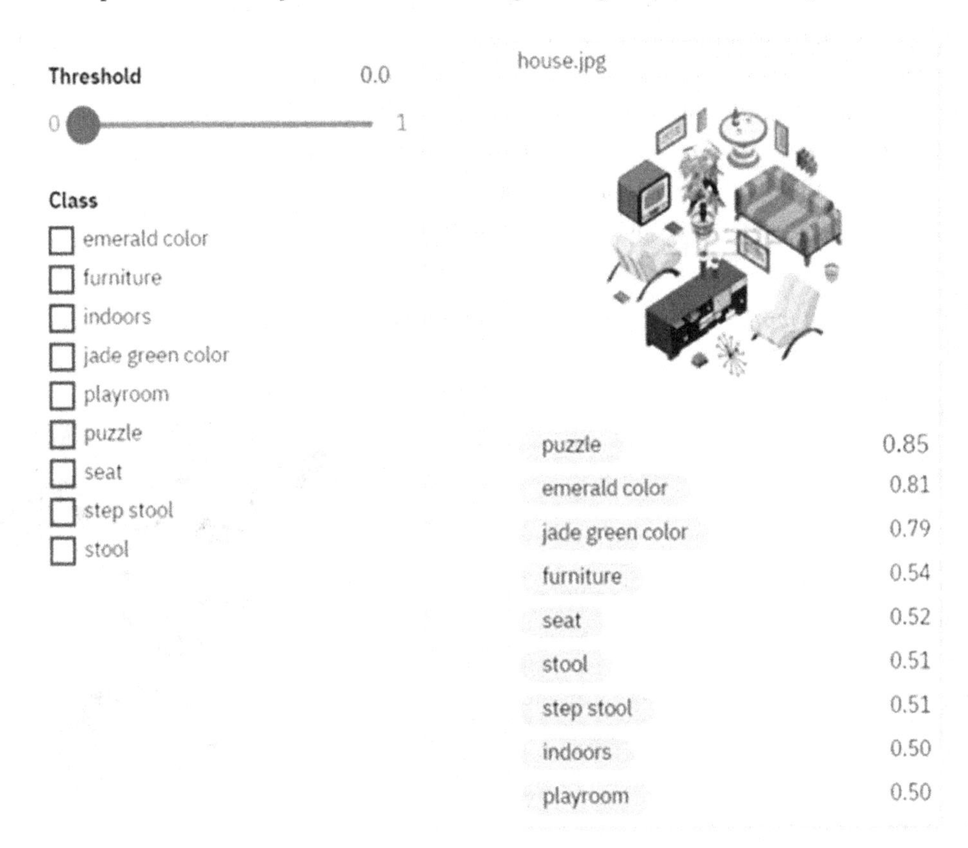

Figure 17. Extrapolation model for Facial Image recognition in AI-ML System

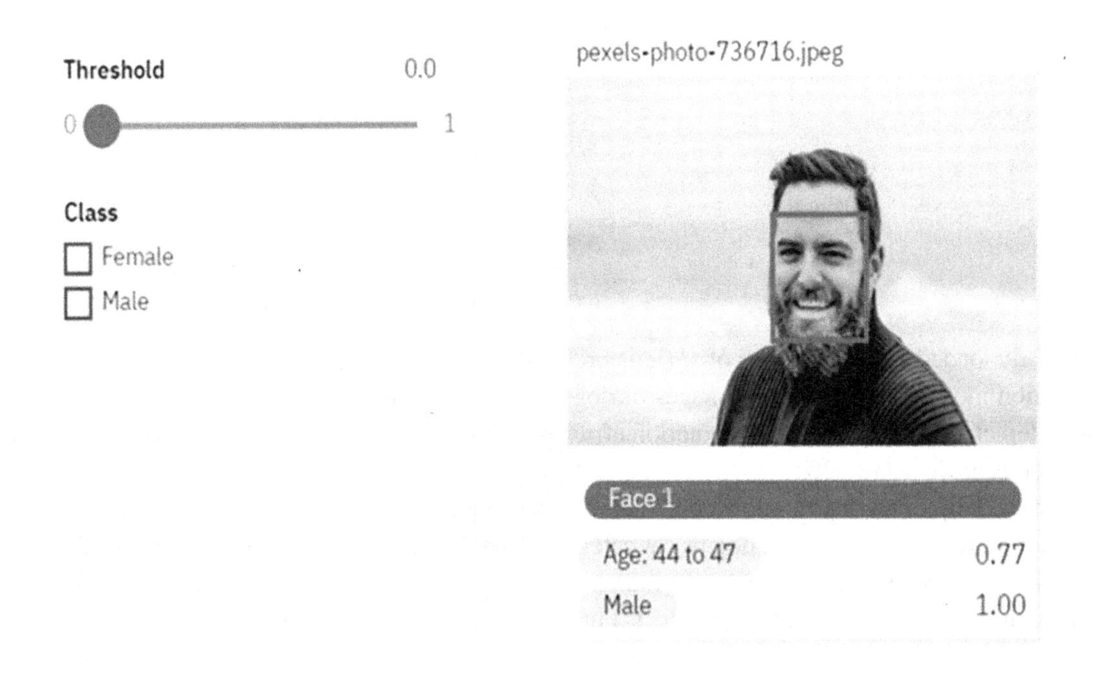

are common in both wrap and sandwich. Both sandwich and wrap class comes under the parent class of snack food; hence all these predictions are made efficaciously with the maximum threshold value of 1. It has been concluded that the system is not a catastrophe model as with the learning ability, the system also identifies the possibility of other option like folded flat bread or flatbed sandwich. Hence this system function with the evidence of Artificial Intelligence to think and Machine learning to learn from its experience. Hence it is not an over trained model even though it produces the maximum threshold points for prediction parameters with the efficiency of 100%

Figure 18. Extrapolation model for Food Image recognition in AI-ML System

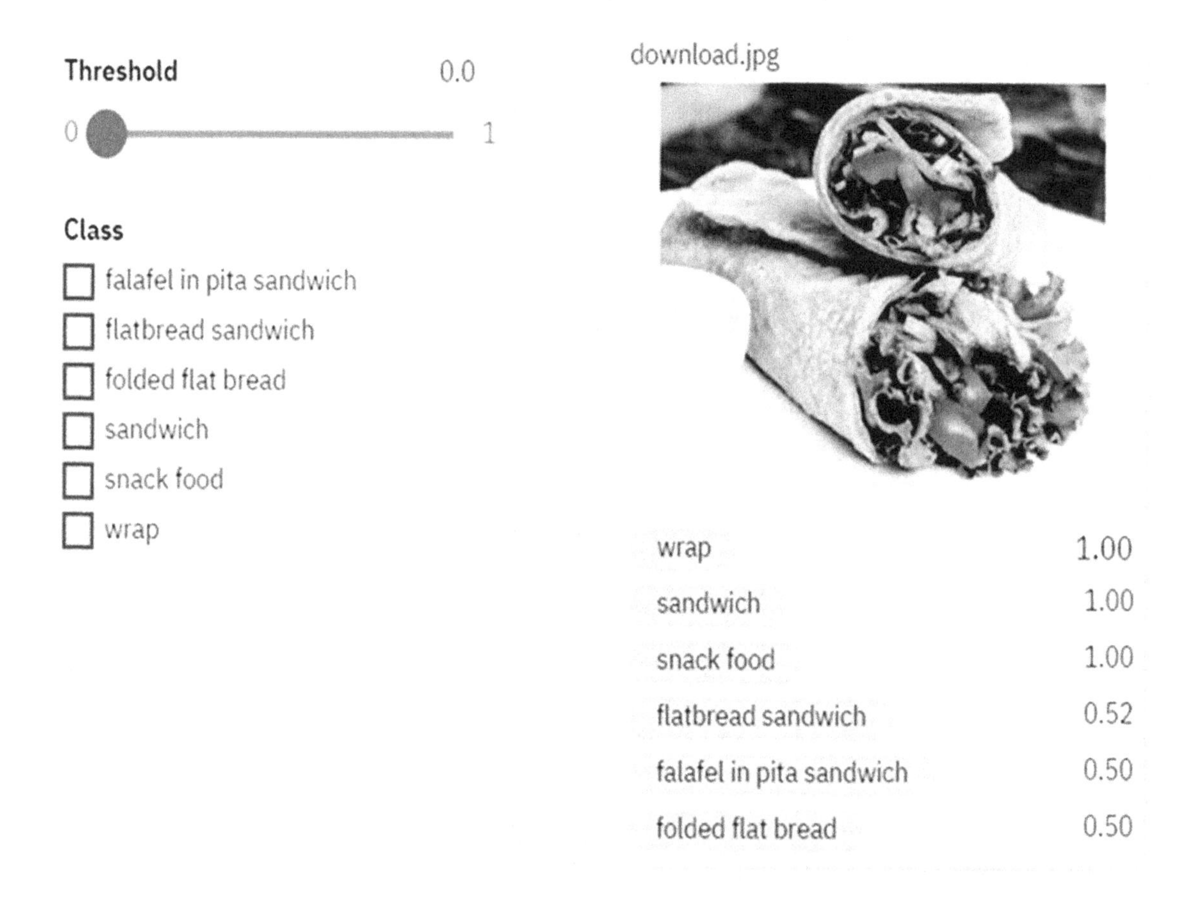

Already loaded training images of various breed of dogs and cats are used to train the AI-ML image recognition models. Live example images of dogs and cats are utilised for training the model to be robust as depicted in Figure 19. The prediction efficiency of the image classification by image recognition technique is achieved to 92%.

With the help of the learnt knowledge the system is made ready to identify whether the given input prediction image is either from a dog or cat category. Live image of a dog is fed into the system and the threshold from the learning experience predicts the image as dog with the efficiency of 92% i.e. 0.92 from the threshold prediction range of 0 to 1. Future extension of this research work is to attain the real-

Figure 19. Extrapolation model for Pet Image recognition in AI-ML System

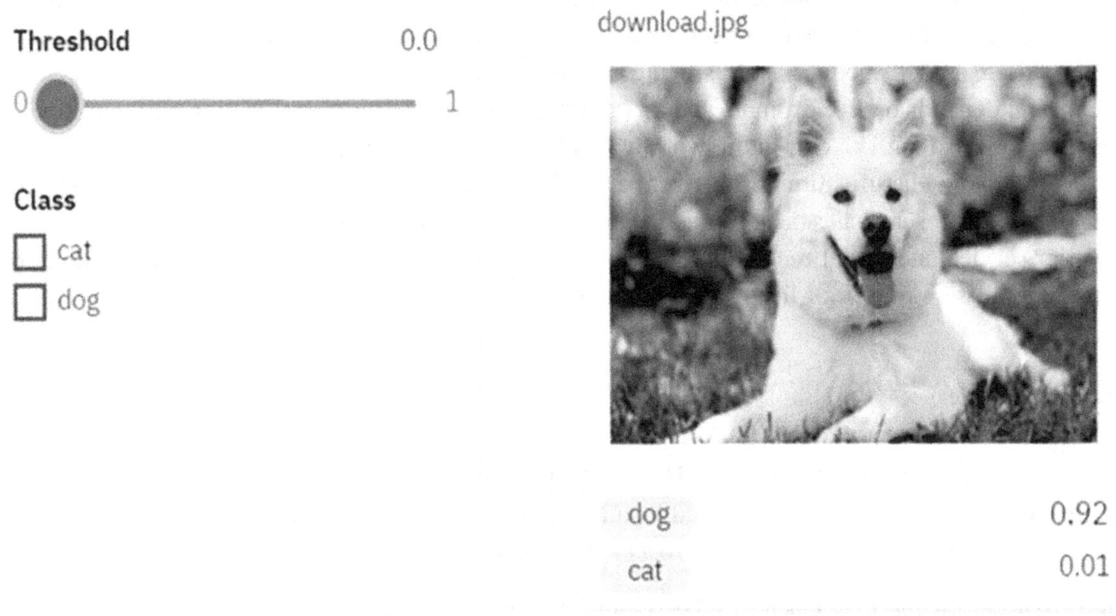

time dataset Internet of Things which are connected to form a WSN. Care to be taken while integrating AI-ML with Internet of Things (IoT) for WSN application which is immune to intermediate attackers, with the implementation of advanced security IoT schema (Jindal et al., 2019; R Tiwari et al., 2019; Goel et al., 2019). Advanced classifiers analysis other than Decision Tree Classifiers (DTC) will be studied for higher accuracy with less redundancy (Bhushan et al., 2017; Bhushan et al., 2018). This research chapter will pave the footprints for implementation of multiple classifier with higher data prediction accuracy for AI-ML-IoT WSN application.

CONCLUSION

Artificial Intelligence and Machine Learning (AI-ML) systems are competent to learn, which permits people, process and things to enhance the performance significantly over time. Highly user interactive Chabot system have been implemented for Natural Language Classifier (NLC) and taxonomy grouping with an efficiency of 99%. The created AI-ML system is said to be well trained when compared with existing system as it depends on live example and it is oriented towards training justification and minimal error prediction of less than 1%. This chapter successfully emphasise the trained and evaluated efficiency of Decision Tree Classifier (DTC) algorithm to implement the AI-ML robust system with area under ROC curve accuracy of 73% and Area under PR accuracy of 91%. Natural language classifier service obtains an accuracy of 99% for predicting the text "temperature " in our dataset.In visual recognition service the pre-trained model, face model obtained 77% accuracy for predicting the age to be 44 to 47 from the given face picture .In Visual recognition service pretrained model, food model obtained maximum accuracy of 52% for predicting the flatbread sandwich in the sandwich picture.In

Visual recognition service user trained model having images of dogs and cats predicts the accuracy of dog to be 92%.The Average prediction accuracy rate of Extrapolation model is 90.8%, for household puzzle extraction, facial i.e. age and gender prediction, food type and pet family live image prediction. The proposed AI-ML model, justifies being efficient and well trained when compared with existing system as it depends on live example and it is oriented towards training justification and minimal error prediction of less than 1%. Our future work would be to obtain the dataset from real-time IOT devices. Planned to obtain classification results from classifiers other than decision tree classifiers with higher accuracy. Future research has been planned to increase the current accuracy obtained in natural language classifier service and visual recognition service. Hence, the proposed AI-ML model is claimed to be robust and fail-safe model for live project implantation.

REFERENCES

Aluri, A., Price, B. S., & McIntyre, N. H. (2019). Using machine learning to cocreate value through dynamic customer engagement in a brand loyalty program. *Journal of Hospitality & Tourism Research (Washington, D.C.), 43*(1), 78–100. doi:10.1177/1096348017753521

Bai, J., Wu, Y., Zhang, J., & Chen, F. (2015). Subset based deep learning for RGB-D object recognition. *Neurocomputing, 165*, 280–292. doi:10.1016/j.neucom.2015.03.017

Bakis, R., Connors, D. P., Dube, P., Kapanipathi, P., Kumar, A., Malioutov, D., & Venkatramani, C. (2017). Performance of natural language classifiers in a question-answering system. *IBM Journal of Research and Development, 61*(4/5), 14–1. doi:10.1147/JRD.2017.2711719

Bhushan, B., & Sahoo, G. (2017). Recent Advances in Attacks, Technical Challenges, Vulnerabilities and Their Countermeasures in Wireless Sensor Networks. *Wireless Personal Communications, 98*(2), 2037–2077. doi:10.100711277-017-4962-0

Bhushan, B., & Sahoo, G. (2018). Routing Protocols in Wireless Sensor Networks. *Computational Intelligence in Sensor Networks Studies in Computational Intelligence*, 215-248. Doi:10.1007/978-3-662-57277-1_10

Bleidorn, W., & Hopwood, C. J. (2019). Using machine learning to advance personality assessment and theory. *Personality and Social Psychology Review, 23*(2), 190–203. doi:10.1177/1088868318772990 PMID:29792115

Cavalcante, I. M., Frazzon, E. M., Forcellini, F. A., & Ivanov, D. (2019). A supervised machine learning approach to data-driven simulation of resilient supplier selection in digital manufacturing. *International Journal of Information Management, 49*, 86–97. doi:10.1016/j.ijinfomgt.2019.03.004

Chen, J. C., Patel, V. M., & Chellappa, R. (2016, March). Unconstrained face verification using deep cnn features. In *2016 IEEE winter conference on applications of computer vision (WACV)* (pp. 1-9). IEEE.

Cruz, F., Twiefel, J., Magg, S., Weber, C., & Wermter, S. (2015, July). Interactive reinforcement learning through speech guidance in a domestic scenario. In *2015 International Joint Conference on Neural Networks (IJCNN)* (pp. 1-8). IEEE. 10.1109/IJCNN.2015.7280477

Dong, C., Loy, C. C., He, K., & Tang, X. (2015). Image super-resolution using deep convolutional networks. *IEEE Transactions on Pattern Analysis and Machine Intelligence*, *38*(2), 295–307. doi:10.1109/TPAMI.2015.2439281 PMID:26761735

Du, Y., Wang, W., & Wang, L. (2015). Hierarchical recurrent neural network for skeleton based action recognition. In *Proceedings of the IEEE conference on computer vision and pattern recognition* (pp. 1110-1118). IEEE.

Dubitzky, W., Lopes, P., Davis, J., & Berrar, D. (2019). The open international soccer database for machine learning. *Machine Learning*, *108*(1), 9–28. doi:10.100710994-018-5726-0

Eitel, A., Springenberg, J. T., Spinello, L., Riedmiller, M., & Burgard, W. (2015, September). Multimodal deep learning for robust RGB-D object recognition. In *2015 IEEE/RSJ International Conference on Intelligent Robots and Systems (IROS)* (pp. 681-687). IEEE. 10.1109/IROS.2015.7353446

Fu, G. S., Levin-Schwartz, Y., Lin, Q. H., & Zhang, D. (2019). Machine Learning for Medical Imaging. *Journal of Healthcare Engineering*. PMID:31183031

Ghahramani, Z. (2015). Probabilistic machine learning and artificial intelligence. *Nature*, *521*(7553), 452–459. doi:10.1038/nature14541 PMID:26017444

Goel, A. K., Rose, A., Gaur, J., & Bhushan, B. (2019). Attacks, Countermeasures and Security Paradigms in IoT. *2019 2nd International Conference on Intelligent Computing, Instrumentation and Control Technologies (ICICICT)*. doi: 10.1109/icicict46008.2019.8993338

Gongal, A., Amatya, S., Karkee, M., Zhang, Q., & Lewis, K. (2015). Sensors and systems for fruit detection and localization: A review. *Computers and Electronics in Agriculture*, *116*, 8–19. doi:10.1016/j.compag.2015.05.021

Jaimahaprabhu, A., Kumar, P., Gangadharan, P. S., & Latha, B. (2019, March). Cloud Analytics based Farming with Predictive Analytics using Artificial Intelligence. In *2019 Fifth International Conference on Science Technology Engineering and Mathematics (ICONSTEM)* (Vol. 1, pp. 65-68). IEEE. 10.1109/ICONSTEM.2019.8918785

Jindal, M., Gupta, J., & Bhushan, B. (2019). Machine learning methods for IoT and their Future Applications. *2019 International Conference on Computing, Communication, and Intelligent Systems (ICCCIS)*. 10.1109/ICCCIS48478.2019.8974551

Khalid, N. S., Shukor, S. A. A., & Syahir, A. F. (2020). Specific Gravity-based of Post-harvest Mangifera indica L. cv. Harumanis for 'Insidious Fruit Rot'(IFR) Detection using Image Processing. In *Computational Science and Technology* (pp. 33–42). Springer. doi:10.1007/978-981-15-0058-9_4

Li, Y., Niu, M., & Zou, Q. (2019). ELM-MHC: An Improved MHC identification method with extreme learning machine algorithm. *Journal of Proteome Research*, *18*(3), 1392–1401. doi:10.1021/acs.jproteome.9b00012 PMID:30698979

Neloy, A. A., Alam, S., Bindu, R. A., & Moni, N. J. (2019, April). Machine Learning based Health Prediction System using IBM Cloud as PaaS. In *2019 3rd International Conference on Trends in Electronics and Informatics (ICOEI)* (pp. 444-450). IEEE. 10.1109/ICOEI.2019.8862754

Perniu, L. M., Moşoi, A. A., Sandu, F., Moraru, S. A., Ungureanu, D. E., Kristály, D. M., & Guşeilă, L. G. (2019, August). Cloud Services for an Active Assisted Living Platform using Wireless Sensors and Mobile Devices. In *2019 International Conference on Sensing and Instrumentation in IoT Era (ISSI)* (pp. 1-6). IEEE. 10.1109/ISSI47111.2019.9043703

Ralston, K., Chen, Y., Isah, H., & Zulkernine, F. (2019). *A Voice Interactive Multilingual Student Support System using IBM Watson.* arXiv preprint arXiv:2001.00471

Solanki, V. K., Cuong, N. H. H., & Lu, Z. J. (2019). Opinion mining: using machine learning techniques. In *Extracting Knowledge From Opinion Mining* (pp. 66–82). IGI Global. doi:10.4018/978-1-5225-6117-0.ch004

Tang, J., Deng, C., & Huang, G. B. (2015). Extreme learning machine for multilayer perceptron. *IEEE Transactions on Neural Networks and Learning Systems*, *27*(4), 809–821. doi:10.1109/TNNLS.2015.2424995 PMID:25966483

Teschendorff, A. E. (2019). Avoiding common pitfalls in machine learning omic data science. *Nature Materials*, *18*(5), 422–427. doi:10.103841563-018-0241-z PMID:30478452

Tiwari, R., Sharma, N., Kaushik, I., Tiwari, A., & Bhushan, B. (2019). Evolution of IoT & Data Analytics using Deep Learning. *2019 International Conference on Computing, Communication, and Intelligent Systems (ICCCIS)*. 10.1109/ICCCIS48478.2019.8974481

Vedaldi, A., & Lenc, K. (2015, October). Matconvnet: Convolutional neural networks for matlab. In *Proceedings of the 23rd ACM international conference on Multimedia* (pp. 689-692). ACM. 10.1145/2733373.2807412

Zhu, Y., Mottaghi, R., Kolve, E., Lim, J. J., Gupta, A., Fei-Fei, L., & Farhadi, A. (2017, May). Target-driven visual navigation in indoor scenes using deep reinforcement learning. In *2017 IEEE international conference on robotics and automation (ICRA)* (pp. 3357-3364). IEEE.

Compilation of References

Abacus. (n.d.). Retrieved May 5, 2020, from https://en.wikipedia.org/wiki/Abacus

Abad, C. L., & Bonilla, R. I. (2007). An analysis on the schemes for detecting and preventing ARP cache poisoning attacks. *Proceedings - International Conference on Distributed Computing Systems*. 10.1109/ICDCSW.2007.19

Abadi, M., Chu, A., Goodfellow, I., McMahan, H. B., Mironov, I., Talwar, K., & Zhang, L. (2016). Deep Learning with Differential Privacy. In *Proceedings of the 2016 ACM SIGSAC Conference on Computer and Communications Security - CCS'16* (pp. 308–318). New York: ACM Press. 10.1145/2976749.2978318

Abbasi, A. A., & Younis, M. (2007). A survey on clustering algorithms for wireless sensor networks. *Computer Communications*, *30*(14-15), 2826–2841. doi:10.1016/j.comcom.2007.05.024

Abdalzaher, M. S., Samy, L., & Muta, O. (2019). Non-zero-sum game-based trust model to enhance wireless sensor networks security for IoT applications. *IET Wireless Sensor Systems*, *9*(4), 218–226. doi:10.1049/iet-wss.2018.5114

Abdulnabi, M., Al-Haiqi, A., Kiah, M. L., Zaidan, A. A., Zaidan, B. B., & Hussain, M. (2017). A distributed framework for health information exchange using smartphone technologies. *Journal of Biomedical Informatics*, *69*, 230–250. doi:10.1016/j.jbi.2017.04.013 PMID:28433825

Aghili, S., Mala, H., & Peris-Lopez, P. (2018). Securing Heterogeneous Wireless Sensor Networks: Breaking and Fixing a Three-Factor Authentication Protocol. *Sensors (Basel)*, *18*(11), 3663. doi:10.339018113663 PMID:30380595

Ahmed, G., Khan, N., Khalid, Z., & Ramer, R. (2009). *Cluster head selection using decision trees for Wireless Sensor Networks*. doi:10.1109/ISSNIP.2008.4761982

Ahmed, M., Taha, A., Hassanien, A. E., & Hassanien, E. (2018). *An Optimized K-Nearest Neighbor Algorithm for Extending Wireless Sensor Network Lifetime*. . doi:10.1007/978-3-319-74690-6_50

Akhand, M. A. H., Ayon, S. I., Shahriyar, S. A., Siddique, N., & Adeli, H. (2020). Discrete Spider Monkey Optimization for Travelling Salesman Problem. *Applied Soft Computing*, *86*, 105887. doi:10.1016/j.asoc.2019.105887

Akkarajitsakul, K., Hossain, E., Niyato, D., & Kim, D. I. (2011). Game theoretic approaches for multiple access in wireless networks: A survey. *IEEE Communications Surveys and Tutorials*, *13*(3), 372–395. doi:10.1109/SURV.2011.122310.000119

Akyildiz, I. F., Melodia, T., & Chowdhury, K. R. (2007). A survey on wireless multimedia sensor networks. *Computer Networks*, *51*(4), 921–960. doi:10.1016/j.comnet.2006.10.002

Akyildiz, I. F., Su, W., Sankarasubramaniam, Y., & Cayirci, E. (2002). A survey on sensor networks. *IEEE Communications Magazine*, *40*(8), 102–114. doi:10.1109/MCOM.2002.1024422

Akyildiz, I. F., Su, W., Sankarasubramaniam, Y., & Cayirci, E. (2002). Wireless Sensor Networks: A Survey. *Computer Networks*, *38*(4), 393–422. doi:10.1016/S1389-1286(01)00302-4

Akyildiz, I. F., Wang, X., & Wang, W. (2005). Wireless mesh networks- a survey. *Computer Networks, 47*(4), 445–487. doi:10.1016/j.comnet.2004.12.001

Alaiad, A., & Zhou, L. (2017). Patients' Adoption of WSN-Based Smart Home Healthcare Systems: An Integrated Model of Facilitators and Barriers. *IEEE Transactions on Professional Communication, 60*(1), 4–23. doi:10.1109/TPC.2016.2632822

Alali, M., Sharef, N. M., Murad, M. A. A., Hamdan, H., & Husin, N. A. (2019). Narrow Convolutional Neural Network for Arabic Dialects Polarity Classification. *IEEE Access: Practical Innovations, Open Solutions, 7*, 96272–96283. doi:10.1109/ACCESS.2019.2929208

Alguliyev, R. M., Aliguliyev, R. M., & Abdullayeva, F. J. (2019). Privacy-preserving deep learning algorithm for big personal data analysis. *Journal of Industrial Information Integration*. doi:10.1016/j.jii.2019.07.002

Ali, B., Mahmood, T., Abbas, M., Hussain, M., Ullah, H., Sarker, A., & Khan, A. (2019). LEACH Robust Routing Approach Applying Machine Learning. *IJCSNS, 19*(6), 18–26.

Alicherry, M., Bhatia, R., & Li, E. (2006). Joint channel assignment and routing for throughput optimization in multi radio wireless mesh networks. *IEEE Journal of Selected Areas in Communication, 24*(11), 1960-1971.

Ali, R., Pal, A. K., Kumari, S., Karuppiah, M., & Conti, M. (2018). A secure user authentication and key-agreement scheme using wireless sensor networks for agriculture monitoring. *Future Generation Computer Systems, 84*, 200–215. doi:10.1016/j.future.2017.06.018

Ali, S., Al-Balushi, T., Nadir, Z., & Hussain, O. K. (2018). Improving the Resilience of Wireless Sensor Networks Against Security Threats: A Survey and Open Research Issues. *International Journal of Technology, 4*(4), 828–839. doi:10.14716/ijtech.v9i4.1526

Aliu, O. G., Imran, A., Imran, M. A., & Evans, B. (2012). A survey of self organisation in future cellular networks. *IEEE Communications Surveys and Tutorials, 15*(1), 336–361. doi:10.1109/SURV.2012.021312.00116

Alliance for Internet Things Innovation. (2019). *AIOTI Vision and Recommendations: European IoT challenges and opportunities*. Retrieved December 11, 2019, from https://aioti.eu/aioti-vision-and-recommendations-european-iot-challenges-and-opportunities-2019-2024/

Allman, M., & Ostermann, S. (1999). *FTP security considerations*. Academic Press.

Almomani, I., Al-Kasasbeh, B., & Al-Akhras, M. (2016, August). WSN-DS: A dataset for intrusion detection systems in wireless sensor networks. *Journal of Sensors, 2016*, 1–16. doi:10.1155/2016/4731953

Alnasir, J. J., & Shanahan, H. P. (2020). The application of hadoop in structural bioinformatics. *Briefings in Bioinformatics, 21*(1), 96–105. PMID:30462158

Alotaibi, M. (2019). Security to wireless sensor networks against malicious attacks using Hamming residue method. *EURASIP Journal on Wireless Communications and Networking, 2019*(1), 8. doi:10.118613638-018-1337-5

Alpaslan, A. H. (2016). Cyberchondria and adolescents. *The International Journal of Social Psychiatry, 62*(7), 679–680. doi:10.1177/0020764016657113 PMID:27358345

Alpaydin, E. (2020). *Introduction to machine learning*. MIT Press.

Alsheikh, M. A., Lin, S., Niyato, D., & Tan, H. P. (2014). Machine learning in wireless sensor networks: Algorithms, strategies, and applications. *IEEE Communications Surveys and Tutorials, 16*(4), 1996–2018. doi:10.1109/COMST.2014.2320099

Althunibat, S., Antonopoulos, A., Kartsakli, E., Granelli, F., & Verikoukis, C. (2016). Countering intelligent-dependent malicious nodes in target detection wireless sensor networks. *IEEE Sensors Journal, 16*(23), 8627–8639. doi:10.1109/JSEN.2016.2606759

Al-Turjman, F., & Alturjman, S. (2018). Context-sensitive access in industrial internet of things (IIoT) healthcare applications. *IEEE Transactions on Industrial Informatics, 14*(6), 2736–2744. doi:10.1109/TII.2018.2808190

Al-Turjman, F., Hasan, M. Z., & Al-Rizzo, H. (2018). Task scheduling in cloud-based survivability applications using swarm optimization in IoT. *Transactions on Emerging Telecommunications Technologies*, 3539.

Aluri, A., Price, B. S., & McIntyre, N. H. (2019). Using machine learning to cocreate value through dynamic customer engagement in a brand loyalty program. *Journal of Hospitality & Tourism Research (Washington, D.C.), 43*(1), 78–100. doi:10.1177/1096348017753521

Alves, T., Das, R., Werth, A., & Morris, T. (2018). Virtualization of SCADA testbeds for cybersecurity research: A modular approach. *Computers & Security, 77*, 531–546. doi:10.1016/j.cose.2018.05.002

Ambigavathi, M., & Sridharan, D. (2018). Energy-Aware Data Aggregation Techniques in Wireless Sensor Network. *Advances in Power Systems and Energy Management Lecture Notes in Electrical Engineering*, 165–173. doi:10.1007/978-981-10-4394-9_17

Amin, R., Islam, S. H., Kumar, N., & Choo, K.-K. R. (2018). An untraceable and anonymous password authentication protocol for heterogeneous wireless sensor networks. *Journal of Network and Computer Applications, 104*, 133–144. doi:10.1016/j.jnca.2017.12.012

Amjad, M., Afzal, M., Umer, T., & Kim, B. (2017). QoS-Aware and Heterogeneously Clustered Routing Protocol for Wireless Sensor Networks. *IEEE Access: Practical Innovations, Open Solutions, 5*, 10250–10262. doi:10.1109/ACCESS.2017.2712662

Anand, C., & Gnanamurthy, R. K. (2016). Localized DoS Attack Detection Architecture for Reliable Data Transmission Over Wireless Sensor Network. *Wireless Personal Communications, 90*(2), 847–859. doi:10.100711277-016-3231-y

Anantharaman, A. (2019). *How Artificial Intelligence is enabling financial inclusion in India*. https://yourstory.com/2019/05/artificial-intelligence-enabling-financial-inclusion

Andrew, J., Mathew, S. S., & Mohit, B. (2019). *A Comprehensive Analysis of Privacy-preserving Techniques in Deep learning based Disease Prediction Systems*. doi:10.1088/1742-6596/1362/1/012070

Andrew, J., & Karthikeyan, J. (2019). Privacy-Preserving Internet of Things: Techniques and Applications. *International Journal of Engineering and Advanced Technology, 8*(6), 3229–3234. doi:10.35940/ijeat.F8830.088619

Andrew, J., & Kathrine, G. J. W. (2018). *An intrusion detection system using correlation, prioritization and clustering techniques to mitigate false alerts* (Vol. 645). Advances in Intelligent Systems and Computing. doi:10.1007/978-981-10-7200-0_23

Andrieu, C., De Freitas, N., Doucet, A., & Jordan, M. I. (2003). An introduction to MCMC for machine learning. *Machine Learning, 50*(1–2), 5–43. doi:10.1023/A:1020281327116

Anguraj, D. K., & Smys, S. (2018). Trust-Based Intrusion Detection and Clustering Approach for Wireless Body Area Networks. *Wireless Personal Communications*. Advance online publication. doi:10.100711277-018-6005-x

Aparna, P., & Kishore, P. V. (2017). An Efficient Medical Image Watermarking Technique in E-healthcare Application Using Hybridization of Compression and Cryptography Algorithm. *Journal of Intelligent Systems, 27*(1), 115–133. doi:10.1515/jisys-2017-0266

Arabi, Z. (2010, June). HERF: A hybrid energy efficient routing using a fuzzy method in wireless sensor networks. In *2010 International Conference on Intelligent and Advanced Systems* (pp. 1-6). IEEE. 10.1109/ICIAS.2010.5716145

Arel, I., Rose, D. C., & Karnowski, T. P. (2010). Deep machine learning-a new frontier in artificial intelligence research. *IEEE Computational Intelligence Magazine*, *5*(4), 13–18. doi:10.1109/MCI.2010.938364

Arita, S., & Nakasato, S. (2017). Fully Homomorphic Encryption for Classification in Machine Learning. *2017 IEEE International Conference on Smart Computing, SMARTCOMP 2017*, 2–5. 10.1109/SMARTCOMP.2017.7947011

Atallah, M. J., & Du, W. (2001). Secure multi-party computational geometry. In Lecture Notes in Computer Science (including subseries Lecture Notes in Artificial Intelligence and Lecture Notes in Bioinformatics) (Vol. 2125, pp. 165–179). Springer Verlag. doi:10.1007/3-540-44634-6_16

Athanasiou, G. N., & Lymberopoulos, D. K. (2016). A comprehensive Reputation mechanism for ubiquitous healthcare environment exploiting cloud model. *2016 38th Annual International Conference of the IEEE Engineering in Medicine and Biology Society (EMBC)*, 5981-5984.

Athanasiou, G., Broustis, I., & Tassiulas, I. (2011). Efficient load-aware channel allocation in wireless mesh networks. *Journal of Computer Networks and Communication*, *31*(7), 1–13. doi:10.1155/2011/972051

Athmani, S., Bilami, A., & Boubiche, D. E. (2019). EDAK: An Efficient Dynamic Authentication and Key Management Mechanism for heterogeneous WSNs. *Future Generation Computer Systems*, *92*, 789–799. doi:10.1016/j.future.2017.10.026

Atoui, I., Makhoul, A., Tawbe, S., Couturier, R., & Hijazi, A. (2016). Tree-Based Data Aggregation Approach in Periodic Sensor Networks Using Correlation Matrix and Polynomial Regression. *2016 IEEE Intl Conference on Computational Science and Engineering (CSE) and IEEE Intl Conference on Embedded and Ubiquitous Computing (EUC) and 15th Intl Symposium on Distributed Computing and Applications for Business Engineering (DCABES)*. 10.1109/CSE-EUC-DCABES.2016.267

Avram, T., Oh, S., & Hariri, S. (2007). Analyzing Attacks in Wireless Ad Hoc Network with Self-Organizing Maps. *Fifth Annual Conference on Communication Networks and Services Research (CNSR 07)*. 10.1109/CNSR.2007.15

Awasthi, A. K., & Srivastava, K. (2013). A Biometric Authentication Scheme for Telecare Medicine Information Systems with Nonce. *Journal of Medical Systems*, *37*(5), 1–4. doi:10.100710916-013-9964-1 PMID:23949846

Ayadi, A., Ghorbel, O., BenSalah, M. S., & Abid, M. (2020). Spatio-temporal correlations for damages identification and localization in water pipeline systems based on WSNs. *Computer Networks*, *171*, 107134. doi:10.1016/j.comnet.2020.107134

Azeta, A.A., Iboroma, D.A., Azeta, V.I., Igbekele, E.O., Fatinikun, D.O., & Ekpunobi, E. (2017). Implementing a medical record system with biometrics authentication in E-health. *2017 IEEE AFRICON*, 979-983.

Badajena, S. N., & Gundimeda, H. (2010). *Self help group bank linkage model and financial inclusion in India*. http://skoch.in

Bai, J., Wu, Y., Zhang, J., & Chen, F. (2015). Subset based deep learning for RGB-D object recognition. *Neurocomputing*, *165*, 280–292. doi:10.1016/j.neucom.2015.03.017

Bakis, R., Connors, D. P., Dube, P., Kapanipathi, P., Kumar, A., Malioutov, D., & Venkatramani, C. (2017). Performance of natural language classifiers in a question-answering system. *IBM Journal of Research and Development*, *61*(4/5), 14–1. doi:10.1147/JRD.2017.2711719

Bakke, A. (2017). *Ethos in E-Health: From Informational to Interactive Websites*. Academic Press.

Bakshi, B., Khorsandi, S., & Capone, A. (2011). On-line joint QoS routing and channel assignment in multi-channel multi-radio wireless mesh networks. *Computer Communications*, *34*(2), 1342–1360. doi:10.1016/j.comcom.2011.02.001

Banafa, A. (2016). *IoT Standardization and Implementation Challenges*. IEEE. https://iot.ieee.org/newsletter/july-2016/iot-standardization-and-implementation-challenges

Banerjee, S., Chunka, C., Sen, S., & Goswami, R. S. (2019). An Enhanced and Secure Biometric Based User Authentication Scheme in Wireless Sensor Networks Using Smart Cards. *Wireless Personal Communications*, *107*(1), 243–270. doi:10.100711277-019-06252-x

Bansal, J. C. (2019). Evolutionary and Swarm Intelligence Algorithms. In J. C. Bansal, P. K. Singh, & N. R. Pal (Eds.), *Evolutionary and Swarm Intelligence Algorithms*. doi:10.1007/978-3-319-91341-4

Bao, F., Chen, I.-R., Chang, M., & Cho, J.-H. (2012, June). Hierarchical trust management for wireless sensor networks and its applications to trust-based routing and intrusion detection. *IEEE eTransactions on Network and Service Management*, *9*(2), 169–183. doi:10.1109/TCOMM.2012.031912.110179

Baranowski, Z., Kleszcz, E., Kothuri, P., Canali, L., Castellotti, R., Marquez, M. M., . . . Duran, J. C. L. (2019). Evolution of the Hadoop Platform and Ecosystem for High Energy Physics. In *EPJ Web of Conferences* (*Vol. 214*, p. 04058). EDP Sciences. 10.1051/epjconf/201921404058

Barbara, A. M., Dobbins, M., Haynes, R. B., Iorio, A., Lavis, J. N., Raina, P., & Levinson, A. J. (2016). The McMaster Optimal Aging Portal: Usability Evaluation of a Unique Evidence-Based Health Information Website. *JMIR Human Factors*, *3*(1), 3. doi:10.2196/humanfactors.4800 PMID:27170443

Barkovskis, N., Salmins, A., Parrilla, F., Ozols, K., & Moreno, M. A. (2016). WSN based on accelerometer, GPS and RSSI measurements for train integrity monitoring. *4th International Conference on Control, Decision and Information Technologies (CoDIT)*.

Basseville, M., Nikiforov, I. V., & ... (1993). *Detection of abrupt changes: theory and application* (Vol. 104). Prentice Hall Englewood Cliffs.

Bauer, M., Boussard, M., Bui, N., De Loof, J., Magerkurth, C., Meissner, S., Nettsträter, A., Stefa, J., Thoma, M., & Walewski, J. W. (2013). *IoT Reference Architecture*. Retrieved from https://link.springer.com/chapter/10.1007/978-3-642-40403-0_8

Bayraktaroglu, E., King, C., Liu, X., Noubir, G., Rajaraman, R., & Thapa, B. (2013). Performance of IEEE 802.11 under jamming. *Mobile Networks and Applications*, *18*(5), 678–696. doi:10.100711036-011-0340-4

Beldad, A., Jong, M. D., & Steehouder, M. F. (2010). How shall I trust the faceless and the intangible? A literature review on the antecedents of online trust. *Computers in Human Behavior*, *26*(5), 857–869. doi:10.1016/j.chb.2010.03.013

Benayache, A., Azeddine, B. S. B., & Pascal, L. H. T. (2019). MsM: A microservice middleware for smart WSN-based IoT application. *Journal of Network and Computer Applications*, *144*(15), 138–154. doi:10.1016/j.jnca.2019.06.015

Bertsekas, D. P. (2005). *Dynamic programming and optimal control* (Vol. 1). Athena Scientific.

Beyens, P., Peeters, M., Steenhaut, K., & Nowe, A. (2005). Routing with compression in wireless sensor networks: a q-learning approach. In *Fifth European workshop on adaptive agents and multi-agent systems (AAMAS 05) (Vol. 8)*. Academic Press.

Beyer, K., Goldstein, J., Ramakrishnan, R., & Shaft, U. (1999). When Is "Nearest Neighbor" Meaningful? *Lecture Notes in Computer Science Database Theory — ICDT'99*, 217–235. doi: . doi:10.1007/3-540-49257-7_15

Bhadre, P., & Gothawal, D. (2014). Detection and blocking of spammers using SPOT detection algorithm. *Networks & Soft Computing (ICNSC), 2014 First International Conference on*, 97-101.

Bhanderi, M. V., & Shah, H. B. (2014). Machine Learning for Wireless Sensor Network: A Review, Challenges and Applications. *Adv. Electron. Electr. Eng*, *4*, 475–486.

Bhattasali, T., & Saeed, K. (2014). Two factor remote authentication in healthcare. *2014 International Conference on Advances in Computing, Communications and Informatics (ICACCI)*. 10.1109/ICACCI.2014.6968594

Bhushan & Sahoo. (2019). A Hybrid Secure and Energy Efficient Cluster Based Intrusion Detection system for Wireless Sensing Environment. *2019 2nd International Conference on Signal Processing and Communication (ICSPC).*

Bhushan, B., & Sahoo, G. (2017). Detection and defense mechanisms against wormhole attacks in wireless sensor networks. *2017 3rd International Conference on Advances in Computing,Communication& Automation (ICACCA)*. DOI: 10.1109/icaccaf.2017.8344730

Bhushan, B., & Sahoo, G. (2018). Routing Protocols in Wireless Sensor Networks. *Computational Intelligence in Sensor Networks Studies in Computational Intelligence,* 215-248. DOI: . doi:10.1007/978-3-662-57277-1_10

Bhushan, B., & Sahoo, G. (2020a). *A Hybrid Secure and Energy Efficient Cluster Based Intrusion Detection system for Wireless Sensing Environment*. Institute of Electrical and Electronics Engineers (IEEE). doi:10.1109/icspc46172.2019.8976509

Bhushan, B., & Sahoo, G. (2020b). Requirements, Protocols, and Security Challenges in Wireless Sensor Networks: An Industrial Perspective. In Handbook of Computer Networks and Cyber Security (pp. 683–713). Springer International Publishing. doi:10.1007/978-3-030-22277-2_27

Bhushan, B., Sahoo, G., & Rai, A. K. (2017). Man-in-the-middle attack in wireless and computer networking — A review. *2017 3rd International Conference on Advances in Computing,Communication & Automation (ICACCA)*. DOI: 10.1109/icaccaf.2017.8344724

Bhushan, B., & Sahoo, G. (2017). A comprehensive survey of secure and energy efficient routing protocols and data collection approaches in wireless sensor networks. *2017 International Conference on Signal Processing and Communication (ICSPC)*. 10.1109/CSPC.2017.8305856

Bhushan, B., & Sahoo, G. (2018). Recent advances in attacks, technical challenges, vulnerabilities and their countermeasures in wireless sensor networks. *Wireless Personal Communications*, *98*(2), 2037–2077. doi:10.100711277-017-4962-0

Bhushan, B., & Sahoo, G. (2019). $$E^{2} SR^{2}$$ E 2 S R 2: An acknowledgement-based mobile sink routing protocol with rechargeable sensors for wireless sensor networks. *Wireless Networks*, *25*(5), 2697–2721. doi:10.100711276-019-01988-7

Bhushan, B., & Sahoo, G. (2019). ISFC-BLS (Intelligent and Secured Fuzzy Clustering Algorithm Using Balanced Load Sub-Cluster Formation) in WSN Environment. *Wireless Personal Communications*. Advance online publication. doi:10.100711277-019-06948-0

Bhushan, B., & Sahoo, G. (2019). Secure Location-Based Aggregator Node Selection Scheme in Wireless Sensor Networks, *Proceedings of ICETIT 2019, Emerging Trends in Information Technology.*

Bhushan, B., & Sahoo, G. (2019). Secure Location-Based Aggregator Node Selection Scheme in Wireless Sensor Networks. *Proceedings of ICETIT 2019 Lecture Notes in Electrical Engineering*, 21–35. DOI: 10.1007/978-3-030-30577-2_2

Bianchi, G. (2000). Performance analysis of the IEEE 802.11 distributed coordination function. *IEEE Journal on Selected Areas in Communications*, *18*(3), 535–547. doi:10.1109/49.840210

Bin, G., Zhe, L., & Ze-Jun, W. (2005, September). A dynamic-cluster energy-aware routing algorithm based on neural structure in the wireless sensor networks. In *The Fifth International Conference on Computer and Information Technology (CIT'05)* (pp. 401-405). IEEE. 10.1109/CIT.2005.8

Biryukov, A., De Cannière, C., Winkler, W. E., Aggarwal, C. C., Kuhn, M., Bouganim, L., ... Smith, S. W. (2011). Differential Privacy. In Encyclopedia of Cryptography and Security (pp. 338–340). Boston, MA: Springer US. doi:10.1007/978-1-4419-5906-5_752

Bleidorn, W., & Hopwood, C. J. (2019). Using machine learning to advance personality assessment and theory. *Personality and Social Psychology Review*, *23*(2), 190–203. doi:10.1177/1088868318772990 PMID:29792115

Borgne, Y.-A. L., Raybaud, S., & Bontempi, G. (2008). Distributed Principal Component Analysis for Wireless Sensor Networks. *Sensors (Basel)*, *8*(8), 4821–4850. doi:10.33908084821 PMID:27873788

Boyan, J. A., & Littman, M. L. (1994). Packet routing in dynamically changing networks: A reinforcement learning approach. In Advances in neural information processing systems (pp. 671-678). Academic Press.

Branch, J. W., Giannella, C., Szymanski, B., Wolff, R., & Kargupta, H. (2013). In-network outlier detection in wireless sensor networks. *Knowledge and Information Systems*, *34*(1), 23–54. doi:10.100710115-011-0474-5

Brinkrolf, J., Göpfert, C., & Hammer, B. (2019). Differential privacy for learning vector quantization. *Neurocomputing*, *342*, 125–136. doi:10.1016/j.neucom.2018.11.095

Brown, J. J. C. J. M. (2020). *Sensor Terminology*. Retrieved February 11, 2020, from https://www.ni.com/en-in/innovations/white-papers/13/sensor-terminology.html

Bruno, R., Conti, M., & Gregori, E. (2007). Throughput analysis and measurements in IEEE 802.11 WLANs with TCP and UDP traffic flows. *IEEE Transactions on Mobile Computing*, *7*(2), 171–186. doi:10.1109/TMC.2007.70718

Buehrer, R. M. (2006). *Synthesis Lectures on Communications*. Morgan & Claypool Publishers.

Bult, K., Burstein, A., Chang, D., Dong, M., Fielding, M., & Kruglick, E., ... Pottie, G. (1996). Low Power Systems for Wireless Microsensors. *International Symposium on Low Power Electronics and Design*. 10.1109/LPE.1996.542724

Butun, I., Morgera, S. D., & Sankar, R. (2014). A Survey of Intrusion Detection Systems in Wireless Sensor Networks. *IEEE Communications Surveys and Tutorials*, *16*(1), 266–282. doi:10.1109/SURV.2013.050113.00191

Cagalj, M., Ganeriwal, S., Aad, I., & Hubaux, J.-P. (2005). On selfish behavior in CSMA/CA networks. *Proceedings IEEE 24th Annual Joint Conference of the IEEE Computer and Communications Societies*, *4*, 2513-2524. 10.1109/INFCOM.2005.1498536

Calkavur, S. (2018). An Image Secret Sharing Method Based on Shamir Secret Sharing. *Current Trends in Computer Sciences & Applications*, *1*(2). Advance online publication. doi:10.32474/CTCSA.2018.01.000106

Cao, X., & Gong, N. Z. (2017). Mitigating evasion attacks to deep neural networks via region-based classification. *ACM International Conference Proceeding Series, Part F1325*, 278–287. 10.1145/3134600.3134606

Car, J., Tan, W. S., Huang, Z., Sloot, P. M., & Franklin, B. D. (2017). eHealth in the future of medications management: Personalisation, monitoring and adherence. *BMC Medicine*, *15*(1), 15. doi:10.118612916-017-0838-0 PMID:28376771

Carr, J. J., & John, M. B. (2002). *Introduction to Biomedical Equipment Technology* (4th ed.). Pearson.

Cavalcante, I. M., Frazzon, E. M., Forcellini, F. A., & Ivanov, D. (2019). A supervised machine learning approach to data-driven simulation of resilient supplier selection in digital manufacturing. *International Journal of Information Management*, *49*, 86–97. doi:10.1016/j.ijinfomgt.2019.03.004

Center, C. C. (1995). IP Spoofing Attacks and Hijacked Terminal Connections. CA-95: 01.

Chakraborty, U. K., Das, S. K., & Abbott, T. E. (2012, June). Energy-efficient routing in hierarchical wireless sensor networks using differential-evolution-based memetic algorithm. In *2012 IEEE Congress on Evolutionary Computation* (pp. 1-8). IEEE. 10.1109/CEC.2012.6252985

Challa, S., Wazid, M., Das, A. K., Kumar, N., Alavalapati, G. R., Yoon, E., & Yoo, K. (2017). Secure Signature-Based Authenticated Key Establishment Scheme for Future IoT Applications. *IEEE Access: Practical Innovations, Open Solutions*, *5*, 3028–3043. doi:10.1109/ACCESS.2017.2676119

Chang, C., & Le, H. (2016). A Provably Secure, Efficient, and Flexible Authentication Scheme for Ad hoc Wireless Sensor Networks. *IEEE Transactions on Wireless Communications*, *15*(1), 357–366. doi:10.1109/TWC.2015.2473165

Chang, C.-Y., Lin, C.-Y., & Kuo, C.-H. (2012). EBDC: An energy-balanced data collection mechanism using a mobile data collector in WSNs. *Sensors (Basel)*, *12*(5), 5850–5871. doi:10.3390120505850 PMID:22778617

Chang, R. K. (2002). Defending against flooding-based distributed denial-of-service attacks: A tutorial. *IEEE Communications Magazine*, *40*(10), 42–51. doi:10.1109/MCOM.2002.1039856

Chatterjee, K. (2019). An Improved Authentication Protocol for Wireless Body Sensor Networks Applied in Healthcare Applications. *Wireless Personal Communications*. Advance online publication. doi:10.100711277-019-07005-6

Chatterjee, S., Roy, S., Das, A. K., Chattopadhyay, S., Kumar, N., & Vasilakos, A. V. (2018). Secure Biometric-Based Authentication Scheme Using Chebyshev Chaotic Map for Multi-Server Environment. *IEEE Transactions on Dependable and Secure Computing*, *15*(5), 824–839. doi:10.1109/TDSC.2016.2616876

Chattopadhyay, A. (2016). *Developing an Innovative Framework for Design and Analysis of Privacy Enhancing Video Surveillance*. Academic Press.

Chattopadhyay, A., & Turkiewicz, K. (2017). Future Directions in Online Healthcare Consumerism Policy Making: Exploring Trust Attributes of Online Healthcare Information. *IEEE Internet Initiative*.

Chattopadhyay, A., Schulz, M.J., Rettler, C., Turkiewicz, K., Fernandez, L., & Ziganshin, A. (2017). Towards a Biometric Authentication-Based Hybrid Trust-Computing Approach for Verification of Provider Profiles in Online Healthcare Information. *2017 IEEE Security and Privacy Workshops (SPW)*, 56-65.

Chaudhary, A., Tiwari, V., & Kumar, A. (2014). A novel intrusion detection system for ad hoc flooding attack using fuzzy logic in mobile ad hoc networks. *Proc. IEEE Recent Adv. Innov. Eng. (ICRAIE),* 1-4. 10.1109/ICRAIE.2014.6909148

Chaudhry, A., Ahmad, N., & Hafez, R. (2012). Improving throughput and fairness by improved channel assignment using topology control based on power control for multi-radio multi-channel wireless mesh networks. *EURASIP Journal on Wireless Communications and Networking*, *2012*(1), 1–25. doi:10.1186/1687-1499-2012-155

Chavan, A., Kurule, D., & Dere, P. (2016). Performance Analysis of AODV and DSDV Routing Protocol in MANET and Modifications in AODV against Black Hole Attack. *Procedia Computer Science*, *79*, 835–844. doi:10.1016/j.procs.2016.03.108

Chen, D., & Varshney, P.K. (2004). QoS Support in Wireless Sensor Networks: A Survey. *Proceedings of the International Conference on Wireless Networks, ICWN'04*, *1*, 227-233.

Chen, J. C., Patel, V. M., & Chellappa, R. (2016, March). Unconstrained face verification using deep cnn features. In *2016 IEEE winter conference on applications of computer vision (WACV)* (pp. 1-9). IEEE.

Chen, J., Su, M., Shen, S., Xiong, H., & Zheng, H. (2019). *POBA-GA : Perturbation optimized black-box adversarial attacks via genetic algorithm*. doi:10.1016/j.cose.2019.04.014

Chen, L., Jung, T., Du, H., Qian, J., Hou, J., & Li, X.-Y. (2018). Crowdlearning: Crowded Deep Learning with Data Privacy. In *2018 15th Annual IEEE International Conference on Sensing, Communication, and Networking (SECON)* (pp. 1–9). IEEE. 10.1109/SAHCN.2018.8397100

Chen, Q., Zhang, D., Guo, M., Deng, Q., & Guo, S. (2010, June). Samr: A self-adaptive mapreduce scheduling algorithm in heterogeneous environment. In *2010 10th IEEE International Conference on Computer and Information Technology* (pp. 2736-2743). IEEE. 10.1109/CIT.2010.458

Chen, R., Park, J.-M., & Bian, K. (2008). Robust distributed spectrum sensing in cognitive radio networks. *INFOCOM 2008. The 27th Conference on Computer Communications,* 1876-1884.

Chen, J., Wang, D., & Zhao, W. (2013). A task scheduling algorithm for Hadoop platform. *Journal of Computers*, *8*(4), 929–936.

Chen, M., Li, Y., Luo, X., Wang, W., Wang, L., & Zhao, W. (2019). A Novel Human Activity Recognition Scheme for Smart Health Using Multilayer Extreme Learning Machine. *IEEE Internet of Things Journal*, *6*(2), 1410–1418. doi:10.1109/JIOT.2018.2856241

Chen, T. Y., Wei, H. W., Wei, M. F., Chen, Y. J., Hsu, T. S., & Shih, W. K. (2013, May). LaSA: A locality-aware scheduling algorithm for Hadoop-MapReduce resource assignment. In *2013 International Conference on Collaboration Technologies and Systems (CTS)* (pp. 342-346). IEEE. 10.1109/CTS.2013.6567252

Chiu, H. S., Yeung, K., & Lui, K. S. (2009). An efficient joint channel assignment and routing protocol for IEEE 802.11-based multi-channel multi-interface mobile ad hoc networks. *IEEE Transactions on Wireless Communications*, *8*(4), 1706–1715. doi:10.1109/TWC.2009.080174

Choi, M., & Paderes, R. E. O. (2015). Biometric Application for Healthcare Records Using Cloud Technology. *2015 8th International Conference on Bio-Science and Bio-Technology (BSBT).* doi: 10.1109/bsbt.2015.16

Chunera, A. (2014). Capability assessment of women self help groups: A study in Nainital district of Uttarakhand. G.B. Pant University of Agriculture and Technology, Pantnagar - 263145 (Uttarakhand).

Ciuonzo, D., De Maio, A., & Rossi, P. S. (2015). A systematic framework for composite hypothesis testing of independent Bernoulli trials. *IEEE Signal Processing Letters*, *22*(9), 1249–1253. doi:10.1109/LSP.2015.2395811

Ciuonzo, D., & Rossi, P. S. (2014). Decision fusion with unknown sensor detection probability. *IEEE Signal Processing Letters*, *21*(2), 208–212. doi:10.1109/LSP.2013.2295054

CNews conducts the Internet of things 2019 conference in Moscow. (2019). Retrieved from https://events.cnews.ru/events/internet_veschei_2018.shtml

Conti, M., Dragoni, N., & Lesyk, V. (2016). A survey of man in the middle attacks. *IEEE Communications Surveys and Tutorials*, *18*(3), 2027–2051. doi:10.1109/COMST.2016.2548426

Crichigno, M., Wu, Y., & Shu, W. (2008). Protocols and Architectures for Channel Assignment in Wireless Mesh Networks. *Ad Hoc Networks*, *6*(7), 1051–1077. doi:10.1016/j.adhoc.2007.10.002

Crosby, G., Pissinou, N., & Gadze, J. (n.d.). A Framework for Trust-based Cluster Head Election in Wireless Sensor Networks. *Second IEEE Workshop on Dependability and Security in Sensor Networks and Systems*. 10.1109/DSSNS.2006.1

Cruz, F., Twiefel, J., Magg, S., Weber, C., & Wermter, S. (2015, July). Interactive reinforcement learning through speech guidance in a domestic scenario. In *2015 International Joint Conference on Neural Networks (IJCNN)* (pp. 1-8). IEEE. 10.1109/IJCNN.2015.7280477

Dai, H.-N., Wang, Q., Dong, L., & Wong, R. (2013). On Eavesdropping Attacks in Wireless Sensor Networks with Directional Antennas. *International Journal of Distributed Sensor Networks*, *9*(8), 1–13. doi:10.1155/2013/760834

Das, A. K. (2011). Analysis and improvement on an efficient biometric-based remote user authentication scheme using smart cards. *IET Information Security*, *5*(3), 145–151. doi:10.1049/iet-ifs.2010.0125

Das, S. K., & Ho, J.-W. (2011). A synopsis on node compromise detection in wireless sensor networks using sequential analysis (Invited Review Article). *Computer Communications*, *34*(17), 2003–2012. doi:10.1016/j.comcom.2011.07.004

Decree of the Government of the Russian Federation of February 8, 2018 No. 127 "On approval of the Rules for categorizing objects of critical information infrastructure of the Russian Federation, as well as the List of indicators of criteria for the significance of objects of critical information infrastructure of the Russian Federation and their values". (2018). *Reference legal system "Consultant Plus".*

Decree of the Government of the Russian Federation of November 16, 2015 No 1236. (2015). https://legalacts.ru/doc/postanovlenie-pravitelstva-rf-ot-16112015-n-1236/

Decree of the President of the Russian Federation of August 16, 2004 No 1085 "Issues of the Federal Service for Technical and Export Control". (2004). Retrieved from https://fstec.ru/

Deebak, B. D., Al-Turjman, F. M., Aloqaily, M., & Alfandi, O. (2019). An Authentic-Based Privacy Preservation Protocol for Smart e-Healthcare Systems in IoT. *IEEE Access: Practical Innovations, Open Solutions*, *7*, 135632–135649. doi:10.1109/ACCESS.2019.2941575

Deebak, B. D., Al-turjman, F., & Mostarda, L. (2019). A Hash-Based RFID Authentication Mechanism for Context-Aware Management in IoT-Based Multimedia Systems. *Sensors (Basel)*, 19. PMID:31487847

Deep Learning. (2020). Retrieved April 15, 2020, from https://en.wikipedia.org/wiki/Deep_learning

Demirkol, I., Ersoy, C., & Alagoz, F. (2006). MAC protocols for wireless sensor networks: A survey. *IEEE Communications Magazine*, *44*(4), 115–121. doi:10.1109/MCOM.2006.1632658

Dempster, A. P., Laird, N. M., & Rubin, D. B. (1977). Maximum likelihood from incomplete data via the EM algorithm. *Journal of the Royal Statistical Society. Series B. Methodological*, *39*(1), 1–22. doi:10.1111/j.2517-6161.1977.tb01600.x

Dener, M. (2018). Comparison of Encryption Algorithms in Wireless Sensor Networks. *ITM Web of Conferences*, *22*, 1-5.

Derryberry, R. T., Gray, S. D., Ionescu, D. M., Mandyam, G., & Raghothaman, B. (2002). Transmit diversity in 3G CDMA systems. *IEEE Communications Magazine*, *40*(4), 68–75. doi:10.1109/35.995853

Devare, M.A.S. (2014). Channel allocation using ARS and BFS-CA analysis in WMN. *International Journal of Advanced Engineering Nano Technology*, 1-6.

Devi, T. K., Sakthivel, P., Pratheep, V., & Priyanka, E. (2020). *120 MHz, 2.2 mW Low Power Phase-Locked Loop using Dual Mode Logic and its Application as Frequency Divider.* Academic Press.

Devi, R. R., & Sujatha, P. (2017). A study on biometric and multi-modal biometric system modules, applications, techniques and challenges. *2017 Conference on Emerging Devices and Smart Systems (ICEDSS)*, 267-271. 10.1109/ICEDSS.2017.8073691

DEWI Project. (2014-2017). *Dependable Embedded Wireless Infrastructure.* http://www.dewiproject.eu/

Dhawan, H., & Waraich, S. (2014). A comparative study on LEACH routing protocol and its variants in wireless sensor networks: A survey. *International Journal of Computers and Applications*, *95*(8).

Di, M., & Joo, E. M. (2007, December). A survey of machine learning in wireless sensor networks from networking and application perspectives. In *2007 6th international conference on information, communications & signal processing* (pp. 1-5). IEEE.

Diaz, P. S., & Sanchez, P. (2016). Simulation of attacks for security in wireless sensor network. *Sensors (Basel), 16*(11), 1932. doi:10.339016111932 PMID:27869710

Diddigi, R. B., Prabuchandran, K. J., & Bhatnagar, S. (2018). Prabuchandran K.J., Shalabh Bhatnagar, "Novel Sensor Scheduling Scheme for Intruder Tracking in Energy Efficient Sensor Networks. *Wireless Communications Letters IEEE, 7*(5), 712–715. doi:10.1109/LWC.2018.2814576

Dierks, T., & Rescorla, E. (2020). *The Transport Layer Security (TLS) Protocol Version 1.3. RFC8446*. Network Working Group.

Digital Financial Inclusion and Consumer Capabilities in India: A handbook for financial service providers. (2017). http://www.ifmrlead.org/wp-content/uploads/2017/09/DFS/IFMR JPM DFS Handbook_PDF.pdf

Dinakaran, K., Adinadh, K. R., Sanjuna, K. R., & Valarmathie, P. (2020). Quality of service (QoS) and priority aware models for adaptive efficient image retrieval in WSN using TBL routing with RLBP features. *Journal of Ambient Intelligence and Humanized Computing*. Advance online publication. doi:10.100712652-020-01793-7

Ding, Y., & Xiao, L. (2011). Channel allocation in multi-channel wireless mesh networks. *Computer Communications, 34*(7), 803–815. doi:10.1016/j.comcom.2010.10.011

Din, I. U., Guizani, M., Hassan, S., Kim, B., Khan, M. K., Atiquzzaman, M., & Ahmed, S. H. (2019). The Internet of Things: A Review of Enabled Technologies and Future Challenges. *IEEE Access: Practical Innovations, Open Solutions, 7*, 7606–7640. doi:10.1109/ACCESS.2018.2886601

Dong, P., Du, X., Zhang, H., & Xu, T. (2016). A detection method for a novel DDoS attack against SDN controllers by vast new low-traffic flows. *Communications (ICC), 2016 IEEE International Conference on*, 1-6.

Dong, S., Agrawal, P., & Sivalingam, K. (2007, November). Reinforcement learning based geographic routing protocol for UWB wireless sensor network. In *IEEE GLOBECOM 2007-IEEE Global Telecommunications Conference* (pp. 652-656). IEEE. 10.1109/GLOCOM.2007.127

Dong, C., Loy, C. C., He, K., & Tang, X. (2015). Image super-resolution using deep convolutional networks. *IEEE Transactions on Pattern Analysis and Machine Intelligence, 38*(2), 295–307. doi:10.1109/TPAMI.2015.2439281 PMID:26761735

Dorigo, M., Birattari, M., & Stutzle, T. (2006). Ant colony optimization. *IEEE Computational Intelligence Magazine, 1*(4), 28–39. doi:10.1109/MCI.2006.329691

Doss, S., Nayyar, A., Suseendran, G., Tanwar, S., Khanna, A., Son, L. H., & Thong, P. H. (2018). APD-JFAD: Accurate Prevention and Detection of Jelly Fish Attack in MANET. *IEEE Access: Practical Innovations, Open Solutions, 6*, 56954–56965. doi:10.1109/ACCESS.2018.2868544

Duarte, M. F., & Eldar, Y. C. (2011). Structured compressed sensing: From theory to applications. *IEEE Transactions on Signal Processing, 59*(9), 4053–4085. doi:10.1109/TSP.2011.2161982

Dubey, N., & Vishwakarma, S. (2016). Cloud Computing in Healthcare International. *Journal of Current Trends in Engineering & Research, 2*(5), 211–216.

Dubitzky, W., Lopes, P., Davis, J., & Berrar, D. (2019). The open international soccer database for machine learning. *Machine Learning, 108*(1), 9–28. doi:10.100710994-018-5726-0

Du, Y., Wang, W., & Wang, L. (2015). Hierarchical recurrent neural network for skeleton based action recognition. In *Proceedings of the IEEE conference on computer vision and pattern recognition* (pp. 1110-1118). IEEE.

Egorova-Förster, A., & Murphy, A. L. (2006, May). A feedback-enhanced learning approach for routing in WSN. In *Communication in Distributed Systems-15. ITG/GI Symposium* (pp. 1-12). VDE.

Eitel, A., Springenberg, J. T., Spinello, L., Riedmiller, M., & Burgard, W. (2015, September). Multimodal deep learning for robust RGB-D object recognition. In *2015 IEEE/RSJ International Conference on Intelligent Robots and Systems (IROS)* (pp. 681-687). IEEE. 10.1109/IROS.2015.7353446

Elappila, M., Chinara, S., & Parhi, D. R. (2020). Survivability Aware Channel Allocation in WSN for IoT applications. *Pervasive and Mobile Computing*, *61*, 101107. doi:10.1016/j.pmcj.2019.101107

El-sayed, T., Badawy, M., & El-Sayed, A. (2019). Impact of Small Files on Hadoop Performance: Literature Survey and Open Points. *Menoufia Journal of Electronic Engineering Research*, *28*(1), 109–120. doi:10.21608/mjeer.2019.62728

El-Semary, A. M., & Diab, H. (2019). BP-AODV: Black-hole Protected AODV Routing Protocol for MANETs Based on Chaotic Map. *IEEE Access: Practical Innovations, Open Solutions*, *7*, 95197–95211. doi:10.1109/ACCESS.2019.2928804

ENACT Project. (2018-2021). *Development, Operation, and Quality Assurance of Trustworthy Smart IoT Systems*. https://www.enact-project.eu/

ENACT. (2018). *D1.1 Use case definition and requirements and validation and evaluation plan* (Grant Agreement N° 780351). ENACT consortium.

ENIAC. (n.d.). Retrieved May 4, 2020, from https://en.wikipedia.org/wiki/ENIAC

Entacher, K. (1998). Bad subsequences of well-known linear congruential pseudorandom number generators. *ACM Transactions on Modeling and Computer Simulation*, *8*(1), 61–70. doi:10.1145/272991.273009

Enz, C. C., El-Hoiydi, A., Decotignie, J.-D., & Peiris, V. (2004). WiseNET: An ultralow-power wireless sensor network solution. *Computer*, *37*(8), 62–70. doi:10.1109/MC.2004.109

Ertin, E. (2007). Gaussian Process Models for Censored Sensor Readings. *2007 IEEE/SP 14th Workshop on Statistical Signal Processing*, 665–669. doi: 10.1109sp.2007.4301342

Esmaeil Zaei, M., Kapil, P., Pelekh, O., & Teimoury Nasab, A. (2018). Does Micro-Credit Empower Women through Self-Help Groups? Evidence from Punjab, Northern India. *Societies (Basel, Switzerland)*, *8*(3), 48. doi:10.3390oc8030048

European Committee for Electrotechnical Standardization. (2010). *CENELEC: EN 50159. Railway applications -communication, signalling and processing systems*. European Union: CENELEC.

European Union Agency for Network and Information Security. (2017). *Baseline Security Recommendations for IoT in the context of Critical Information Infrastructures*. ENISA.

Ever, Y. K. (2020). A secure authentication scheme framework for mobile sinks used in the Internet of Drones application. *Computer Communications*, *155*, 143–149. doi:10.1016/j.comcom.2020.03.009

Fadlullah, Z. M., Tang, F., Mao, B., Kato, N., Akashi, O., Inoue, T., & Mizutani, K. (2017). State-of-the-art deep learning: Evolving machine intelligence toward tomorrow's intelligent network traffic control systems. *IEEE Communications Surveys and Tutorials*, *19*(4), 2432–2455. doi:10.1109/COMST.2017.2707140

Faghihniya, M. J., Hosseini, S. M., & Tahmasebi, M. (2016). Security upgrade against RREQ flooding attack by using balance index on vehicular ad hoc network. *Wireless Networks*, *23*(6), 1863–1874. doi:10.100711276-016-1259-2

Fakhet, W., El Khediri, S., Dallali, A., & Kachouri, A. (2017, October). New K-means algorithm for clustering in wireless sensor networks. In *2017 International Conference on Internet of Things, Embedded Systems and Communications (IINTEC)* (pp. 67-71). IEEE. 10.1109/IINTEC.2017.8325915

Fan, O., Cheng, H., Lan, Y., Zhang, Y., Yin, X., Hu, J., Peng, X., Wang, G., & Chen, S. (2019). Automatic delivery and recovery system of Wireless Sensor Networks (WSN) nodes based on UAV for agricultural applications. *Computers and Electronics in Agriculture, 162*, 31–43. doi:10.1016/j.compag.2019.03.025

Farjamnia, G., Gasimov, Y., & Kazimov, C. (2019). Review of the Techniques against the Wormhole Attacks on Wireless Sensor Networks. *Wireless Personal Communications, 105*(4), 1561–1584. doi:10.100711277-019-06160-0

Fasoulakis, M., Tsiropoulou, E. E., & Papavassiliou, S. (2019). Satisfy instead of maximize: Improving operation efficiency in wireless communication networks. *Computer Networks, 159*, 135–146. doi:10.1016/j.comnet.2019.05.012

Fazeldehkordi, E., Amiri, I. S., & Akanbi, O. A. (2016). *A study of black hole attack solutions on Aodv routing protocol in Manet*. Elsevier.

Federal Law of December 27, 2002 No 184-FZ "On Technical Regulation", *Collection of Legislation of the Russian Federation,* December 30, 2002, No. 52 (part 1), Art. 5140. (2002). https://ivprom.ru/lib/245/

Ferjani, A. A., Liouane, N., & Kacem, I. (2016). Task allocation for wireless sensor network using logic gate-based evolutionary algorithm. *2016 International Conference on Control, Decision and Information Technologies (CoDIT)*, 654–658. 10.1109/CoDIT.2016.7593640

Fernandez, M. F., & Aridgides, T. (2003). Measures for evaluating sea mine identification processing performance and the enhancements provided by fusing multisensor/multiprocess data via an M-out-of-N voting scheme. *Detection and Remediation Technologies for Mines and Minelike Targets VIII, 5089*, 425-437.

Forster, A. (2007, December). Machine learning techniques applied to wireless ad-hoc networks: Guide and survey. In *2007 3rd international conference on intelligent sensors, sensor networks and information* (pp. 365-370). IEEE.

Forster, A., & Murphy, A. L. (2009). CLIQUE: Role-Free Clustering with Q-Learning for Wireless Sensor Networks. *2009 29th IEEE International Conference on Distributed Computing Systems.* doi: 10.1109/icdcs.2009.43

Förster, A., Murphy, A. L., Schiller, J., & Terfloth, K. (2008, October). An efficient implementation of reinforcement learning based routing on real WSN hardware. In *2008 IEEE International Conference on Wireless and Mobile Computing, Networking and Communications* (pp. 247-252). IEEE. 10.1109/WiMob.2008.99

Fragkiadakis, A. G., Tragos, E. Z., & Askoxylakis, I. G. (2013). A survey on security threats and detection techniques in cognitive radio networks. *IEEE Communications Surveys and Tutorials, 15*(1), 428–445. doi:10.1109/SURV.2011.122211.00162

Friday, H., Wah, Y., Al-garadi, M. A., & Rita, U. (2018). Deep learning algorithms for human activity recognition using mobile and wearable sensor networks : State of the art and research challenges. *Expert Systems with Applications, 105*, 233–261. doi:10.1016/j.eswa.2018.03.056

Fudenberg, D., & Maskin, E. (1986). The folk theorem in repeated games with discounting or with incomplete information. *Econometrica, 54*(3), 533–554. doi:10.2307/1911307

Fudenberg, D., & Maskin, E. (1991). On the dispensability of public randomization in discounted repeated games. *Journal of Economic Theory, 53*(2), 428–438. doi:10.1016/0022-0531(91)90163-X

Fu, G. S., Levin-Schwartz, Y., Lin, Q. H., & Zhang, D. (2019). Machine Learning for Medical Imaging. *Journal of Healthcare Engineering*. PMID:31183031

Fu, T.-Y., Peng, W.-C., & Lee, W.-C. (2010). Parallelizing Itinerary-Based KNN Query Processing in Wireless Sensor Networks. *IEEE Transactions on Knowledge and Data Engineering*, *22*(5), 711–729. doi:10.1109/TKDE.2009.146

Gai, K., Qiu, M., Ming, Z., Zhao, H., & Qiu, L. (2017). Spoofing-Jamming Attack Strategy Using Optimal Power Distributions in Wireless Smart Grid Networks. *IEEE Transactions on Smart Grid*, *8*(5), 2431–2439. doi:10.1109/TSG.2017.2664043

Gálvez, J. J., & Ruiz, P. M. (2013). Efficient rate allocation, routing and channel assignment in wireless mesh networks supporting dynamic traffic flows. *Ad Hoc Networks*, *11*(6), 1765–1781. doi:10.1016/j.adhoc.2013.04.002

Galvez, J. J., & Ruiz, P. M. (2015). Joint link rate allocation, routing and channel assignment in multi-rate multi-channel wireless networks. *Ad Hoc Networks*, *29*(c), 78–98. doi:10.1016/j.adhoc.2015.02.002

Gandomi, A., Reshadi, M., Movaghar, A., & Khademzadeh, A. (2019). HybSMRP: A hybrid scheduling algorithm in Hadoop MapReduce framework. *Journal of Big Data*, *6*(1), 106. doi:10.118640537-019-0253-9

Ganeriwal, S., Balzano, L. K., & Srivastava, M. B. (2008). Reputation-based framework for high integrity sensor networks. *ACM Transactions on Sensor Networks*, *4*(3), 1–37. doi:10.1145/1362542.1362546

Gara, F., Saad, L. B., & Ayed, R. B. (2017). An intrusion detection system for selective forwarding attack in IPv6-based mobile WSNs. *2017 13th International Wireless Communications and Mobile Computing Conference (IWCMC)*, 276-281.

Gentry, C. (2009). *Fully Homomorphic Encryption Using Ideal Lattices*. Academic Press.

Ghahramani, Z. (2015). Probabilistic machine learning and artificial intelligence. *Nature*, *521*(7553), 452–459. doi:10.1038/nature14541 PMID:26017444

Ghazvini, M., Movahedinia, N., Jamshidi, K., & Moghim, N. (2013). Game theory applications in CSMA methods. *IEEE Communications Surveys and Tutorials*, *15*(3), 1062–1087. doi:10.1109/SURV.2012.111412.00167

Gispan, L., Leshem, A., & Beery, Y. (2017). Decentralized estimation of regression coefficients in sensor networks. *Digital Signal Processing*, *68*, 16–23. doi:10.1016/j.dsp.2017.05.005

Glushkova, D., Jovanovic, P., & Abelló, A. (2019). Mapreduce performance model for Hadoop 2. x. *Information Systems*, *79*, 32–43. doi:10.1016/j.is.2017.11.006

Goel, A. K., Rose, A., Gaur, J., & Bhushan, B. (2019). Attacks, Countermeasures and Security Paradigms in IoT. *2019 2nd International Conference on Intelligent Computing, Instrumentation and Control Technologies (ICICICT)*. doi:10.1109/icicict46008.2019.8993338

Goertzel, K. (2016). Legal Liability for Bad Software. *Crosstalk*, *2016*(September/October). http://static1.1.sqspcdn.com/static/f/702523/27213494/1472233517737/201609 – Goertzel.pdf

Gong, M., Feng, J., & Xie, Y. (2020). *Privacy-enhanced multi-party deep learning*. Academic Press.

Gongal, A., Amatya, S., Karkee, M., Zhang, Q., & Lewis, K. (2015). Sensors and systems for fruit detection and localization: A review. *Computers and Electronics in Agriculture*, *116*, 8–19. doi:10.1016/j.compag.2015.05.021

Gong, M. X., Midkiff, S. F., & Mao, S. (2009). On-demand routing and channel assignment in multi-channel mobile ad hoc networks. *Ad Hoc Networks*, *7*(1), 63–78. doi:10.1016/j.adhoc.2007.11.011

Good Practices for Security of IoT in the context of Smart Manufacturing. (2018). https://www.enisa.europa.eu/publications/good-practices-for-security-of-iot?fbclid=IwAR1q-chv88kZRsIESHtGTEwbA0Mbx8mb9hV1Euqy-Y--IHVYvLuFhGuvi6o

Gope, P., & Hwang, T. (2016). A Realistic Lightweight Anonymous Authentication Protocol for Securing Real-Time Application Data Access in Wireless Sensor Networks. *IEEE Transactions on Industrial Electronics*, *63*(11), 7124–7132. doi:10.1109/TIE.2016.2585081

Gope, P., Lee, J., & Quek, T. Q. (2018). Lightweight and Practical Anonymous Authentication Protocol for RFID Systems Using Physically Unclonable Functions. *IEEE Transactions on Information Forensics and Security*, *13*(11), 2831–2843. doi:10.1109/TIFS.2018.2832849

GOST R 54593-2011. National standard of the Russian Federation. Information Technology. Free software. General Provisions (approved and enforced by the Order of Rosstandart of December 06, 2011 No 718-St). (2011). *Reference legal system "Consultant Plus"*.

GOST R ISO / IEC 15408-3-2013 "Information technology. Methods and means of security. Criteria for assessing the security of information technology. Requirements for trusting OS security. (2013). http://docs.cntd.ru/document/1200105711

Grosse, K., Manoharan, P., Papernot, N., Backes, M., & McDaniel, P. (2017). *On the (Statistical) Detection of Adversarial Examples*. Academic Press.

Gui, T., Ma, C., Wang, F., & Wilkins, D. E. (2016, March). Survey on swarm intelligence based routing protocols for wireless sensor networks: An extensive study. In 2016 IEEE international conference on industrial technology (ICIT) (pp. 1944-1949). IEEE.

Guidelines for the implementation and use of the industrial Internet of things to optimize control (supervisory) activities "(approved by the minutes of the meeting of the Project Committee of 09.11.2017 No 73 (13)). (2017). Retrieved from https://legalacts.ru/doc/metodicheskie-rekomendatsii-po-vnedreniiu-i-ispolzovaniiu-promyshlennogo-interneta-veshchei/

Guo, F., Zhao, Q., Li, X., Kuang, X., Zhang, J., Han, Y., & an Tan, Y. (2019). Detecting adversarial examples via prediction difference for deep neural networks. *Information Sciences*, *501*, 182–192. doi:10.1016/j.ins.2019.05.084

Guo, Y., Zhang, Y., Mi, Z., Yang, Y., & Obaidat, M. S. (2019). Distributed task allocation algorithm based on connected dominating set for WSANs. *Ad Hoc Networks*, *89*, 107–118. doi:10.1016/j.adhoc.2019.03.006

Gupta, O., & Raskar, R. (2018). Distributed learning of deep neural network over multiple agents. *Journal of Network and Computer Applications*, *116*, 1–8. doi:10.1016/j.jnca.2018.05.003

Gupta, S. S., Chattopadhyay, A., Sinha, K., Maitra, S., & Sinha, B. P. (2013). High-Performance Hardware Implementation for RC4 Stream Cipher. *IEEE Transactions on Computers*, *62*(4), 730–743. doi:10.1109/TC.2012.19

Haghighat, M., Zonouz, S. A., & Abdel-Mottaleb, M. (2015). CloudID: Trustworthy cloud-based and cross-enterprise biometric identification. *Expert Systems with Applications*, *42*(21), 7905–7916. doi:10.1016/j.eswa.2015.06.025

Haibo, B. L., Sohraby, L. K., & Wang, C. (2010). Future internet services and applications. *IEEE Network*, *24*(4), 4–5. doi:10.1109/MNET.2010.5510911

Haidar, A. M., Saleh, N., Itani, W., & Shirahama, H. (2015). Toward a neural network computing: A novel NN-SRAM. *Proc. NOLTA*, 672-675.

Ha, M., Byun, Y., Kim, J., Lee, J., Lee, Y., & Lee, S. (2019). Selective Deep Convolutional Neural Network for Low Cost Distorted Image Classification. *IEEE Access: Practical Innovations, Open Solutions*, *7*, 133030–133042. doi:10.1109/ACCESS.2019.2939781

Hameed, K., Khan, A., Ahmed, M., Alavalapati, G. R., & Rathore, M. M. (2018). Towards a formally verified zero watermarking scheme for data integrity in the Internet of Things based-wireless sensor networks. *Future Generation Computer Systems*, *82*, 274–289. doi:10.1016/j.future.2017.12.009

Hamouda, Y. E. M. (2019). Modified random bit climbing (λ-MRBC) for task mapping and scheduling in wireless sensor networks. *Jordanian Journal of Computers and Information Technology*, *5*(1), 17–33. doi:10.5455/jjcit.71-1541688581

Han, W., Tian, Z., Huang, Z., Huang, D., & Jia, Y. (2019). Quantitative Assessment of Wireless Connected Intelligent Robot Swarms Network Security Situation. *IEEE Access: Practical Innovations, Open Solutions*, *7*, 134293–134300. doi:10.1109/ACCESS.2019.2940822

Harb, H., Makhoul, A., & Couturier, R. (2015). An Enhanced K-Means and ANOVA-Based Clustering Approach for Similarity Aggregation in Underwater Wireless Sensor Networks. *IEEE Sensors Journal*, *15*(10), 5483–5493. doi:10.1109/JSEN.2015.2443380

Harris, B., & Hunt, R. (1999). TCP/IP security threats and attack methods. *Computer Communications*, *22*(10), 885–897. doi:10.1016/S0140-3664(99)00064-X

Hashemi, A. S., & Mozaffari, S. (2019). Secure deep neural networks using adversarial image generation and training with Noise-GAN. *Computers & Security*, *86*, 372–387. doi:10.1016/j.cose.2019.06.012

Hastings, N. E., & McLean, P. A. (1996, March). TCP/IP spoofing fundamentals. In *Conference Proceedings of the 1996 IEEE Fifteenth Annual International Phoenix Conference on Computers and Communications* (pp. 218-224). IEEE. 10.1109/PCCC.1996.493637

Hecht-Nielsen, R. (1992). Theory of the Backpropagation Neural Network. In Neural Networks for Perception (pp. 65–93). doi:10.1016/b978-0-12-741252-8.50010-8

He, D., Chan, S., & Guizani, M. (2017). Cyber Security Analysis and Protection of Wireless Sensor Networks for Smart Grid Monitoring. *IEEE Wireless Communications*, *24*(6), 98–103. doi:10.1109/MWC.2017.1600283WC

He, D., & Zeadally, S. (2015). An Analysis of RFID Authentication Schemes for Internet of Things in Healthcare Environment Using Elliptic Curve Cryptography. *IEEE Internet of Things Journal*, *2*(1), 72–83. doi:10.1109/JIOT.2014.2360121

He, H., Zhu, Z., & Makinen, E. (2009). A Neural Network Model to Minimize the Connected Dominating Set for Self-Configuration of Wireless Sensor Networks. *IEEE Transactions on Neural Networks*, *20*(6), 973–982. doi:10.1109/TNN.2009.2015088 PMID:19398401

Hennebert, C., & Dos Santos, J. (2014). Security protocols and privacy issues into 6LoWPAN stack: A synthesis. *IEEE Internet of Things Journal*, *1*(5), 384–398. doi:10.1109/JIOT.2014.2359538

Hernandez-Leal, P., Kaisers, M., Baarslag, T., & de Cote, E. M. (2017). *A Survey of Learning in Multiagent Environments: Dealing with Non-Stationarity.* arXiv preprint arXiv:1707.09183

Ho, J.-W., Wright, M., & Das, S. K. (2011). Fast detection of mobile replica node attacks in wireless sensor networks using sequential hypothesis testing. *IEEE Transactions on Mobile Computing*, *10*(6), 767–782. doi:10.1109/TMC.2010.213

Hsieh, C. J., Chang, K. W., Lin, C. J., Keerthi, S. S., & Sundararajan, S. (2008). A dual coordinate descent method for large-scale linear SVM. In *Proceedings of the 25th International Conference on Machine Learning* (pp. 408–415). 10.1145/1390156.1390208

Hsu, R. C., Liu, C.-T., Wang, K.-C., & Lee, W.-M. (2009). QoS-Aware Power Management for Energy Harvesting Wireless Sensor Network Utilizing Reinforcement Learning. *2009 International Conference on Computational Science and Engineering.* 10.1109/CSE.2009.83

Huang, H., Savkin, A. V., Ding, M., & Huang, C. (2019). Mobile robots in wireless sensor networks: A survey on tasks. *Computer Networks*, *148*, 1–19. doi:10.1016/j.comnet.2018.10.018

Huang, P., Li, B., Guo, L., Jin, Z., & Chen, Y. L. (2016). A Robust and Reusable ECG-Based Authentication and Data Encryption Scheme for eHealth Systems. *2016 IEEE Global Communications Conference (GLOBECOM)*, 1-6. 10.1109/GLOCOM.2016.7841541

Huang, W., Wang, H., Zhang, Y., & Zhang, S. (2019). A novel cluster computing technique based on signal clustering and analytic hierarchy model using hadoop. *Cluster Computing*, *22*(6), 13077–13084. doi:10.100710586-017-1205-9

Hugh, T. (2020). *What Are Cyber Threats: How They Affect You and What to Do About Them.* Retrieved from https://preyproject.com/blog/en/what-are-cyber-threats-how-they-affect-you-what-to-do-about-them/

Hull, T. E., & Dobell, A. R. (1962). Random number generators. *SIAM Review*, *4*(3), 230–254. doi:10.1137/1004061

Hussain, S., & Matin, A.W., Jodrey, & Hussain, S. (2006). Hierarchical Cluster-based Routing in Wireless Sensor Networks. *J. Netw. Acad. Publisher*, *2*(5), 87–97.

IEEE. (2016). *IEEE Standard for Information technology--Telecommunications and information exchange between systems Local and metropolitan area networks--Specific requirements - Part 11: Wireless LAN Medium Access Control (MAC) and Physical Layer (PHY) Specifications.*

ILO. (2014). *The role of economic and social councils and similar institutions in promoting social protection floors for all through social dialogue. Report of the ILO-AICESIS Conference.*

Industrial Revolution. (n.d.). Retrieved May 4, 2020, from https://en.wikipedia.org/wiki/Industrial_Revolution

Information Age. (n.d.). Retrieved May 4, 2020, from https://en.wikipedia.org/wiki/Information_Age

Information protection measures in state information systems. (2014). Retrieved from https://fstec.ru/component/attachments/download/675

InSecTT Project (2020-2023). Intelligent Secure Trustable Things.

IoT-2019: main trends in the Russian market. (2019). Retrieved from https://iot.ru/promyshlennost/iot-2019-glavnye-trendy-na-rossiyskom-rynke

Ishmanov, F., & Bin Zikria, Y. (2017). Trust mechanisms to secure routing in wireless sensor networks: Current state of the research and open research issues. *Journal of Sensors*, *2017*, 2017. doi:10.1155/2017/4724852

ISO/IEC 19464:2014 (2019). Information technology – Advanced Message Queuing Protocol (MQP) v1.0. ISO/IEC Information Technology.

ISO/IEC 20922:2016 (2016). Information technology - Message Queuing Telemetry Transport (MQTT) v3.1.1. ISO/IEC JTC 1 Information Technology.

ISO/IEC 29182-1:2013 (2013). Information technology - Sensor networks: Sensor Network Reference Architecture (SNRA). International Standard ISO/IEC.

Issac, T., Silas, S., & Rajsingh, E. B. (2019b). Role assignment in lighting application for a greener and safer smart city. *2019 3rd International Conference on Computing and Communications Technologies (ICCCT)*, 1–8. 10.1109/ICCCT2.2019.8824914

Issac, T., Silas, S., & Rajsingh, E. B. (2019a). Luminaire aware centralized outdoor illumination role assignment scheme: A smart city perspective. In J. D. Peter, A. H. Alavi, & B. Javadi (Eds.), *Advances in Big Data and Cloud Computing* (pp. 443–456). doi:10.1007/978-981-13-1882-5_38

Issac, T., Silas, S., & Rajsingh, E. B. (2020a). Dynamic and static system modelling with simulations of an eco-friendly smart lighting system. In D. Peter & S. L. Fernandes (Eds.), *Systems Simulation and Modeling for Cloud Computing and Big Data Applications* (pp. 81–97). doi:10.1016/B978-0-12-819779-0.00005-8

Issac, T., Silas, S., & Rajsingh, E. B. (2020b). Investigations on PSO based task assignment algorithms for heterogeneous wireless sensor network. *2nd International Conference on Signal Processing and Communication (ICSPC)*, 89–93. 10.1109/ICSPC46172.2019.8976850

J, A., Karthikeyan, J., & Jebastin, J. (2019). Privacy Preserving Big Data Publication On Cloud Using Mondrian Anonymization Techniques and Deep Neural Networks. In *2019 5th International Conference on Advanced Computing & Communication Systems (ICACCS)* (pp. 722–727). IEEE. doi:10.1109/ICACCS.2019.8728384

Jacobsen, R. H. (2019). *On Fully Homomorphic Encryption for Privacy-Preserving Deep Learning*. Academic Press.

Jahan, S., Chowdhury, M., & Islam, R. (2017). Robust fingerprint verification for enhancing security in healthcare system. *2017 International Conference on Image and Vision Computing New Zealand (IVCNZ)*. 10.1109/IVCNZ.2017.8402502

Jaimahaprabhu, A., Kumar, P., Gangadharan, P. S., & Latha, B. (2019, March). Cloud Analytics based Farming with Predictive Analytics using Artificial Intelligence. In *2019 Fifth International Conference on Science Technology Engineering and Mathematics (ICONSTEM)* (Vol. 1, pp. 65-68). IEEE. 10.1109/ICONSTEM.2019.8918785

Jain, A.K.; Hong, L. & Bolle, R. (1997). On-line fingerprint verification. *IEEE Transactions on Pattern Analysis and Machine Intelligence, 19*, 302-314.

Jain, A. K., Nandakumar, K., & Ross, A. (2016). 50 years of biometric research: Accomplishments, challenges, and opportunities. *Pattern Recognition Letters, 79*, 80–105. doi:10.1016/j.patrec.2015.12.013

Jaitly, S., Malhotra, H., & Bhushan, B. (2017). Security vulnerabilities and countermeasures against jamming attacks in Wireless Sensor Networks: A survey. *2017 International Conference on Computer, Communications and Electronics (Comptelix)*. 10.1109/COMPTELIX.2017.8004033

Jamshidi, M., Shaltooki, A. A., Zadeh, Z. D., & Darwesh, A. M. (2018). A Dynamic ID Assignment Mechanism to Defend Against Node Replication Attack in Static Wireless Sensor Networks. JOIV. *International Journal on Informatics Visualization, 3*(1). Advance online publication. doi:10.30630/joiv.3.1.161

Janakiram, D., Kumar, A., & V., A. M. R. (2006). Outlier Detection in Wireless Sensor Networks using Bayesian Belief Networks. *2006 1st International Conference on Communication Systems Software & Middleware*. doi:10.1109/comswa.2006.1665221

Janakiram, D., Reddy, V. A., & Kumar, A. P. (2006, January). Outlier detection in wireless sensor networks using Bayesian belief networks. In *2006 1st International Conference on Communication Systems Software & Middleware* (pp. 1-6). IEEE. 10.1109/COMSWA.2006.1665221

Jaradat, T., Benhaddou, D., Balakrishnan, M., & Al-Fuqaha, A. (2013, July). Energy efficient cross-layer routing protocol in wireless sensor networks based on fuzzy logic. In *2013 9th International Wireless Communications and Mobile Computing Conference (IWCMC)* (pp. 177-182). IEEE. 10.1109/IWCMC.2013.6583555

Jardosh, A. P., Ramachandran, K. N., Almerith, K. C., & Belding-Royer, E. M. (2005). Understanding congestion in IEEE 802.11b wireless networks. *Proceedings of the 5th ACM SIGCOMM Conference on Internet Measurement*, 279-292. 10.1145/1330107.1330140

Jarecki, S., & Krawczyk, H. (1995). Or : How to Cope With Perpetual Leakage. *Communication*, 1–22.

Jaswal, G., Nigam, A., & Nath, R. (2019). Finger Biometrics for e-Health Security. Handbook of Multimedia Information Security. doi:10.1007/978-3-030-15887-3_28

Jedari, B., Xia, F., & Ning, Z. (2018). A survey on human-centric communications in non-cooperative wireless relay networks. *IEEE Communications Surveys and Tutorials*, *20*(2), 914–944. doi:10.1109/COMST.2018.2791428

Jelicic, V., Magno, M., Brunelli, D., Bilas, V., & Benini, L. (2014). Benefits of Wake-Up Radio in Energy-Efficient Multimodal Surveillance Wireless Sensor Network. *IEEE Sensors Journal*, *14*(9), 3210–3220. doi:10.1109/JSEN.2014.2326799

Jia, S., Ma, L., & Qin, D. (2019). Research on Low Energy Consumption Distributed Fault Detection Mechanism in Wireless Sensor Network. *Publisher: IEEE. China Communications*, *16*(3).

Jin, J., Luo, J., Song, A., Dong, F., & Xiong, R. (2011, May). Bar: An efficient data locality driven task scheduling algorithm for cloud computing. In *2011 11th IEEE/ACM International Symposium on Cluster, Cloud and Grid Computing* (pp. 295-304). IEEE.

Jindal, M., Gupta, J., & Bhushan, B. (2020). *Machine learning methods for IoT and their Future Applications*. Institute of Electrical and Electronics Engineers (IEEE). doi:10.1109/icccis48478.2019.8974551

Jindal, M., Gupta, J., & Bhushan, B. (2019). Machine learning methods for IoT and their Future Applications. *2019 International Conference on Computing, Communication, and Intelligent Systems (ICCCIS)*. 10.1109/ICCCIS48478.2019.8974551

Jin, Y., Jin, J., Gluhak, A., Moessner, K., & Palaniswami, M. (2012). An Intelligent Task Allocation Scheme for Multihop Wireless Networks. *IEEE Transactions on Parallel and Distributed Systems*, *23*(3), 444–451. doi:10.1109/TPDS.2011.172

Joseph, J., & Varghese, T. (2014). *Role of Financial Inclusion in the Development of Indian Economy*. www.iiste.org

Joshi, V. K. (2014). Financial Inclusion : Urban-Poor in India. *SCMS Journal of Indian Management*, *11*(4), 29–37.

Kadam, K., & Srivastava, N. (2012, March). Application of machine learning (reinforcement learning) for routing in Wireless Sensor Networks (WSNs). In *2012 1st International Symposium on Physics and Technology of Sensors (ISPTS-1)* (pp. 349-352). IEEE.

Kalaivani, K., & Sivakumar, R. (2016). A Novel Fuzzy Based Bio-Key Management scheme for Medical Data Security. *Journal of Electrical Engineering & Technology*, *11*(5), 1509–1518. doi:10.5370/JEET.2016.11.5.1509

Kalid, N., Zaidan, A. A., Zaidan, B. B., Salman, O. H., Hashim, M., Albahri, O. S., & Albahri, A. S. (2018). Based on Real Time Remote Health Monitoring Systems: A New Approach for Prioritization "Large Scales Data" Patients with Chronic Heart Diseases Using Body Sensors and Communication Technology. *Journal of Medical Systems*, *42*(4), 1–37. doi:10.100710916-018-0916-7 PMID:29500683

Kaliyannan, G. V., Palanisamy, S. V., Priyanka, E. B., Thangavel, S., Sivaraj, S., & Rathanasamy, R. (2020). Investigation on sol-gel based coatings application in energy sector–A review. *Materials Today: Proceedings*. Advance online publication. doi:10.1016/j.matpr.2020.03.484

Kamboj, S. (2014). Financial inclusion and growth of Indian economy: An empirical analysis. *International Journal of Business and Management*, *2*(9), 175–179.

Kandpal, V. (2020). Role of regulators in intensifying financial access to the untouched segment of society in developing country. *Corporate Governance and Organizational Behavior Review*, *4*(1), 8–14.

Kaplantzis, S., Shilton, A., Mani, N., & Sekercioglu, Y. A. (2007). Detecting Selective Forwarding Attacks in Wireless Sensor Networks using Support Vector Machines. *2007 3rd International Conference on Intelligent Sensors, Sensor Networks and Information*. doi: 10.1109/issnip.2007.4496866

Kapoor, R., Gupta, R., Kumar, R., & Jha, S. (2019). New scheme for underwater acoustically wireless transmission using direct sequence code division multiple access in MIMO systems. *Wireless Networks*, *25*(8), 4541–4553. doi:10.100711276-018-1750-z

Karakurt, Y., Yıldız, H. U., & Tavlı, B. (2018). The impact of mitigation of eavesdropping on wireless sensor networks lifetime. *Proceedings of 26th Signal Processing and Communications Applications Conference (SIU)*, 1-4. 10.1109/SIU.2018.8404252

Karp, B., & Kung, H. T. (2000, August). GPSR: Greedy perimeter stateless routing for wireless networks. In *Proceedings of the 6th annual international conference on Mobile computing and networking* (pp. 243-254). 10.1145/345910.345953

Karthik, N., & Ananthanarayana, V. S. (2017). A Hybrid Trust Management Scheme for Wireless Sensor Networks. *Wireless Personal Communications*, *97*(4), 5137–5170. doi:10.100711277-017-4772-4

Kasongo, S. M., & Sun, Y. (2019). A Deep Learning Method With Filter Based Feature Engineering for Wireless Intrusion Detection System. *IEEE Access: Practical Innovations, Open Solutions*, *7*, 38597–38607. doi:10.1109/ACCESS.2019.2905633

Kaspersky Security Network. (2019). https://securelist.ru/it-threat-evolution-q1-2019-statistics/94021/

Kaur, H. (2014). *A study on role of financial institutions in financial inclusion with special reference to micro financing in North India*. Punjabi University.

Kaur, M., Sarangal, M., & Nayyar, A. (2014). *Simulation of Jelly Fish Periodic Attack in Mobile Ad hoc Networks. International Journal of Computer Trends and Technology* , 15.

Kaushik, I., & Sharma, N. (2020). Black Hole Attack and Its Security Measure in Wireless Sensors Networks. Advances in Intelligent Systems and Computing Handbook of Wireless Sensor Networks: Issues and Challenges in Current Scenarios, 401–416. doi:10.1007/978-3-030-40305-8_20

Kaushik, I., Sharma, N., & Singh, N. (2019). Intrusion Detection and Security System for Black-hole Attack. *2019 2nd International Conference on Signal Processing and Communication (ICSPC)*. doi: 10.1109/icspc46172.2019.8976797

Kavianpour, A., & Anderson, M. C. (2017, June). An Overview of Wireless Network Security. In *2017 IEEE 4th International Conference on Cyber Security and Cloud Computing (CSCloud)* (pp. 306-309). IEEE. 10.1109/CSCloud.2017.45

Kay, S. M. (1993). *Fundamentals of Statistical Signal Processing: Estimation Theory*. Prentice-Hall, Inc.

Kennedy, J. (2006). Swarm intelligence. In *Handbook of nature-inspired and innovative computing* (pp. 187–219). Springer. doi:10.1007/0-387-27705-6_6

Kerry Maletsky. (2015). *RSA vs ECC Comparison for Embedded Systems. Technical Report*. Atmel Corporation.

Khalaf, O. I., Abdulsahib, G. M., & Sadik, M. (2018). A Modified Algorithm for Improving Lifetime WSN. *Journal of Engineering and Applied Sciences (Asian Research Publishing Network)*, *13*, 9277–9282.

Khalid, A., Paul, G., & Chattopadhyay, A. (2017). RC4-AccSuite: A Hardware Acceleration Suite for RC4-Like Stream Ciphers. IEEE Transactions on Very Large Scale Integration (VLSI). *Systems*, *25*(3), 1072–1084. doi:10.1109/tvlsi.2016.2606554

Khalid, N. S., Shukor, S. A. A., & Syahir, A. F. (2020). Specific Gravity-based of Post-harvest Mangifera indica L. cv. Harumanis for 'Insidious Fruit Rot'(IFR) Detection using Image Processing. In *Computational Science and Technology* (pp. 33–42). Springer. doi:10.1007/978-981-15-0058-9_4

Khan, F., Memon, S., & Jokhio, S. H. (2016). Support vector machine based energy aware routing in wireless sensor networks. *2016 2nd International Conference on Robotics and Artificial Intelligence (ICRAI)*. doi: 10.1109/icrai.2016.7791218

Khan, F., Memon, S., & Jokhio, S. H. (2016, November). Support vector machine based energy aware routing in wireless sensor networks. In *2016 2nd International Conference on Robotics and Artificial Intelligence (ICRAI)* (pp. 1-4). IEEE. 10.1109/ICRAI.2016.7791218

Khan, M. A., & Salah, K. (2018). IoT security: Review blockchain solutions and open challenges. *Future Generation Computer Systems*, *82*, 395–411. doi:10.1016/j.future.2017.11.022

Khanna, N., & Sachdeva, M. (2019). A comprehensive taxonomy of schemes to detect and mitigate black-hole attack and its variants in MANETs. *Computer Science Review*, *32*, 24–44. doi:10.1016/j.cosrev.2019.03.001

Khanvilkar, A. S. (2016). *Challenges faced by Microfinance Institutions*. https://www.nelito.com/blog/challenges-faced-microfinance-institutions.html

Khan, Z. A., & Samad, A. (2017). A study of machine learning in wireless sensor network. *Int. J. Comput. Netw. Appl*, *4*(4), 105–112. doi:10.22247/ijcna/2017/49122

Kho, J., Rogers, A., & Jennings, N. R. (2009). Decentralized control of adaptive sampling in wireless sensor networks. *ACM Transactions on Sensor Networks*, *5*(3), 1–35. doi:10.1145/1525856.1525857

Khosrowjerdi, M. (2016). A review of theory-driven models of trust in the online health context. *IFLA Journal*, *42*(3), 189–206. doi:10.1177/0340035216659299

Kieyzun, A., Guo, P. J., Jayaraman, K., & Ernst, M. D. (2009, May). Automatic creation of SQL injection and cross-site scripting attacks. In *Proceedings of the 31st International Conference on Software Engineering* (pp. 199-209). IEEE Computer Society. 10.1109/ICSE.2009.5070521

Kim, D. S., & Chung, Y. J. (2006, June). Self-organization routing protocol supporting mobile nodes for wireless sensor network. In First International Multi-Symposiums on Computer and Computational Sciences (IMSCCS'06) (Vol. 2, pp. 622-626). IEEE. doi:10.1109/IMSCCS.2006.265

Kinalis, A., Nikoletseas, S., Patroumpa, D., & Rolim, J. (2014). Biased sink mobility with adaptive stop times for low latency data collection in sensor networks. *Information Fusion*, *15*(1), 56–63. doi:10.1016/j.inffus.2012.04.003

Kingma, D. P., & Ba, J. (2015). Adam: A Method for Stochastic Optimization. *3rd International Conference on Learning Representations, ICLR.*

Konorski, J. (2006). A game-theoretic study of CSMA/CA under a backoff attack. *IEEE/ACM Transactions on Networking*, *14*(6), 1167–1178. doi:10.1109/TNET.2006.886298

Kos, A., Milutinović, V., & Umek, A. (2019). Challenges in wireless communication for connected sensors and wearable devices used in sport biofeedback applications. *Future Generation Computer Systems*, *92*, 582–592. doi:10.1016/j.future.2018.03.032

Krasteva, V., Jekova, I., & Abächerli, R. (2017). Biometric verification by cross-correlation analysis of 12-lead ECG patterns: Ranking of the most reliable peripheral and chest leads. *Journal of Electrocardiology*, *50*(6), 847–854. doi:10.1016/j.jelectrocard.2017.08.021 PMID:28916172

Krishnamurthy, B., & Wills, C. E. (2009). *On the Leakage of Personally Identifiable Information Via Online Social Networks*. Retrieved from https://www.facebook.com/profile.php?

Krishnan, R., & Perumal, G. (2018). Family-Based Algorithm for Recovering from Node Failure in WSN. In D. Reddy Edla, P. Lingras, & K. Venkatanareshbabu (Eds.), *Advances in Machine Learning and Data Science. Advances in Intelligent Systems and Computing* (Vol. 705). Springer. doi:10.1007/978-981-10-8569-7_31

Krishnan, R., & Perumal, G. (2019). H2B2H protocol for addressing link failure in WSN. *Cluster Computing*, *22*(S4, Suppl 4), 9687–9696. doi:10.100710586-017-1355-9

Kulkarni, N., Prasad, N. R., & Prasad, R. (2017). Q-MOHRA: QoS Assured Multi-objective Hybrid Routing Algorithm for Heterogeneous WSN. *Wireless Personal Communications*, *100*(2), 255–266. doi:10.100711277-017-5064-8

Kumar, S. (2014). Improving WSN Routing and Security with an Artificial Intelligence Approach. In DWAI@ AI* IA (pp. 62-71). Academic Press.

Kumar, S., Sharma, R., & Vans, E. (2016). Localization for Wireless Sensor Networks: A Neural Network Approach. *International Journal of Computer Networks & Communications, 8*, 61-71. doi:10.5121/ijcnc.2016.8105

Kumar, D. P., Amgoth, T., & Annavarapu, C. S. R. (2019). Machine learning algorithms for wireless sensor networks: A survey. *Information Fusion*, *49*, 1–25. doi:10.1016/j.inffus.2018.09.013

Kumar, D., Aseri, T. C., & Patel, R. B. (2009). EEHC: Energy efficient heterogeneous clustered scheme for wireless sensor networks. *Elsevier Comput. Commun.*, *32*(4), 662–667. doi:10.1016/j.comcom.2008.11.025

Kumar, D., Chand, S., & Kumar, B. (2018). Cryptanalysis and improvement of an authentication protocol for wireless sensor networks applications like safety monitoring in coal mines. *Journal of Ambient Intelligence and Humanized Computing*, *10*(2), 641–660. doi:10.100712652-018-0712-8

Kumar, V.; Kumar, S. (2016). Mahila Bank in India- A Catalyst for Economic Empowerment of Women (A study of Women Customers in Delhi & Semi Urban Area). *Conference Proceedings of Confernce on Stand-up India - International Conference on "Entrepreneurship and Women Empowerment", THE NATIONAL INSTITUTE OF ENTREPRENEURSHIP AND SMALL BUSINESS DEVELOPMENT (NIESBUD)*, 9–38.

Kuzmanovic, A., & Knightly, E. W. (2006). Low-rate TCP-targeted denial of service attacks and counter strategies. *IEEE/ACM Transactions on Networking*, *14*(4), 683–696. doi:10.1109/TNET.2006.880180

Kwon, H., Yoon, H., & Park, K. W. (2019). Selective poisoning attack on deep neural networks. *Symmetry*, *11*(7), 1–13. doi:10.3390ym11070892

L'Ecuyer, P., & Simard, R. (1999). Beware of linear congruential generators with multipliers of the form a=±2 q±2 r. *ACM Transactions on Mathematical Software*, *25*(3), 367–374. doi:10.1145/326147.326156

Labovitz, C., Ahuja, A., Bose, A., & Jahanian, F. (2000). Delayed Internet routing convergence. *Computer Communication Review*, *30*(4), 175–187. doi:10.1145/347057.347428

Lahiri Chavan, A., Arora, S., Kumar, A., & Koppula, P. (2009). How Mobile Money Can Drive Financial Inclusion for Women at the Bottom of the Pyramid (BOP) in Indian Urban Centers. In N. Aykin (Ed.), *Internationalization, Design and Global Development* (pp. 475–484). Springer Berlin Heidelberg. doi:10.1007/978-3-642-02767-3_53

Lai, T. L. (1988). Nearly Optimal Sequential Tests of Composite Hypotheses. *Annals of Statistics*, *16*(2), 856–886. doi:10.1214/aos/1176350840

Lai, T. L. (2001). Sequential analysis: Some classical problems and new challenges. *Statistica Sinica*, *11*, 303–351.

Lalwani, S., Sharma, H., Chandra, S., Kusum, S., Jagdish, D., & Bansal, C. (2019). A Survey on Parallel Particle Swarm Optimization Algorithms. *Arabian Journal for Science and Engineering*, *44*(4), 2899–2923. doi:10.100713369-018-03713-6

Lamport, L. (1981). Password authentication with insecure communication. *Communications of the ACM, 24*(11), 770–772. doi:10.1145/358790.358797

Lang, J. (2015). *The Elliptic Curve Diffie-Hellman*. ECDH.

Lan, K.-C., & Wei, M.-Z. (2017). A Compressibility-Based Clustering Algorithm for Hierarchical Compressive Data Gathering. *IEEE Sensors Journal, 17*(8), 2550–2562. doi:10.1109/JSEN.2017.2669081

Laskov, P., Schäfer, C., Kotenko, I., & Müller, K.-R. (2004). Intrusion Detection in Unlabeled Data with Quarter-sphere Support Vector Machines. *PIK - Praxis Der Informationsverarbeitung Und Kommunikation, 27*(4), 228–236. doi:10.1515/piko.2004.228

Le Duc, D. T. L. T., Zalyubovskiy, V. V., Kim, D. S., & Choo, H. (2017). Collision-tolerant broadcast scheduling in duty-cycled wireless sensor networks. *Journal of Parallel and Distributed Computing, 100*, 42–56. doi:10.1016/j.jpdc.2016.10.006

Lee, E., Park, S., Yu, F., Choi, Y., Jin, M.-S., & Kim, S.-H.(2008). A predictable mobility-based data dissemination protocol for wireless sensor networks. *Proc. 22nd Int. Conf. Adv. Inf. Netw. Appl.*, 741–747. 10.1109/AINA.2008.139

Lee, S., & Chung, T. (2005). Data Aggregation for Wireless Sensor Networks Using Self-organizing Map. *Lecture Notes in Computer Science Artificial Intelligence and Simulation*, 508–517. doi:10.1007/978-3-540-30583-5_54

Lee, J.-S., & Kao, T.-Y. (2016). An Improved Three-Layer Low-Energy Adaptive Clustering Hierarchy for Wireless Sensor Networks. *IEEE Internet of Things Journal, 3*(6), 951–958. doi:10.1109/JIOT.2016.2530682

Lee, S. B., Ahn, G. S., & Campbell, A. T. (2001). Improving UDP and TCP performance in mobile ad hoc networks with INSIGNIA. *IEEE Communications Magazine, 39*(6), 156–165. doi:10.1109/35.925684

Lee, S., & Chung, T. (2004, October). Data aggregation for wireless sensor networks using self-organizing map. In *International Conference on AI, Simulation, and Planning in High Autonomy Systems* (pp. 508-517). Springer.

Li, G., Liu, X., & Wang, C. (2010). A sequential mesh test based selective forwarding attack detection scheme in wireless sensor networks. *Networking, Sensing and Control (ICNSC), 2010 International Conference on*, 554-558.

Li, W., Yi, P., Wu, Y., Pan, L., & Li, J. (2014). A New Intrusion Detection System Based on KNN Classification Algorithm in Wireless Sensor Network. *Journal of Electrical and Computer Engineering*, 1–8. doi:10.1155/2014/240217

Li, Y., Zhang, H., Bermudez, C., Chen, Y., Landman, B. A., & Vorobeychik, Y. (2019). Anatomical context protects deep learning from adversarial perturbations in medical imaging. *Neurocomputing*. doi:10.1016/j.neucom.2019.10.085

Liang, X., Chen, M., Xiao, Y., Balasingham, I., & Leung, V. C. (2009). A novel cooperative communication protocol for QoS provisioning in wireless sensor networks. *2009 5th International Conference on Testbeds and Research Infrastructures for the Development of Networks & Communities and Workshops*. doi: 10.1109/tridentcom.2009.4976244

Liang, X., Balasingham, I., & Byun, S. S. (2008, October). A multi-agent reinforcement learning based routing protocol for wireless sensor networks. In *2008 IEEE International Symposium on Wireless Communication Systems* (pp. 552-557). IEEE. 10.1109/ISWCS.2008.4726117

Li, C., Chen, C., Lee, C., Weng, C., & Chen, C. (2018). A novel three-party password-based authenticated key exchange protocol with user anonymity based on chaotic maps. *Soft Computing, 22*(8), 2495–2506. doi:10.100700500-017-2504-z

Li, C., Lee, C., Weng, C., & Chen, C. (2018). Towards secure authenticating of cache in the reader for RFID-based IoT systems. *Peer-to-Peer Networking and Applications, 11*(1), 198–208. doi:10.100712083-017-0564-6

Li, C., Wu, T., Chen, C., Lee, C., & Chen, C. (2017). An Efficient User Authentication and User Anonymity Scheme with Provably Security for IoT-Based Medical Care System. *Sensors (Basel)*, *17*(7), 17. doi:10.339017071482 PMID:28644381

Li, D., Wong, K., Hu, Y. H., & Sayeed, A. (2002). Detection, classification, and tracking of targets. *IEEE Signal Processing Magazine*, *19*(2), 17–29. doi:10.1109/79.985674

Li, J., Zhang, W., Kumari, S., Choo, K.-K. R., & Hogrefe, D. (2018). Security analysis and improvement of a mutual authentication and key agreement solution for wireless sensor networks using chaotic maps. *Transactions on Emerging Telecommunications Technologies*, *29*(6), e3295. Advance online publication. doi:10.1002/ett.3295

Lina, Deshmukh, & Potgantwar. (2015). Ensuring an early recognition and avoidance of the vampire attacks in WSN using routing loops. *Advance Computing Conference (IACC) 2015 IEEE International*, 61-66.

Lin, P., Qiao, C., & Wang, X. (2004). Medium access control with a dynamic duty cycle for sensor networks. *Wireless Communications and Networking Conference, 2004. WCNC. 2004 IEEE, 3*, 1534-1539. 10.1109/WCNC.2004.1311671

Lin, S., Kalogeraki, V., Gunopulos, D., & Lonardi, S. (2006). Online Information Compression in Sensor Networks. *2006 IEEE International Conference on Communications*. 10.1109/ICC.2006.255237

Lin, Y., Wan, K., Zhang, B., Liu, Y., & Li, X. (2016). An enhanced biometric-based three factors user authentication scheme for multi-server environments. *International Journal of Security and Its Applications*, *10*(1), 315–328. doi:10.14257/ijsia.2016.10.1.29

Li, P., Li, J., Huang, Z., Li, T., Gao, C.-Z., Yiu, S.-M., & Chen, K. (2017). Multi-key privacy-preserving deep learning in cloud computing. *Future Generation Computer Systems*, *74*, 76–85. doi:10.1016/j.future.2017.02.006

Li, P., Li, T., Ye, H., Li, J., Chen, X., & Xiang, Y. (2018). Privacy-preserving machine learning with multiple data providers. *Future Generation Computer Systems*, *87*, 341–350. doi:10.1016/j.future.2018.04.076

Liu, C., Li, B., Vorobeychik, Y., & Oprea, A. (2017). Robust linear regression against training data poisoning. *AISec 2017 - Proceedings of the 10th ACM Workshop on Artificial Intelligence and Security, Co-Located with CCS 2017*, 91–102. 10.1145/3128572.3140447

Liu, Y., Chen, X., Liu, C., & Song, D. (2016). *Delving into Transferable Adversarial Examples and Black-box Attacks*. Retrieved from https://arxiv.org/abs/1611.02770

Liu, F., & Bai, Y. (2012). An overview of topology control mechanism in multi-radio multi-channel wireless mesh networks. *EURASIP Journal on Wireless Communications and Networking*, *2012*(1), 1–12. doi:10.1186/1687-1499-2012-324

Liu, H., Meng, Z., Wang, H., & Xu, M. (2016). Spatio-temporal variation analysis of soil temperature based on wireless sensor network. *International Journal of Agricultural and Biological Engineering*, *9*(6), 131–138.

Liu, M., Vemuri, B. C., Amari, S.-I., & Nielsen, F. (2012). Shape Retrieval Using Hierarchical Total Bregman Soft Clustering. *IEEE Transactions on Pattern Analysis and Machine Intelligence*, *34*(12), 2407–2419. doi:10.1109/TPAMI.2012.44 PMID:22331859

Liu, Q., Yin, J., Leung, V. C. M., & Cai, Z. (2013). FADE: Forwarding Assessment Based Detection of Collaborative Grey Hole Attacks in WMNs. *IEEE Transactions on Wireless Communications*, *12*(10), 5124–5137. doi:10.1109/TWC.2013.121906

Liu, T., Xu, N., Liu, Q., Wang, Y., & Wen, W. (2019). A System-level Perspective to Understand the Vulnerability of Deep Learning Systems. *Proceedings of the Asia and South Pacific Design Automation Conference, ASP-DAC*, 544–549. 10.1145/3287624.3288751

Liu, X., Xiong, N., Zhang, N., Liu, A., Shen, H., & Huang, C. (2018). A Trust With Abstract Information Verified Routing Scheme for Cyber-Physical Network. *IEEE Access: Practical Innovations, Open Solutions*, *6*, 3882–3898. doi:10.1109/ACCESS.2018.2799681

Liu, X., Zhao, H., Yang, H., & Li, X. (2013). SinkTrail: A proactive data reporting protocol for wireless sensor networks. *IEEE Transactions on Computers*, *62*(1), 151–162. doi:10.1109/TC.2011.207

Liu, Z., Liu, W., Ma, Q., Liu, G., Zhang, L., Fang, L., & Sheng, V. S. (2019). Security cooperation model based on topology control and time synchronization for wireless sensor networks. *Journal of Communications and Networks (Seoul)*, *21*(5), 469–480. doi:10.1109/JCN.2019.000041

Li, Y., Niu, M., & Zou, Q. (2019). ELM-MHC: An Improved MHC identification method with extreme learning machine algorithm. *Journal of Proteome Research*, *18*(3), 1392–1401. doi:10.1021/acs.jproteome.9b00012 PMID:30698979

Li, Y., Wang, Y., & Li, D. (2019). Privacy-preserving lightweight face recognition. *Neurocomputing*, *363*, 212–222. doi:10.1016/j.neucom.2019.07.039

Lodeiro-Santiago, M., Caballero-Gil, C., & Caballero-Gil, P. (2017). Collaborative SQL-injections detection system with machine learning. *ACM International Conference Proceeding Series*. 10.1145/3109761.3158395

Lokhande, M. A. (2014). Financial inclusion initiatives in India-A review. *Intercontinental Journal Of Finance Research Review*, *2*(12), 22–33.

Luca, C., Antonio, M., & Dario, M. (2019). IoT Manager: An open-source IoT framework for smart cities. *Journal of Systems Architecture*, *98*(September), 413–423. doi:10.1016/j.sysarc.2019.04.003

Luo, H., Zhang, G., Liu, Y., & Das, S. K. (2018). Adaptive Routing in Wireless Sensor Networks. *Adaptation and Cross Layer Design in Wireless Networks*, 263–299. doi:10.1201/9781315219813-9

Lu, Y., Li, L., Peng, H., & Yang, Y. (2016). An Energy Efficient Mutual Authentication and Key Agreement Scheme Preserving Anonymity for Wireless Sensor Networks. *Sensors (Basel)*, *16*(6), 837. doi:10.339016060837 PMID:27338382

Luzuriaga, J. E., Perez, P., Boronat, P., Cano, J. C., Calafate, C., & Manzoni, P. (2015). A comparative evaluation of AMQP and MQTT protocols over unstable and mobile networks. Valencia: Department of Computer Engineering (Universitat Politecnica de Valencia). doi:10.1109/CCNC.2015.7158101

Lv, L., Ding, Z., Chen, J., & Al-Dhahir, N. (2019). Design of Secure NOMA Against Full-Duplex Proactive Eavesdropping. *IEEE Wireless Communications Letters*, *8*(4), 1090–1094. doi:10.1109/LWC.2019.2907852

Macua, S. V., Belanovic, P., & Zazo, S. (2010, June). Consensus-based distributed principal component analysis in wireless sensor networks. In *2010 IEEE 11th International Workshop on Signal Processing Advances in Wireless Communications (SPAWC)* (pp. 1-5). IEEE.

Madhu. (2014). BhartiyaMahila Bank (BMB) An empowerment to Indian Wome. *Radix International Journal of Banking, Finance and Accounting*, *3*(11).

Madsen, T. K., Fitzek, F. H., Prasad, R., & Schulte, G. (2005). Connectivity Probability of Wireless Ad Hoc Networks: Definition, Evaluation, Comparison. *Wireless Personal Communications*, *35*(1-2), 135–151. doi:10.100711277-005-8745-7

Maheswari, C., Priyanka, E. B., & Meenakshipriya, B. (2017). Fractional-order PIλDμ controller tuned by coefficient diagram method and particle swarm optimization algorithms for SO2 emission control process. *Proceedings of the Institution of Mechanical Engineers. Part I, Journal of Systems and Control Engineering*, *231*(8), 587–599. doi:10.1177/0959651817711626

Maheswari, C., Priyanka, E. B., Thangavel, S., & Parameswari, P. (2018). Development of Unmanned Guided Vehicle for Material Handling Automation for Industry 4.0. *International Journal of Recent Technology and Engineering*, *7*(4S), 428–432.

Mailath, G. J., & Samuelson, L. (2006). *Repeated games and reputations: long-run relationships*. Oxford university press. doi:10.1093/acprof:oso/9780195300796.001.0001

Malik, M., Neshatpour, K., Rafatirad, S., Joshi, R. V., Mohsenin, T., Ghasemzadeh, H., & Homayoun, H. (2019). Big vs little core for energy-efficient hadoop computing. *Journal of Parallel and Distributed Computing*, *129*, 110–124. doi:10.1016/j.jpdc.2018.02.017

Malone, D., Duffy, K., & Leith, D. (2007). Modeling the 802.11 distributed coordination function in nonsaturated heterogeneous conditions. *IEEE/ACM Transactions on Networking*, *15*(1), 159–172. doi:10.1109/TNET.2006.890136

Ma, M., Yang, Y., & Zhao, M. (2013). Tour planning for mobile data-gathering mechanisms in wireless sensor networks. *IEEE Transactions on Vehicular Technology*, *62*(4), 1472–1483. doi:10.1109/TVT.2012.2229309

Mao, Q., Hu, F., & Hao, Q. (2018). Deep learning for intelligent wireless networks: A comprehensive survey. *IEEE Communications Surveys and Tutorials*, *20*(4), 2595–2621. doi:10.1109/COMST.2018.2846401

Marina, M. K., Das, S. R., & Sunramaniam, A. P. (2010). A topology control approach for utilizing multiple channels in multi-radio wireless mesh networks. *Computer Networks*, *54*(2), 241–256. doi:10.1016/j.comnet.2009.05.015

Masiero, R., Quer, G., Munaretto, D., Rossi, M., Widmer, J., & Zorzi, M. (2009). Data Acquisition through Joint Compressive Sensing and Principal Component Analysis. *GLOBECOM 2009 - 2009 IEEE Global Telecommunications Conference*. doi: 10.1109/glocom.2009.5425458

Masiero, R., Quer, G., Munaretto, D., Rossi, M., Widmer, J., & Zorzi, M. (2009, November). Data acquisition through joint compressive sensing and principal component analysis. In *GLOBECOM 2009-2009 IEEE Global Telecommunications Conference* (pp. 1-6). IEEE. 10.1109/GLOCOM.2009.5425458

Maurya, A., & Jain, V. K. (2016). Fuzzy based energy efficient sensor network protocol for precision agriculture. *Computers and Electronics in Agriculture*, *130*, 20–37. doi:10.1016/j.compag.2016.09.016

Ma, X., Chen, X., & Zhang, X. (2019). Non-interactive privacy-preserving neural network prediction. *Information Sciences*, *481*, 507–519. doi:10.1016/j.ins.2018.12.015

Ma, X., Zhang, F., Chen, X., & Shen, J. (2018). Privacy preserving multi-party computation delegation for deep learning in cloud computing. *Information Sciences*, *459*, 103–116. doi:10.1016/j.ins.2018.05.005

Ma, Y., Wang, Y., Yang, J., Miao, Y., & Li, W. (2017). Big Health Application System based on Health Internet of Things and Big Data. *IEEE Access: Practical Innovations, Open Solutions*, *5*, 7885–7897. doi:10.1109/ACCESS.2016.2638449

Mbogo, M. (2013). The Impact of Mobile Payments on the Success and Growth of Micro-Business: The Case of M-PESA in Kenya. *Journal of Language. Technology & Entrepreneurship in Africa*, *53*(9), 1689–1699. doi:10.1017/CBO9781107415324.004

McKinsey. (2013). *Retail banking in Asia: Actionable insights for new opportunities*. https://goo.gl/4QjSLw

Mehmood, A., Khan, S., Shams, B., & Lloret, J. (2015). Energy-efficient multi-level and distance-aware clustering mechanism for WSNs. *International Journal of Communication Systems*, *28*(5), 972–989. doi:10.1002/dac.2720

Mekonnen, T., Porambage, P., Harjula, E., & Ylianttila, M. (2017). Energy Consumption Analysis of High Quality Multi-Tier Wireless Multimedia Sensor Network. *IEEE Access: Practical Innovations, Open Solutions*, *5*, 15848–15858. doi:10.1109/ACCESS.2017.2737078

Mellouli, D., Hamdani, T. M., Sanchez-Medina, J. J., Ayed, M. B., & Alimi, A. M. (2019). Morphological Convolutional Neural Network Architecture for Digit Recognition. *IEEE Transactions on Neural Networks and Learning Systems*, *30*(9), 2876–2885. doi:10.1109/TNNLS.2018.2890334 PMID:30676985

Melnikov, A., & Zeilenga, K. (2018). *Simple Authentication and Security Layer (SASL). RFC4422.* Network Working Group.

Mihaylov, M., Tuyls, K., & Nowé, A. (2010). Decentralized Learning in Wireless Sensor Networks. Adaptive and Learning Agents Lecture Notes in Computer Science, 60–73. doi:10.1007/978-3-642-11814-2_4

Milton, M. A. A. (2018). *Evaluation of Momentum Diverse Input Iterative Fast Gradient Sign Method (M-DI2-FGSM) Based Attack Method on MCS 2018 Adversarial Attacks on Black Box Face Recognition System.* Academic Press.

Ming, T., Zu, J., Zhusong, L., Aniello, C., & Francesco, P. (2018, January). Multi-layer cloud architectural model and ontology-based security service framework for IoT -based smart homes. *Future Generation Computer Systems*, *78*(Part 3), 1040–1051. doi:10.1016/j.future.2016.11.011

Mishra, V., & Agarwal, M. (2019). *Effect of Redundant Work on Machine.* SSRN 3444784

Mishra, S., Sarkar, U., Taraphder, S., Datta, S., Swain, D., Saikhom, R., Panda, S., & Laishram, M. (2017). Principal Component Analysis. *International Journal of Livestock Research*, *1*. Advance online publication. doi:10.5455/ijlr.20170415115235

Mishra, V., Agarwal, M., & Puri, N. (2018). Comprehensive and Comparative Analysis of Neural Network. *International Journal of Computers and Applications*, *2*(8).

Misra, S., & Vaish, A. (2011). Reputation-based role assignment for role-based access control in wireless sensor networks. *Computer Communications*, *34*(3), 281–294. Advance online publication. doi:10.1016/j.comcom.2010.02.013

Mnih, V., Kavukcuoglu, K., Silver, D., Graves, A., Antonoglou, I., Wierstra, D., & Riedmiller, M. (2013). *Playing atari with deep reinforcement learning.* arXiv preprint arXiv:1312.5602

Mnih, V., Kavukcuoglu, K., Silver, D., Rusu, A. A., Veness, J., Bellemare, M. G., Graves, A., Riedmiller, M., Fidjeland, A. K., Ostrovski, G., Petersen, S., Beattie, C., Sadik, A., Antonoglou, I., King, H., Kumaran, D., Wierstra, D., Legg, S., & Hassabis, D. (2015). Human-level control through deep reinforcement learning. *Nature*, *518*(7540), 529–533. doi:10.1038/nature14236 PMID:25719670

Mo, F., Shamsabadi, A. S., Katevas, K., Cavallaro, A., & Haddadi, H. (2019). *Towards Characterizing and Limiting Information Exposure in DNN Layers.* doi:10.1145/3319535.3363279

Mogaibel, H. A., Othman, M., Subramaniam, S., & Hamid, N. A. W. A. (2012). On-demand channel reservation scheme for common traffic in wireless mesh networks. *Journal of Network and Computer Applications*, *35*(4), 132–151. doi:10.1016/j.jnca.2012.01.017

Mohan, J., Kanagasabai, A., & Pandu, V. (2016). *Advances in Biometrics for Secure Human Authentication System: Biometric Authentication System.* Academic Press.

Mohsenian Rad, A. H., & Wong, V. W. S. (2007). Joint logical topology design, interface assignment, channel allocation and routing for multichannel wireless mesh networks. *IEEE Transactions on Wireless Communications*, *6*(12), 4432–4440. doi:10.1109/TWC.2007.060312

Momani, M., Challa, S., & Al-Hmouz, R. (2010). Bayesian Fusion Algorithm for Inferring Trust in Wireless Sensor Networks. *JNW*, *5*(7), 815–822. doi:10.4304/jnw.5.7.815-822

Mpitziopoulos, A., Gavalas, D., Konstantopoulos, C., & Pantziou, G. (2009). A survey on jamming attacks and countermeasures in WSNs. *IEEE Communications Surveys and Tutorials*, *11*(4), 42–56. doi:10.1109/SURV.2009.090404

Mughal, M. A., Luo, X., Ullah, A., Ullah, S., & Mahmood, Z. (2018). A Lightweight Digital Signature Based Security Scheme for Human-Centered Internet of Things. *IEEE Access: Practical Innovations, Open Solutions*, *6*, 31630–31643. doi:10.1109/ACCESS.2018.2844406

Muhammad, K., Sugeng, K. A., & Murfi, H. (2018). Machine Learning with Partially Homomorphic Encrypted Data. *Journal of Physics: Conference Series*, *1108*(1), 012112. Advance online publication. doi:10.1088/1742-6596/1108/1/012112

Munir, A., Gordon-Ross, A., & Ranka, S. (2014). Multi-core embedded wireless sensor networks: Architecture and applications. *IEEE Transactions on Parallel and Distributed Systems*, *25*(6), 1553–1562. doi:10.1109/TPDS.2013.219

Munns, G., Sawyer, W., & Cole, B. (2013). *Exemplary teachers of students in poverty*. Routledge. doi:10.4324/9780203076408

NABARD. (2018). *Status Of MicroFinance in India 2018-19*. Author.

Nakagami, M., Fortes, J. A., & Yamaguchi, S. (2019, July). Job-Aware Optimization of File Placement in Hadoop. In *2019 IEEE 43rd Annual Computer Software and Applications Conference (COMPSAC)* (Vol. 2, pp. 664-669). IEEE. 10.1109/COMPSAC.2019.10284

Nayak, P., & Devulapalli, A. (2016). A Fuzzy Logic-Based Clustering Algorithm for WSN to Extend the Network Lifetime. *IEEE Sensors Journal*, *16*(1), 137–144. doi:10.1109/JSEN.2015.2472970

Ndiaye, M., Hancke, G. P., & Abu-Mahfouz, A. M. (2017). Software defined networking for improved wireless sensor network management: A survey. *Sensors (Basel)*, *17*(5), 1031. doi:10.339017051031 PMID:28471390

Nehra, N. K., Kumar, M., & Patel, R. B. (2009, December). Neural network based energy efficient clustering and routing in wireless sensor networks. In *2009 First International Conference on Networks & Communications* (pp. 34-39). IEEE. 10.1109/NetCoM.2009.56

Neloy, A. A., Alam, S., Bindu, R. A., & Moni, N. J. (2019, April). Machine Learning based Health Prediction System using IBM Cloud as PaaS. In *2019 3rd International Conference on Trends in Electronics and Informatics (ICOEI)* (pp. 444-450). IEEE. 10.1109/ICOEI.2019.8862754

Neyman, J., & Pearson, E. S. (1933). IX. On the problem of the most efficient tests of statistical hypotheses. *Philosophical Transactions of the Royal Society of London. Series A, Mathematical and Physical Sciences*, *231*, 289–337.

Nguyen, P., Simon, T., Halem, M., Chapman, D., & Le, Q. (2012, November). A hybrid scheduling algorithm for data intensive workloads in a mapreduce environment. In *2012 IEEE Fifth International Conference on Utility and Cloud Computing* (pp. 161-167). IEEE. 10.1109/UCC.2012.32

Niranjan. Pandey, S., & Ganz, A. (2006). Design and evaluation of multichannel multirate wireless networks. *Mobile Networks and Applications*, *11*(5), 697–709. doi:10.100711036-006-7796-7

Nita, M. C., Pop, F., Voicu, C., Dobre, C., & Xhafa, F. (2015). MOMTH: Multi-objective scheduling algorithm of many tasks in Hadoop. *Cluster Computing*, *18*(3), 1011–1024. doi:10.100710586-015-0454-8

Niu, R., & Varshney, P. K. (2005). Decision fusion in a wireless sensor network with a random number of sensors. *Acoustics, Speech, and Signal Processing, 2005. Proceedings.(ICASSP'05). IEEE International Conference on, 4.*

Niu, R., & Varshney, P. K. (2008). Performance analysis of distributed detection in a random sensor field. *IEEE Transactions on Signal Processing*, *56*(1), 339–349. doi:10.1109/TSP.2007.906770

Ochola, E., Mejaele, L., Eloff, M., & Poll, J. V. D. (2017). Manet Reactive Routing Protocols Node Mobility Variation Effect in Analysing the Impact of Black Hole Attack. *SAIEE Africa Research Journal*, *108*(2), 80–92. doi:10.23919/SAIEE.2017.8531629

Okdem, S., & Karaboga, D. (2009). Routing in wireless sensor networks using an ant colony optimization (ACO) router chip. *Sensors (Basel)*, *9*(2), 909–921. doi:10.339090200909

Onaolapo, A. R. (2015). Effects of Financial Inclusion on the Economic Growth of Nigeria. *International Journal of Business and Management Review*, *3*(8), 11–28.

Order of the FSTEC of Russia dated February 11, 2013 No 17 "On approval of the requirements for the protection of information not constituting state secrets contained in state information systems". (2013). Retrieved from https://fstec.ru/normotvorcheskaya/akty/53-prikazy/702

Order of the Government of the Russian Federation of August 28, 2019 N 1911-r "On approval of the Concept of creating a unified cloud platform". (2019). *Reference legal system "Consultant Plus"*.

Order of the Ministry of Communications of Russia of December 31, 2015 No 622 "On approval of the rules for using the classifier of programs for electronic computers and databases". (2015). Retrieved from https://legalacts.ru/doc/prikaz-minkomsvjazi-rossii-ot-31122015-n-622/

Osborne, M. A., Roberts, S. J., Rogers, A., Ramchurn, S. D., & Jennings, N. R. (2008). Towards Real-Time Information Processing of Sensor Network Data Using Computationally Efficient Multi-output Gaussian Processes. *2008 International Conference on Information Processing in Sensor Networks (Ipsn 2008)*. 10.1109/IPSN.2008.25

Osisanwo, F. Y., Akinsola, J. E. T., Awodele, O., Hinmikaiye, J. O., Olakanmi, O., & Akinjobi, J. (2017). Supervised machine learning algorithms: Classification and comparison. *International Journal of Computer Trends and Technology*, *48*(3), 128–138. doi:10.14445/22312803/IJCTT-V48P126

Otoum, S., Kantarci, B., & Mouftah, H. T. (2019). On the feasibility of deep learning in sensor network intrusion detection. *IEEE Networking Letters*, *1*(2), 68–71. doi:10.1109/LNET.2019.2901792

Ouferhat, N., & Mellouk, A. (2007). A QoS Scheduler Packets for Wireless Sensor Networks. *2007 IEEE/ACS International Conference on Computer Systems and Applications*. 10.1109/AICCSA.2007.370885

Pakize, S. R. (2014). A comprehensive view of Hadoop MapReduce scheduling algorithms. *International Journal of Computer Networks & Communications Security*, *2*(9), 308–317.

Pan, L., & Li, J. (2010). K-Nearest Neighbor Based Missing Data Estimation Algorithm in Wireless Sensor Networks. *Wireless Sensor Network.*, *2*(02), 115–122. doi:10.4236/wsn.2010.22016

Panos, C., Ntantogian, C., Malliaros, S., & Xenakis, C. (2017). Analyzing, quantifying, and detecting the black-hole attack in infrastructure-less networks. *Computer Networks*, *113*, 94–110. doi:10.1016/j.comnet.2016.12.006

Pant, M., Deep, K., Bansal, J. C., Das, K. N., & Nagar, A. K. (2018). Soft computing for problem solving (SocProS 2015). *International Journal of System Assurance Engineering and Management*, *9*(1), 1–1. doi:10.100713198-018-0713-1

Paradis, L., & Han, Q. (2007). A Survey of Fault Management in Wireless Sensor Networks. *Journal of Network and Systems Management*, *15*(2), 171–190. doi:10.100710922-007-9062-0

Park, J. C., & Kasera, S. K. (2007, March). *Securing Ad Hoc wireless networks against data injection attacks using firewalls. In 2007 IEEE Wireless Communications and Networking Conference*. IEEE.

Parras, J., & Zazo, S. (2018). Wireless Networks under a Backoff Attack: A Game Theoretical Perspective. *Sensors (Basel), 18*(2), 404. doi:10.339018020404 PMID:29385752

Parras, J., & Zazo, S. (2019a). Learning attack mechanisms in Wireless Sensor Networks using Markov Decision Processes. *Expert Systems with Applications, 122*, 376–387. doi:10.1016/j.eswa.2019.01.023

Parras, J., & Zazo, S. (2019b). Repeated Game Analysis of a CSMA/CA Network under a Backoff Attack. *Sensors (Basel), 19*(24), 5393. doi:10.339019245393 PMID:31817778

Parras, J., & Zazo, S. (2019c). Using one class SVM to counter intelligent attacks against an SPRT defense mechanism. *Ad Hoc Networks, 94*, 101946. doi:10.1016/j.adhoc.2019.101946

Passport of the national project. "National Program" Digital Economy of the Russian Federation "(approved by the Presidium of the Presidential Council for Strategic Development and National Projects, Minutes dated June 06, 2019 No 7). (2019). https://legalacts.ru/doc/pasport-natsionalnogo-proekta-natsionalnaja-programma-tsifrovaja-ekonomika-rossiiskoi-federatsii/

Patel, H. P., & Chaudhari, M. B. (2013). A Time Space Cryptography Hashing Solution for Prevention Jellyfish Reordering Attack in Wireless Adhoc Network. *International Conference on Computing Communication and Networking Technologies (ICCCNT)*. 10.1109/ICCCNT.2013.6726689

Pathan, A. S. K., & Hong, C. S. (2007, August). A secure energy-efficient routing protocol for WSN. In *International symposium on parallel and distributed processing and applications* (pp. 407-418). Springer. 10.1007/978-3-540-74742-0_38

Patil, M., & Biradar, R. C. (2012, December). A survey on routing protocols in wireless sensor networks. In *2012 18th IEEE International Conference on Networks (ICON)* (pp. 86-91). IEEE. 10.1109/ICON.2012.6506539

Patil, C. M., & Gowda, S. (2017). An Approach for Secure Identification and Authentication for Biometrics using Iris. *2017 International Conference on Current Trends in Computer, Electrical, Electronics and Communication (CTCEEC)*, 421-424. 10.1109/CTCEEC.2017.8455148

Patil, S., & Chaudhari, S. (2016). DoS Attack Prevention Technique in Wireless Sensor Networks. *Procedia Computer Science, 79*, 715–721. doi:10.1016/j.procs.2016.03.094

Pauer, F., Göbel, J., Storf, H., Litzkendorf, S., Babac, A., Frank, M., LYeshrs, V., Schauer, F., Schmidtke, J., Biehl, L., Wagner, T. O., YESckert, F., Schulenburg, J. G., & Hartz, T. (2016). Adopting Quality Criteria for Websites Providing Medical Information About Rare Diseases. *Interactive Journal of Medical Research, 5*(3), 5. doi:10.2196/ijmr.5822 PMID:27562540

Păun, G. (2000). Computing with Membranes. *Journal of Computer and System Sciences, 61*(1), 108–143. doi:10.1006/jcss.1999.1693

Payal, A., Rai, C. S., & Reddy, B. R. (2015). Analysis of some feedforward artificial neural network training algorithms for developing localization framework in wireless sensor networks. *Wireless Personal Communications, 82*(4), 2519–2536. doi:10.100711277-015-2362-x

Perniu, L. M., Moşoi, A. A., Sandu, F., Moraru, S. A., Ungureanu, D. E., Kristály, D. M., & Guşeilă, L. G. (2019, August). Cloud Services for an Active Assisted Living Platform using Wireless Sensors and Mobile Devices. In *2019 International Conference on Sensing and Instrumentation in IoT Era (ISSI)* (pp. 1-6). IEEE. 10.1109/ISSI47111.2019.9043703

Perrig, A., Stankovic, J., & Wagner, D. (2004). *Security in wireless sensor networks*. Academic Press.

Perrig, A., Stankovic, J., & Wagner, D. (2004). Security in Wireless Sensor Networks. *Communications of the ACM, 47*(6), 53. doi:10.1145/990680.990707

Pham, T. N. D., & Yeo, C. K. (2016). Detecting Colluding Black-hole and Greyhole Attacks in Delay Tolerant Networks. *IEEE Transactions on Mobile Computing, 15*(5), 1116–1129. doi:10.1109/TMC.2015.2456895

Phan, N. H., Vu, M. N., Liu, Y., Jin, R., Dou, D., Wu, X., & Thai, M. T. (2019). Heterogeneous Gaussian mechanism: Preserving differential privacy in deep learning with provable robustness. *IJCAI International Joint Conference on Artificial Intelligence,* 4753–4759. 10.24963/ijcai.2019/660

Phan, N., Wu, X., Hu, H., & Dou, D. (2017). Adaptive Laplace Mechanism: Differential Privacy Preservation in Deep Learning. In *2017 IEEE International Conference on Data Mining (ICDM)* (pp. 385–394). IEEE. 10.1109/ICDM.2017.48

Phong, L. T., & Phuong, T. T. (2019). Privacy-Preserving Deep Learning via Weight Transmission. *IEEE Transactions on Information Forensics and Security*, 1–1. doi:10.1109/TIFS.2019.2911169

Pilloni, V., Navaratnam, P., Vural, S., Atzori, L., & Tafazolli, R. (2014). TAN: A distributed algorithm for dynamic task assignment in WSNs. *IEEE Sensors Journal, 14*(4), 1266–1279. doi:10.1109/JSEN.2013.2294540

Pinto, A. R., Montez, C., Araújo, G., Vasques, F., & Portugal, P. (2014). An approach to implement data-fusion techniques in wireless sensor networks using genetic machine learning algorithms. *Information Fusion, 15*, 90–101. doi:10.1016/j.inffus.2013.05.003

Pirbhulal, S., Zhang, H., Mukhopadhyay, S. C., Li, C., Wang, Y., Li, G., Wu, W., & Zhang, Y. (2015). An Efficient Biometric-Based Algorithm Using Heart Rate Variability for Securing Body Sensor Networks. *Sensors (Basel), 15*(7), 15067–15089. doi:10.3390150715067 PMID:26131666

Pirbhulal, S., Zhang, H., Wu, W., Mukhopadhyay, S. C., & Zhang, Y. (2018). Heartbeats Based Biometric Random Binary Sequences Generation to Secure Wireless Body Sensor Networks. *IEEE Transactions on Biomedical Engineering, 65*(12), 2751–2759. doi:10.1109/TBME.2018.2815155 PMID:29993429

Plyler, M., Haas, S., & Nagarajan, G. (2010). *Community Level Economic Effects of M-PESA in Kenya: Initial Findings.* Financial Services Assessment - Iris Center, University of Maryland College Park. http://www.fsassessment.umd.edu/publications/pdfs/Kenya-MPESA-Community.pdf

Pooja & Chauhan. (2017). Review on Security attacks and Countermeasures in Wireless Sensor Networks. *International Journal of Advanced Research in Computer Science, 8*(5), 1275–1283.

Poor, H. V., & Hadjiliadis, O. (2009). *Quickest detection* (Vol. 40). Cambridge University Press Cambridge.

Porkodi, S., & Aravazhi, D. D. (2013). Role of Micro Finance and Self Help Groups in Financial Inclusion. *International Journal of Marketing, 2*(3), 137–149. http://indianresearchjournals.com/pdf/IJMFSMR/2013/March/13.pdf

Portillo, C., Martinez-Bauset, J., Pla, V., & Casares-Giner, V. (2020). Modeling of Duty-Cycled MAC Protocols for Heterogeneous WSN with Priorities. *Electronics (Basel), 9*(3), 467. doi:10.3390/electronics9030467

Pragya, M., Arya, K. V., & Pal, S. H. (2017). Intrusion Detection System Against Colluding Misbehavior in MANETs. *Wireless Personal Communications, 100*(2), 491–503. doi:10.100711277-017-5094-2

Prajapati, J., & Jain, S. C. (2018, April). Machine learning techniques and challenges in wireless sensor networks. In *2018 Second International Conference on Inventive Communication and Computational Technologies (ICICCT)* (pp. 233-238). IEEE. 10.1109/ICICCT.2018.8473187

Prasad, E. (2009, Dec.). Growth in Asia. *Finance & Development*, 19–22.

Pratheep, V. G., Priyanka, E. B., & Prasad, P. H. (2019, October). Characterization and Analysis of Natural Fibre-Rice Husk with Wood Plastic Composites. *IOP Conference Series. Materials Science and Engineering*, *561*(1), 012066. doi:10.1088/1757-899X/561/1/012066

Pratheep, V. G., Priyanka, E. B., Thangavel, S., Gousanal, J. J., Antony, P. B., & Kavin, E. D. (2020). Investigation and analysis of corn cob, coir pith with wood plastic composites. *Materials Today: Proceedings*. Advance online publication. doi:10.1016/j.matpr.2020.02.288

Preeth, S. S. L., Dhanalakshmi, R., Kumar, R., & Shakeel, P. M. (2018). An adaptive fuzzy rule based energy efficient clustering and immune-inspired routing protocol for WSN-assisted IoT system. *Journal of Ambient Intelligence and Humanized Computing*, 1–13. doi:10.100712652-018-1154-z

Premarathne, U.S., Abuadbba, A., Alabdulatif, A., Khalil, I., Tari, Z., Zomaya, A.Y., & Buyya, R. (2015). *Hybrid Cryptographic Access Control for Cloud based Electronic Health Records Systems*. Academic Press.

Priyanka, E. B., & Thangavel, S. (2018b). Advanced Enhancement Model of Bionics Fish Cilia MEMS Vector Hydrophone-Systematic Analysis Review. *International Journal of Applied Physics, 3.*

Priyanka, E. B., & Thangavel, S. (2018c). Optimization of Large-Scale Solar Hot Water System Using Non-traditional Optimization Technique. *J Environ Sci Allied Res, 2018*, 5-14.

Priyanka, E. B., Maheswari, C., & Thangavel, S. (2020b). A smart-integrated IoT module for intelligent transportation in oil industry. *International Journal of Numerical Modelling: Electronic Networks, Devices and Fields*, e2731.

Priyanka, E. B., Krishnamurthy, K., & Maheswari, C. (2016, November). Remote monitoring and control of pressure and flow in oil pipelines transport system using PLC based controller. In *2016 Online International Conference on Green Engineering and Technologies (IC-GET)* (pp. 1-6). IEEE. 10.1109/GET.2016.7916754

Priyanka, E. B., Maheswari, C., Ponnibala, M., & Thangavel, S. (2019b). SCADA Based Remote Monitoring and Control of Pressure & Flow in Fluid Transport System Using IMC-PID Controller. *Advances in Systems Science and Applications*, *19*(3), 140–162.

Priyanka, E. B., Maheswari, C., & Thangavel, S. (2018d). IoT based field parameters monitoring and control in press shop assembly. *Internet of Things*, *3*, 1–11. doi:10.1016/j.iot.2018.09.004

Priyanka, E. B., Maheswari, C., & Thangavel, S. (2018e). Online monitoring and control of flow rate in oil pipelines transportation system by using PLC based Fuzzy-PID Controller. *Flow Measurement and Instrumentation*, *62*, 144–151. doi:10.1016/j.flowmeasinst.2018.02.010

Priyanka, E. B., Maheswari, C., & Thangavel, S. (2018f, December). Proactive Decision Making Based IoT Framework for an Oil Pipeline Transportation System. In *International conference on Computer Networks, Big data and IoT* (pp. 108-119). Springer.

Priyanka, E. B., Maheswari, C., & Thangavel, S. (2019a). Remote monitoring and control of LQR-PI controller parameters for an oil pipeline transport system. *Proceedings of the Institution of Mechanical Engineers. Part I, Journal of Systems and Control Engineering*, *233*(6), 597–608. doi:10.1177/0959651818803183

Priyanka, E. B., Maheswari, C., Thangavel, S., & Bala, M. P. (2020a). Integrating IoT with LQR-PID controller for online surveillance and control of flow and pressure in fluid transportation system. *Journal of Industrial Information Integration*, *17*, 100127. doi:10.1016/j.jii.2020.100127

Priyanka, E. B., & Subramaniam, T. (2018a). Fuzzy Logic Forge Filter Weave Pattern Recognition Analysis on Fabric Texture. *International Journal of Electrical and Electronic Science*, *5*(3), 63–70.

Pruthi, V., Mittal, K., Sharma, N., & Kaushik, I. (2019). Network Layers Threats & its Countermeasures in WSNs. *2019 International Conference on Computing, Communication, and Intelligent Systems (ICCCIS)*. 10.1109/ICCCIS48478.2019.8974523

Public declaration of the goals and objectives of the federal service for supervision in the field of communication, information technologies and mass communications (Roskomnadzor). (2019). https: Rkn.gov.rn

Puri, D., & Bhushan, B. (2019). Enhancement of security and energy efficiency in WSNs: Machine Learning to the rescue. *2019 International Conference on Computing, Communication, and Intelligent Systems (ICCCIS)*. 10.1109/ICCCIS48478.2019.8974465

Qin, Zhu, Ni, Gu, & Zhu. (2019). Blockchain for the IoT and industrial IoT: A review. *IoT*. .2019.100081 doi:10.1016/j.IoT

Qu, Y., Ng, B., & Seah, W. (2016). A survey of routing and channel assignment in multi-channel multi-radio (WMNs). *Journal of Network and Computer Applications*, *65*(C), 120–130. doi:10.1016/j.jnca.2016.02.017

Rahaman, M. F., & Khan, M. Z. (2018). Low-Complexity Optimal Hard Decision Fusion Under the Neyman—Pearson Criterion. *IEEE Signal Processing Letters*, *25*(3), 353–357. doi:10.1109/LSP.2017.2766245

Rahman, A., & Gburzynski, P. (2006). Hidden problems with the hidden node problem. *23rd Biennial Symposium on Communications*, 270-273. 10.1109/BSC.2006.1644620

Rajagopalan, R., & Varshney, P. K. (2006). *Data aggregation techniques in sensor networks: A survey*. Academic Press.

Rajeshkumar, G., & Valluvan, K. R. (2016). An Energy Aware Trust Based Intrusion Detection System with Adaptive Acknowledgement for Wireless Sensor Network. *Wireless Personal Communications*, *94*(4), 1993–2007. doi:10.100711277-016-3349-y

Rajeshwari, M. S. (2015). A Study on Issues and Challenges of Women Empowerment in India. *Journal of Business and Management*, *17*(4).

Ralston, K., Chen, Y., Isah, H., & Zulkernine, F. (2019). *A Voice Interactive Multilingual Student Support System using IBM Watson*. arXiv preprint arXiv:2001.00471

Ranade, A. (2017). *Role of "Fintech" in Financial inclusion and new business models*. Academic Press.

Rani, Y., & Rohil, H. (2013). A study of hierarchical clustering algorithm. *ter S & on Te SIT-2*, 113.

Raniwala, A., & Chiueh, T. (2005). Architecture and algorithms for an IEEE 802.11-based multi-channel wireless mesh networks. *Proceedings 24th Annual Joint Conference of the IEEE Computer and Communications Societies*, *3*, 2223-2234.

Raniwala, A., Gopalan, K., & Chiueh, T. (2004). Centralized channel assignment and routing algorithms for multi-channel wireless mesh networks. *Computer Communications*, *8*, 50–65.

Rasooli, A., & Down, D. G. (2011, November). An adaptive scheduling algorithm for dynamic heterogeneous Hadoop systems. In *Proceedings of the 2011 Conference of the Center for Advanced Studies on Collaborative Research* (pp. 30-44). Academic Press.

Rawat, P., Singh, K. D., Chaouchi, H., & Bonnin, J. M. (2014). Wireless sensor networks: A survey on recent developments and potential synergies. *The Journal of Supercomputing*, *68*(1), 1–48. doi:10.100711227-013-1021-9

Reith, R. N., Schneider, T., & Tkachenko, O. (2019). *Efficiently Stealing your Machine Learning Models*. doi:10.1145/3338498.3358646

Ren, J., Zhang, Y., Zhang, K., & Shen, X. (2016). Adaptive and Channel-Aware Detection of Selective Forwarding Attacks in Wireless Sensor Networks. *IEEE Transactions on Wireless Communications, 15*(5), 3718–3731. doi:10.1109/TWC.2016.2526601

Revathi, M., & Deva Priya, S. (2014). Channel allocation for interference analysis in wireless networks. *Proceedings of ICSEM'14 – 2nd International Conference on Science, Engineering and Management*, 1-7.

Riggio, R., Rasheed, T., Testi, S., Granelli, F., & Chlamtac. (2011). Interfeerence and traffic aware channel assignment in wifi-based wireless mesh networks. *Ad Hoc Networks, 9*(5), 864-875.

Rojas-Suarez, L. (2016, Oct.). Financial Inclusion in Latin America: Facts and Obstacles. SSRN Electronic Journal. doi:10.2139srn.2860875

Rom'an, Fern'andez-Gago, & L'opez. (2009). Trust and Reputation Systems for Wireless Sensor Networks. In Security and Privacy in Mobile and Wireless Networking. Troubador Publishing Ltd.

Römer, K., Frank, C., Marrón, P. J., & Becker, C. (2004). Generic role assignment for wireless sensor networks. *Proceedings of the 11th Workshop on ACM SIGOPS European Workshop: Beyond the PC - EW11*, 2. 10.1145/1133572.1133588

Rooshenas, A., Rabiee, H. R., Movaghar, A., & Naderi, M. Y. (2010). Reducing the data transmission in Wireless Sensor Networks using the Principal Component Analysis. *2010 Sixth International Conference on Intelligent Sensors, Sensor Networks and Information Processing*. 10.1109/ISSNIP.2010.5706781

Roozeboom, C. L., Hopcroft, M. A., Smith, W. S., Sim, J. Y., Wickeraad, D. A., Hartwell, P. G., & Pruitt, B. L. (2013). Integrated Multifunctional Environmental Sensors. *Journal of Microelectromechanical Systems, 22*(3), 779–793. doi:10.1109/JMEMS.2013.2245400

Roy, A., Memon, N., & Ross, A. (2017). MasterPrint: Exploring the Vulnerability of Partial Fingerprint-Based Authentication Systems. *IEEE Transactions on Information Forensics and Security, 12*(9), 2013–2025. doi:10.1109/TIFS.2017.2691658

Roy, M. A., & Wheeler, D. (2006). A Survey of Micro-Enterprise in Urban West Africa: Drivers Shaping the Sector. *Development in Practice, 16*(5), 452–464. doi:10.1080/09614520600792432

Roy, S., Chatterjee, S., Das, A. K., Chattopadhyay, S., Kumari, S., & Jo, M. (2018). Chaotic Map-Based Anonymous User Authentication Scheme with User Biometrics and Fuzzy Extractor for Crowdsourcing Internet of Things. *IEEE Internet of Things Journal, 5*(4), 2884–2895. doi:10.1109/JIOT.2017.2714179

Russell, S. J., & Norvig, P. (2016). *Artificial intelligence: a modern approach*. Pearson Education Limited.

Sachi, N., Mohanty, C. R., Sheeba, R., Deepak, G. K., Shankar, S. K., & Lakshmanaprabu, A. K. (2019). An efficient Lightweight integrated Blockchain (ELIB) model for IoT security and privacy. *Future Generation Computer Systems*. Advance online publication. doi:10.1016/j.future.2019.09.09.050

Sae-Bae, N., Wu, J., Memon, N., Konrad, J., & Ishwar, P. (2019). Emerging NUI-Based Methods for User Authentication: A New Taxonomy and Survey. *IEEE Transactions on Biometrics, Behavior, and Identity Science, 1*(1), 5–31. doi:10.1109/TBIOM.2019.2893297

Sagar, A. K., & Lobiyal, D. K. (2015). Probabilistic Intrusion Detection in Randomly Deployed Wireless Sensor Networks. *Wireless Personal Communications, 84*(2), 1017–1037. doi:10.100711277-015-2673-y

Saleem, M., Di Caro, G. A., & Farooq, M. (2011). Swarm intelligence based routing protocol for wireless sensor networks: Survey and future directions. *Information Sciences, 181*(20), 4597–4624. doi:10.1016/j.ins.2010.07.005

Saleh, N., Itani, W., Haidar, A., & Nassar, H. (2015). A Novel Scheme to Reduce the Energy Consumption of Wireless Sensor Networks. *Int. J. Enhanced Res. Sci. Technol. Eng.*, *4*(5), 190–195.

Saleh, N., Kassem, A., & Haidar, A. M. (2018). Energy-Efficient Architecture for Wireless Sensor Networks in Healthcare Applications. *IEEE Access: Practical Innovations, Open Solutions*, *6*, 6478–6486. doi:10.1109/ACCESS.2018.2789918

Sarangi, S., & Thankchan, B. (2012). A novel routing algorithm for wireless sensor network using particle swarm optimization. *IOSR Journal of Computer Engineering*, *4*(1), 26–30. doi:10.9790/0661-0412630

Sarasvathi, V., & Iyengar, N. (2012). Centralized rank-based channel assignment for multi-radio multi-channel wireless mesh networks. *2nd international conference on Computer, Communication, Control and Information Technology*, 182-186. 10.1016/j.protcy.2012.05.027

Sarella, V. R., Reddy, P. P., Rao, S. K., & Padala, P. (2017). EEECARP: Efficient Energy Clustering Adaptive Routing Procedure for Wireless Sensor Networks. *Journal of Global Information Management*, *25*(4), 125–138. doi:10.4018/JGIM.2017100108

Sarkar, S. (2017). Results on significant anomalies of state values after key scheduling algorithm in RC4. *IET Information Security*, *11*(5), 267–272. doi:10.1049/iet-ifs.2016.0451

Sassi, S., & Goaied, M. (2013). Financial development, ICT diffusion and economic growth: Lessons from MENA region. *Telecommunications Policy*, *37*(4), 252–261. doi:10.1016/j.telpol.2012.12.004

Savaglio, C., Pace, P., Aloi, G., Liotta, A., & Fortino, G. (2019). Lightweight Reinforcement Learning for Energy Efficient Communications in Wireless Sensor Networks. *IEEE Access: Practical Innovations, Open Solutions*, *7*, 29355–29364. doi:10.1109/ACCESS.2019.2902371

Schuba, C. L., Krsul, I. V., Kuhn, M. G., Spafford, E. H., Sundaram, A., & Zamboni, D. (1997, May). Analysis of a denial of service attack on TCP. In *Proceedings. 1997 IEEE Symposium on Security and Privacy (Cat. No. 97CB36097)* (pp. 208-223). IEEE. 10.1109/SECPRI.1997.601338

Schweitzer, N., Stulman, A., Margalit, R. D., & Shabtai, A. (2017). Contradiction Based Gray-Hole Attack Minimization for Ad-Hoc Networks. *IEEE Transactions on Mobile Computing*, *16*(8), 2174–2183. doi:10.1109/TMC.2016.2622707

SCOTT Project. (2017-2020). *Secure COnnected Trustable Things*. https://scottproject.eu/

SCOTT Project. (2018). *D23.2 Dependable Communication and Safety Building Blocks.* Call H2020-ECSEL-2016-2-IA-two-stage - Innovation Actions (IA). Secure COnnected Trustable Things.

Seah, M. W. M., Tham, C.-K., Srinivasan, V., & Xin, A. (2007). Achieving Coverage through Distributed Reinforcement Learning in Wireless Sensor Networks. *2007 3rd International Conference on Intelligent Sensors, Sensor Networks and Information.* doi: 10.1109/issnip.2007.4496881

Second Industrial Revolution. (n.d.). Retrieved from https://en.wikipedia.org/wiki/Second_Industrial_Revolution

Sethi, T. S., & Kantardzic, M. (2018). Neurocomputing Data driven exploratory attacks on black box classifiers in adversarial domains. *Neurocomputing*, *289*, 129–143. doi:10.1016/j.neucom.2018.02.007

Shafahi, A., Ronny Huang, W., Najibi, M., Suciu, O., Studer, C., Dumitras, T., & Goldstein, T. (2018). Poison frogs! Targeted clean-label poisoning attacks on neural networks. Advances in Neural Information Processing Systems, 6103–6113.

Shah, P., & Dubhashi, M. (2015). Review Paper on Financial Inclusion-The Means of Inclusive Growth. *Chanakya International Journal of Business Research*, *1*(1), 37. doi:10.15410/cijbr/2015/v1i1/61403

Shakya, K. (2016). *Microfinance and Woman Empowerment*. Academic Press.

Shankar, R. K., Priyanka, E. B., & Saravanan, B. (2015). Performance analysis of gasoline direct injection in two-stroke spark-ignition engines. *Int J Adv Res Electr Electron Instrum Eng*, *4*, 4940–4947.

Sharawi, M., Saroit, I. A., El-Mahdy, H., & Emary, E. (2013). Routing wireless sensor networks based on soft computing paradigms: Survey. *International Journal on Soft Computing Artificial Intelligence and Applications (Commerce, Calif.)*, *2*(4), 21–36.

Sharma, M., Tandon, A., Narayan, S., & Bhushan, B. (2017). Classification and analysis of security attacks in WSNs and IEEE 802.15.4 standards: A survey. *2017 3rd International Conference on Advances in Computing,Communication& Automation (ICACCA).* DOI: 10.1109/icaccaf.2017.8344727

Sharma, N., Kaushik, I., Singh, N., & Kumar, R. (2019). Performance Measurement Using Different Shortest Path Techniques in Wireless Sensor Network. *2019 2nd International Conference on Signal Processing and Communication (ICSPC).* doi: 10.1109/icspc46172.2019.8976618

Sharma, S., & Chen, K. (2018). *Image Disguising for Privacy-preserving Deep Learning*. doi:10.1145/3243734.3278511

Sharma, V. K., Shukla, S. S. P., & Singh, V. (2012, December). A tailored Q-Learning for routing in wireless sensor networks. In *2012 2nd IEEE International Conference on Parallel, Distributed and Grid Computing* (pp. 663-668). IEEE. 10.1109/PDGC.2012.6449899

Sharma, D. (2016). Nexus between financial inclusion and economic growth: Evidence from the emerging Indian economy. *Journal of Financial Economic Policy*, *8*(1), 13–36. doi:10.1108/JFEP-01-2015-0004

Sharma, V., Rai, S., & Dev, A. (2012). A comprehensive study of artificial neural networks. *International Journal of Advanced Research in Computer Science and Software Engineering*, *2*(10).

Shei, Y., & Su, Y. T. (2008). A sequential test based cooperative spectrum sensing scheme for cognitive radios. *Personal, Indoor and Mobile Radio Communications, 2008. PIMRC 2008. IEEE 19th International Symposium on*, 1-5.

She, W., Liu, Q., Tian, Z., Chen, J. S., Wang, B., & Liu, W. (2019). Blockchain Trust Model for Malicious Node Detection in Wireless Sensor Networks. *IEEE Access: Practical Innovations, Open Solutions*, *7*, 38947–38956. doi:10.1109/ACCESS.2019.2902811

Shibahara, T., Yamanishi, K., Takata, Y., Chiba, D., Akiyama, M., Yagi, T., ... Murata, M. (2017, May). Malicious URL sequence detection using event de-noising convolutional neural network. In *2017 IEEE International Conference on Communications (ICC)* (pp. 1-7). IEEE. 10.1109/ICC.2017.7996831

Shin, M., Lee, S., & Kim, Y. A. (2006). Distributed channel assignment for multi-radio wireless networks. *IEEE International Conference on Mobile Ad Hoc and Sensor Systems*, 417-426. 10.1109/MOBHOC.2006.278582

Shin, M., Ma, J., Mishra, A., & Arbaugh, W. A. (2006). Wireless network security and interworking. *Proceedings of the IEEE*, *94*(2), 455–466. doi:10.1109/JPROC.2005.862322

Shivani, R., & Hunar. (2017). Survey on cluster based data aggregation in wireless sensor network. *International Journal of Advanced Research in Computer Science*, *8*(5), 311–314.

Shnidman, D. A. (1998). Binary integration for Swerling target fluctuations. *IEEE Transactions on Aerospace and Electronic Systems*, *34*(3), 1043–1053. doi:10.1109/7.705926

Shokri, R., & Shmatikov, V. (2015, October). Privacy-preserving deep learning. In *Proceedings of the 22nd ACM SIGSAC conference on computer and communications security* (pp. 1310-1321). Academic Press.

Shokri, R., & Shmatikov, V. (2015). Privacy-Preserving Deep Learning. In *Proceedings of the 22nd ACM SIGSAC Conference on Computer and Communications Security - CCS '15* (pp. 1310–1321). New York: ACM Press. 10.1145/2810103.2813687

Shomron, G., & Weiser, U. (2018). Spatial Correlation and Value Prediction in Convolutional Neural Networks. *IEEE Computer Architecture Letters*, *18*(1), 10–13. doi:10.1109/LCA.2018.2890236

Siddiqui, M. S., & Hong, C. S. (2007). Security issues in wireless mesh networks. *Proceedings of the International Conference on Multimedia and Ubiquitous Engineering*, 717-722.

Significant lack in awareness of IoT security among IT leaders, study finds. (2018). Retrieved from https://www.IoT technews.com/news/2018/nov/21/significant-lack-awareness-IoT -security-among-it-and-security-leaders-study-finds/

Simon, E., Henry, D., & Morris, J. S. (2015). *IoT – Enabled Analytic Applications Revolutionize Supply Chain Planning and Execution*. November 2015, IDC #259697.

Singal, H., & Kohli, S. (2016). Mitigating Information Trust: Taking the Edge off Health Websites. *International Journal of Technoethics*, *7*(1), 16–33. doi:10.4018/IJT.2016010102

Singh, C., & Naik, G. (2018). Financial Inclusion After PMJDY: A Case Study of Gubbi Taluk, Tumkur. *SSRN Electronic Journal*. doi:10.2139srn.3151257

Singh, A. (2014). Leadership in female SHGs: Traits/abilities, situational or forced? *The International Journal of Sociology and Social Policy*, *34*(3/4), 247–262. doi:10.1108/IJSSP-10-2013-0110

Singh, A., Rathkanthiwar, S., & Kakde, S. (2016, April). Energy efficient routing of WSN using particle swarm optimization and V-LEACH protocol. In *2016 International Conference on Communication and Signal Processing (ICCSP)* (pp. 2078-2082). IEEE. 10.1109/ICCSP.2016.7754544

Singh, K., & Kaur, J. (2017). Machine learning based link cost estimation for routing optimization in wireless sensor networks. *Adv. Wirel. Mob. Commun*, *10*, 39–49.

Singh, P., , & Tewari, P., & Verma, D. (2017). Status of Self Help Groups in Uttarakhand. *International Journal of Basic and Agricultural Research*, *15*(3), 242–247.

Singh, S., & Saini, H. S. (2018). Security approaches for data aggregation in Wireless Sensor Networks against Sybil Attack. *Proceedings of Second International Conference on Inventive Communication and Computational Technologies (ICICCT)*, 190-193. 10.1109/ICICCT.2018.8473091

Sinha, Jha, Rai, & Bhushan. (2017). Security vulnerabilities, attacks and countermeasures in wireless sensor networks at various layers of OSI reference model: A survey. *Proceedings of IEEE International Conference on Signal Processing and Communication (ICSPC)*, 288-293.

Sinha, P., Rai, A. K., & Bhushan, B. (2019). Information Security threats and attacks with conceivable counteraction. *2019 2nd International Conference on Intelligent Computing, Instrumentation and Control Technologies (ICICICT)*. doi: 10.1109/icicict46008.2019.8993384

Sinha, P., Jha, V. K., Rai, A. K., & Bhushan, B. (2017). Security vulnerabilities, attacks and countermeasures in wireless sensor networks at various layers of OSI reference model: A survey. *2017 International Conference on Signal Processing and Communication (ICSPC)*. 10.1109/CSPC.2017.8305855

Si, W., Selvakennedy, S., & Zomaya, A. Y. (2010). An overview of channel assignment methods for multi-radio multi-channel wireless mesh networks. *Journal of Parallel and Distributed Computing*, *70*(5), 505–524. doi:10.1016/j.jpdc.2009.09.011

Skalli, H., Ghosh, S., Das, S. K., Lenzini, L., & Conti, M. (2007). Channel assignment strategies for multi-radio wireless mesh networks: Issues and solutions. *IEEE Communications Magazine*, *45*(11), 86–95. doi:10.1109/MCOM.2007.4378326

Smaragdakis, G., Matta, I., & Bestavros, A. (2004). SEP: a stable election protocol for clustered heterogeneous wireless sensor networks. *Proc. Int. Workshop Sensor and Actor Network Protocols and Applications*, 251-261.

Snow, A., Rastogi, P., & Weckman, G. (2005). Assessing dependability of wireless networks using neural networks. *MILCOM 2005 - 2005 IEEE Military Communications Conference*. doi: . doi:10.1109/milcom.2005.1606090

Sodhro, A. H., Pirbhulal, S., Qaraqe, M., Lohano, S., Sodhro, G. H., Junejo, N. U., & Luo, Z. (2018). Power Control Algorithms for Media Transmission in Remote Healthcare Systems. *IEEE Access: Practical Innovations, Open Solutions*, *6*, 42384–42393. doi:10.1109/ACCESS.2018.2859205

Sodhro, A. H., Pirbhulal, S., & Sangaiah, A. K. (2018). Convergence of IoT and product lifecycle management in medical health care. *Future Generation Computer Systems*, *86*, 380–391. doi:10.1016/j.future.2018.03.052

Sodhro, A. H., Sangaiah, A. K., Pirphulal, S., Sekhari, A., & Ouzrout, Y. (2018). Green media-aware medical IoT system. *Multimedia Tools and Applications*, *78*(3), 3045–3064. doi:10.100711042-018-5634-0

Sohraby, K., Minoli, D., & Znati, T. (2007). *Wireless Sensor Network Technology, Protocols, and Applications*. John Wiley & Sons. doi:10.1002/047011276X

Solanki, V. K., Cuong, N. H. H., & Lu, Z. J. (2019). Opinion mining: using machine learning techniques. In *Extracting Knowledge From Opinion Mining* (pp. 66–82). IGI Global. doi:10.4018/978-1-5225-6117-0.ch004

Song, X., Wang, C., Gao, J., & Hu, X. (2013). DLRDG: Distributed linear regression-based hierarchical data gathering framework in wireless sensor network. *Neural Computing & Applications*, *23*(7-8), 1999–2013. doi:10.100700521-012-1248-z

Soni, S., & Bhushan, B. (2019). Use of Machine Learning algorithms for designing efficient cyber security solutions. In *2019 2nd International Conference on Intelligent Computing, Instrumentation and Control Technologies, ICICICT 2019* (pp. 1496–1501). Institute of Electrical and Electronics Engineers Inc. 10.1109/ICICICT46008.2019.8993253

Soni, P., Kumar, A., & Hafizul Islam, S. K. (2019). An improved three-factor authentication scheme for patient monitoring using WSN in remote health-care system. *Computer Methods and Programs in Biomedicine*, *182*, 105054. doi:10.1016/j.cmpb.2019.105054 PMID:31499422

Son, J., Lee, J., & Seo, S. (2009). Topological Key Hierarchy for Energy-Efficient Group Key Management in Wireless Sensor Networks. *Wireless Personal Communications*, *52*(2), 359–382. doi:10.100711277-008-9653-4

Sridhar, S., Guo, J., & Jha, S. (2009). Channel assignment in multi-radio wireless mesh networks: a graph-theoretic approach. *Proceedings of the First international Conference on Communication Systems and Networks*, 180-189. 10.1109/COMSNETS.2009.4808856

Sriram, M., & Sundaram, N. (2016). Financial inclusion in India: A review. *International Journal of Applied Engineering Research*, *11*(3), 1575–1578.

Steinhardt, J., Koh, P. W., & Liang, P. (2017). Certified defenses for data poisoning attacks. Advances in Neural Information Processing Systems, 3518–3530.

Subramaniam, T., & Bhaskaran, P. (2019). Local intelligence for remote surveillance and control of flow in fluid transportation system. *Journal*, *74*(1), 15–21. http://iieta. org/journals/ama_c

Subramanian, A., Buddhikot, M., & Miller, S. (2006). Interference aware routing in multi-radio wireless mesh networks. *2nd IEEE Workshop on Wireless Mesh Networks*, 55-63. 10.1109/WIMESH.2006.288620

Subramanian, A., Gupta, H., Das, S. R., & Cao, J. (2008). Minimum interference channel assignment in multi radio wireless mesh networks. *IEEE Transactions on Mobile Computing*, *7*(12), 1459–1473. doi:10.1109/TMC.2008.70

Sugihara, R., & Gupta, R. K. (2010). Optimal speed control of mobile node for data collection in sensor networks. *IEEE Transactions on Mobile Computing*, *9*(1), 127–139. doi:10.1109/TMC.2009.113

Sun, X., He, C., & Lu, Y. (2012, December). ESAMR: An enhanced self-adaptive mapreduce scheduling algorithm. In *2012 IEEE 18th International Conference on Parallel and Distributed Systems* (pp. 148-155). IEEE.

Sun, T., Hu, G., Yang, G., & Jia, J. (2015). Real-time and clock-shared rainfall monitoring with a wireless sensor network. *Computers and Electronics in Agriculture*, *119*, 1–11. doi:10.1016/j.compag.2015.09.023

Sutton, R. S., & Barto, A. G. (1998). *Reinforcement learning: An introduction.* MIT Press.

Svozil, D., Kvasnicka, V., & Pospichal, J. (1997). Introduction to multi-layer feed-forward neural networks. *Chemometrics and Intelligent Laboratory Systems*, *39*(1), 43–62. doi:10.1016/S0169-7439(97)00061-0

Tandon, A., Lim, T. J., & Tefek, U. (2019). Sentinel based malicious relay detection in wireless IoT networks. *Journal of Communications and Networks (Seoul)*, *21*(5), 458–468. doi:10.1109/JCN.2019.000049

Tang, J., Deng, C., & Huang, G. B. (2015). Extreme learning machine for multilayer perceptron. *IEEE Transactions on Neural Networks and Learning Systems*, *27*(4), 809–821. doi:10.1109/TNNLS.2015.2424995 PMID:25966483

Tariq, H., Al-Sahaf, H., & Welch, I. (2019, December). Modelling and Prediction of Resource Utilization of Hadoop Clusters: A Machine Learning Approach. In *Proceedings of the 12th IEEE/ACM International Conference on Utility and Cloud Computing* (pp. 93-100). 10.1145/3344341.3368821

Tashtarian, F., Haghighat, A. T., Honary, M. T., & Shokrzadeh, H. (2007, September). A new energy-efficient clustering algorithm for wireless sensor networks. In *2007 15th International Conference on Software, Telecommunications and Computer Networks* (pp. 1-6). IEEE.

Teschendorff, A. E. (2019). Avoiding common pitfalls in machine learning omic data science. *Nature Materials*, *18*(5), 422–427. doi:10.103841563-018-0241-z PMID:30478452

Thangaramya, K., Kulothungan, K., Logambigai, R., Selvi, M., Ganapathy, S., & Kannan, A. (2019). Energy aware cluster and neuro-fuzzy based routing algorithm for wireless sensor networks in IoT. *Computer Networks*, *151*, 211–223. doi:10.1016/j.comnet.2019.01.024

The Determination of the Supreme Court of the Russian Federation of February 20, 2018 No 306-KG17-22931 in the case of No A65-25932 / 2016. (2018). https://legalacts.ru/sud/opredelenie-verkhovnogo-suda-rf-ot-20022018-n-306-kg17-22931-po-delu-n-a65-259322016/

The Forecast of scientific and technological development of the Russian Federation for the period until 2030. (2014). Retrieved from https://www.garant.ru/products/ipo/prime/doc/70484380/

The 'Internet of Things' will be the world's most massive device market and save companies billions of dollars. (2015). Retrieved from https://www.businessinsider.com/the-internet-of-things-market-growth-and-trends-2015-2

The Ministry of Communications approved the concept of the development of the Internet of things in Russia. (2019). Retrieved from https://iot.ru/promyshlennost/minkomsvyaz-odobrila-kontseptsiyu-razvitiya-interneta-veshchey-v-rossii

The National Program. "Digital Economy of the Russian Federation" (approved by the Presidium of the Presidential Council for Strategic Development and National Projects, Minutes No. 7 dated June 4, 2019). (2019). https://digital.gov.ru/ru/activity/directions/858/

The Order of the Ministry of Information Technologies and Communications of the Russian Federation of September 27, 2007. (2007). On the approval of the requirements for the organizational-technical support of sustainable operation of the communication network. *Reference legal system "Consultant Plus".*

The Resolution of the Government of the Russian Federation of October 29, 2019 No 1382 "The rules for granting subsidies from the federal budget to achieve the results of the federal project "Information Security" of the national program "Digital Economy of the Russian Federation". (2019). *Reference legal system "Consultant Plus".*

Thilagavathi, Saveetha, Willson Joseph, Anusree, & Roshni Menon. (2018). Certain Applications of Sensor Network in the Field of IoT, Big Data Analysis and Cloud. *Journal of Advanced Research in Dynamical & Control Systems, 10*, 2012–2015.

Thrun, S., Burgard, W., & Fox, D. (2005). *Probabilistic robotics*. MIT Press.

Titus, I., Rajsingh, E. B., & Silas, S. (2015). E-Health Navigator : A Tool for Selection of Desired Health Care Services. *Advances in Computer Science and Information Technology (ACSIT), 2*(14), 54–58. Retrieved from https://www.krishisanskriti.org/Publication.html

Titus, I., Silas, S., & Rajsingh, E. B. (2016). Investigations on task and role assignment protocols in Wireless Sensor Network. *Journal of Theoretical and Applied Information Technology, 89*(1), 209–219. http://www.jatit.org/volumes/Vol89No1/21Vol89No1.pdf

Tiwari, R., Sharma, N., Kaushik, I., Tiwari, A., & Bhushan, B. (2019). Evolution of IoT & Data Analytics using Deep Learning. *2019 International Conference on Computing, Communication, and Intelligent Systems (ICCCIS)*. 10.1109/ICCCIS48478.2019.8974481

Tkach, I., & Edan, Y. (2020). Heterogeneous Distributed Bees Algorithm. In Automation, Collaboration, & E-Services: Vol. 7. Distributed Heterogeneous Multi Sensor Task Allocation Systems (7th ed., pp. 34–27). doi:10.1007/978-3-030-34735-2

Toledo, A. L., & Wang, X. (2007). Robust detection of selfish misbehavior in wireless networks. *IEEE Journal on Selected Areas in Communications, 25*(6), 1124–1134. doi:10.1109/JSAC.2007.070807

Tomić, I., & McCann, J. A. (2017). A survey of potential security issues in existing wireless sensor network protocols. *IEEE Internet of Things Journal, 4*(6), 1910–1923. doi:10.1109/JIOT.2017.2749883

Townsend, L. (2018). *Wireless Sensor Network Clustering with Machine Learning*. Academic Press.

Tran, D., & Nguyen, T. (2008). Localization In Wireless Sensor Networks Based on Support Vector Machines. Parallel and Distributed Systems. *IEEE Transactions on., 19*, 981–994. doi:10.1109/TPDS.2007.70800

Tripathi, M., Gaur, M. S., & Laxmi, V. (2013). Comparing the impact of black hole and gray hole attack on LEACH in WSN. *Procedia Computer Science, 19*, 1101–1107. doi:10.1016/j.procs.2013.06.155

Tripathy, A., Pradhan, S. K., Tripathy, A. R., & Nayak, A. K. (2019). A New Hybrid Cryptography Technique in Wireless Sensor Network. *International Journal of Innovative Technology and Exploring Engineering, 8*(10), 121–131. doi:10.35940/ijitee.I8736.0881019

Tseng, F.-H., Chou, L.-D., & Chao, H.-C. (2011). A survey of black hole attacks in wireless mobile ad hoc networks. *Human-Centric Computing and Information Sciences, 1*(1), 4. doi:10.1186/2192-1962-1-4

Turner, S., & Polk, T. (2018). *Prohibiting Secure Sockets Layer (SSL) Version 2.0. RFC6176*. Network Working Group.

Ullah, I., Chen, J., Su, X., Esposito, C., & Choi, C. (2019). Localization and Detection of Targets in Underwater Wireless Sensor Using Distance and Angle Based Algorithms. *IEEE Access: Practical Innovations, Open Solutions, 7*, 45693–45704. doi:10.1109/ACCESS.2019.2909133

V., & Agarwal, M. (2019). Intelligent Handwritten Digit Recognition Based on Multiple Parameters using CNN. *International Journal of Computer Sciences and Engineering, 7*(5), 636–641.

Valerio, V. D., Presti, F. L., Petrioli, C., Picari, L., Spaccini, D., & Basagni, S. (2019). CARMA: Channel-Aware Reinforcement Learning-Based Multi-Path Adaptive Routing for Underwater Wireless Sensor Networks. *IEEE Journal on Selected Areas in Communications, 37*(11), 2634–2647. doi:10.1109/JSAC.2019.2933968

Vamsi, P. R., & Kant, K. (2014). A lightweight sybil attack detection framework for wireless sensor networks. *Contemporary computing (IC3), 2014 Seventh International conference on*, 387-393.

Van Dam, T., & Langendoen, K. (2003). An adaptive energy-efficient MAC protocol for wireless sensor networks. *Proceedings of the 1st international conference on Embedded networked sensor systems*, 171-180. 10.1145/958491.958512

Varghese, G., Fingerhut, J. A., & Bonomi, F. (2006). *Detecting evasion attacks at high speeds without reassembly*. Association for Computing Machinery (ACM). doi:10.1145/1159913.1159951

Vedaldi, A., & Lenc, K. (2015, October). Matconvnet: Convolutional neural networks for matlab. In *Proceedings of the 23rd ACM international conference on Multimedia* (pp. 689-692). ACM. 10.1145/2733373.2807412

Venkataramana, S., Sekhar, B. V. D. S., Deshai, N., Chakravarthy, V. V. S. S. S., & Rao, S. K. (2019, May). Efficient time reducing and energy saving routing algorithm for wireless sensor network. *Journal of Physics: Conference Series, 1228*(1), 012002. doi:10.1088/1742-6596/1228/1/012002

Venkatesan, V.P., & Senthamaraikannan, K. (2018). *A Comprehensive Survey on Various Biometric Systems*. Academic Press.

Venus, M., Amir, M. R., Aso, M. D., & Amir, S. (2019). Trust based recommendation systems in IoT: a systematic literature review. *Human-Centric Computing and Information Sciences, 9*, 21. 01901838 doi:10.118613673

Vikhar, P. A. (2016, December). Evolutionary algorithms: A critical review and its future prospects. In *2016 International conference on global trends in signal processing, information computing and communication (ICGTSPICC)* (pp. 261-265). IEEE. 10.1109/ICGTSPICC.2016.7955308

Vimalapriya, M. D., Vignesh, B. S., & Sandhya, S. (2019). *Energy-Centric Route Planning using Machine Learning Algorithm for Data Intensive Secure Multi-Sink Sensor Networks. International Journal of Innovative Technology and Exploring Engineering*. IJITEE.

Vupputuri, S., Rachuri, K. K., & Murthy, C. S. R. (2010). Using mobile data collectors to improve network lifetime of wireless sensor networks with reliability constraints. *Journal of Parallel and Distributed Computing, 70*(7), 767–778. doi:10.1016/j.jpdc.2010.03.010

Wald, A. (1945). Statistical decision functions which minimize the maximum risk. *Annals of Mathematics, 46*(2), 265–280. doi:10.2307/1969022

Wald, A. (1973). *Sequential analysis*. Courier Corporation.

Wang, C., Xu, G., & Sun, J. (2017). An Enhanced Three-Factor User Authentication Scheme Using Elliptic Curve Cryptosystem for Wireless Sensor Networks. *Sensors (Basel), 17*(12), 2946. doi:10.339017122946 PMID:29257066

Wang, F., Xu, G., & Xu, G. (2019). A Provably Secure Anonymous Biometrics-Based Authentication Scheme for Wireless Sensor Networks Using Chaotic Map. *IEEE Access: Practical Innovations, Open Solutions*, *7*, 101596–101608. doi:10.1109/ACCESS.2019.2930542

Wang, K., Li, W. T., Deng, L., Lyu, Q., Zheng, Y. Q., Yi, S. L., Xie, R., Ma, Y., & He, S. (2018). Rapid detection of chlorophyll content and distribution in citrus orchards based on low-altitude remote sensing and bio-sensors. *International Journal of Agricultural and Biological Engineering*, *11*(2), 164–169. doi:10.25165/j.ijabe.20181102.3189

Wang, K., Yuan, L., Miyazaki, T., Zeng, D., Guo, S., & Sun, Y. (2017). Strategic antieavesdropping game for physical layer security in wireless cooperative networks. *IEEE Transactions on Vehicular Technology*, *66*(10), 9448–9457. doi:10.1109/TVT.2017.2703305

Wang, N., Qi, X.-G., Duan, L., Jiang, H., & Liu, X. (2014). Clustering-based routing algorithm in wireless sensor networks with mobile sink. *J. Netw.*, *9*(9), 2376–2383. doi:10.4304/jnw.9.9.2376-2383

Wang, W., Sun, Y., Li, H., & Han, Z. (2010). Cross-layer attack and defense in cognitive radio networks. In *Global Telecommunications Conference (GLOBECOM 2010), 2010 IEEE* (pp. 1-6). IEEE. 10.1109/GLOCOM.2010.5684069

Wang, X., Hou, R., Zhu, Y., Zhang, J., & Meng, D. (2019). NPUFort: A secure architecture of DNN accelerator against model inversion attack. *ACM International Conference on Computing Frontiers 2019, CF 2019 - Proceedings*, 190–196. 10.1145/3310273.3323070

Wang, Y., Martonosi, M., & Peh, L. S. (2006, May). A supervised learning approach for routing optimizations in wireless sensor networks. In *Proceedings of the 2nd international workshop on Multi-hop ad hoc networks: from theory to reality* (pp. 79-86). 10.1145/1132983.1132997

Wang, Y., Martonosi, M., & Peh, L.-S. (2007). Predicting link quality using supervised learning in wireless sensor networks. *Mobile Computing and Communications Review*, *11*(3), 71–83. doi:10.1145/1317425.1317434

Wang, Z., Song, G., & Gao, C. (2019). An Isolation-Based Distributed Outlier Detection Framework Using Nearest Neighbor Ensembles for Wireless Sensor Networks. *IEEE Access: Practical Innovations, Open Solutions*, *7*, 96319–96333. doi:10.1109/ACCESS.2019.2929581

Waspmote Technical guide. (2018). Retrieved April 24, 2018, from Libelium Comunicaciones Distribuidas S.L website: http://www.libelium.com/downloads/documentation/waspmote_plug_and_sense_technical_guide.pdf

Watro, R.J., Kong, D., Cuti, S., Gardiner, C., Lynn, C., & Kruus, P. (2004). TinyPK: securing sensor networks with public key technology. *SASN '04.*

Wazid, M., Das, A. K., Khan, M. M., Al-Ghaiheb, A. A., Kumar, N., & Vasilakos, A. V. (2017). Secure Authentication Scheme for Medicine Anti-Counterfeiting System in IoT Environment. *IEEE Internet of Things Journal*, *4*(5), 1634–1646. doi:10.1109/JIOT.2017.2706752

Wazid, M., Das, A. K., Odelu, V., Kumar, N., Conti, M., & Jo, M. (2018). Design of Secure User Authenticated Key Management Protocol for Generic IoT Networks. *IEEE Internet of Things Journal*, *5*(1), 269–282. doi:10.1109/JIOT.2017.2780232

Wong, K. H., Zheng, Y., Cao, J., & Wang, S. (2006). A dynamic user authentication scheme for wireless sensor networks. *IEEE International Conference on Sensor Networks, Ubiquitous, and Trustworthy Computing (SUTC'06)*, 1, 8. 10.1109/SUTC.2006.1636182

Wood, A., & Stankovic, J. (2002). Denial of service in sensor networks. *Computer*, *35*(10), 54–62. doi:10.1109/MC.2002.1039518

Wu, J., Yu, Y., Song, T., & Hu, J. (2018). Sequential 0/1 for Cooperative Spectrum Sensing in the Presence of Strategic Byzantine Attack. *IEEE Wireless Communications Letters.*

Wu, S., Zhu, Y., & Pu, Q. (2012). Robust smart-cards-based user authentication scheme with user anonymity. *Security and Communication Networks, 5*(2), 236–248. doi:10.1002ec.315

X2Rail-1. (2016). *D7.1 Analysis of existing lines and economic models.* European Union: Shift2Rail.

X2Rail-1. (2016-2020). *Start-up activities for Advanced Signalling and Automation Systems.* https://projects.shift2rail.org/s2r_ip2_n.aspx?p=X2RAIL-1

X2Rail-1. (2018). D3.1 User & System Requirements (Telecommunications).

X2Rail-1. (2018). D7.2 Railway requirements and Standards application conditions.

Xia, J., Li, C., Lai, X., Lai, S., Zhu, F., Deng, D., & Fan, L. (2020). Cache-aided mobile edge computing for B5G wireless communication networks. *EURASIP Journal on Wireless Communications and Networking, 2020*(1), 15. doi:10.118613638-019-1612-0

Xianging, F., & Yulin, S. (2007). Improvement on LEACH protocol of wireless sensor networks. *Proc. Int. Conf. Sensor Technologies and Applications*, 260-264.

Xiao, H., Biggio, B., Brown, G., Fumera, G., Eckert, C., & Roli, F. (2015). Is feature selection secure against training data poisoning? *32nd International Conference on Machine Learning, ICML 2015, 2*, 1689–1698.

Xiao, Y., Xing, C., Zhang, T., & Zhao, Z. (2019). An Intrusion Detection Model Based on Feature Reduction and Convolutional Neural Networks. *IEEE Access: Practical Innovations, Open Solutions, 7*, 42210–42219. doi:10.1109/ACCESS.2019.2904620

Xia, Y., Wang, L., Zhao, Q., & Zhang, G. (2011). Research on job scheduling algorithm in hadoop. *Journal of Computer Information Systems, 7*(16), 5769–5775.

Xie, H., Yan, Z., Yao, Z., & Atiquzzaman, M. (2018). Data Collection for Security Measurement in Wireless Sensor Networks: A Survey. *IEEE Internet of Things Journal, 6*(2), 2205–2224. doi:10.1109/JIOT.2018.2883403

Xiong, N., Zhang, L., Zhang, W., Vasilakos, A., & Imran, M. (2017). Design and Analysis of an Efficient Energy Algorithm in Wireless Social Sensor Networks. *Sensors (Basel), 17*(10), 2166. doi:10.339017102166 PMID:28934171

Yadav, R., Varma, S., Malaviya, N., & others. (2009). A survey of MAC protocols for wireless sensor networks. *UbiCC Journal, 4*, 827-833.

Yadav, M., Das, A., Sharma, S., & Tiwari, A. (2015). Understanding teachers' concerns about inclusive education. *Asia Pacific Education Review, 16*(4), 653–662. Advance online publication. doi:10.100712564-015-9405-6

Yang, H., Qin, G., & Ye, L. (2019). Combined Wireless Network Intrusion Detection Model Based on Deep Learning. *IEEE Access: Practical Innovations, Open Solutions, 7*, 82624–82632. doi:10.1109/ACCESS.2019.2923814

Yang, H., & Wang, F. (2019). Wireless Network Intrusion Detection Based on Improved Convolutional Neural Network. *IEEE Access: Practical Innovations, Open Solutions, 7*, 64366–64374. doi:10.1109/ACCESS.2019.2917299

Yang, J., Zhang, H., Ling, Y., Pan, C., & Sun, W. (2014). Task allocation for wireless sensor network using modified binary particle swarm optimization. *IEEE Sensors Journal, 14*(3), 882–892. doi:10.1109/JSEN.2013.2290433

Yang, K. (2014). *Wireless sensor networks.* Springer. doi:10.1007/978-1-4471-5505-8

Yang, M. S. (1993). A survey of fuzzy clustering. *Mathematical and Computer Modelling*, *18*(11), 1–16. doi:10.1016/0895-7177(93)90202-A

Yang, Z., Meratnia, N., & Havinga, P. (2008). An online outlier detection technique for wireless sensor networks using unsupervised quarter-sphere support vector machine. *2008 International Conference on Intelligent Sensors, Sensor Networks and Information Processing*. 10.1109/ISSNIP.2008.4761978

Yan, J. F., Gao, Y., & Yang, L. (2011, July). Ant colony optimization for wireless sensor networks routing. In *2011 International Conference on Machine Learning and Cybernetics* (Vol. 1, pp. 400-403). IEEE. 10.1109/ICMLC.2011.6016670

Yavuz, A. A., Mudgerikar, A., Singla, A., Papapanagiotou, I., & Bertino, E. (2017). Real-Time Digital Signatures for Time-Critical Networks. *IEEE Transactions on Information Forensics and Security*, *12*(11), 2627–2639. doi:10.1109/TIFS.2017.2716911

Yeh, H., Chen, T., Liu, P., Kim, T., & Wei, H. (2011). A Secured Authentication Protocol for Wireless Sensor Networks Using Elliptic Curves Cryptography. *Sensors (Basel)*, *11*(5), 4767–4779. doi:10.3390110504767 PMID:22163874

Ye, L. (2020). Study on embedded system in monitoring of intelligent city pipeline network. *Computer Communications*, *153*, 451–458. doi:10.1016/j.comcom.2020.02.004

Ye, W., Heidemann, J., & Estrin, D. (2004). Medium access control with coordinated adaptive sleeping for wireless sensor networks. *IEEE/ACM Transactions on Networking*, *12*(3), 493–506. doi:10.1109/TNET.2004.828953

Ye, Z., Wen, T., Liu, Z., Song, X., & Fu, C. (2017). An Efficient Dynamic Trust Evaluation Model for Wireless Sensor Networks. *Journal of Sensors*, *2017*, 1–16. doi:10.1155/2017/7864671

Yoon, E., & Kim, C. (2013). Advanced Biometric-Based User Authentication Scheme for Wireless Sensor Networks. *Sensor Letters*, *11*(9), 1836–1843. doi:10.1166l.2013.3014

Yu, Y., & Prasanna, V. K. (2005). Energy-Balanced Task Allocation for Collaborative Processing in Wireless Sensor Networks. *Mobile Networks and Applications*, *10*(1/2), 115–131. doi:10.1023/B:MONE.0000048550.31717.c5

Zadeh, L. A. (1999). Fuzzy logic= computing with words. In *Computing with Words in Information/Intelligent Systems 1* (pp. 3–23). Physica. doi:10.1007/978-3-7908-1873-4_1

Zhang, X., Bao, H., Ye, J., Yan, K., & Zhang, H. (2013). A data gathering scheme for WSN/WSAN based on partitioning algorithm and mobile sinks. *Proc. 10th IEEE Int. Conf. High Perform. Comput. Commun., IEEE Int. Conf. Embedded Ubiquitous Comput.*, 1968–1973.

Zhang, J., Ren, F., Gao, S., Yang, H., & Lin, C. (2015). Dynamic Routing for Data Integrity and Delay Differentiated Services in Wireless Sensor Networks. *IEEE Transactions on Mobile Computing*, *14*(2), 328–343. doi:10.1109/TMC.2014.2313576

Zhang, K., Chuai, G., Gao, W., Liu, X., Maimaiti, S., & Si, Z. (2019). A new method for traffic forecasting in urban wireless communication network. *EURASIP Journal on Wireless Communications and Networking*, *2019*(1), 66. doi:10.118613638-019-1392-6

Zhang, L., Ding, G., Wu, Q., Zou, Y., Han, Z., & Wang, J. (2015). Byzantine attack and defense in cognitive radio networks: A survey. *IEEE Communications Surveys and Tutorials*, *17*(3), 1342–1363. doi:10.1109/COMST.2015.2422735

Zhang, Q., Yin, Y., Zhan, D., & Peng, J. (2014). A Novel Serial Multimodal Biometrics Framework Based on Semisupervised Learning Techniques. *IEEE Transactions on Information Forensics and Security*, *9*(10), 1681–1694. doi:10.1109/TIFS.2014.2346703

Zhang, Y., Meratnia, N., & Havinga, P. J. (2013). Distributed online outlier detection in wireless sensor networks using ellipsoidal support vector machine. *Ad Hoc Networks, 11*(3), 1062–1074. doi:10.1016/j.adhoc.2012.11.001

Zhao, C., Zhang, W., Yang, Y., & Yao, S. (2015). Treelet-Based Clustered Compressive Data Aggregation for Wireless Sensor Networks. *IEEE Transactions on Vehicular Technology, 64*(9), 4257–4267. doi:10.1109/TVT.2014.2361250

Zhao, H., Chen, C., Hu, J., & Qin, J. (2015). Securing Body Sensor Networks with Biometric Methods: A New Key Negotiation Method and a Key Sampling Method for Linear Interpolation Encryption. *International Journal of Distributed Sensor Networks, 11*(8), 11. doi:10.1155/2015/764919

Zhao, M., & Yang, Y. (2012). Bounded relay hop mobile data gathering in wireless sensor networks. *IEEE Transactions on Computers, 61*(2), 265–277. doi:10.1109/TC.2010.219

Zhao, P., & Hoi, S. C. (2013, August). Cost-sensitive online active learning with application to malicious URL detection. In *Proceedings of the 19th ACM SIGKDD international conference on Knowledge discovery and data mining* (pp. 919-927). 10.1145/2487575.2487647

Zhao, W., Liu, D., & Jiang, Y. (2009, January). Distributed neural network routing algorithm based on global information of wireless sensor network. In *2009 WRI International Conference on Communications and Mobile Computing* (Vol. 1, pp. 552-555). IEEE. 10.1109/CMC.2009.103

Zharova, A., Elin, V., & Panfilov, P. (2017). Personal Data in Cloud. Russia Experience. *Proceedings of the 28th DAAAM International Symposium*, 1136-1142. DOI: 10.2507/28th.daaam.proceedings

Zharova, A. (2020, June). The protect mobile user data in Russia. *Iranian Journal of Electrical and Computer Engineering, 10*(3), 3184–3192. doi:10.11591/ijece.v10i3.pp3184-3192

Zhong, Y. (2016, August). The analysis of cases based on decision tree. In *2016 7th IEEE International Conference on Software Engineering and Service Science (ICSESS)* (pp. 142-147). IEEE.

Zhou, W., Marshall, A., & Gu, Q. (2006, April). A novel classification scheme for 802.11 WLAN active attacking traffic patterns. In *IEEE Wireless Communications and Networking Conference, 2006. WCNC 2006.* (Vol. 2, pp. 623-628). IEEE. 10.1109/WCNC.2006.1683541

Zhu, Y., Mottaghi, R., Kolve, E., Lim, J. J., Gupta, A., Fei-Fei, L., & Farhadi, A. (2017, May). Target-driven visual navigation in indoor scenes using deep reinforcement learning. In *2017 IEEE international conference on robotics and automation (ICRA)* (pp. 3357-3364). IEEE.

Zhu, W., Cao, J., & Raynal, M. (2018). Energy-Efficient Composite Event Detection in Wireless Sensor Networks. *IEEE Communications Letters, 22*(1), 177–180. doi:10.1109/LCOMM.2017.2764458

Zou, K., Ouyang, Y., Niu, C., & Zou, Y. (2012, September). Simulation of Malicious Nodes Detection Based on Machine Learing for WSN. In *International Conference on Information Computing and Applications* (pp. 492-499). Springer. 10.1007/978-3-642-34038-3_68

Zou, Y., Zhu, J., Wang, X., & Hanzo, L. (2016). A survey on wireless security: Technical challenges, recent advances, and future trends. *Proceedings of the IEEE, 104*(9), 1727–1765. doi:10.1109/JPROC.2016.2558521

Zungeru, A. M., Ang, L. M., & Seng, K. P. (2012). *Performance evaluation of ant-based routing protocols for wireless sensor networks.* arXiv preprint arXiv:1206.5938

About the Contributors

K. Martin Sagayam received his PhD in Electronics and Communication Engineering (Signal image processing using machine learning algorithms) from Karunya University. He received Master of Engineering in Communication Systems from Anna University, Chennai and received Bachelor of Enigeering in Electronics and Communication Engineering from Anna University. Currently, he is working as Assistant Professor in the Department of ECE, Karunya Institute Technology and Sciences, Coimbatore, India. He has authored/ co-authored more than 30 referred International Journals. He has also presented more than 20 papers in reputed international and national conferences. He has authored 10 book chapters with reputed international publishers like Elsevier, Springer, IGI Global and CRC press. He has editor in the reputed book and book series with reputed international publishers like IGI Global, and CRC Press. He has reviewed research articles from the journal of Signal, Image and Video Processing, Intelligent Decision Technologies, International Journal of Engineering Research & Technology, Pattern analysis and its applications and Informing Science Institute. He is an active member of professional bodies such as Engineering and Scientific Research Groups, International Society of Promising Computer Engineers, Copernicus, Scientific Engineering Research Corporation, International Association of Computer Science and Information Technology, International Association of Engineers, Indian Society of Electronics and Communication Engineering and Orcid. His area of interest includes Communication systems, signal and image processing, machine learning and virtual reality.

Bharat Bhushan is an Assistant Professor of Computer Science and Engineering (CSE) Department at HMRITM, GGSIP University, New Delhi, India. Currently, he is also serving as a Managing Director at Gate villa, Allahabad, U.P, India. He is an alumnus as well as a Ph.D. scholar of Birla Institute of Technology, Mesra. He received his Undergraduate Degree (B-Tech in Computer Science and Engineering) with Distinction in 2012 and received his Postgraduate Degree (M-Tech in Information Security) with Distinction in 2015 from Birla Institute of Technology, Mesra, India. He earned numerous international certifications such as Cisco Certified Network Associate (CCNA), Cisco Certified Entry Networking Technician (CCENT), Microsoft Certified Technology Specialist (MCTS), Microsoft Certified IT Professional (MCITP) and Cisco Certified Network Professional Trained (CCNP). In the last three years, he has Published more than 50 research papers in various renowned International conferences and SCI indexed journals including Wireless Networks (Springer), Wireless Personal Communications (Springer) and Emerging Transactions on Telecommunications (Wiley). He has contributed with several book chapters in various books and is currently in the process of editing 7 books from the most famed publishers like Elsevier, IGI Global, and CRC Press. He has served as a Reviewer/Editorial Board Member for several reputed international journals including IEEE Access, IEEE Communication Surveys and Tutorials and

Wireless Personal Communication (Springer). He has also served as Speaker and Session Chair at more than 15 National and International Conferences. In the past, he worked as a Network Engineer in HCL Infosystems Ltd., Noida. He has qualified GATE exams for successive years and gained the highest percentile of 98.48 in GATE 2013. He is also a member of numerous renowned bodies including IEEE, IAENG, CSTA, SCIEI, IAE and UACEE.

Diana Andrushia is working as Assistant Professor in the department of Electronics and Communication Engineering at Karunya Institute of Technology and Sciences, Coimbatore, India. She received her Bachelor of Engineering degree in Electronics and Communication Engineering with first class from Anna University, Chennai, India in 2006 and her Master of Engineering degree in Applied Electronics with first class from Anna University, Chennai, India in 2008. She holds PhD in information and Communication Engineering from Anna University, Chennai, India. Dr.A.Diana Andrushia has 12 years of research experience in the field of computer vision and image understanding. She has published 35 research papers in the reputed journals and conferences. She has guided 10 MTECH students and 25 batch of BTECH students in the field computer vision and its applications. She is currently working in the field of machine learning, pattern recognition, structural health monitoring and deep neural networks.

Victor Hugo C. de Albuquerque (Senior Member of IEEE) has a Ph.D. in Mechanical Engineering with emphasis on Materials from the Federal University of Paraíba (UFPB, 2010), an MSc in Teleinformatics Engineering from the Federal University of Ceará (UFC, 2007), and he graduated in Mechatronics Technology at the Federal Center of Technological Education of Ceará (CEFETCE, 2006). He is currently a Full Professor of the Graduate Program in Applied Informatics, and coordinator of the Laboratory of Industrial Informatics, Electronics and Health at the University of Fortaleza (UNIFOR). Data Science Director at the Superintendency for Research and Public Safety Strategy of Ceará State (SUPESP/CE), Brazil. He has experience in Computer Systems, mainly in the research fields of: Applied Computing, Intelligent Systems, Visualization and Interaction, with specific interest in Pattern Recognition, Artificial Intelligence, Image Processing and Analysis, as well as Automation with respect to biological signal/image processing, image segmentation, biomedical circuits and human/brain-machine interaction, including Augmented and Virtual Reality Simulation Modeling for animals and humans. Additionally, he has research at the microstructural characterization field through the combination of non-destructive techniques with signal/image processing and analysis and pattern recognition. Prof. Victor is the leader of the Industrial Informatics, Electronics and Health Research Group. He is Editor-in-Chief of the Journal of Artificial Intelligence and Systems and Associate Editor of the IEEE Access, Applied Soft Computing, Frontiers in Communications and Networks, Computational Intelligence and Neuroscience, Journal of Nanomedicine and Nanotechnology Research, and Journal of Mechatronics Engineering, and he has been Lead Guest Editor of several high-reputed journals, and TPC member of many international conferences.

* * *

Manisha Agarwal is an Associate Professor in Banasthali Vidyapith.She completed her PhD from Banasthali Vidyapith.

Vikash Kumar Agarwal completed his BE from Bapuji Institute of Engineering and Technology, Davangere, and M.Tech. from BIT Mesra, Ranchi. He is currently working as Assistant Professor in RTC Institute of Technology, Ranchi. His research area includes wireless sensor networks, internet of things and blockchain technology.

David Batista is a Bachelor of Science + Master of Science in Telecommunication engineering specialized in Communication Systems. Innovation System Engineer with experience on international projects, with technical knowledge on satellite communications, rail transport solutions, ICT and emerging technologies as V2X Communications and IoT.

Elijah Blessing Rajsingh is the Registrar of Karunya Institute of Technology and Sciences and Professor in the Department of Computer Science and Engineering. He completed his Ph.D. degree in Information and Communication Engineering from Anna University, India in 2005. His research areas include network security, mobile computing, wireless and ad hoc networks. He is an Associate Editor for International Journal of Computers & Applications, Taylor and Francis and member of the editorial review board for many peer reviewed journals. He is a member of IEEE, ISTE and has served as International advisory board member for various International conferences.

Alvino Rock C. completed his UG Engineering Studies in the field of Computer Science and Engineering. His field of research interest's include Data Science, Big Data, Cloud Computing and Networking.

Jon Colado has a degree in Telecommunication engineering specialized in Communication Systems. Innovation System Engineer with experience on international projects, with technical knowledge on Transportation, ICT and emerging technologies as V2X Communications, IoT and Cloud/Fog computing.

Tushar Dhiman is pursuing his Bachelors in Technology in Computer Science from Guru Indraprastha University, Delhi. His research interests include Machine Learning, AI.

Ramesh G. is working as Professor in the Department of Information Technology at KLN College of Engineering, Madurai, India. He has completed Under Graduation in Electrical and Electronics Engineering, Post Graduation in Computer Science and Engineering and Ph.D in Routing - Optical networks. He published papers in many international journals and conference proceedings. He organized many workshops, seminar and FDPs. He has applied Indian patent for IOT based product for reaping the fruit from the tree. He completed consultancy work from TNEB and undergoing funded IOT based project. He is the life time member of IEI and ISTE.

Vidit Gulyani is currently pursuing his graduation degree in Bachelors of Technology in Department of Computer Science and Engineering from HMR institute of technology and management. He's keenly interested in Software Development. His Research interest include Machine Learning, 5G and blockchain.

Chiung Ching Ho is a Senior Lecturer at the Department of Computing and Information Systems, School of Science and Technology, Sunway University. I am currently working on location intelligence, text mining, and health informatics. I have a keen interest in application of interpretable and explainable machine learning can be applied in the domains of health and finance.

Osamah Ibrahim Khalaf is Senior Engineering and Telecommunications Lecturer in Al-Nahrain University/College of Information Engineering. He has hold 10 years of university-level teaching experience in computer science and network technology and has a strong CV about research activities in computer science and information technology projects. He has had many published articles indexed in (ISI/Thomson Reuters) and has also participated and presented at numerous international conferences. He has a patent and has received several medals and awards due to his innovative work and research activities. He has good skills in software engineering including experience with: .Net, SQL development, database management, mobile applications design, mobile techniques, Java development, android development, and IOS mobile development, Cloud system and computations, website design. His brilliant personal Strengths are in highly self-motivated team player who can work independently with minimum supervision, strong leadership skills, and outgoing personality. In 2004, he got his B.Sc. in software engineering field from Al_Rafidain University College in Iraq. Then in 2007, he got his M. Sc. in computer engineering field from Belarussian National Technical University. After that, he got his PhD in 2017 in the field of computer networks from faculty of computer systems & software engineering -University Malaysia, Pahang. He has overseas Work experiences in University in Binary University in Malaysia and University Malaysia Pahang.

Titus Issac received his B.E degree in Information Technology from Karunya Institute of Technology and Sciences India, in 2010 and M. Tech degree in Network and Internet Engineering from Karunya Institute of Technology, India, in 2012. He was a Senior Research Fellow for Indian Council of Medical Research, Government of India, through Extramural Project Scheme during 2013-16. He is currently a research student in the Department of Computer Science and Engineering, Karunya Institute of Technology and Sciences. His research interests include WSN, IOT, Edge and Fog computing. He is a reviewer in Journal of supercomputing.

Karthikeyan J, received his Ph.D. degree from Vellore Institute of Technology (VIT University), India in 2013. He has obtained his Bachelor of Science (B.Sc.) and Master of Computer Applications (M.C.A.) in the years 2005 and 2010 from VIT University, India. At present, he is serving as an assistant professor in the Department of Software and Systems Engineering, School of Information Technology and Engineering, VIT University, India. He is currently guiding number of research scholars. He has authored and co-authored several research articles, book chapters, and conference contributions. His research area includes machine learning, big data security, privacy-preserving data publication, deep learning, etc.

Roopa JayaSingh J. did her UG and PG in Karunya University. Completed her PhD degree in Information and Communication Engineering under Anna University, Coimbatore. Her research focused on Biomedical Signal Processing. She is currently working as Assistant Professor in Electronics and Communication Engineering, Karunya University, Coimbatore. Her total teaching experience of 14 Plus years. Her field of Interest is Biomedical Engineering, Data Science and Communication Engineering.

T. Jemima Jebaseeli is currently working as an Assistant Professor in Karunya Institute of Technology and Sciences, Coimbatore. Her research interests include Medical Image Processing, Image Fusion and Image Compression. She has published various papers in reputed international journals and conferences.

Sergio Jiménez Gómez has a Masters [MEng] in Telecommunication engineering specialized in Communication Systems. Innovation System Engineer with experience on international projects, with technical knowledge on Transportation, ICT and emerging technologies as V2X Communications, IoT and Cloud/Fog computing.

C. Willson Joseph is working at Sahrdaya College of Engineering and technology. He is at present pursuing his research in the area of Internet of Things Security at Karunya Institute of Technology and Sciences.

Rajkumar K. received B.E (CSE) from Madurai Kamaraj University in 2004, M.E (CSE) from Anna University in 2007 and Ph.D from Anna University in 2019. He has 15 years of teaching experience in the departments of CSE and IT. Currently he is working as Assistant professor in department of Information Technology at PSNA College of Engineering and Technology. He is a life member of ISTE. He has published many SCI and Scopus indexed journals. His area of interest includes Computer Network, Wireless sensor network, etc.

Vinay Kandpal holds Ph.D. in Management from Department of Management Studies, Kumaun University, Nainital. He is honours graduate in Commerce from University of Calcutta and did his MBA with dual specialization in Finance & Marketing. He is pursuing D Lit from Kumaun University Nainital on A Study of Social and Economic Implications of Financial Inclusion in Uttarakhand. He is Assistant Professor in Department of General Management, School of Business, University of Petroleum & Energy Studies, Dehradun. He has over 13 years of experience in Academics. He has published 35 research papers in the areas of topics like Banking, Digitalization, Smart Cities, CSR, Corporate Governance and Infrastructure Finance in leading refereed and indexed Journals. His fields of Teaching and Research Interest are Financial Accounting, Management Accounting & Cost Accounting, Working Capital Management, Capital Market, Mutual Fund, Financial Management, Banking, Financial Inclusion and Financial Institutions. He is a member of All India Management Association Indian Accounting Association and Indian Commerce Association. He has published 4 books in the area of Accounting & Finance.

Aditya Khamparia is working as an Associate Professor in Department of Computer Science and Engineering at Lovely Professional University, Punjab, India. He has about eight years of teaching experience and his research interests are semantic information processing and semantic web, e-learning, cognitive science and artificial thinking, soft computing, neural network and data mining.

Naghma Khatoon is currently working as Assistant Professor in Faculty of Computing and Information Technology, Usha Martin University, Ranchi. Her expertise include WSN, MANET and IoT. She has completed her B.Sc. IT from Ranchi University and M.Sc. IT and Ph.D. from Birla Institute of Technology, Mesra.

Nitish Kumar completed his B.Tech in Computer Technology from Priyadarshini college of Engineering, Nagpur and M.Tech in Information Technology from Birla institute of technology, Mesra, India. He is currently working as Assistant Professor in RTC Institute of Technology, Ranchi.

Santosh Kumar received his Ph.D. in Computer Science & Engineering at Birla Institute of Technology, Mesra, Ranchi. He is an Assistant Professor at the Department of Computer Science and Information Technology at ITER, Siksha 'O' Anusandhan University, Bhubaneswar. His research interests are machine learning, evolutionary computation, and data mining method aiming at diagnosis and disease prediction. He has published numerous research papers at national and international journals, conference proceedings as well as chapters of books.

Daniel Maldonado is a Telecommunications Engineer working as Innovation Systems Engineer at transportation department at Indra. With strong technical knowledge on emerging technologies as V2X Communications, IoT, Industrial IoT, Cloud/Fog computing and Cybersecurity.

Vidushi Mishra is pursuing her PhD from Banasthali Vidyapith. She is a member of ACM. She qualified NET and GATE.

Gabriel Mujica is an assistant professor and research member in the Center of Industrial Electronics at the Universidad Politecnica de Madrid, where he is mainly involved in the area of Networked Embedded Systems and Wireless Sensor Networks (WSN), and where he got the PhD degree on Industrial Electronics engineering. He has participated in different national and European research projects (including Horizon 2020 funded projects) related to the development and optimization of WSN as well as the integration of heterogeneous hardware, software and communication technologies for wireless distributed systems. This way, he has authored several contributions in high-impact conferences and journals in the area of Internet of Things. He has collaborated in the organization of research tutorials, special issues and seminars and as a reviewer for several international conferences and journals (IEEE, Springer). Moreover, he was a visiting researcher at Trinity College Dublin within the area of IoT technologies for smart and sustainable cities. Currently, his main research interests are related to multi-hop distributed networks, hardware-software co-design and protocols for embedded systems in smart urban and industrial application contexts.

J. Andrew Onesimu, is an Assistant Professor in the Department of Computer Science and Engineering (CSE) at Karunya Institute of Technology and Sciences (Deemed Univerisity), Coimbatore, India. He is also pursuing a Ph.D. degree specializing in Big Data Security at Vellore Institute of Technology (VIT University), Vellore, India. He received his Bachelor of Engineering (B.E.) in CSE with First Class in 2011 and received his Master of Engineering (M.E.) degree in CSE with First Class in 2013 from Anna University, Chennai, India. He has earned certifications in Big Data Analytics and other notable core areas. He has attended several workshops and faculty development programs to develop his skills to the recent trends. He has published research articles in Springer and other Scopus indexed journals. He also has presented papers in many international and national conferences. He has given guest lectures on the topic "Machine Learning Techniques" also has organized workshops. He is a member of the International Association of Engineers (IAENG). He has guided many UG/PG students. He has 6 years of teaching experience with excellent academic records for both UG and PG degree students. His research interest includes privacy-preserving techniques, big data security, machine learning, deep learning, and blockchain technologies.

Juan Parras received his Bachelor in Telecommunications Engineering from Universidad de Jaén in 2014, his Master in Telecommunications Engineering from Universidad Politécnica de Madrid (UPM) in 2016 and his Ph.D. in Telecommunications Engineering from UPM in 2020. His research interests include audio and radio signal processing, game theory and deep reinforcement learning applied to communications networks.

Francisco Parrilla Ayuso has a Masters [MEng] in Telecommunication engineering (COIT Associated Member: nº 18.895), Masters in Bussiness Administration (MBA) and Masters in Development and Management of International Research, Development and Innovation Projects and Actions [UPM]. Innovator, Senior Manager (PMP/PRINCE2/ITIL and agile methodologies) with extensive experience managing complex international projects and teams, with a strong technical knowledge on Transportation, ICT and emerging technologies as V2X Communications, IoT, Cloud/Fog computing, Big Data and Machine Learning.

Jorge Portilla received the M.Sc.degree in Physics from the Universidad Complutense de Madrid, Madrid, Spain, on 2003, and the Ph.D.degree in Electronic Engineering from Universidad Polítecnica de Madrid (UPM), Madrid, Spain, in 2010. He is currently an Assistant Professor Tenured at Universidad Politécnica de Madrid. He carries out his research activity within the Centro de Electrónica Industrial. His research interests are focused on Wireless Sensor Networks, Internet of Things, Digital Embedded Systems and Reconfigurable FPGA-based embedded systems. He has participated in more than 30 funded research projects, including European Union FP7 and H2020 projects and Spain Government funded projects, as well as private industry funded projects, mainly related to Wireless Sensor Networks and Internet of Things. He has numerous publications in prestigious international conferences as well as in journals with impact factor.

E. B. Priyanka received her ME in Mechatronics Engineering from Kongu Engineering College in 2016. In 2018 she was appointed as Senior Research Fellow by CSIR under Research and Development group at the Center for Research in Kongu Engineering College. She writes and presents widely on the field like IoT, Data analysis, Industrial automation and control.

Jeba Kumar R. J. S. completed his Undergraduate Professional Engineering Studies in the field of Electronics and Communication Engineering. He is closely associated with the Research and Development activities in the School of Electrical Science and Industry collaboration, Karunya University. His wide spectrum of research interest's include Artificial Intelligence, Machine Learning, Data Science, Communication Engineering, Network Engineering and Instrumentation Engineering.

Robin D. Sebastian, 3rd year Computer Sciences student at Karunya Institute of Technology and Sciences. Robin is specialized in Deep Learning and regularly attends some courses regarding it. Robin regularly updates himself in the latest technology. Robin was inspired by Artificial Intelligence and made his career in the field of Deep Learning by doing many kinds of research in privacy and security issues. Robin always believes in positivity even if any of the things go negatively. Robin goes for cycling or waters plants whenever he feels stressed.

Jaime Señor is a student of the MSc programs in Industrial Engineering and Industrial Electronics at Universidad Politécnica de Madrid.

Mini Sharma is currently working as State project manager, Rural development department, Government of Jharkhand, India. She has completed her Bachelor of Engineering (Computer science and engineering) from SSGMCE Shegaon, Amravati University. She had completed her Master of Engineering (Software Engineering) from Birla Institute of Technology, Mesra, India in 2008. She has rich experience of 17 years in the field of research. She has published several research papers in various international conferences and journals. Her research area includes wireless sensor networks, internet of things and blockchain technology.

Salaja Silas is currently working as Associate Professor in the Department of Computer Science and Engineering, Karunya Institute of Technology and Sciences. She received her B.E. Degree in computer science and engineering from the University of Madras in 2000. She received her M.E. Degree from Madurai Kamaraj University, India in 2002. She completed her PhD in computer science and engineering in 2012. Her research interests include computer networks, pervasive computing, and data analytics. She is a reviewer in Journal of supercomputing and Journal of Cluster computing.

S. Thangavel received his ME in Mechatronics Engineering from Kongu Engineering College in 2014. In 2015 he was appointed as Assistant Professor at the Mechatronics engineering department in Kongu Engineering College. He writes and presents widely on the field like MEMS, Electrical machines, Industrial automation and control.

Ebinezer V. received his B. Tech degree in Information Technology and M.E degree in Computer Science and Engineering from Anna University, Chennai in the year of 2009 and 2012. He is pursuing PhD at Anna University, Chennai. He is currently working as Assistant professor in the department of Computer Science and Engineering, Karunya Institute of Technology and Sciences, Coimbatore Tamilnadu, India. He has published many research papers in the various International / National Conferences and Journals. His area of interests include Cloud computing, Body Area Networks, Data Structures and distributed systems etc.

Jegathesan V. has completed his Bachelor Degree in Electrical and Electronics Engineering and Masters Degree in Applied Electronics from Bharathiar University, Coimbatore and Doctorate Degree in Electrical Engineering from Anna University, Chennai. He is currently working as Associate Professor, Department of Electrical and Electronics Engineering, Karunya Institute of Technology and Sciences, Coimbatore, India. He has more than 20 years of teaching and Research experience. His research has included Static Power Converters, Inverters, AC Motor Drives, Power Quality Enhancement by eliminating the lower order harmonics in Voltage Source Inverter. He has published more than 30 papers in International/National Journals and 25 papers in International/National Conferences. He is a life member of Indian Society for Technical Education and International Association of Engineers.

Jeyalakshmi V. (Member ISTE, IETE) received the B.E and M.E degrees with First Rank holder from the Madurai kamaraj University and Ph.D degree in information and Communication from Anna University, Tamilnadu, India. Dr. Jeyalakshmi Veeranan, working as a Professor in Electronics and

Communication Engineering at the Anna University, Chennai, Tamilnadu, India. She has more than 20 years of teaching experience. She has published more than 40 papers in journals and 22 papers in conferences. She was delivered more video lecture modules and principal investigator involved in MOOC course funded by MHRD at EMRC during 2016. Her research interests are Digitization, wireless sensor networks, IOT, VLSI, Signal Processing, Mobile Communication, Networking and MEMS.

D. Venkatesa Prabu received his ME in Mechatronics Engineering from Kongu Engineering College in 2016. Now he was appointed as Research Fellow by CSIR under Research and Development group at the Center for Research in Kongu Engineering College. He writes and presents widely on the field like Robotics and control.

D. Samuel Joshua Viswas, is currently pursuing his Bachelor's in Computer Science and Engineering at Karunya Institute of Technology and Sciences, Coimbatore, India. He is also a student member for Association of Computing Machinery (ACM) and has research experience in the field of Machine Learning and Deep Learning and its application to the field of Natural Sciences, Medical Imaging and Cyber Security. He has carried out research work along with distinguished faculty from Karunya Institute of Technology and Sciences and also from Ben-Gurion University of Negev, Israel. His current research collaboration is based on application of Machine Learning in the field of drug discovery along with faculty from Ben-Gurion University of the Negev. He has received various medals in International Informatics Olympiad and has completed certification courses from India's premier academic and research & development institutions such as Society for Electronic Transactions and Security (SETS), Chennai, India and from Indian Institute of Technology, Kanpur, India. He has participated in various workshops in fields such as advancements in surgery by Lancet, Internet of things, Machine Learning and Deep Learning. He has also received certification from extended learning programs such as HarvardX, University of Michigan and has been a member of National Cadet Corps, National Service Scheme and Youth Red Cross.

Santiago Zazo is Dr. Engineer by the Universidad Politécnica de Madrid (UPM) in 1995. In 1998 he joined UPM, where currently he is Professor in Signal Theory and Communications. His main research activities are in the field of Signal Processing. More recently, he has been mostly focused on distributed optimization, optimum control, game theory and reinforcement learning. He is author / coauthor of more than 40 journal papers and about 200 conference papers.

Index

E

F

G

H

I

L

M

N

P

Q

R

S

T

U

W